Progress and Trends in Microbial Engineering

Progress and Trends in Microbial Engineering

Editor: Ralph Becker

Callisto Reference

www.callistoreference.com

Callisto Reference,
118-35 Queens Blvd., Suite 400,
Forest Hills, NY 11375, USA

Visit us on the World Wide Web at:
www.callistoreference.com

ISBN: 978-1-63239-919-9 (Hardback)

Printed in China.

Cataloging-in-Publication Data

Progress and trends in microbial engineering / edited by Ralph Becker.
p. cm.
Includes bibliographical references and index.
ISBN 978-1-63239-919-9
1. Microbial biotechnology. 2. Microorganisms. 3. Bioengineering.
4. Microbiological chemistry. I. Becker, Ralph.
TP248.27.M53 P76 2018
660.63--dc23

Table of Contents

Preface

The purpose of the book is to provide a glimpse into the dynamics and to present opinions and studies of some of the scientists engaged in the development of new ideas in the field from very different standpoints. This book will prove useful to students and researchers owing to its high content quality.

Microbial engineering is the study of microorganisms and microbial ecology. Microbial engineering has applications in varied areas such as industrial fermentation, microbial degradation, medicine, etc. The study of microbial cultures helps in understanding bacterial and viral resistance against antibiotics and antimicrobials. This book traces the progress of this field and highlights some of its key concepts. It includes some of the vital pieces of work being conducted across the world, on various topics related to microbial engineering. Scientists and students actively engaged in this field will find this book full of crucial and unexplored concepts. This book attempts to assist those with a goal of delving into the field of microbial engineering.

At the end, I would like to appreciate all the efforts made by the authors in completing their chapters professionally. I express my deepest gratitude to all of them for contributing to this book by sharing their valuable works. A special thanks to my family and friends for their constant support in this journey.

Editor

1

Simultaneous Detection of Forbidden Chemical Residues in Milk Using Dual-Label Time-Resolved Reverse Competitive Chemiluminescent Immunoassay Based on Amine Group Functionalized Surface

Dongdong Zhang[1]◔, Xiaoqi Tao[1,2]◔, Haiyang Jiang[1], Kai Wen[1], Jianzhong Shen[1], Xingyuan Cao[1]*

1 College of Veterinary Medicine, China Agricultural University, Beijing, China, **2** College of Food Science, Southwest University, Chongqing, China

Abstract

In this study, a sensitive dual-label time-resolved reverse competitive chemiluminescent immunoassay was developed for simultaneous detection of chloramphenicol (CAP) and clenbuterol (CLE) in milk. The strategy was performed based on the distinction of the kinetic characteristics of horseradish peroxidase (HRP) and alkaline phosphatase (ALP) in chemiluminesecence (CL) systems and different orders of magnitude in HRP CL value for CAP and ALP CL value for CLE in the chemiluminescent immunoassay. Capture antibodies were covalently bound to the amine group functionalized chemiluminescent microtiter plate (MTP) for efficient binding of detection antibodies for the enzymes labeled CAP (HRP-CAP) and CLE (ALP-CLE). The CL signals were recorded at different time points by the automatic luminometers with significant distinction in the dynamic curves. When we considered the ALP CL value (about 10^5) of CLE as background for HRP CL signal value (about 10^7) of CAP, there was no interaction from ALP CL background of CLE and the differentiation of CAP and CLE can be easily achieved. The 50% inhibition concentration (IC_{50}) values of CAP and CLE in milk samples were 0.00501 μg L^{-1} and 0.0128 μg L^{-1}, with the ranges from 0.0003 μg L^{-1} to 0.0912 μg L^{-1} and from 0.00385 μg L^{-1} to 0.125 μg L^{-1}, respectively. The developed method is more sensitive and of less duration than the commercial ELISA kits, suitable for simultaneous screening of CAP and CLE.

Editor: Sabato D'Auria, CNR, Italy

Funding: This study was supported by Technology Pillar Program in the Twentieth Five-Year Plan Period (2011BAK10B01, 2011BAZ0319816). The funders had no role in study design, data collection and analysis, decision to publish, or preparation of the manuscript.

Competing Interests: The authors have declared that no competing interests exist.

* Email: 06046@cau.edu.cn

◔ These authors contributed equally to this work.

Introduction

Feeding of drugs and chemicals to food animals (pigs, cattle, and goats) can leave residues in meat or milk, offal and other parts of the animals. With food poisoning cases and toxic effects arising from consumption of these residues and the emergence of drug resistant bacteria over recent years, there is a need to control the feeding of drugs and chemicals to the food animals [1–3]. A variety of analytical methods to detect and qualify drugs and chemicals in food matrices have been reported, such as gas chromatography–mass spectrometry (GC–MS), liquid chromatography (LC) with an iron trap detector and LC–MS, enzyme linked immunoassay (ELISA), surface plasmon resonance-based biosensor, immunoassay based on gold nanoparticles and magnetic beads, quantum dot-based lateral flow immunoassay, electrochemical biosensor, amperometric immunosensor and piezoelectric immunosensor [4–13]. Briefly, current trends for detection chemicals are divided to two directions: confirmatory methods such as GC-MS or LC-MS which could differentiate the chemicals and make multi-component detection, and rapid screening methods which satisfied the requirement of rapid and 'Yes/NO' at the level of interest for in-field controls. Monitoring the growing number of antibiotics, hormones, endocrine disrupting chemicals, toxins in animal-derived food samples places a high demand on rapid, cheap and reliable screening methods. Recently, increasing attention has been paid to the development of various multiplexed immunoassay (MIA) in a single run to achieve multi-residue determination and serve specific analytical purposes, which has been demonstrated to be promising in food safety, clinical diagnosis and environmental monitoring [14–17]. In this study, the goat anti-rabbit and goat anti-mouse immunoglobulins were bound covalently to the 3-aminopropyltriethoxysilane (APTES)-functionalized microtiter plates (MTP) in a leach-proof fashion using 1-ethyl-3-(3-dimethylaminopropyl) carbodiimide hydrochloride (EDC) and N-hydroxysulfosuccinimide (SNHS) to enhance the sensitivity of the MIA and reduce assay duration [18–20].

Chloramphenicol (CAP), a broad spectrum antibiotic, is frequently employed in animal production for its excellent antibacterial and pharmacokinetic properties. However, in humans it leads to hematotoxic side effects [2], in particular CAP-induced aplastic anaemia for which a dosage-effect relationship has not yet been established, leading to a prohibition of CAP for the treatment of food-producing animals [21]. Clenbuterol (CLE), a representative of the class of synthesized β_2-adrenergic agonists, is abused illegally for food-producing animals as "lean meat agent", which have potential hazard to human health. Hence, the development of an immunoassay format with enhanced analytical performance is essential for the precise simultaneous detection of trace CAP and CLE.

In this study, we have developed a sensitive rapid dual-label time-resolved reverse competitive chemiluminescent immunoassay (DLTRRC-CIA) for simultaneous determination of trace CAP and CLE applicable in milk. Figure 1 introduces the new assay design, which incorporates the advantages of different properties of chemiluminescence (CL) reaction (horseradish peroxidase (HRP) and alkaline phosphatase (ALP)) and sensitivity improvement of covalent binding. To date, the DLTRRC-CIA is the most sensitive assay with the least assay time for the simultaneous determination of trace CAP and CLE.

Materials and Methods

Materials and Reagents

(a) Standards—CAP (99% purity, Sigma-Aldrich, St. Louis, MO, USA); florfenicol (FF, 99%), florfenicol amine (FFA, 97.6%) and thiamphenicol (TAP, 97.6% purity) were puchased from Schering-Plough Corp. (Kenilworth, NJ, USA); CLE, salbutamol (SAL), ractopamine (RAC), sulfadiazine (SUL), ciprofloxacin (CIP), penicillin (PEN) were purchased from Shanghai Caienfu Technology Co. Ltd. (Shanghai, China).

(b) Analytical grade regents—EDC, SNHS, 2-(N-morpholino) ethanesulfonic acid (MES, pH 4.7), potassium hydroxide (KOH) and 3-APTES were purchased from Sigma (St. Louis, MO, USA). All other chemicals and organic solvents were of reagent grade and were from Beijing Chemical Co. (Beijing, China).

(c) The chemiluminescence Super Signal substrate solution was purchased from Pierce (Rockford, IL, USA). The Visiglo Plus AP Chemiluminescent Substrate was obtained from Invitrogen (NY, USA).

(d) The anti-CAP polyclonal antibodies (PAbs) (rabbit source), anti-CLE monoclonal antibody (MAb) (mouse source), HRP labeled CAP (CAP-HRP), ALP labeled CLE (CLE-ALP) were obtained from WDWK Biotech Co. (Beijing, China). Both of commercial conventional CAP ELISA kit and CLE ELISA kit were purchased from WDWK Biotech Co. (Beijing, China) and R-Biopharm (Beijing, China), respectively. Goat anti-rabbit immunoglobulins and goat anti-mouse immunoglobulins were from Sigma-Aldrich (St, Louis. MO, USA).

(e) Buffers and solutions were prepared in Milli-Q deionised water (DIW). Coating buffer (CB, pH 9.6) was made with 1.59 g Na_2CO_3 and 2.93 g $NaHCO_3$ in 1 L of purified water. Blocking

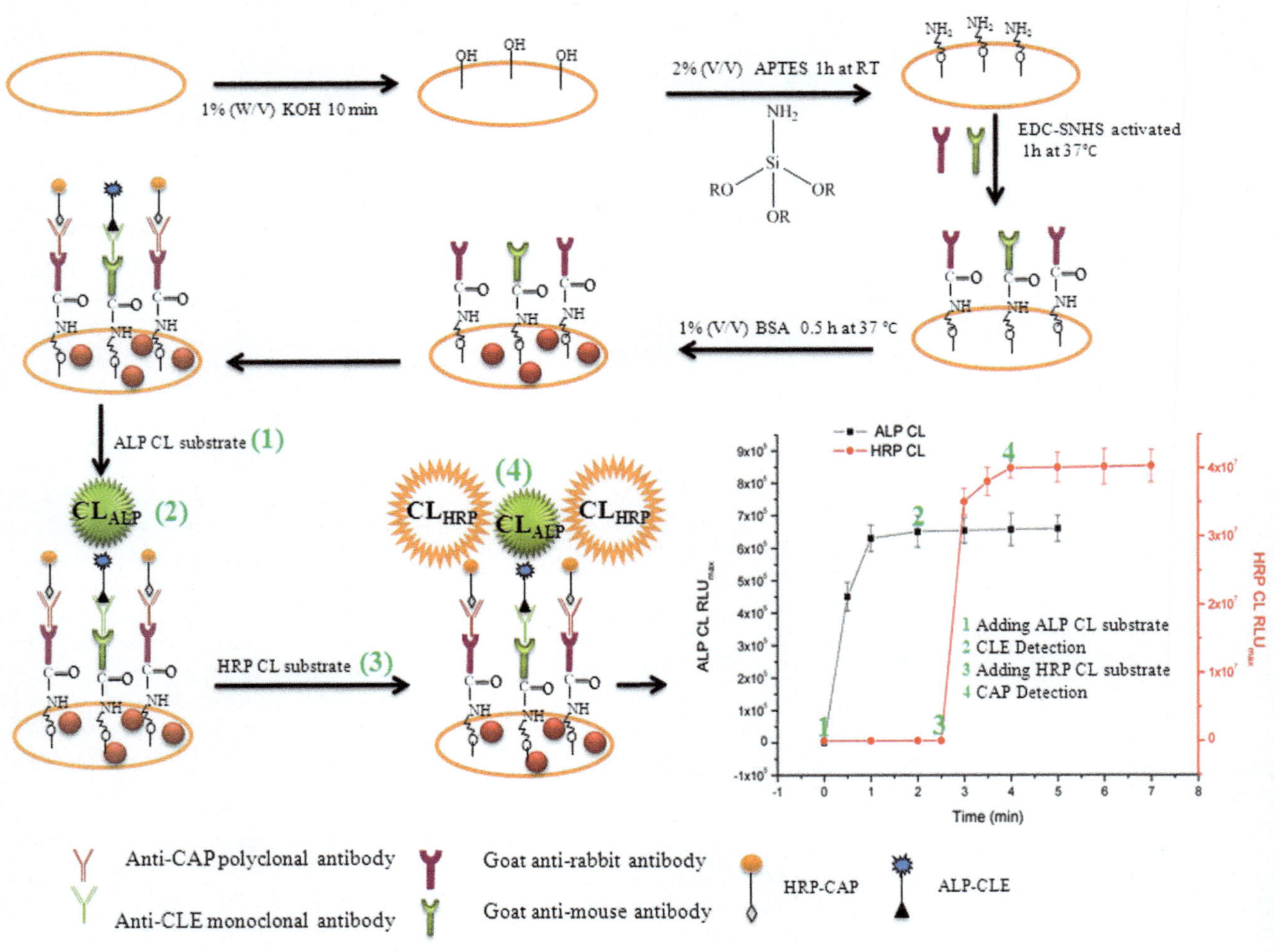

Figure 1. Schematic representation of DLTRRC-CIA for quantitative determination of CAP and CLE.

buffer was prepared by 0.01 M sodium phosphate-buffered saline (PBS) with 1% BSA, pH 7.4. 0.01 M Phosphate-buffered saline (PBS, pH 7.4) was prepared by dissolving 8.0 g NaCl, 0.2 g KCl, 0.24 g KH_2PO_4, and 3.63 g $Na_2HPO_4 \cdot 12H_2O$ in 1 L of purified water. PBST contained 0.01 M PBS with 0.05% Tween-20. A 0.02 M sodium phosphate (PB, pH 7.2) was 1.1 g $NaH_2PO_4 \cdot 2H_2O$ and 5.16 g $Na_2HPO_4 \cdot 12H_2O$ in 1 L of purified water. Solution Carrez A was 0.36 M $K_4Fe(CN)_6 \cdot 3H_2O$ and solution Carrez B was 1.04 M $ZnSO_4 \cdot 7H_2O$. The dilutions of KOH and 3-APTES were diluted in DIW. EDC and SNHS were reconstituted in 0.1 M MES buffer, pH 4.7.

Apparatus

Chemiluminescence was measured with Veritas Microplate Luminometer (Turner BioSystems, Sunny Vale, CA, USA). The colorimetric-ELISA was made by Sunrise microtiter plate reader (TECAN, Groedig, Austria). Transparent 96-well microtiter ELISA plates for colorimetric assay and 96-well chemiluminescent white opaque MTP were purchased from Costar (Cambridge, MA, USA). All buffers were prepared using Milli-Q H_2O system (18 MΩ/cm) (EMD Millipore Corporation, Belleria, MA, USA).

Covalent Binding of Capture Antibodies to the APTES-functionalized MTP

Each well of the 96-well MTP was treated with 100 μL of 1.0% (w/v) KOH at 37°C for 10 min followed by five DIW washes (300 μL/well). The KOH-treated wells were then functionalized with amino groups by incubating with 100 μL of 2% (v/v) APTES at room temperature (RT) for 1 h inside the fume cabinet. The amine-functionalized MTP was subsequently washed five times with DIW in order to remove excess unbound APTES from the surface. The 990 μL of goat anti-rabbit (2 μg mL^{-1}) and goat anti-mouse immunoglobulins (2 μg mL^{-1}) mixture (capture antibodies) was incubated with 10 μL of a pre-mixed solution of EDC (4 mg mL^{-1}) and SNHS (11 mg mL^{-1}) for 15 min at 37°C. The 100 μL of resulting EDC-SNHS cross-linked goat anti-rabbit and goat anti-mouse immunoglobulins solution was added to each of the APTES-functionalized wells and incubated for 1 h at 37°C. The goat anti-rabbit and goat anti-mouse immunoglobulins -bound wells were then washed five times with PBS (300 μL/well).

The Procedure of DLTRRC-CIA for CAP and CLE

The goat anti-rabbit and goat anti-mouse immunoglobulins -bound MTP wells were blocked with 200 μL/well of blocking buffer at 37°C for 0.5 h, subsequently washed five times with PBST. The goat anti-rabbit and goat anti-mouse immunoglobulins -bound MTP wells were incubated with mixture of anti-CAP PAbs (1:40000 dilution in PBST) and anti-CLE MAb (1:40000 dilution in PBST) for 30 min at 37°C, and subsequently washed five times with PBST. Then 100 μL/well of standard (cocktail of CAP and CLE) in 0.02 M PB or sample solution was added, followed 50 μL/well mixture of CAP-HRP (1:40000 dilution in PBST) and CLE-ALP (1:10000 dilution in PBST). The competitive reaction was allowed to take place for 30 min at RT. After washing five times, the CL signal was measured using a chemiluminesence reader at 2 min after automatic addition of 100 μL/well Visiglo Plus AP Chemiluminescent Substrate in injector A and the results were expressed in relative light units (RLU). At 2.5 min, the HRP activity was revealed by automatically adding 100 μL/well of a freshly prepared substrate mixture of Super Signal substrate solution in injector B. The CL signal was measured using a chemiluminesence reader at 4 min and the results were expressed in RLU.

Data Analysis

Standards and samples were run in quadruplicate wells, and mean chemiluminescence intensity values were divided by RLU_{max} (chemiluminescence intensity in the absence of analyte).The ratio is defined as B/B_0. Standard curves were obtained by plotting B/B_0 against the logarithm of analyte concentration and fitted to a four-parameter logistic equation using Origin (version 8.0, Microcal, Northampton, MA, USA) software packages

$$y = \{(A-D)/[1+(x/C)^B]\}+D$$

where A is the asymptotic maximum 1, B is the curve slope at the inflection point, C is the x value at the inflection point (corresponding to the analyte concentration that reduces RLU_{max}

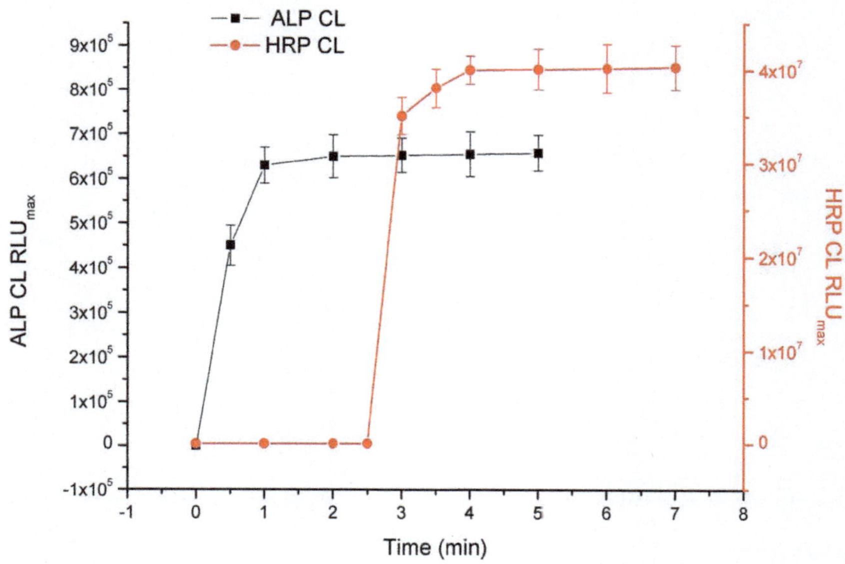

Figure 2. Kinetic measurement of chemiluminescence (CL) output intensity (RLU) for Super Signal and Visiglo Plus substrates catalyzed by HRP and ALP in the developed DLTRRC-CIA.

Table 1. A comparative analysis of the developed DLTRRC-CIA with various immunoassay formats and commercial kits for CAP and CLE detection in standard solution.

Manufacturer	Antibody binding	Sensitivity (μg L^{-1})	Assay duration (h)	Refer to
			CAP	
Developed DLTRRC-CIA	Covalently-bound	0.00501	5	Reported here
commercial CAP ELISA kits (WDWK Biotech)	Passively adsorbed	0.084	18	http://yu18710047045.foodmate.net/sell/index.php?itemid=27609
commercial CAP ELISA kits (R-Biopharm)	Passively adsorbed	0.082	18	http://www.xygen.com/pdfs/mycotoxins/antibiotics/Glucoronid-Cap.pdf
Direct competitive CL-ELISA	Passively adsorbed	0.017	18	[24]
Indirect competitive CL-ELISA	Passively adsorbed	0.13	>18	[25]
			CLE	
Developed DLTRRC-CIA	Covalently-bound	0.0128	5	Reported here
commercial CLE ELISA kits (WDWK Biotech)	Passively adsorbed	0.493	18	http://yu18710047045.foodmate.net/sell/index.php?itemid=21866
commercial CLE ELISA kits (R-Biopharm)	Passively adsorbed	0.472	18	http://www.bioon.com.cn/reagent/show_product.asp?id=934563
ELISA	Passively adsorbed	0.3	>18	[26]
Time-resolved CL immunoassay	Passively adsorbed	0.5	>18	[17]

to 50%), and D is the asymptotic minimum ($RLU_{background\ signal}/RLU_{max}$).

Recovery and Precision

Standard solution were added into the blank milk samples, known to be free of CAP, and CLE, to yield CAP/CLE concentrations of 0.001/0, 0.0075/0, 0.04/0, 0/0.04, 0/0.020, 0/0.10, 0.001/0.004, 0.0075/0.020, 0.040/0.10 μg L^{-1}, respectively. Measured concentrations with five times in duplicate for each sample to assess accuracy and precision.

Analysis of Field Milk Samples

Forty whole cow milk samples with pakages were purchased from retail outlets in Beijing. Each sample was divided into three portions in brown polystyrene bottle, which would be analyzed by developed DLTRRC-CIA, conventional ELISA kits. All of the samples were stored at −20°C until use.

Sample Preparation

For extraction of CAP and CLE from milk, added 500 μL Carrez A and 500 μL Carrez B to 10 mL milk, mixed thoroughly, then centrifuged for 10 min at 4000 g in 4°C. Transferred 4.4 mL of aqueous supernatant to a new tube, adjusted the pH to 11 with 1 M NaOH and thoroughly mixed with 8.0 mL ethyl acetate for 10 min in a new tube. Centrifuged at 4000 g for 10 min, 4 mL of organic supernatant was transferred to a new tube and dried by nitrogen at 60°C. The residue was dissolved in 2 mL of 0.02 M PB. The sample solution could be used for determination.

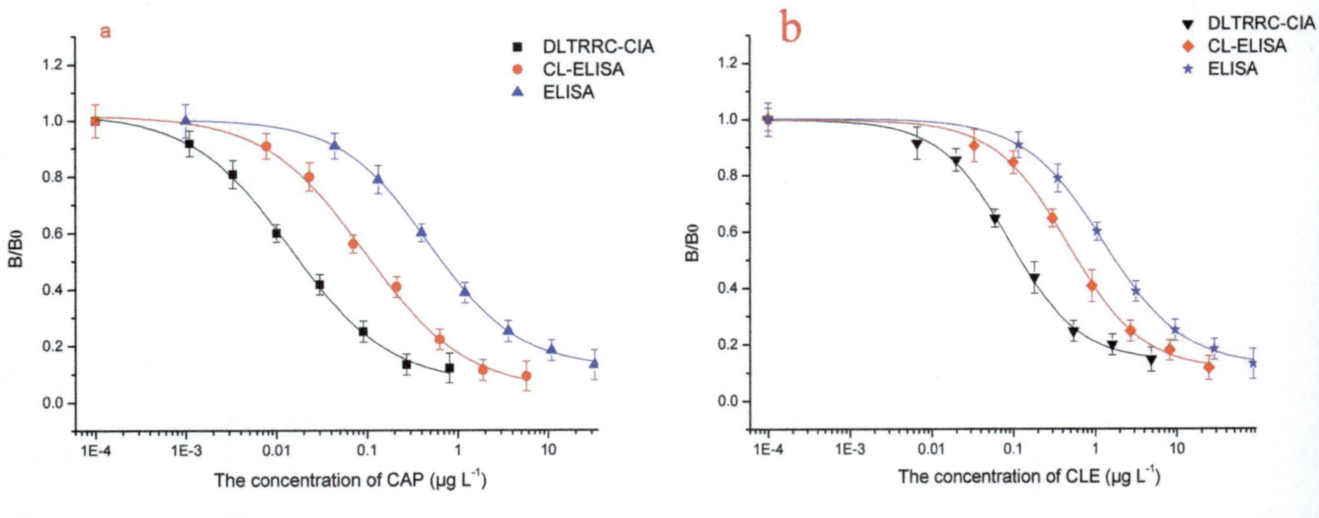

Figure 3. Normalized standard curve by developed DLTRRC-CIA for CAP (a) and CLE (b) under optimized conditions compared to the standard curve obtained by reverse competitive CL-ELISA and traditional ELISA method.

Table 2. CR of CAP and CLE in DLTRRC-CIA with some structurally related and unrelated compounds.

Compound	Structure	IC_{50} (μg L^{-1})	CR (%)
CAP	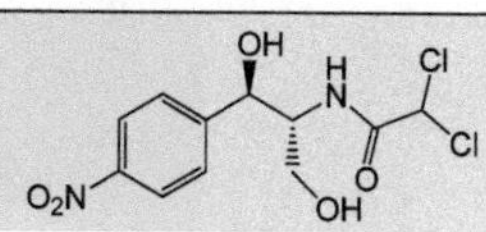	0.00501	100
TAP	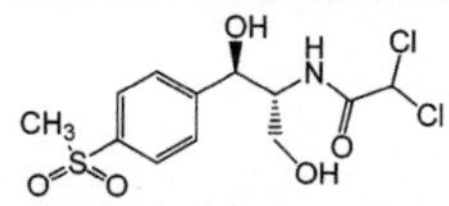	>1000	<0.1
FF	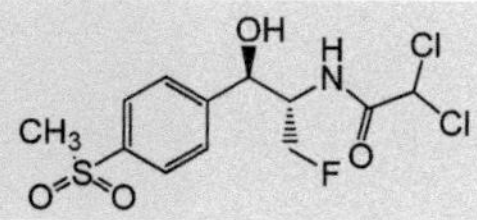	>1000	<0.1
FFA	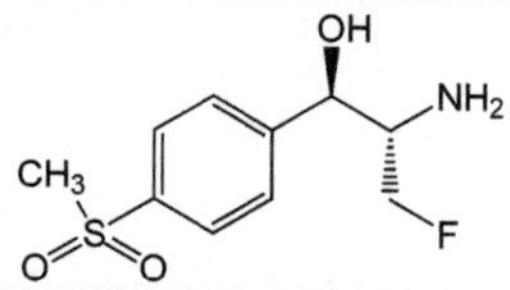	>1000	<0.1
CLE	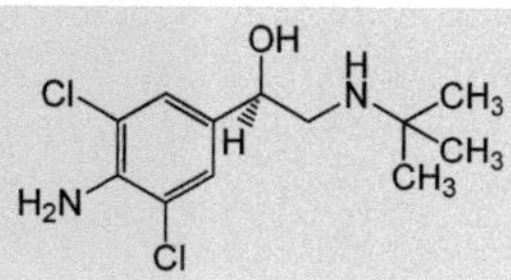	0.0128	100
SAL	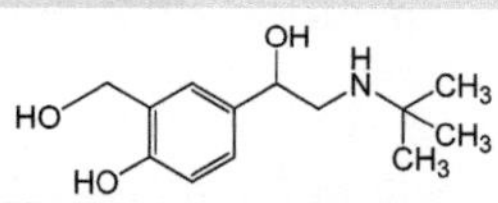	0.41	3.1
RAC	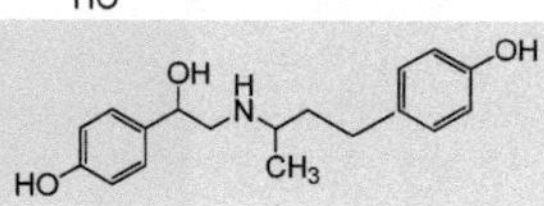	>1000	<0.1
SUL	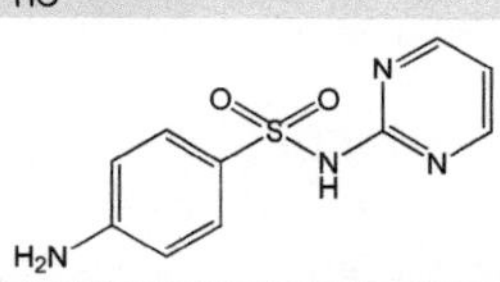	>1000	<0.1
CIP	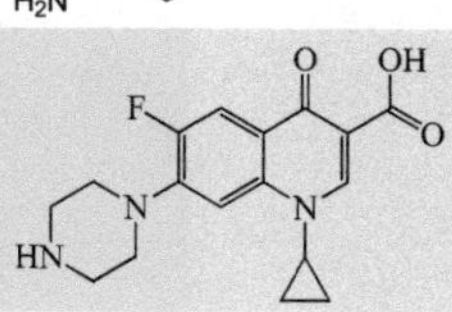	>1000	<0.1
PEN	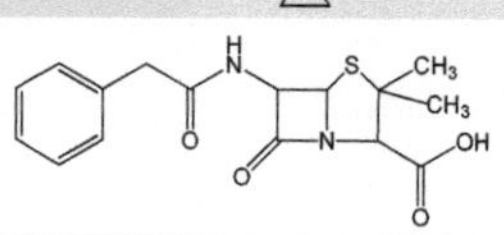	>1000	<0.1

Results and Discussion

DLTRRC-CIA Strategy Based on Time-Resolved CL and Covalent Binding of Capture Antibodies

The DLTRRC-CIA for CAP and CLE based on time-resolved CL and covalent binding of capture antibodies is illustrated in Figure 1. HRP and ALP were adopted as the signal probes to tag CAP and CLE, respectively, due to their different CL kinetic characteristics. With a competitive immunoassay format, the HRP- and ALP-tagged immunocomplexes (goat anti-rabbit immunoglobulins—anti-CAP PAbs—CAP-HRP and goat anti-mouse immunoglobulins—anti-CLE MAb—CLE-ALP) were formed in the well of the MTP. The two CL signals were successively triggered by adding the two different CL substrates (ALP and HRP CL solutions were prepared in different injections in the automatic luminometer).

The CL signal of ALP-tagged immunocomplex reached the maximum value at 2 min after injection of ALP CL substrate in injector A and showed a steady plateau in the whole measured period (Figure 2). The CL signal of the HRP-tagged immunocomplex reached the maximum value at 4 min after injection of HRP CL substrate in injector B (at 2.5 min), and showed a steady plateau in the whole measured period. Then the signal for CLE was detected at 2 min after adding ALP CL substrate (injector A) into the well. The signal for CAP was collected at 4 min after adding HRP CL substrate (injector B) into the wells at 2.5 min.

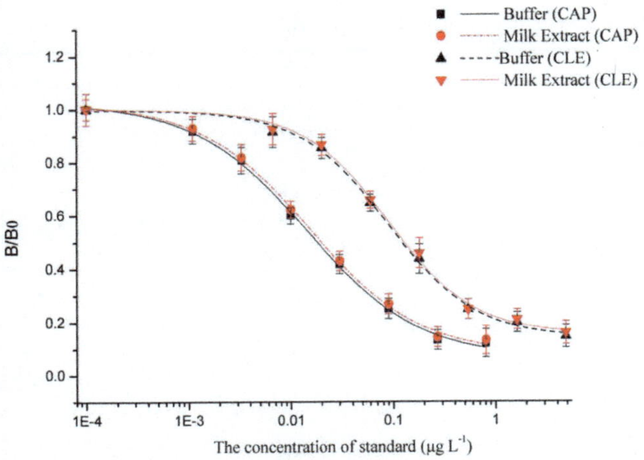

Figure 4. Inhibition curves of CAP and CLE in buffer and milk extract.

When the chemiluminesence reader collected the CL signal at 4 min, there were two chemiluminescent systems—HRP CL system for CAP and ALP CL system for CLE, and the CL value was the sum of HRP CL for CAP and ALP CL for CLE. But the ALP value of $RLU_{max\text{-}CLE}$ was only 1/61 of $RLU_{max\text{-}CAP}$ and the effect of ALP CL signal for CLE (0.65 million) on HRP CL signal for CAP (40 million) was negligible (Figure S1a). Thus we detected the signals of CAP at 2 min and CLE at 4 min, after the automatic injection of ALP CL substrate at 0 min and HRP CL substrate at 2.5 min. When CAP and CLE were both absent, the maximum HRP CL signal for CAP was 61 times of maximum ALP CL signal for CLE. Due to the distinguishable gap for HRP and ALP signals, the cross-talk resulting from the mixed CL reaction systems was effectively avoided (Materials S1 and Materials S2).

The APTES-functionalized bioanalytical platforms have been widely used for the leach-proof immobilization of biomolecules as they enable high immobilization density, long-term stability, high reproducibility, and less biofouling [19,20,22,23]. In the present study, the covalent crosslinking of antibodies in a leach-proof manner on APTES-functionalized bioanalytical platform leads to high functional antibody immobilization density, which results in highly sensitive analyte detection and saving the assay duration (Table 1).

Performance of DLTRRC-CIA in Standard Solutions

Under the optimal conditions, the CL responses decreased linearly with the increase in the concentrations of CAP and CLE since a competitive immunoassay mode was employed. The developed DLTRRC-CIA detected CAP and CLE in the range of 0.0003–0.0912 μg L^{-1} (R^2 = 0.9971) and 0.00385–0.125 μg L^{-1} (R^2 = 0.9974), with 50% inhibition concentration (IC_{50}) values of 0.00501 and 0.0128 μg L^{-1}, respectively (Figure 3). The sensitivity represented by IC_{50} of the developed DLTRRC-CIA for CAP were more than 9.2 times and 93 times greater than reverse competitive CL-ELISA (IC_{50} = 0.046 μg L^{-1}) and traditional ELISA method with colorimetric detector (IC_{50} = 0.47 μg L^{-1}) developed by our own, respectively (Figure 3a). Meanwhile, the sensitivity of the developed DLTRRC-CIA for CLE were more than 8 times and 96 times greater than reverse competitive CL-ELISA (IC_{50} = 0.11 μg L^{-1}) and traditional ELISA method with colorimetric detector (IC_{50} = 1.25 μg L^{-1}) developed by our own, respectively (Figure 3b). The detection limits represented by 10% inhibition concentration (IC_{10}) values for CAP and CLE were 0.0003 and 0.001 μg L^{-1}, respectively.

TAP, FF, FFA, SAL and RAC, structurally related with CAP or CLE, were selected for cross-reactivity (CR) experiments to evaluate the specificity of anti-CAP PAbs and anti-CLE MAb. No significant CR of anti-CAP PAbs to other amphenicols and β-adrenergic agonist was observed (Table 2). Meanwhile, no amphenicols and β-adrenergic agonists at concentration up to more than 1000 μg L^{-1} showed binding with the anti-CLE MAb. Furthermore, to help define the specificity of the two antibodies, structurally unrelated drugs including SUL, CIP and PEN were also tested. No CR was observed.

Comparison of the Developed DLTRRC-CIA for CAP and CLE with Other Immunoassays in Standard Solutions

The analytical comparison of various immunoassays for CAP and CLE detection is important to analyze their suitability for screening the residue in field samples. The developed DLTRRC-CIA decreased the overall assay duration significantly by more

Table 3. Recovery of spiked CAP and CLE in milk.

Drug	Spiked concentration (μg L^{-1})	Measured (μg L^{-1})	Measured (μg L^{-1})	Recovery	Recovery
	CAP/CLE	CAP	CLE	CAP	CLE
	0.001/0	0.0009±0.00005[a]	ND[b]	90.0%	ND
	0.0075/0	0.0067±0.0006	ND	89.3%	ND
	0.040/0	0.0366±0.0032	ND	91.5%	ND
CAP/CLE	0/0.004	ND	0.0035±0.0004	ND	87.5%
	0/0.020	ND	0.0179±0.002	ND	89.5%
	0/0.10	ND	0.085±0.026	ND	85.0%
	0.001/0.004	0.00085±0.0006	0.0036±0.0003	85.5%	89.4%
	0.0075/0.020	0.0068±0.0006	0.0177±0.0015	90.7%	88.6%
	0.040/0.10	0.0363±0.0034	0.089±0.0076	90.75%	89.0%

[a]Each value was repeated five times.
[b]Not detectable, <LOD.

Table 4. Determination of milk samples collected from retail outlets in Beijing by the DLTRRC-CIA and traditional ELISA kit.

Sample	DLTRRC-CIA		Conventional ELISA kit (WDWK Biotech)		Conventional ELISA kit (R-Biopharm)	
	(μg L^{-1})		(μg L^{-1})		(μg L^{-1})	
	CAP	CLE	CAP	CLE	CAP	CLE
S2	0.074[a]	ND	0.072	ND [b]	0.076	ND
S4	ND	0.046	ND	ND	ND	0.045
S5	0.089	0.021	0.091	ND	0.090	ND
S11	ND	0.060	ND	ND	ND	0.058
S16	0.082	ND	0.079	ND	0.078	ND
S17	ND	0.060	ND	0.062	ND	0.067
S19	0.091	ND	0.11	ND	0.12	ND
S25	0.083	0.056	0.089	ND	0.090	0.054
S33	0.079	0.065	0.082	0.061	0.076	0.069
S1,S3, S6–S10, S12–S15, S18, S20–S24, S26–S32, S34–S40			ND			

[a]Each was determined with 3 repeats.
[b]ND not detectable.

than 3-fold i.e. from 18 h (using the CL-ELISA procedure based on passively adsorbed anti-CAP PAbs, as employed in the commercial CAP kit) to about 5 h (Table 1). Its analytical sensitivity for CAP (0.00501 μg L^{-1}) was 16-fold better than that of the commercial CAP ELISA kits (WDWK Biotech: 0.084 μg L^{-1}; R-Biopharm: 0.082 μg L^{-1}) and for CLE (0.0128 μg L^{-1}) was about 36-fold better than that of the commercial CLE ELISA kits (WDWK Biotech: 0.493 μg L^{-1}; R-Biopharm: 0.472 μg L^{-1}). Moreover, it has better analytical performance, i.e. better analytical sensitivity with lower limit of detection (LOD), than our previously developed competitive direct CL-ELISA formats for CAP [24] and newly reported time-resolved chemiluminescence immunoassay for CLE [17], where the anti-CAP PAbs and anti-CLE MAb were passively adsorbed to the MTP, respectively.

Matrix Effect

To apply a new method in real sample analysis, a matrix effect is an important issue to be considered, especially in animal tissues due to the complicated matrix. In this study, the established DLTRRC-CIA was used to determine CAP and CLE in field milk samples. When determining the matrix effects, interferences are quantified by comparing a standard inhibition curve in buffer with that generated in the milk extract matrix known to be free of CAP and CLE. Extracting with ethyl acetate, drying by nitrogen and dissolving in the same buffer with the standard curve buffer were adopted to overcome matrix interference. The two group curves for either of CAP and CLE are superposable, indicating that the matrix effect is not significant (Figure 4). Then the samples can be analyzed using the standard curve instead of the matrix curve.

Application in Real Samples

Precision and recovery. Some spiked milk samples were detected using the developed DLTRRC-CIA method to evaluate the application potential. Five blank samples were spiked with CAP and CLE standards at different known amounts, prior to the recovery tests. As seen in Table 3, the recoveries of the spiked CAP and CLE were in the range of 85.5–91.5% and 83.7–89.6%, respectively. The coefficients of variation (CVs) for the detections of the two analytes were all less than 8.9%, indicating acceptable accuracy of the proposed method.

Analysis of CAP and CLE in Field Milk Samples. To evaluate the determination capability of the developed DLTRRC-CIA in milk samples, 40 field samples were analyzed by the developed DLTRRC-CIA, conventional ELISA kits (Table 4). The results demonstrated that the developed DLTRRC-CIA could simultaneously screen CAP and CLE in the incurred samples as the ELISA kits did. Thereafter, the developed DLTRRC-CIA was reliable for the simultaneous screening of trace CAP and CLE residues in milk samples.

Conclusion

A novel sensitive DLTRRC-CIA was developed for simultaneous detection of CAP and CLE in milk based on distinguishable CL kinetic characteristics and covalent binding of capture antibodies. Since the adding substrate and detect time windows for HRP and ALP were different, and CL values for CAP (about 10^7) and CLE (about 10^5) were not in the same orders, the differentiation of CAP and CLE can be easily achieved. Goat anti-rabbit and goat anti-mouse immunoglobulins were covalently bound to the APTES-functionalized MTP, enhancing the sensitivity of the CIA for CAP and CLE, and reducing the assay duration. There was almost no cross-interaction resulting from the mixed CL reaction systems, testified by the recovery test. This strategy shows outstanding advantages such as low cost, high sensitivity, short assay duration and high efficiency. The recovery test and analysis in field samples demonstrates potential utility of the developed DLTRRC-CIA for clinical simultaneous screening of CAP and CLE in milk.

Author Contributions

Conceived and designed the experiments: XQT DDZ HYJ. Performed the experiments: XQT KW. Analyzed the data: XQT. Contributed reagents/materials/analysis tools: XYC JZS. Wrote the paper: XQT DDZ. Provided the instruments and money: JZS XYC.

References

1. Wu ML, Deng JF, Chen Y, Chu WL, Hung DZ, et al. (2013) Late diagnosis of an outbreak of leanness-enhancing agent–related food poisoning. Am J Emerg Med 31: 1501–1503.
2. Festing MFW, Diamanti P, Turton JA (2001) Strain differences in haematological response to chloroamphenicol succinate in mice: implications for toxicological research. Food Chem Toxicol 39: 375–383.
3. Davis F, Higson SPJ (2010) Label-free immunochemistry approach to detect and identity antibiotics in milk. Pediatr Res 67: 476–480.
4. Shen JZ, Xia X, Jiang HY, Li C, Li JC, et al. (2009) Determination of chloramphenicol, thiamphenicol, florfenicol, and florfenicol amine in poultry and porcine muscle and liver by gas chromatography-negative chemical ionization mass spectrometry. J Chromatogr B 877: 1523–1529.
5. Moragues F, Igualada C, León N (2012) Validation of the determination of chloramphenicol residues in animal feed by liquid chromatography with an Ion Trap Detector based on European decision 2002/657/EC. Food Anal Method 5: 416–421.
6. Capriotti AL, Cavaliere C, Colapicchioni V, Piovesana S, Samperi R, et al. (2013) Analytical strategies based on chromatography–mass spectrometry for the determination of estrogen-mimicking compounds in food. J Chromatogr A 1313: 62–77.
7. Sai N, Chen Y, Liu N, Yu GG, Su P, et al. (2010) A sensitive immunoassay based on direct hapten coated format and biotin–streptavidin system for the detection of chloramphenicol. Talanta 82: 1113–1121.
8. Guo XW (2012) Surface plasmon resonance based biosensor technique: a review. J. Biophotonics 5: 483–501.
9. Tao X, Jiang H, Yu XZ, Zhu JH, Wang X, et al. (2013) An ultrasensitive chemiluminescence immunoassay of chloramphenicol based on gold nanoparticles and magnetic beads. Drug Test Anal 5: 346–352.
10. Berlina AN, Taranova NA, Zherdev AV, Vengerov YY, Dzantiev BB (2012) Quantum dot-based lateral flow immunoassay for detection of chloramphenicol in milk. Anal Bioanal Chem 405: 4997–5000.
11. Huet C, Delahaut P, Fodey T, Haughey SA, Elliott C, et al. (2010) Advances in biosensor-based analysis for antimicrobial residues in foods. Trend Anal Chem 29: 1281–1294.
12. Li S, Zhang J, Deng WP, Shen XZ, Song SP, et al. (2013) A Highly Sensitive amperometric immunosensor for clenbuterol detection in livestock urine. Electroanal 25: 867–873.
13. Holford TRJ, Davis F, Higson SPJ (2012) Recent trends in antibody based sensors. Biosens. Bioelectron 34: 12–24.
14. Zhang MZ, Wang MZ, Chen ZL, Fang JH, Fang MM, et al. (2009) Development of a colloidal gold-based lateral-flow immunoassay for the rapid simultaneous detection of clenbuterol and ractopamine in swine urine. Anal Bioanal Chem 395: 2591–2599.
15. Ge L, Yan JX, Song XR, Yan M, Ge SG, et al. (2012) Three-dimensional paper-based electrochemiluminescence immunodevice for multiplexed measurement of biomarkers and point-of-care testing. Biomaterials 33: 1024–1031.
16. Zhou Y, Tian XL, Li YS, Zhang YY, Yang L, et al. (2011) A versatile and highly sensitive probe for Hg (II), Pb (II) and Cd (II) detection individually and totally in water samples. Biosens Bioelectron 30: 310–314.
17. Han J, Gao HF, Wang WW, Wang ZX, Fu ZF (2013) Time-resolved chemiluminescence strategy for multiplexed immunoassay of clenbuterol and ractopamine. Biosens Bioelectron 48: 39–42.
18. Vashist SK, Dixit CK, MacCraith BD, O'Kennedy R (2011) Effect of antibody immobilization strategies on the analytical performance of a surface plasmon resonance-based immunoassay. Analyst 136: 4431–4436.
19. Vashist SK (2012) Comparison of 1-Ethyl-3-(3-Dimethylaminopropyl) Carbodiimide Based Strategies to Crosslink Antibodies on Amine-Functionalized Platforms for Immunodiagnostic Applications. Diagnostics 2: 23–33.
20. Vashist SK (2012) A sub-picogram sensitive rapid chemiluminescent immunoassay for the detection of human fetuin A. Biosens Bioelectron 40: 297–302.
21. Stolker AAM, Brinkman UAT (2005) Analytical strategies for residue analysis of veterinary drugs and growth-promoting agents in food-producing animals – a review. J Chromatogr A 1067: 15–53.
22. Zheng D, Vashist SK, Al-Rubeaan K, Luong JHT, Sheu FS (2012) Rapid and simple preparation of a reagentless glucose electrochemical biosensor. Analyst 137: 3800–3805.
23. Zheng D, Vashist SK, Al-Rubeaan K, Luong JHT, Sheu FS (2012) Mediatorless amperometric glucose biosensing using 3-aminopropyltriethoxysilane-functionalized grapheme. Talanta 99: 22–28.
24. Tao XQ, Jiang HY, Zhu JH, Shi WM, Wang ZH, et al. (2013) An ultrasensitive chemiluminescent ELISA for determination of chloramphenicol in milk, milk powder, honey, eggs and chicken muscle. Food Agri Immuno 25: 137–148.
25. Zhang SX, Zhang Z, Shi WM, Eremin SA, Shen JZ (2006) Development of a chemiluminescent ELISA for determining chloramphenicol in chicken muscle. J. Agri. Food Chem 54: 5718–5722.
26. Ren XF, Zhang FM, Chen FJ, Yang TB (2009) Development of a sensitive monoclonal antibody-based ELISA for the detection of clenbuterol in animal tissues. Food Agri Immunol 20: 333–344.

2

Development and Customization of a Color-Coded Microbeads-Based Assay for Drug Resistance in HIV-1 Reverse Transcriptase

Lijun Gu[1,2], Ai Kawana-Tachikawa[3], Teiichiro Shiino[4], Hitomi Nakamura[5,6], Michiko Koga[3,6], Tadashi Kikuchi[3,6], Eisuke Adachi[6], Tomohiko Koibuchi[6], Takaomi Ishida[1,2], George F. Gao[7], Masaki Matsushita[8], Wataru Sugiura[4], Aikichi Iwamoto[1,3,5,6], Noriaki Hosoya[3,5]*

1 Research Center for Asian Infectious Diseases, the Institute of Medical Science, the University of Tokyo, Tokyo, Japan, **2** Japan-China Joint Laboratory of Molecular Immunology and Molecular Microbiology, Institute of Microbiology, Chinese Academy of Sciences, Beijing, P. R. China, **3** Division of Infectious Diseases, Advanced Clinical Research Center, the Institute of Medical Science, the University of Tokyo, Tokyo, Japan, **4** AIDS Research Center, National Institute of Infectious Diseases, Tokyo, Japan, **5** Department of Infectious Disease Control, International Research Center for Infectious Diseases, the Institute of Medical Science, the University of Tokyo, Tokyo, Japan, **6** Division of Infectious Diseases and Applied Immunology, Research Hospital, The Institute of Medical Science, The University of Tokyo, Tokyo, Japan, **7** CAS Key Laboratory of Pathogenic Microbiology and Immunology, Institute of Microbiology, Chinese Academy of Sciences, Beijing, P. R. China, **8** Biotech Research and Development, Wakunaga Pharmaceutical Corporation, Hiroshima, Japan

Abstract

Background: Drug resistance (DR) of HIV-1 can be examined genotypically or phenotypically. Although sequencing is the gold standard of the genotypic resistance testing (GRT), high-throughput GRT targeted to the codons responsible for DR may be more appropriate for epidemiological studies and public health research.

Methods: We used a Japanese database to design and synthesize sequence-specific oligonucleotide probes (SSOP) for the detection of wild-type sequences and 6 DR mutations in the clade B HIV-1 reverse transcriptase region. We coupled SSOP to microbeads of the Luminex 100 xMAP system and developed a GRT based on the polymerase chain reaction (PCR)-SSOP-Luminex method.

Results: Sixteen oligoprobes for discriminating DR mutations from wild-type sequences at 6 loci were designed and synthesized, and their sensitivity and specificity were confirmed using isogenic plasmids. The PCR-SSOP-Luminex DR assay was then compared to direct sequencing using 74 plasma specimens from treatment-naïve patients or those on failing treatment. In the majority of specimens, the results of the PCR-SSOP-Luminex DR assay were concordant with sequencing results: 62/74 (83.8%) for M41, 43/74 (58.1%) for K65, 70/74 (94.6%) for K70, 55/73 (75.3%) for K103, 63/73 (86.3%) for M184 and 68/73 (93.2%) for T215. There were a number of specimens without any positive signals, especially for K65. The nucleotide position of A2723G, A2747G and C2750T were frequent polymorphisms for the wild-type amino acids K65, K66 and D67, respectively, and 14 specimens had the D67N mutation encoded by G2748A. We synthesized 14 additional oligoprobes for K65, and the sensitivity for K65 loci improved from 43/74 (58.1%) to 68/74 (91.9%).

Conclusions: We developed a rapid high-throughput assay for clade B HIV-1 DR mutations, which could be customized by synthesizing oligoprobes suitable for the circulating viruses. The assay could be a useful tool especially for public health research in both resource-rich and resource-limited settings.

Editor: Nicolas Sluis-Cremer, University of Pittsburgh, United States of America

Funding: This work was supported in part by a contract research fund from the Ministry of Education, Culture, Sports, Science and Technology (MEXT) for Program of Japan Initiative for Global Research Network on Infectious Diseases (10005010)(AI); Global COE Program (Center of Education and Research for Advanced Genome-Based Medicine - For personalized medicine and the control of worldwide infectious diseases -) of MEXT (F06)(AI); JSPS KAKENHI (25293226)(AKT); JSPS KAKENHI (24790437, 26860300)(NH);Grants for AIDS research from the Ministry of Health, Labor, and Welfare of Japan (H24-AIDS-IPPAN-008)(AKT); Grants for AIDS research from the Ministry of Health, Labor, and Welfare of Japan (H25-AIDS-IPPAN-006)(NH); Research on international cooperation in medical science, Research on global health issues, Health and Labour Science Research Grants, the Ministry of Health, Labor, and Welfare of Japan (H25-KOKUI-SITEI-001)(AI). The funders had no role in study design, data collection and analysis, decision to publish, or preparation of the manuscript.

Competing Interests: A.I. has received grant support from Toyama Chemical Co. Ltd., astellas, ViiV Healthcare K.K., MSD K.K., Baxter through the University of Tokyo. A.I. has received speaker's honoraria/payment for manuscript from Eiken Chemical Co. Ltd., astellas, Toyama Chemical Co. Ltd, Torii Pharmaceutical Co. Ltd., Takeda Pharmaceutical Co. Ltd. and MSD. MM is an employee of Wakunaga Pharmaceutical Corporation which keeps a patent on PCR amplification sequence-specific oligonucleotide probes (SSOP) method. For the remaining authors none were declared.

* Email: hnori@ims.u-tokyo.ac.jp

Introduction

Since combination antiretroviral therapy (cART) was introduced, the prognosis of patients with HIV-1 infection has improved dramatically [1,2]. In resource-rich settings, new classes, new drugs or new formulations of previously-known classes of antiretroviral drugs (ARV) have been introduced continuously for clinical use. Nucleoside/nucleotide reverse transcriptase inhibitor (NRTI) resistance has declined over time in resource-rich settings, presumably reflecting the improvement of treatment regimens [3,4]. Rates of transmitted HIV-1 drug resistance (DR) have remained limited also in resource-limited settings; however, limitation of the first-line and subsequent regimens would be a concern. cART consisting of two NRTIs and one non-nucleoside reverse transcriptase inhibitors (NNRTI), most often zidovudine (AZT) + lamivudine (3TC) or stavudine (d4T) + 3TC plus nevirapine (NVP) or efavirenz (EFV), has been widely used as the treatment regimen in the resource-limited settings [5,6]; consequently, DR might become a larger public health challenge in the developing countries.

DR can be examined genotypically or phenotypically [7] (http://www.aidsmap.com/pdf/Resistance-tests/page/1044559/). Although sequencing is the gold standard of the genotypic resistance testing (GRT), high-throughput GRT targeted to the codons responsible for DR may be more convenient and suitable for public health research [8,9]. We applied the PCR-SSOP-Luminex method [10–12] to an HIV-1 GRT. As an initial approach, we focused on designing an assay for six major DR mutations: M41L, K65R, K70R, K103N, M184V and T215Y/F. M41L, K70R, T215F/Y are examples of thymidine analogue mutations (TAMs) and associated with AZT and d4T [13] (HIV Drug resistance database, Stanford University, http://hivdb.stanford.edu/index.html). K65R is associated multi-nucleoside and nucleotide DR. Although K65R is selected by nucleotide reverse transcriptase inhibitor tenofovir (TDF) usually, it can be selected by d4T. K103N is highly associated with EFV and NVP resistance. The K103N mutation reduces susceptibility to NVP by 50-flod, and EFV by 20-fold. M184V is highly associated with 3TC and emtricitabine (FTC) resistance, and reduce the susceptibility to 3TC by 200-fold. The monitoring of these six DR mutations should be important for molecular epidemiologic study estimating the efficacy of anti-HIV drugs especially in resource limited settings. We synthesized the oligonucleotides for the primers and probes based on the Japanese data base on reverse transcriptase mutations. In order to validate the initial assay system and examine the flexibility for customization, we focused on the clade B HIV-1 which is most prevalent in Japan. Here we report the results of the comparison between sequencing and the PCR-SSOP-Luminex assay using the specimens of a Japanese cohort.

Methods

PCR-SSOP-Luminex assay

HIV-1 DR genotyping described here is based on the reverse SSOP method coupled with a microsphere beads array platform (Luminex Corporation, Austin, TX, USA). Briefly, the method involves PCR amplification by biotinylated primers, hybridization to nucleotide probes coupled to microbeads, detection of the bound PCR products by streptavidin-phycoerythrin (SAPE) reaction, and quantitation by measurement of median fluorescence intensity (MFI).

Color-coded microbeads were coupled to oligoprobes derived from DR mutations or conserved sequences in HIV-1 RT coding region. These synthesized probes were modified at the 5′-end with a terminal amino group and covalently bound to the carboxylated fluorescent microbeads using ethylene dichloride (EDC), following the procedures recommended by the manufacturer (Wakunaga Pharmaceutical Co. Ltd, Hiroshima, Japan). Briefly, 6.25×10^5 carboxylated microbeads were suspended in 50 μl of 0.1 M MES (2-(N-morpholino) ethane sulfonic acid, pH 4.5 (Dojindo Laboratories, Kumamoto Techno Research Park, Kumamoto, Japan). Afterwards, 0.5 μM of amine-substituted oligonucleotide probes was added, followed by 100 mg/ml EDC (1-Ethyl-3-(3-dimethylaminopropyl) carbodiimido hydrochloride) (Pierce Biotechnology, Rockford, IL, USA), and the mixture was incubated in the dark for 30 min at 25°C. The EDC addition and incubation were repeated twice and the microbeads were washed once with 0.02% Tween-20 and once with 0.1% SDS. After the final wash, the pellets were resuspended in 50 μl TE buffer (pH 8.0), and counted on a hemocytometer. The concentration of fluorescence-labeled microbeads coupled to oligonucleotide probes (oligobeads) was adjusted to 8000–12000/μl, and oligobeads were stored at 4°C in the dark.

Five-microliter aliquots of the 5′-biotinlabeled amplified DNA were added to wells in a 96-well PCR tray containing 5 μl/well of denaturation solution, and allowed to denature for 5 min at RT. Hybridization mixture was prepared using oligobeads stocks, SAPE and hybridization solution, according to the manufacturer's instructions (Wakunaga Pharmaceutical Co. Ltd, Hiroshima, Japan). Twenty-five-microliter aliquots of hybridization mixture containing 500 each sequence-specific oligobeads were added to each well. The amplicons were hybridized at 55°C for 30 min using the thermal cycler. Hybridized amplicons were washed twice with 75 μl of wash solution in each well. Reaction outcomes were measured by the Luminex 100 flow cytometer that is equipped with two types of lasers. The bead populations were detected and identified using the 635 nm laser. The SAPE fluorescence of the biotin labeled amplicons that had hybridized to the oligobeads was quantitated using the 532 nm laser. The MFI of SAPE was used to quantify the amount of DNA bound to the oligobeads. Assays were performed in triplicate.

Site-directed mutagenesis and plasmid construction

To assess the feasibility of the assay system, we chose cloned SF2 genome (GenBank accession number K02007) as a template. Synthesized PCR fragments with DR mutations created by site-directed mutagenesis were inserted into the SF2 genome. The numbering system used to refer to DR mutations was based on HXB2 genome (HXB2 location 2485–3309, GenBank accession number K03455) sequences. The SF2 and HXB2 genome had the same sequence in the regions covered by the probes.

Plasmid p9B/R7 from the HIV-1 SF2 strain was a kind gift of Dr. T. Shioda (Osaka University, Osaka, Japan) [14,15]. An *Xho* I- *BamH* I fragment including the pol gene from p9B/R7 was subcloned into pBluescript II SK (+) (Stratagene, La Jolla, CA, USA). As shown in Fig. 1A, various DR mutations were introduced into the *Xho* I- *BamH* I fragment of the plasmid by site-directed mutagenesis using oligonucleotides and PCR methods as previously described [16], and were confirmed by sequencing.

Clinical specimens

Sixty subjects were selected from among patients participating in an ongoing study on microbes at an HIV outpatient clinic affiliated with the Institute of Medical Science, the University of Tokyo (IMSUT). The study was approved by the internal review board of IMSUT (20-31-1120), and all subjects provided written informed consent. The median HIV-1 RNA and CD4 cells count

A

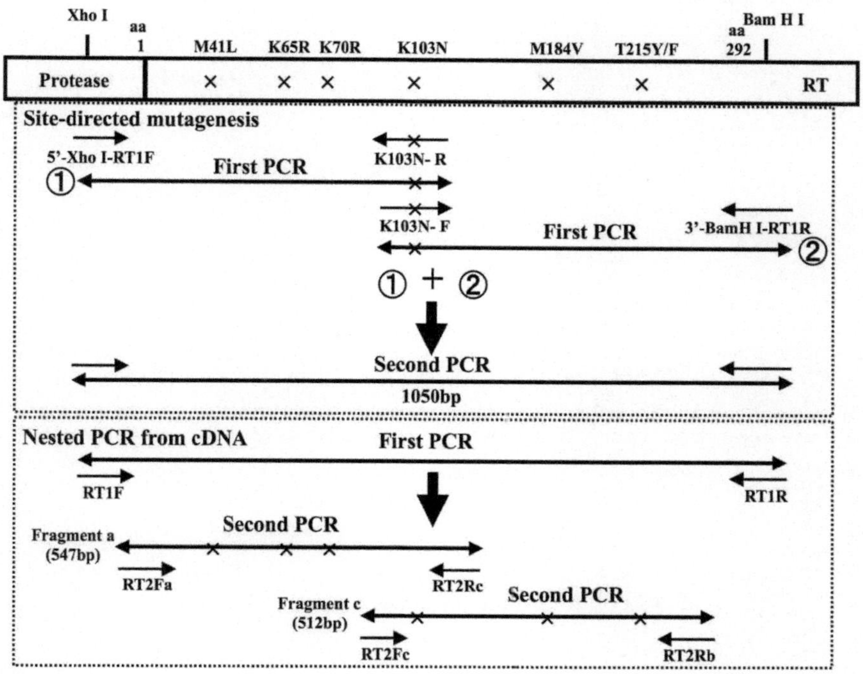

B

PCR primers

RT1F: 5'-ATGATAGGGGGAATTGGAGGTTT- 3'

RT1R: 5'- TACTTCTGTTAGTGCTTTGGTTCC- 3'

RT2Fa: 5'- GACCTACACCTGTCAACATAATTGG- 3'

RT2Rc: 5'- TGGAATATTGCTGGTGATCC- 3'

RT2Fc: 5'-AACTCAAGACTTCTGGGAAGT- 3'

RT2Rb: 5'-CAGTCCAGCTGTCTTTTTCTGGC- 3'

Sequencing primers

T7: 5′-TAATACGACTCACTATAGGG-3′

Rev: 5′-CAGGAAACAGCTATGAC-3′

Figure 1. Schematic representation of PCR amplification and sequences of primers for PCR and sequencing. (A) Top: Site-directed mutagenesis using oligonucleotide is shown using K103N as an example. Desired mutations in the reverse transcriptase gene were engineered in two PCR fragments, then incorporated into a larger fragment (1050 bp, HXB2 location 2388–3425) by the second PCR, and cloned into pBluescript II SK (+) at *XhoI-BamHI* sites. Bottom: Negative strand cDNA was synthesized from patients' plasma. After the first strand PCR using RT1F and RT1R as primers, Fragment a or Fragment c were amplified by nested PCR. (B) Primer sequences for PCR amplification and sequencing.

at sampling were 4.15 (range 2.60–5.88) log 10 copies/ml and 264 (range: 9–902) cells/μl, respectively. Seventy-four specimens from 60 patients were analyzed in this study. Forty-eight patients contributed one plasma specimen, 10 patients contributed two plasma specimens from different time points, and two patients contributed three separate plasma specimens. Twenty-two specimens were from patients who were treatment-naïve when the plasma specimens were obtained. The remaining 52 specimens were from patients on failing treatment including NRTI at the time of sample collection.

Viral RNA extraction, cDNA synthesis, and PCR amplification

Viral RNA was extracted from 140 μl plasma using QIAamp Viral RNA Mini Kit (QIAGEN, Valencia, CA, USA) and eluted

Table 1. Design of oligoprobes based on clade B HIV-1 sequences from Japanese surveillance database.

Locus	genetic code	Frequency(%)	Name of probe	Oligoprobes	Nucleotide sequence (5′-3′)	100% match isolates of database
M41					TGT ACA GAA ATG GAA	
	M41-ATG	795/795(100)	M41M-ATG	GTACAGAAATGGAA	— --- --- --- ---	618/868(71.2%)
	M41L-TTG	109/180(60.6)	M41L-TTG	GTACAGAATTGGAA	— --- --- T— ---	
	M41L-CTG	69/180(38.3)	M41L-CTG	GTACAGAACTGGAA	— --- --- C— ---	
	M41L-TTA	1/180(0.6)				
	M41L-TTR	1/180(0.6)				
K65					ATA AAG AAA AAA GAC AGT	
	K65-AAA	961/1012(95.0)	K65K-AAA	ATAAAGAAAAAAGACAG	--- --- --- --- --- —	516/870(59.3%)
	K65-AAG	30/1012(3.0)				
	K65-AAR	21/1012(2.1)				
	K65R-AGA	7/7(100)	K65R-AGA	ATAAAGAGAAAAGACAG	--- --- -G- --- --- —	
	K65R-AGG	0/7(0.0)				
K70					GAC AGT ACT AAA TGG AGA	
	K70-AAA	849/885(95.9)	K70K-AAA-1	GACAGTACTAAATG	--- --- --- --- —	621/769(80.8%)
			K70K-AAA-2	GTACTAAATGGAGA	--- --- --- --- ---	
	K70-AAG	17/885(1.9)				
	K70-AAR	19/885(2.1)				
	K70R-AGA	85/88(96.6)	K70R-AGA	GTACTAGATGGAGA	--- --- -G- --- ---	
	K70R-AGG	0/88(0.0)	K70R-AGG	GTACTAGGTGGAGA	--- --- -GG --- ---	
	K70R-AGR	3/88(3.4)				
K103					TTA AAA AAG AAA AAA TCA GTA	
	K103-AAA	815/867(94.0)	K103K-AAA	AAAAAAGAAAAAATCAG	- --- --- --- --- --- -	558/785(71.1%)
	K103-AAG	25/867(2.9)				
	K103-AAR	27/867(3.1)				
	K103N-AAC	112/128(87.5)	K103N-AAC	AAAAAAGAACAAATCAG	- --- --- —C --- --- -	
	K103N-AAT	3/128(2.3)	K103N-AAT	AAAAAAGAATAAATCAGT	- --- --- —T --- --- —	
	K103N-AAY	13/128(10.2)				
M184					TAT CAA TAC ATG GAT GAT	
	M184-ATG	659/659(100)	M184M-ATG	TATCAATACATGGATG	--- --- --- --- --- -	761/888(85.7%)
	M184V-GTG	295/335(88.1)	M184V-GTG	TATCAATACGTGGATG	--- --- --- G— --- -	
	M184V-GTA	14/335(4.2)				
	M184V-GTR	26/335(7.8)				
T215					TGG GGA TTT ACC ACA CCA	
	T215-ACC	692/724(95.6)	T215T-ACC-1	GGGGATTTACCACA	— --- --- --- ---	583/813(71.7%)
			T215T-ACC-2	GGGGATTTACCACACCA	— --- --- --- --- ---	
	T215-ACT	13/724(1.8)				
	T215-ACA	7/724(1.0)				
	T215-ACG	3/724(0.4)				
	T215-ACY	8/724(1.1)				
	T215-ACM	1/724(0.1)				
	T215F-TTC	54/54(100)	T215F-TTC	GGGGATTTTTCACAC	— --- --- TT- ---	
	T215Y-TAC	182/185(98.4)	T215Y-TAC	GGGGATTTTACACAC	— --- --- TA- --- -	
	T215Y-TAT	2/185(1.1)				
	T215Y-TAY	1/185(0.5)				
Standard probes			S2582	ATTAAAGCCAGGAATGGAT		
			S2693	AAAAATTGGGCCTGAAAAT		
			S3230	CCTTTGGATGGGTTATGAA		
			S3252	CATCCTGATAAATGGACAG		

Table 1. Cont.

Locus	genetic code	Frequency(%)	Name of probe	Oligoprobes	Nucleotide sequence (5′-3′)	100% match isolates of database
			S3243	TATGAACTCCATCCTGA		

R: Mixed base of A and G; Y: Mixed base of C and T; M: Mixed base of C and A.

with 60 μl AVE buffer. For cDNA synthesis, 55 μl of RNA solution was mixed with 5 μl of 100 pmol/μl random primer (TaKaRa Bio, Otsu, Shiga, Japan) or specific primer, RT1R (Fig. 1B), and 5 μl of 10 mM dNTPs, and denatured at 70°C for 10 min. The RT mixture containing 20 μl of 5× First-Strand buffer, 5 μl of 0.1 M DTT, 5 μl of RNaseOUT Recombinant RNase inhibitor (40 U/μl; Invitrogen, Carlsbad, California, USA) and 5 μl of SuperScript III RT (200 U/μl, Invitrogen) was added to the 65 μl denatured viral RNA-primer-dNTP mixture. The reaction mixture (100 μl final volume) was incubated at 25°C for 5 min for annealing and then at 55°C for 60 min for reverse transcription. The reaction was inactivated by heating at 70°C for 15 min.

RT gene fragments were amplified by nested PCR from cDNAs or by single PCR from plasmids. For the first reaction a 1050 bp fragment from the RT coding region was amplified from 5 μl aliquots of cDNAs using RT1F and RT1R as outer primers in a reaction mixture containing 50 μl of 1×Prime STAR buffer, 0.2 mM of each dNTPs, 0.5 μM of each primer, and 0.5 μl Prime STAR HS DNA Polymerase (25 U/μl, TaKaRa Bio, Otsu, Shiga, Japan). Amplification conditions consisted of 35 cycles denaturation at 98°C for 10 s, annealing at 55°C for 5 s, and extension at 72°C for 30 s.

For the second reaction of the nested PCR and for the single PCR from plasmids (0.1 pg), RT coding fragments were amplified in two PCR fragments using two 5′ biotinylated primer sets, PS1 and PS2, as described previously [17]. The PS1 primer set produced a 547 bp amplicon that was used to detect M41L, K65R, K70R, and K103N, and the PS2 primer set produced a 512 bp amplicon that was used to detect K103N, M184V, and T215Y/F (Fig. 1A). The reaction mixture used 5 μl of the first PCR products or 0.1 pg of plasmid DNA in a final volume of 50 μl, as described above, with amplification conditions as follows: 25 cycles (second nested PCR) or 35 cycles (plasmid DNA amplification) of denaturation at 98°C for 10 s, annealing at 55°C for 5 s, and extension at 72°C for 1 min. The PCR amplicons were used for Luminex detection or direct sequencing.

Sequencing

PCR products were purified with the QIAquick PCR Purification Kit (QIAGEN, Valencia, CA, USA) and were directly sequenced in both directions using ABI 3130xl genetic analyzer (Applied Biosystems, Foster City, CA, USA) and Big Dye terminator V3.1 cycle sequencing kit (Applied Biosystems). In the case of sequence ambiguity due to a coexistence of multiple nucleotides, we confirmed the sequence by cloning and sequencing. For cloning and sequencing, the purified PCR fragments were phosphorylated using T4 polynucleotide kinase (TaKaRa Bio, Otsu, Shiga, Japan) and inserted into the dephosphorylated *EcoR*V restriction site of pBluescript II SK(+). Inserts were sequenced using T7 and Rev universal primers (Fig. 1B).

HIV-1 Japanese surveillance database

In Japan 10 university hospitals, 5 medical centers, 5 public health laboratories, and the National Institute of Infectious Diseases are collaborating in the surveillance of newly diagnosed HIV/AIDS cases. HIV/AIDS patients with both acute and chronic infections, newly diagnosed at these centers since January 2003 were enrolled [18]. Prevalence of DR codons in these patients was determined by analysis of sequences of clade B HIV-1 reverse transcriptase positions 1–240 amino acids.

Statistical analysis

GraphPad Prism 5.0 software (GraphPad Software Inc., San Diego, CA) was used for statistical data analysis. Statistical significance was defined as $P<0.05$.

Results

Design of oligoprobes based on the database of clade B HIV-1 sequences in Japan

Based on the frequency of the codon usage in the Japanese surveillance database for amino acids M41, K65, K70, K103, M184 and T215 in RT gene, we designed the nucleotide sequence of 18 oligoprobes for DR mutation and 5 standard probes (Table 1). We adopted nucleotide sequences of HXB2 strain for the flanking sequences. Synthesized oligoprobes could cover 71.2%, 59.3%, 80.8%, 71.1%, 85.7% and 71.7% of M41, K65, K70, K103, M184 and T215, respectively (Table 1). Five standard probes were designed in the conserved region of RT gene and used as the assay control.

Evaluation of PCR-SSOP-Luminex DR assay using cloned HIV-1

We examined the sensitivity and specificity of the PCR-SSOP-Luminex DR assay using cloned HIV-1. The test fragments were amplified (Fig. 2A), and the amplicons were hybridized to the 16 oligobeads. We performed three independent assays with triplicate hybridization and detection in each assay. The mean positive signal and standard deviation were 5237±1398 (Fig. 2B). The CV% of positive signal and standard deviation were 10.1% ±10.7. The mean of negative signal and standard deviation were 131.2±69.4. The CV% of negative signal and standard deviation were 21.6±23.6 (mean ± S.D.). MFIs of hybridization signals were clearly high only when the oligoprobes matched the mutations in the fragments (Fig. 2B). These data confirmed that the assay system could discriminate one base mismatch at M41, K65, K70, K103, M184 and T215 codons in the plasmid-probe model system.

Next, we examined the sensitivity to detect a particular sequence in a mixture for each DR-related site. The plasmids carrying the wild type and mutant sequences were mixed at various ratios. In samples containing only the wild type sequences, the mean background signal (% ±2SD) was 2.0% ±1.2, 4.1%

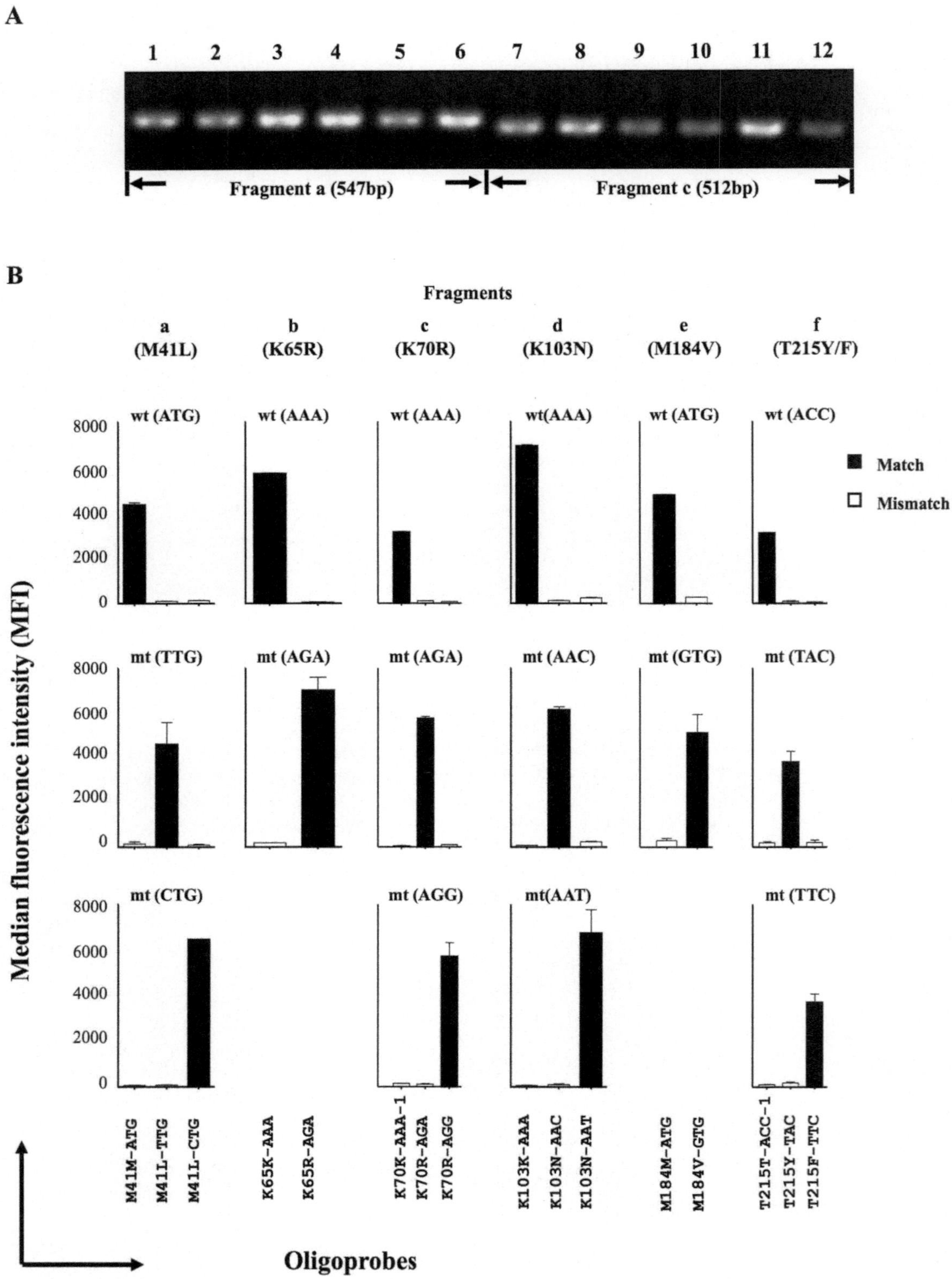

Figure 2. Validation of PCR-SSOP-Luminex assay using plasmids as templates. (A) Agarose gel electrophoresis of amplified fragments. Lanes 1–6: Fragment a (547 bp). Lanes 7–14: Fragment c (512 bp). 1, wild type; 2, M41L-TTG; 3, M41L-CTG; 4, K65R-AGA; 5, 70R-AGA; 6, 70R-AGG; 7, wild type; 8, 103N-AAC; 9, K103N-AAT; 10, M184V-GTG; 11, T215Y-TAC; 12, T215F-TTC. (B) Median fluorescence intensities (MFIs). The plasmid in the test sample is indicated on the top of each panel. Oligoprobes used for detection are indicated at the bottom. Matched results are shown as black bars, mismatched results as white bars. Assays were performed in triplicate. The mean ± standard deviation is shown at the top of each bar.

±2.6, 3.3% ±1.2, 4.6% ±1.8, 6.2% ±3.2 and 3.3% ±0.4 at M41, K65, K70, K103, M184 and T215, respectively (Fig. 3B). The signal for the mutant was judged positive when "% signal" from the mutant oligobeads exceeded the mean + 2SD; 3.2%,

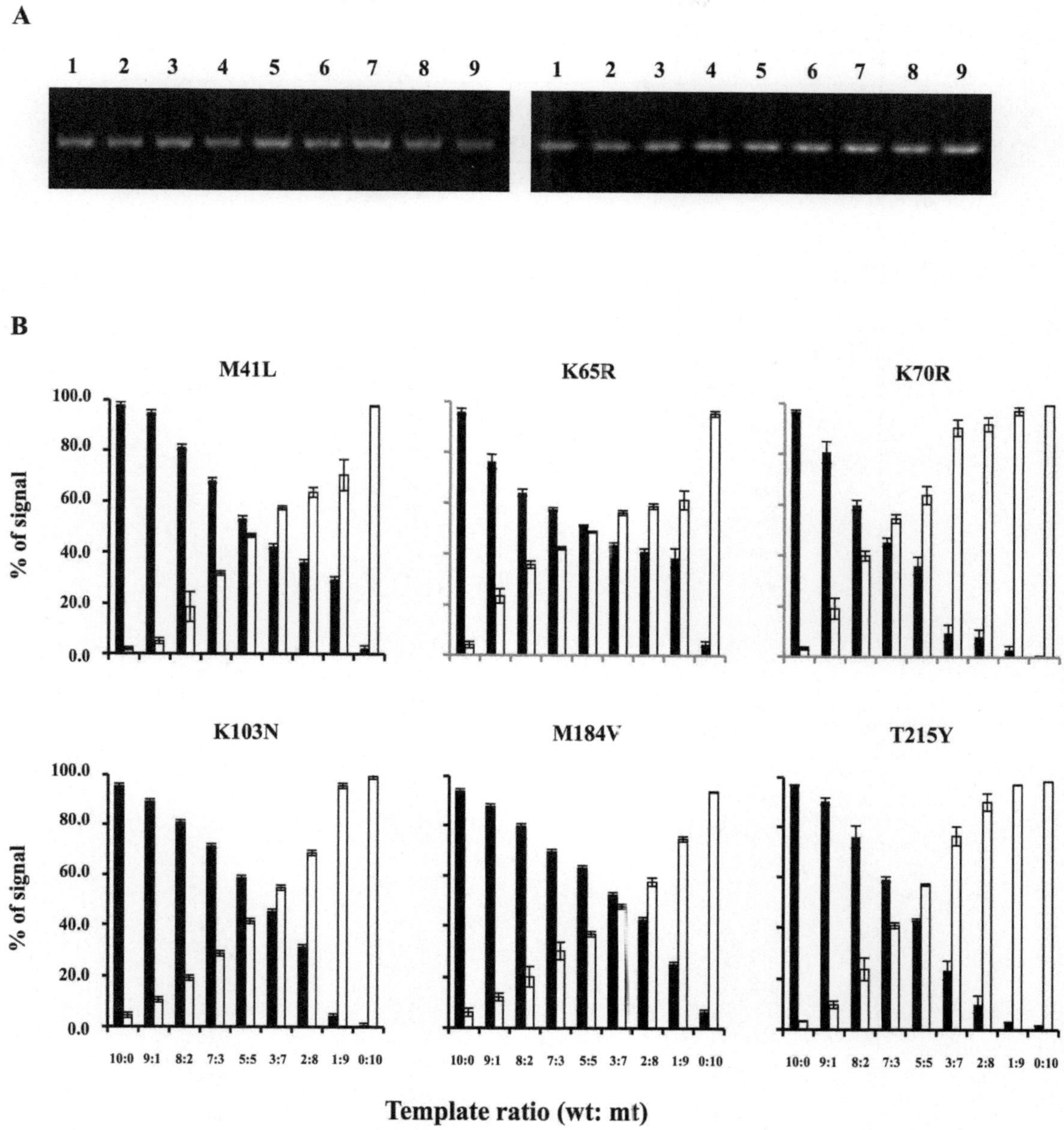

Figure 3. Assay sensitivity in a mixture. (A) Agarose gel electrophoresis of mixtures of amplified fragments. Left panel: Fragment a (547 bp). Right panel: Fragment c (512 bp). Wild-type:mutant ratio; Lanes 1 (10:0); 2 (9:1); 3 (8:2); 4 (7:3); 5 (5:5); 6 (3:7); 7 (2:8); 8 (1:9); 9 (0:10). (B) Signals from wild type probes (black bars) and mutant probes (white bars) in each mixture. "% of signal" was calculated by "MFI of wild type or mutant signal" divided by "MFI of wild type plus mutant signal" and multiplied by 100. Triplicate experiments were performed three times. "% of signal" is shown with standard deviations.

6.7%, 4.5%, 6.4%, 9.4%, 3.7% at M41, K65, K70, K103, M184 and T215, respectively. Actual signals from the mutant oligoprobes at 9:1 (wild type:mutant) mixture at these sites was 5.1%, 23.6%, 19.2%, 10.7%, 12.3%, 9.8%, respectively. Therefore, we infer that the assay can detect 10% DR mutants in the population. There was a big variation (5.1 to 23.6%) in detection of 10% mixtures. We suppose that the difference in minor variant detection could be caused by the melting temperature (GC content and length). Although we compared the GC content and probe length, we could not find a reasonable explanation from this list (data not shown).

Identification of DR mutations in clinical specimens by sequencing

In order to determine the sequence, we amplified the RT gene in two separate fragments from frozen plasma (Fig. 1A). We succeeded to amplify fragment a (containing M41, K65, K70) from all 74 specimens, but failed to amplify fragment c (containing K103, M184, T215) in one patient. Infection with clade B HIV-1 was confirmed by phylogenetic analysis of the RT gene. DR mutations were found at codons M41L (n = 22), K65R (n = 3), K70R (n = 10), K103N (n = 7), M184V (n = 21) and T215Y/F (n = 22) in 40 specimens (Table 2).

Table 2. Comparison of the results between sequencing and PCR-SSOP-Luminex assay.

Position	Aminoacid	Codons	Number of specimens (percentage)	
			Direct sequencing	PCR-SSOP-Luminex
41	Met	ATG	52 (70.3)	41 (55.4)
	Leu	TTG	16 (21.6)	16 (21.6)
		CTG	5 (6.8)	5 (6.8)
	Met, Leu mix		1 (1.4)	0
	No reaction		-	12 (16.2)
65	Lys	AAA	64 (86.5)	41 (55.4)
		AAG	4 (5.4)	0
		AAA, AAG mix	3 (4.1)	2 (2.7)
	Arg	AGA	3 (4.1)	0
	No reaction		-	31 (41.9)
70	Lys	AAA	62 (83.8)	61 (82.4)
		AAG	2 (2.7)	0
	Arg	AGA	8 (10.8)	8 (10.8)
		AGG	1 (1.4)	1 (1.4)
	Lys, Arg mix		1 (1.4)	0
	No reaction		-	4 (5.4)
103	Lys	AAA	60 (82.2)	50 (68.5)
		AAG	1 (1.4)	0
		AAA, AAG mix	1 (1.4)	0
	Asn	AAC	4 (5.5)	2 (2.7)
		AAT	2 (2.7)	2 (2.7)
		AAC, AAT mix	1 (1.4)	1 (1.4)
	Arg	AGA	4 (5.5)	0
	No reaction		-	18 (24.7)
184	Met	ATG	52 (71.2)	46 (63.0)
	Val	GTG	19 (26.0)	16 (21.9)
		GTA	1 (1.4)	0
	Met, Val mix		1 (1.4)	1 (1.4)
	No reaction		-	10 (13.7)
215	Thr	ACC	48 (65.8)	46 (63.0)
	Tyr	TAC	16 (21.9)	15 (20.5)
	Phe	TTC	6 (8.2)	6 (8.2)
	Thr, Tyr mix		1 (1.4)	1 (1.4)
	Other		2 (2.7)	0
	No reaction		-	5 (6.8)

Identification of DR mutations in clinical specimens by the PCR-SSOP-Luminex DR assay

We performed the PCR-SSOP-Luminex DR assay on 74 specimens whose DR mutations had been sequenced (Fig. 4). The lowest MFI of five standard probes (MFI = 837.5) was assumed as the cut off value for the positive signal. We synthesized two additional wild-type oligoprobes (K70K-AAA-2 and T215T-ACC-2), since the original oligobeads (K70K-AAA-1 and T215T-ACC-1) gave marginal signals in some specimens (Table 1 and Fig. 4). By the modification of the position of the target codons and the length of flanking sequences, we could obtain higher MFI signals. Successful determination of the genotypes was 62/74 (83.8%), 43/74 (58.1%), 70/74 (94.6%), 55/73 (75.3%), 63/73 (86.3%) and 68/73 (93.2%) for M41, K65, K70, K103, M184 and T215, respectively. The median of background signal (without sample) was MFI = 178, and median negative signal from the patients sample was MFI = 181 (interquartile range (IQR) = 101–307). PCR-SSOP-Luminex assays create very low background and non-specific signal from negative samples. When the genotype was successfully determined by the PCR-SSOP-Luminex assay, the results were always concordant with those of sequencing (Table 2). We inferred that the failure to determine the genotype was due to sequence diversities. Therefore, we decided to customize the assay according to the sequences around K65 which had the lowest success rate (58.1%).

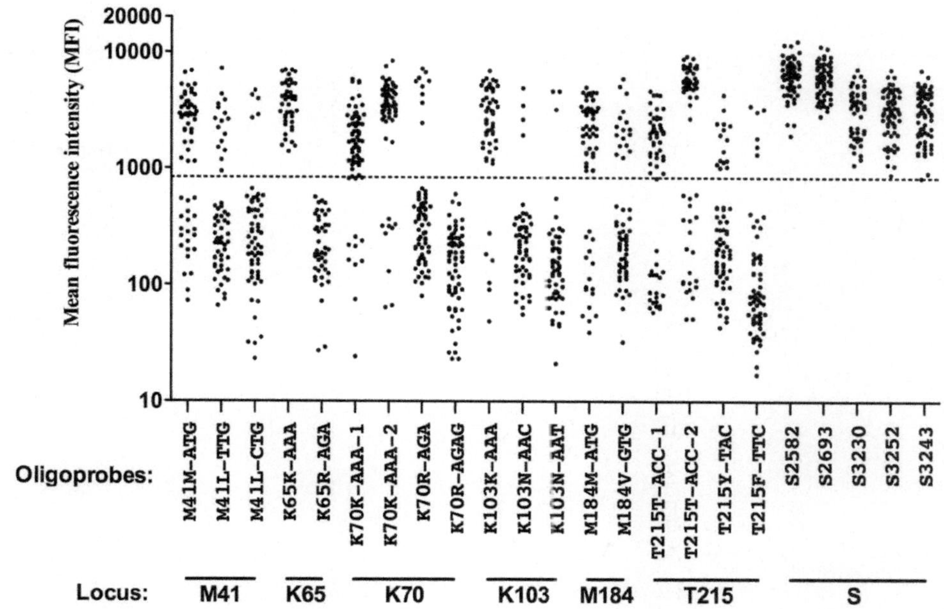

Figure 4. PCR-SSOP-Luminex DR assay of clinical samples. Results of 18 probes for 6 DR loci and 5 standard probes are shown. Each dot represents the mean of triplicates. Dashed line indicates the cut-off value, the lowest MFI among 5 standard probes (MFI = 837.5).

Customization of oligoprobes to detect genetic diversity around K65

In 31 specimens without signals, the sequence around K65 was very diverse, especially at codons K65, K66 and K67 (Table 3). The nucleotide position of A2723G, A2747G and C2750T were frequent polymorphism for the wild type amino acid K65, K66 and D67, respectively. Fourteen specimens had a G2748A mutation that led to D67N, a thymidine analogue mutation [19]. Based on these results we synthesized 10 additional oligoprobes (Table 4).

The specificity of the newly added oligoprobes was confirmed by the plasmids carrying the mutation (data not shown). Customization of the assay decreased the number of specimens without signals from 31 to 6 (Fig. 5).

Discussion

We developed a PCR-SSOP-Luminex DR assay that can identify 6 clinically important DR mutations for NRTI and NNRTI in a single well. To simplify the development, we chose clade B virus and focused on the following mutations in the RT region: M41L, K65R, K70R, K103N, M184V and T215Y/F. We designed a series of capture probes according to the database of the Japanese patients. MFI from hybridization with the corresponding probes was greater than 20-fold signal-to-noise ratio in plasmid experiments (Fig. 2), at least 2-fold signal-to-noise ratio using clinical samples (Fig. 4). The initial positive reaction was as low as 58.1% (43/74) in the highly polymorphic K65 region. The use of additional probes designed to match sequences in the patients' specimens improved the detection rate to 91.9% (68/74), demonstrating that the PCR-SSOP-Luminex assay can be customized to reflect sequences of the viruses prevalent in a given environment.

Transmission of viral strains with major DR mutations can reduce the efficacy of first-line regimens. Since the first report of a horizontal transmission of HIV-1 harboring a zidovudine-resistant mutation [20], 5% to 15% of treatment-naïve, HIV-1-infected individuals harbored the viruses with DR mutations in early 2000s in resource-rich settings [21–24]. It was suggested that transmission of drug resistance in the resource-rich settings can remain stable and at a low level [25]. However, in Japan where the transmission of drug-resistant viruses has historically been low, there seems to be an increasing trend [18,26]. Rates of transmitted HIV drug resistance has remained limited also in resource-limited settings (http://www.who.int/hiv/topics/drugresistance/en/), however, limitation on the first line and subsequent regimens would be a concern. Continued surveillance of drug resistant HIV-1 is warranted.

There are some multiplex strategies to detect the single-nucleotide differences, LigAmp assay [27,28], Nanostring assays [29], oligonucleotide ligation assay-based DNA chip [30], AS-PCR [31,32]. These assays provide substantial improvements in their detection sensitivity over conventional sequencing-based assays, however major limitation of these assays could be detecting one or few DR mutations at a time. PCR-SSOP-Liminex assay should be able to accommodate more DR mutations than the others.

One limitation of PCR-SSOP-Luminex assay is the assay sensitivity caused by diversity of HIV. We were able to detect DR mutations that constituted 10% of the mixture in the isogenic system using plasmids. Even after successful amplification, we could not get signal in considerable number of patient's specimens. By comparison with cloning and sequencing, we estimate that at least 20% mutant was necessary in the patient's specimens to be detected by the PCR-SSOP-Luminex assay. The sensitivity of detection by Sanger sequencing has been reported to be ~20% [33,34]. Sanger sequencing and PCR-SSOP-Luminex had approximately the same detection sensitivity on patient materials. In Sanger sequencing, nucleotide sequences are determined by the wave height. When the virus in the plasma is a mixture of the wild type and a mutant, each nucleotide is displayed as two waves in the same locus with different height according to the fraction in the sample. In the case of PCR-SSOP-Luminex, the wild type or mutant nucleotides are detected independently by the signal bound to the proper probes. Therefore, as we showed partly in this paper, the mutant detection by PCR-SSOP-Luminex assay could

Table 3. Sequence diversity around K65 in the 31 specimens.

Amino acid (HXB2-wt)	I63 K64 K65 K66 D67 S68	
Nucleotide (HXB2-wt)	ATA AAG AAA AAA GAC AGT	Number of specimens
	2747	
K66K-AAG	--- --- --- —G --- —	10
	2747 2748	
K66K-AAG/D67N-AAC	--- --- --- —G A— —	5
	2748	
D67N-AAC	--- --- --- --- A— —	3
	2750	
D67D-GAT	--- --- --- --- —T —	1
	2748	
K65K-AAG/D67N-AAC	--- --- —G --- A— —	2
	2748	
K65K-AAR/D67N-AAC	--- --- —R --- A— —[a]	1
	2747 2748	
K65K-AAG/K66K-AAG/D67N-AAC	--- --- —G —G A— —	1
	2741 2748	
K65K-AAG/D67N-AAC/K64K-AAA	--- —A —G --- A— —	1
	2747 2750	
K66K-AAG/D67D-GAT	--- --- --- —G —T —	1
	2748 2750	
D67N-AAT	--- --- --- --- A-T —	1
	2750 2751	
D67D-GAT/S68G-GGT	--- --- --- --- —T G-	2
	2747	
K65R-AGA/K66K-AAG	--- --- -G- —G --- —	1
	2750	
K65R-AGA/D67D-GAT	--- --- -G- --- —T —	2

[a]R: Mixed base of A and G.

be improved by further customization to the circulating viruses. In our paper, we used maximum 23 oligoprobes in one tube or well. The evaluation of assay volume etc. would be necessary to determine actually possible maximum number of ologibrobes in

Table 4. Design of additional oligoprobes based on clinical samples.

Locus	Name of probe	Oligoprobes	Nucleotide sequence (5′-3′)
K65			ATA AAG AAA AAA GAC AGT ACT
	K66K-AAG	ATAAAGAAAAAGGACAG	--- --- --- —G --- —
	K65R-AGA/K66K-AAG	ATAAAGAGAAAGGACAG	--- --- -G- —G --- —
	K66K-AAG/D67N-AAC	ATAAAGAAAAAGAACAG	--- --- --- —G A— ---
	K65R-AGA/K66K-AAG/D67N-AAC	ATAAAGAGAAAGAACAG	--- --- -G- —G A— —
	D67N-AAC	ATAAAGAAAAAAAACAGT	--- --- --- --- A— ---
	K65R-AGA/D67N-AAC	ATAAAGAGAAAAAACAG	--- --- -G- --- A— ---
	D67D-GAT	ATAAAGAAAAAAGATAGT	--- --- --- --- —T ---
	K65R-AGA/D67D-GAT	ATAAAGAGAAAAGATAGT	--- --- -G- --- —T ---
	K65K-AAG	ATAAAGAAGAAAGACAG	--- --- —G --- --- —
	K65K-AAG/D67N-AAC	ATAAAGAAGAAAAACAGTA	--- --- —G --- A— --- -

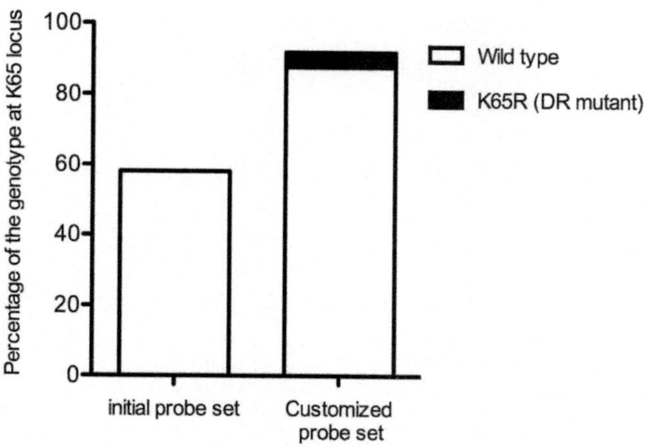

Figure 5. Improvement of the detection by the additional probes at K65 locus. The ordinate shows the percentage of the genotype at K65 locus determined by the PCR-SSOP-Luminex DR assay. One hundred % is the sample number (74) successfully amplified by PCR.

one tube. 100 color-coded beads are available, theoretically, 100 probes could be applied (http://www.luminexcorp.com/Products/Instruments/Luminex100200/). Since the wild type or mutant nucleotides are detected as the signal of hybridized probes for each in the PCR-SSOP-Luminex assay, inclusion of multiplex probes based on the codon usages for the DR mutations and the sequence variations in the flanking region of circulating viruses would improve the detection. According to our results, we could develop, validate and customized PCR-SSOP-Luminex assay for detecting DR mutations at six positions in HIV-1 RT gene. The study numbers are very limited and we worked only on subtype B HIV-1 in this article. HIV-1 diversity is notoriously huge and sequences in an individual could be more diverse than acutely expanding viruses in the field [35]. Diagnostic use in individual patients is far from the actual application at present. We believe the assay would be more suitable for molecular epidemiological studies detecting regional trends of HIV-1 DR mutation over time.

PCR-SSOP-Luminex assay were widely used for the detection of papillomavirus, influenza surveillance etc. in developing countries [36,37]. Furthermore, application of Luminex technology to other pathogen (*C. difficile*, *Norovirus*, *E. coli* or *Salmonella*, *Rotavirus A*, *Campylobacter*, *Shigella* etc.) were reported recently [38]. It would be feasible to expect that the use of Luminex technology will be wide-spread in the near future. Furthermore, the application of PCR-SSOP-Luminex DR assay to non-clade B HIV-1 is currently under development. Future studies include refinement of the assay for use with specimens co-infected with HIV-1 and HBV.

Conclusions

We have developed a rapid high-throughput assay for DR testing. The assay can be customized by adding oligoprobes suitable for the circulating viruses. The assay may turn out to be a useful method especially for public health research in both resource-rich and resource-limited settings.

Acknowledgments

The authors thank Dr. Barbara Rutledge for discussion and editing English.

Author Contributions

Conceived and designed the experiments: LG NH AKT AI. Performed the experiments: LG NH AKT. Analyzed the data: LG NH TS WS. Contributed reagents/materials/analysis tools: MM. Contributed to the writing of the manuscript: LG NH AI. Provided clinical information and patient care: HN MK TK EA TK AI. Discussion: TI GFG.

References

1. Palella FJ Jr, Delaney KM, Moorman AC, Loveless MO, Fuhrer J, et al. (1998) Declining morbidity and mortality among patients with advanced human immunodeficiency virus infection. HIV Outpatient Study Investigators. N Engl J Med 338: 853–860.
2. Samji H, Cescon A, Hogg RS, Modur SP, Althoff KN, et al. (2013) Closing the Gap: Increases in Life Expectancy among Treated HIV-Positive Individuals in the United States and Canada. PLoS One 8: e81355.
3. Frentz D, Boucher CA, van de Vijver DA (2012) Temporal changes in the epidemiology of transmission of drug-resistant HIV-1 across the world. AIDS Rev 14: 17–27.
4. WHO (2012) WHO HIV Drug Resistance Report 2012. WHO Press.
5. WHO (2013) Consolidated guidelines on the use of antiretroviral drugs for treating and preventing HIV infection: Recommendations for a public health approach. WHO Press.
6. Sungkanuparph S, Techasathit W, Utaipiboon C, Chasombat S, Bhakeecheep S, et al. (2010) Thai national guidelines for antiretroviral therapy in HIV-1 infected adults and adolescents 2010. Asian Biomedicine 4: 515–528.
7. Clavel F, Hance AJ (2004) HIV drug resistance. N Engl J Med 350: 1023–1035.
8. Wainberg MA, Friedland G (1998) Public health implications of antiretroviral therapy and HIV drug resistance. Jama 279: 1977–1983.
9. Shafer RW, Rhee SY, Pillay D, Miller V, Sandstrom P, et al. (2007) HIV-1 protease and reverse transcriptase mutations for drug resistance surveillance. Aids 21: 215–223.
10. Balajee SA, Sigler L, Brandt ME (2007) DNA and the classical way: identification of medically important molds in the 21st century. Med Mycol 45: 475–490.
11. Dunbar SA (2006) Applications of Luminex xMAP technology for rapid, high-throughput multiplexed nucleic acid detection. Clin Chim Acta 363: 71–82.
12. Itoh Y, Mizuki N, Shimada T, Azuma F, Itakura M, et al. (2005) High-throughput DNA typing of HLA-A, -B, -C, and -DRB1 loci by a PCR-SSOP-Luminex method in the Japanese population. Immunogenetics 57: 717–729.
13. Wensing AM, Calvez V, Gunthard HF, Johnson VA, Paredes R, et al. (2014) 2014 update of the drug resistance mutations in HIV-1. Top Antivir Med 22: 642–650.
14. Cheng-Mayer C, Quiroga M, Tung JW, Dina D, Levy JA (1990) Viral determinants of human immunodeficiency virus type 1 T-cell or macrophage tropism, cytopathogenicity, and CD4 antigen modulation. J Virol 64: 4390–4398
15. Shioda T, Levy JA, Cheng-Mayer C (1991) Macrophage and T cell-line tropisms of HIV-1 are determined by specific regions of the envelope gp120 gene. Nature 349: 167–169.
16. Ho SN, Hunt HD, Horton RM, Pullen JK, Pease LR (1989) Site-directed mutagenesis by overlap extension using the polymerase chain reaction. Gene 77: 51–59.
17. Koga I, Odawara T, Matsuda M, Sugiura W, Goto M, et al. (2006) Analysis of HIV-1 sequences before and after co-infecting syphilis. Microbes Infect 8: 2872–2879.
18. Hattori J, Shiino T, Gatanaga H, Yoshida S, Watanabe D, et al. (2010) Trends in transmitted drug-resistant HIV-1 and demographic characteristics of newly diagnosed patients: nationwide surveillance from 2003 to 2008 in Japan. Antiviral Res 88: 72–79.
19. Johnson VA, Calvez V, Gunthard HF, Paredes R, Pillay D, et al. (2013) Update of the drug resistance mutations in HIV-1: March 2013. Top Antivir Med 21: 6–14.
20. Erice A, Mayers DL, Strike DG, Sannerud KJ, McCutchan FE, et al. (1993) Brief report: primary infection with zidovudine-resistant human immunodeficiency virus type 1. N Engl J Med 328: 1163–1165.
21. Cane P, Chrystie I, Dunn D, Evans B, Geretti AM, et al. (2005) Time trends in primary resistance to HIV drugs in the United Kingdom: multicentre observational study. Bmj 331: 1368.
22. Descamps D, Chaix ML, Andre P, Brodard V, Cottalorda J, et al. (2005) French national sentinel survey of antiretroviral drug resistance in patients with HIV-1 primary infection and in antiretroviral-naive chronically infected patients in 2001–2002. J Acquir Immune Defic Syndr 38: 545–552.

23. Weinstock HS, Zaidi I, Heneine W, Bennett D, Garcia-Lerma JG, et al. (2004) The epidemiology of antiretroviral drug resistance among drug-naive HIV-1-infected persons in 10 US cities. J Infect Dis 189: 2174–2180.
24. Wensing AM, van de Vijver DA, Angarano G, Asjo B, Balotta C, et al. (2005) Prevalence of drug-resistant HIV-1 variants in untreated individuals in Europe: implications for clinical management. J Infect Dis 192: 958–966.
25. Yerly S, von Wyl V, Ledergerber B, Boni J, Schupbach J, et al. (2007) Transmission of HIV-1 drug resistance in Switzerland: a 10-year molecular epidemiology survey. Aids 21: 2223–2229.
26. Gatanaga H, Ibe S, Matsuda M, Yoshida S, Asagi T, et al. (2007) Drug-resistant HIV-1 prevalence in patients newly diagnosed with HIV/AIDS in Japan. Antiviral Res 75: 75–82.
27. Shi C, Eshleman SH, Jones D, Fukushima N, Hua L, et al. (2004) LigAmp for sensitive detection of single-nucleotide differences. Nat Methods 1: 141–147.
28. Church JD, Towler WI, Hoover DR, Hudelson SE, Kumwenda N, et al. (2008) Comparison of LigAmp and an ASPCR assay for detection and quantification of K103N-containing HIV variants. AIDS Res Hum Retroviruses 24: 595–605.
29. Geiss GK, Bumgarner RE, Birditt B, Dahl T, Dowidar N, et al. (2008) Direct multiplexed measurement of gene expression with color-coded probe pairs. Nat Biotechnol 26: 317–325.
30. Deng JY, Zhang XE, Mang Y, Zhang ZP, Zhou YF, et al. (2004) Oligonucleotide ligation assay-based DNA chip for multiplex detection of single nucleotide polymorphism. Biosens Bioelectron 19: 1277–1283.
31. Boltz VF, Ambrose Z, Kearney MF, Shao W, Kewalramani VN, et al. (2012) Ultrasensitive allele-specific PCR reveals rare preexisting drug-resistant variants and a large replicating virus population in macaques infected with a simian immunodeficiency virus containing human immunodeficiency virus reverse transcriptase. J Virol 86: 12525–12530.
32. Rowley CF, Boutwell CL, Lockman S, Essex M (2008) Improvement in allele-specific PCR assay with the use of polymorphism-specific primers for the analysis of minor variant drug resistance in HIV-1 subtype C. J Virol Methods 149: 69–75.
33. Zagordi O, Klein R, Daumer M, Beerenwinkel N (2010) Error correction of next-generation sequencing data and reliable estimation of HIV quasispecies. Nucleic Acids Res 38: 7400–7409.
34. Mohamed S, Ravet S, Camus C, Khiri H, Olive D, et al. (2014) Clinical and analytical relevance of NNRTIs minority mutations on viral failure in HIV-1 infected patients. J Med Virol 86: 394–403.
35. Walker BD, Korber BT (2001) Immune control of HIV: the obstacles of HLA and viral diversity. Nat Immunol 2: 473–475.
36. Jiang HL, Zhu HH, Zhou LF, Chen F, Chen Z (2006) Genotyping of human papillomavirus in cervical lesions by L1 consensus PCR and the Luminex xMAP system. J Med Microbiol 55: 715–720.
37. Vongphrachanh P, Simmerman JM, Phonekeo D, Pansayavong V, Sisouk T, et al. (2010) An early report from newly established laboratory-based influenza surveillance in Lao PDR. Influenza Other Respir Viruses 4: 47–52.
38. Claas EC, Burnham CA, Mazzulli T, Templeton K, Topin F (2013) Performance of the xTAG(R) gastrointestinal pathogen panel, a multiplex molecular assay for simultaneous detection of bacterial, viral, and parasitic causes of infectious gastroenteritis. J Microbiol Biotechnol 23: 1041–1045.

Molecular Epidemiology and Clinical Characteristics of Drug-Resistant *Mycobacterium tuberculosis* in a Tuberculosis Referral Hospital in China

Qi Wang[1☉], Susanna K. P. Lau[2☉], Fei Liu[1☉], Yanlin Zhao[3], Hong Min Li[4], Bing Xi Li[1], Yong Liang Hu[4], Patrick C. Y. Woo[2]*, Cui Hua Liu[1]*

1 CAS Key Laboratory of Pathogenic Microbiology and Immunology, Institute of Microbiology, Chinese Academy of Sciences, Beijing, China, **2** State Key Laboratory of Emerging Infectious Diseases, Department of Microbiology, The University of Hong Kong, Hong Kong Special Administrative Region, China, **3** National Center for Tuberculosis Control and Prevention, Chinese Center for Disease Control and Prevention, Beijing, China, **4** Institute for Tuberculosis Research, the 309 Hospital, Beijing, China

Abstract

Background: Despite the large number of drug-resistant tuberculosis (TB) cases in China, few studies have comprehensively analyzed the drug resistance-associated gene mutations and genotypes in relation to the clinical characteristics of *M. tuberculosis* (Mtb) isolates.

Methodology/Principal Findings: We thus analyzed the phenotypic and genotypic drug resistance profiles of 115 Mtb clinical isolates recovered from a tuberculosis referral hospital in Beijing, China. We also performed genotyping by 28 loci MIRU-VNTR analysis. Socio-demographic and clinical data were retrieved from medical records and analyzed. In total, 78 types of mutations (including 42 previously reported and 36 newly identified ones) were identified in 115 Mtb clinical isolates. There was significant correlation between phenotypic and genotypic drug resistance rates for first-line anti-TB drugs ($P<0.001$). Genotyping revealed 101 MIRU-VNTR types, with 20 isolates (17.4%) being clustered and 95 isolates (82.6%) having unique genotypes. Higher proportion of re-treatment cases was observed among patients with clustered isolates than those with unique MIRU-VNTR genotypes (75.0% vs. 41.1%). Moreover, clinical epidemiological links were identified among patients infected by Mtb strains belonging to the same clusters, suggesting a potential of transmission among patients.

Conclusions/Significance: Our study provided information on novel potential drug resistance-associated mutations in Mtb. In addition, the genotyping data from our study suggested that enforcement of the implementation of genotyping in diagnostic routines would provide important information for better monitor and control of TB transmission.

Editor: Qian Gao, Fudan University, China

Funding: This study was supported by the National Basic Research Programs of China (2014CB744400 and 2012CB518700), National Natural Science Foundation of China (81371769), the Ministry of Health and the Ministry of Science and Technology, China (2013ZX10003006 and 2012ZX10005007-011), the Chinese Academy of Sciences (KJZD-EW-L02), and the Research Fund for the Control of Infectious Diseases (Commissioned study) of the Health, Welfare and Food Bureau of the Hong Kong SAR Government (HK-09-01-10). The funders had no role in study design, data collection and analysis, decision to publish, or preparation of the manuscript.

Competing Interests: The authors have declared that no competing interests exist.

* Email: liucuihua@im.ac.cn (CHL); pcywoo@hkucc.hku.hk (PCYW)

☉ These authors contributed equally to this work.

Introduction

Tuberculosis (TB) remains a major infectious and deadly disease in the world. In addition, multidrug resistance (MDR) and extensively drug-resistance (XDR) pose a more serious problem for TB control. WHO reported that about 8.6 million people developed TB, and 1.3 million died from the disease in 2012. India and China alone accounted for 26% and 12% of global cases, respectively. In addition, the global estimate of the burden of MDR-TB was 300, 000 cases among notified TB patients in 2012. India and China were the two countries estimated to have the largest numbers of TB patients with MDR-TB [1]. Therefore, there is an urgent need to better understand the molecular epidemiology and clinical characteristics of drug-resistant *M. tuberculosis* (Mtb) isolates, so as to provide knowledge for rapid molecular diagnosis for and better management of drug-resistant TB.

Despite the high TB burden and large number of drug-resistant TB cases in China, relatively few studies have performed comprehensive analysis of drug resistance-associated mutations of clinical Mtb isolates, their genotypes, as well as their association with clinical characteristics. We previously observed high rates of

MDR- and XDR-TB among patients from a TB referral hospital in Beijing, China [2]. In addition, we established a TBDReaMDB database-coupled automatic analysis for drug resistance-associated mutations in Mtb isolates [3]. In this study, we further applied this method to perform comprehensive analysis of the drug resistance-associated mutations (genotypic drug resistance profiles) for a larger sample of Mtb clinical isolates from hospitalized TB patients. Molecular genotyping is a useful tool for analyzing Mtb strain diversity and transmission patterns, especially in high incidence settings such as TB referral hospitals. Genotyping methods based on variable number of tandem repeats (VNTR) of mycobacterial interspersed repetitive units (MIRU) have become popular methods used for TB epidemiological analysis [4,5]. We thus also evaluated the genotypic population structure for the Mtb isolates based on 28 loci MIRU-VNTR genotyping analysis and its correlation with drug resistance associated mutations and clinical characteristics and epidemiology. Our study added to the growing list of potential TB drug resistance-associated mutations, which could be invaluable in applying MDR- and XDR-TB molecular detection tests. In addition, the genotyping data from our study suggested that enforcement of the implementation of genotyping in diagnostic routines would provide important information for better monitor and control of TB transmission.

Materials and Methods

Ethics statement

The investigation protocols used in this study were approved by the institutional ethics committee of the 309 Hospital, Beijing, China. Written informed consent for Mtb isolates to be collected as well as for their information to be stored in the hospital database for research purposes was provided by participants. Written informed consent was obtained from the next of kin, caretakers, or guardians on the behalf of the minors/children participants involved in this study. Permission for using the information in the medical records of the patients for research purposes was obtained from the 309 Hospital. The Institute ethics committee of the 309 Hospital and the Institute of Microbiology, Chinese Academy of Sciences approved this study.

Mtb isolates, drug susceptibility testing and detection of drug resistance-associated mutations

Mtb isolates used in this study were obtained from TB patients being treated in the 309 Hospital, a TB referral hospital located in Beijing, China over a 3-year period (January 1, 2009–December 31, 2011). A total of 3860 culture positive TB patients were diagnosed and treated in the 309 Hospital during 2009–2011. Of these, 150 non-repetitive Mtb isolates were randomly selected to obtain pure cultures for further confirmation by p-nitrobenzoic acid and thiophene carboxylic acid hydrazine resistance tests as well as 16S rDNA sequencing analysis. Thirty-five isolates were excluded as a result of contaminated cultures or ambiguous sequencing results. The remaining 115 isolates were then subjected to DST, sequencing for drug resistance-associated gene mutations, and genotyping analysis. Epidemiologic and clinical data of the patients were extracted from the subjects' medical records. Mtb isolates were cultured using the BACTEC 960 system (Becton Dickinson Diagnostic Systems, Sparks, MD, USA) according to the manufacturer's instructions. The drug susceptibility testing (DST) were conducted according to the WHO guidelines as described previously [2]. Identification of drug resistance-associated mutations of Mtb isolates was conducted as described previously [3]. Written informed consent was obtained from participants.

MIRU-VNTR genotyping of Mtb isolates

Deletion-targeted multiplex PCR (DTM-PCR) method was used initially to identify the Beijing strains [6]. The Mtb isolates were further genotyped by a 28 loci MIRU-VNTR method based on the 24 loci described by Supply et al. [4,7] as well as the four hyper variable loci (1982, 3232, 3820, and 4120) described by Gao et al. [8] and Supply et al. [9]. The amplicons were evaluated on the 2% standard agarose gels by using a 100-bp DNA ladder (Takara), and sizing of the various VNTR alleles was done with the Peak Scanner Software v1.0 (PE Applied Biosystems). The number of repeats at each locus of MIRUs was determined and numerical values were assigned accordingly. Genetic clusters were defined as a group of two or more isolates exhibiting the same MIRU-VNTR pattern. The reference strain H37Rv was run as an additional control.

Phylogenetic analysis

The phylogenetic tree based on the 28 loci MIRU-VNTR data was constructed using PAUP 4.0b software [10] with UPGMA (unweighted pair-group method with arithmetic means) method.

Statistical analyses

All data were analyzed using the SPSS software (15.0 version). Comparisons of categorical variables were performed using the Pearson Chi-square test to compare different groups. A P value of <0.05 was considered to be statistically significant.

Results

General characteristics of the study subjects

A total of 115 non-repetitive Mtb isolates collected during 2009–2011 period were included in this study. The TB patients from whom the Mtb isolates were obtained included 53 susceptible TB cases, 21 MDR-TB cases, 17 XDR-TB cases, and 24 other types of TB cases (drug-resistant TB cases excluding MDR- and XDR-TB). All cases were HIV negative. The median age was 37 (range 14–86). Sixty-nine (60.0%) patients were male and 46 (40.0%) were female. Sixty-one (53.0%) patients were new cases and 54 (47.0%) were re-treatment cases. More detiled information on socio-demographic and clinical characteristics of the patients is shown in Table S1.

Drug resistance-associated gene mutations in Mtb isolates

We identified 78 types of mutations in drug resistance-associated genes, including 42 previously reported ones (Table 1) and 36 newly identified ones (Table 2 and Table 3). More detailed information on the PCR primers used for as well as the mutations identified in drug resistance-associated loci in all Mtb isolates were listed in Table S2 and Table S3, respectively. There was significant correlation between phenotypic and genotypic drug resistance rates for first-line anti-TB drugs ($P<0.001$ for isoniazid, rifampicin, streptomycin, ethambutol and pyrazinamide by chi-square test). The percentages of genotypic resistant isolates among phenotypic resistant isolates for first-line anti-TB drugs varied between 40.0% to 95.7%. There was also significant correlation between some of the second-line anti-TB drugs ($P<0.001$ for ofloxacin/levofloxacin and kanamycin by chi-square test). The percentages of genotypic resistant isolates among phenotypic resistant isolates for second-line anti-TB drugs varied between 0% to 41.5% (Table S4).

Tables 1. Previously reported mutations identified in drug resistance-associated loci in *M. tuberculosis* isolates.

Genes	Base Mutations	Amino acid mutations	Drugs[a]	No. of genotypic resistant isolates/No. of phenotypic resistant isolates	No. of genotypic resistant isolates/No. of phenotypic susceptible isolates	*P* value	Mutation categories
Rv0341 (*iniB*)	CTGGTGT CGGCG665 (del)	-	INH	3/47	0/68	0.035	Drug resistance mutation
Rv1483 (*mabA*)	C-15T	-	INH	1/47	0/68	0.227	Phylogenetic informative mutation
Rv1483 (*mabA*)	T-8C	-	INH	3/47	0/68	0.035	Drug resistance mutation
Rv1908c (*katG*)	G944C	S315T	INH	14/47	0/68	<0.001	Drug resistance mutation
Rv1908c (*katG*)	G1389T	R463L	INH	43/47	0/68	<0.001	Drug resistance mutation
Rv2247 (*accD6*)	A686G	D229G	INH	43/47	0/68	<0.001	Drug resistance mutation
Rv2428 (*ahpC*)	G-51A	-	INH	1/47	0/68	0.227	Phylogenetic informative mutation
Rv2428 (*ahpC*)	C-52T	-	INH	1/47	0/68	0.227	Phylogenetic informative mutation
Rv2428 (*ahpC*)	G-48A	-	INH	1/47	0/68	0.227	Phylogenetic informative mutation
Rv0667 (*rpoB*)	T1289C	L511P	RMP	2/39	0/76	0.046	Drug resistance mutation
Rv0667 (*rpoB*)	A1291G	S512G	RMP	1/39	0/76	0.161	Phylogenetic informative mutation
Rv0667 (*rpoB*)	A1304G	D516G	RMP	1/39	0/76	0.161	Phylogenetic informative mutation
Rv0667 (*rpoB*)	A1304T	D516V	RMP	1/39	0/76	0.161	Drug resistance mutation[b]
Rv0667 (*rpoB*)	A1304C	D516A	RMP	1/39	0/76	0.161	Phylogenetic informative mutation
Rv0667 (*rpoB*)	C1333G	H526D	RMP	3/39	0/76	0.046	Drug resistance mutation
Rv0667 (*rpoB*)	A1334G	H526R	RMP	1/39	0/76	0.161	Phylogenetic informative mutation
Rv0667 (*rpoB*)	C1349T	S531L	RMP	13/39	0/76	<0.001	Drug resistance mutation
Rv0682 (*rpsL*)	A128G	K43R	SM	13/37	0/78	<0.001	Drug resistance mutation
Rv0682 (*rpsL*)	A263G	K88R	SM	3/37	0/78	0.011	Drug resistance mutation
Rv3919c (*gidB*)	C413T	A138V	SM	1/37	0/78	0.145	Phylogenetic informative mutation
Rv3919c (*gidB*)	A276C	E92D	SM	37/37	59/78	0.001	Phylogenetic informative mutation[c]
Rv3794 (*embA*)	C-12T	-	EMB	2/34	0/81	0.028	Drug resistance mutation
Rv3794 (*embA*)	G1049A	G350D	EMB	1/34	0/81	0.121	Phylogenetic informative mutation
Rv3795 (*embB*)	A916G	M306V	EMB	6/34	0/81	<0.001	Drug resistance mutation
Rv3795 (embB)	G918A	M306I	EMB	1/34	0/81	0.121	Phylogenetic informative mutation
Rv3795 (*embB*)	G982T	D328Y	EMB	1/34	0/81	0.121	Phylogenetic informative mutation
Rv3795 (*embB*)	G1217C	G406A	EMB	3/34	2/81	0.127	Phylogenetic informative mutation
Rv3795 (*embB*)	G1216A	G406S	EMB	1/34	0/81	0.121	Phylogenetic informative mutation
Rv3795 (*embB*)	G1217A	G406D	EMB	1/34	0/81	0.121	Phylogenetic informative mutation
Rv3795 (*embB*)	G1412C	R471P	EMB	0/34	1/81	0.515	Phylogenetic informative mutation
Rv3795 (*embB*)	C1489A	Q497K	EMB	1/34	0/81	0.121	Phylogenetic informative mutation

Table 1. Cont.

Genes	Base Mutations	Amino acid mutations	Drugs[a]	No. of genotypic resistant isolates/No. of phenotypic resistant isolates	No. of genotypic resistant isolates/No. of phenotypic susceptible isolates	*P* value	Mutation categories
Rv3795 (*embB*)	A1490G	Q497R	EMB	0/34	1/81	0.515	Phylogenetic informative mutation
Rv2043c (*pncA*)	T384G	V128G	PZA	3/15	0/100	<0.001	Drug resistance mutation
Rv2043c (*pncA*)	GG391 (ins)	-	PZA	2/15	0/100	<0.001	Drug resistance mutation
Rv2043c (*pncA*)	C227T	T76I	PZA	1/15	0/100	0.010	Drug resistance mutation
Rv0006 (*gyrA*)	A281G	D94G	OFX, LVX	10/41	0/74	<0.001	Drug resistance mutation
Rv0006 (*gyrA*)	G284C	S95T	OFX, LVX	41/41	67/74	0.042	Phylogenetic informative mutation[d]
Rv0006 (*gyrA*)	C269T	A90V	OFX, LVX	5/41	0/74	0.002	Drug resistance mutation
Rv0006 (*gyrA*)	G280A	D94N	OFX, LVX	1/41	0/74	0.177	Phylogenetic informative mutation
Rv0006 (*gyrA*)	G280T	D94Y	OFX, LVX	1/41	0/74	0.177	Phylogenetic informative mutation
Rvnr01 (rrs)	G1332A	-	KAN	2/20	1/95	0.023	Drug resistance mutation
Rvnr01 (*rrs*)	A1401G	-	KAN	2/20	0/95	0.002	Drug resistance mutation

[a]INH, isoniazid; RMP, rifampicin; SM, streptomycin; EMB, ethambutol; PZA, pyrazinamide; OFX, ofloxacin; LVX, levofloxacin; KAN, kanamycin.
[b]Previously known to be associated with drug resistance (Luo T et al., Antimicrobial Agents and Chemotherapy, 2010).
[c]Previously known not associated with drug resistance (Spies FS et al., Journal of Clinical Microbiology, 2011).
[d]Previously known not associated with drug resistance (Giannoni F et al., Antimicrobial Agents and Chemotherapy, 2005).

Table 2. Newly identified mutations which happened alone in drug resistance-associated loci in *M. tuberculosis* isolates.

Genes	Base mutations	Amino acid mutations	Drugs[a]	No. of genotypic resistant isolates/No. of phenotypic resistant isolates	No. of Genotypic resistant isolates/No. of phenotypic susceptible isolates	*P* value	Mutation categories
Rv0667 (*rpoB*)	A1198G	T481A	RMP	1/39	0/76	0.161	Phylogenetic informative mutation
Rv0667 (*rpoB*)	A1081 (del)		RMP	3/39	0/76	0.014	Potential drug-resistant mutation
Rvnr01 (rrs)	A908C	-	SM	3/37	0/78	0.011	Potential drug-resistant mutation
Rv3919c (*gidB*)	C356A	A119D	SM	4/37	0/78	0.003	Potential drug resistance mutation
Rv3919c (*gidB*)	G579C	A193	SM	4/37	0/78	0.003	Potential drug resistance mutation
Rv3795 (*embB*)	C584T	P195L	EMB	0/34	1/81	0.515	Phylogenetic informative mutation
Rv3795 (*embB*)	G940C	G314R	EMB	1/34	0/81	0.121	Phylogenetic informative mutation
Rv3795 (*embB*)	A386G	N129S	EMB	1/34	0/81	0.121	Phylogenetic informative mutation
Rv3795 (*embB*)	A1602T	D534	EMB	1/34	0/81	0.121	Phylogenetic informative mutation
Rv3795 (*embB*)	A1687C	I563L	EMB	1/34	0/81	0.121	Phylogenetic informative mutation
Rv3795 (*embB*)	G2247 (del)		EMB	1/34	0/81	0.121	Phylogenetic informative mutation
Rv3795 (*embB*)	G2067A	A609	EMB	1/34	0/81	0.121	Phylogenetic informative mutation
Rv2043c (*pncA*)	A28(del)		PZA	1/15	0/100	0.010	Potential drug resistance mutation
Rv2043c (*pncA*)	G232T	G78C	PZA	0/15	1/100	0.697	Phylogenetic informative mutation

[a]INH, isoniazid; RMP, rifampicin; SM, streptomycin; EMB, ethambutol; PZA, pyrazinamide; OFX, ofloxacin; LVX, levofloxacin; KAN, kanamycin; ETH, ethionamide.

Table 3. Newly identified mutations which happened together with other known resistance mutations in drug resistance-associated loci in *M. tuberculosis* isolates.

Genes	Base mutations	Amino acid mutations	Drugs[a]	No. of Genotypic resistant isolates/No. of phenotypic resistant isolates	No. of genotypic resistant isolates/No. of phenotypic susceptible isolates	*P* value	Mutation categories
Rv0340	G444A	G148	INH	1/47	0/68	0.227	Phylogenetic informative mutation
Rv1592c	T70 (del)	-	INH	47/47	68/68	No data	Reference specific mutation
Rv1908c (*katG*)	T511G	D171A	INH	1/47	0/68	0.227	Phylogenetic informative mutation
Rv1908c (*katG*)	A566C	C189G	INH	1/47	0/68	0.227	Phylogenetic informative mutation
Rv2247 (*accD6*)	G747C	T249	INH	1/47	0/68	0.227	Phylogenetic informative mutation
Rv2247 (*accD6*)	T759C	D253	INH	1/47	0/68	0.227	Phylogenetic informative mutation
Rv2247 (*accD6*)	G662C	S221T	INH	1/47	0/68	0.227	Phylogenetic informative mutation
Rv2247 (*accD6*)	CG761AC	A254D	INH	1/47	0/68	0.227	Phylogenetic informative mutation
Rv2428 (*ahpC*)	G119A	S40N	INH	1/47	0/68	0.227	Phylogenetic informative mutation
Rv2428 (*ahpC*)	C-81T	-	INH	1/47	0/68	0.227	Phylogenetic informative mutation
Rv2846c (*efpA*)	T368 (del)	-	INH	1/47	0/68	0.227	Phylogenetic informative mutation
Rv0667 (rpoB)	C1819A	R688S	RMP	1/39	0/76	0.161	Phylogenetic informative mutation
Rv0667 (*rpoB*)	A1789(del)		RMP	4/39	0/76	0.004	Potential drug resistant mutation
Rv3919c (*gidB*)	T615C	A205	SM	21/37	59/78	0.012	Potential drug resistant mutation?
Rv3919c (*gidB*)	G299A	S100F	SM	37/37	78/78	No data	Reference specific mutation
Rv3793 (*embC*)	C2379T	D793	EMB	0/34	1/79	0.515	Phylogenetic informative mutation
Rv3793 (*embC*)	C2781T	R927	EMB	34/34	81/81	No data	Reference specific mutation
Rv3794 (*embA*)	C228T	C76	EMB	34/34	79/81	0.355	Phylogenetic informative mutation
Rv3795 (*embB*)	G524C	G175A	EMB	1/34	0/81	0.121	Phylogenetic informative mutation
Rv0006 (*gyrA*)	G61C	E21Q	OFX, LVX	41/41	68/74	0.061	Phylogenetic informative mutation
Rv1694 (*tlyA*)	A33G	L11	KAN	20/20	93/95	0.513	Phylogenetic informative mutation
Rv3854c (*ethA*)	A1080C	Q360H	ETH	17/17	98/98	No data	Reference specific mutation

[a]INH, isoniazid; RMP, rifampicin; SM, streptomycin; EMB, ethambutol; PZA, pyrazinamide; OFX, ofloxacin; LVX, levofloxacin; KAN, kanamycin; ETH, ethionamide.

Figure 1. UPGMA tree based on the 28 loci MIRU-VNTR data of 115 *M. tuberculosis* isolates. Note: M = MDR; X = XDR.

MIRU-VNTR genotype profiles of the Mtb isolates

Initially, we applied the DTM PCR method to identify the Beijing genotype strains, and the majority of the studied isolates belonged to Beijing genotype (96.5%, 111/115), which limits the usefulness of spoligotyping. We thus further performed 28 loci MIRU-VNTR analyses to genotype all the isolates. The detailed information on the PCR primers used as well as the 28 loci MIRU-VNTR genotype profiles for each isolates are listed in Table S2 and Table S5, respectively. A total of 101 MIRU-VNTR genotypes were detected. Twenty isolates (17.4%) were clustered and 95 isolates (82.6%) had unique genotypes (Fig. 1 and Table S6).

Association between clinical characteristics of the patients and MIRU-VNTR genotype clustering

Comparison of demographic and clinical characteristics of the patients with clustered vs. unique genotype patterns are shown in Table 4. We observed higher proportion of re-treatment cases among patients with clustered isolates than those with unique MIRU-VNTR genotypes (75.0% vs. 41.1%). However, there was no obvious association between the drug resistance as well as several other characteristics of patients (such as gender, age, underlying diseases, and geographic location, etc.) with the clustered MIRU-VNTR genotypes of the corresponding Mtb isolates. By further examining the clinical data of the clustered isolates, we noticed that two groups of clustered isolates had clinical epidemiological links. Specifically, Mtb 1056 (susceptible)

Table 4. Demographic and clinical characteristics of the patients for clustered vs. unique patterns.

Characteristics	Clustered n = 20 (%)	Unique patterns n = 95 (%)	*P* value
Gender			0.132
Male	9 (45.0)	60 (63.2)	
Female	11 (55.0)	35 (36.8)	
Age group, years			0.722
<14	0	1 (1.1)	
15–29	6 (30.0)	38 (40.0)	
30–44	5 (25.0)	20 (21.1)	
45–59	4 (20.0)	15 (15.8)	
60–74	3 (15.0)	18 (18.9)	
>75	2 (10.0)	3 (3.2)	
Marital status			0.554
Married	16 (80.0)	70 (73.7)	
Single	4 (20.0)	25 (26.3)	
Residence situation			0.623
Beijing resident	5 (25.0)	29 (30.5)	
Migrant	15 (75.0)	66 (69.5)	
Ethnicity			0.210
The largest group (Han)	20 (100.0)	88 (92.6)	
Ethnic groups	0	7 (7.4)	
Geographic location			0.295
East China	2 (10.0)	7 (7.4)	
South China	0	0	
North China	11 (55.0)	72 (75.8)	
Central China	2 (10.0)	4 (4.2)	
Northeast China	4 (20.0)	7 (7.4)	
Southwest China	1 (5.0)	2 (2.1)	
Northwest China	0	3 (3.2)	
TB treatment history			0.006
New cases	5 (25.0)	56 (58.9)	
Re-treatment cases	15 (75.0)	39 (41.1)	
Underlying diseases			
Diabetes mellitus	4 (20.0)	14 (14.7)	0.556
Hypertension	2 (10.0)	9 (9.5)	0.942
Abnormal liver function	5 (25.0)	12 (12.6)	0.157
Chronic obstructive pulmonary disease	1 (5.0)	4 (4.2)	0.875
Sites of TB			0.107
Extrapulmonary TB	0	14 (14.7)	
Pulmonary TB	12 (60.0)	58 (61.1)	
Pulmonary and extrapulmonary TB	8 (40.0)	23 (24.2)	
Radiological findings at onset			0.066
Non-cavitary	6 (30.0)	50 (52.6)	
Cavitary disease	14 (70.0)	45 (47.4)	
Drug resistance types			0.110
Susceptible	13 (65.0)	40 (42.1)	
MDR	1 (5.0)	20 (21.1)	
XDR	4 (20.0)	13 (13.7)	
Other types	2 (10.0)	22 (23.2)	
Treatment outcome			0.083

Table 4. Cont.

Characteristics	Clustered n=20 (%)	Unique patterns n=95 (%)	*P* value
Cure	15 (75.0)	79 (83.2)	
Died	1 (5.0)	0	
No data	4 (20.0)	16 (16.8)	

and 1059 (XDR) were isolated from patients having overlapping hospitalization dates in Tuberculosis ward 2, while Mtb 1053 (XDR) and 1057 (XDR) were obtained from patients having overlapping hospitalization dates in Tuberculosis ward 3. The more detailed epidemiological and clinical information of the patients from whom those clustered isolates were isolated were shown in Supplementary Table S7.

Phylogenetic analysis of Mtb isolates

A total of 38,355 bp covering drug resistance-associated loci in Mtb were used for analysis as described in a previous study [3]. From the UPGMA tree based on the 28 loci MIRU-VNTR data of Mtb isolates, we observed that clustered isolates contained both susceptible and drug-resistant isolates (Fig. 1).

Discussion

Early detection of drug resistance in Mtb is important for the successful treatment of TB. New tools for rapid and cost-effective diagnosis of drug resistance in Mtb isolates are urgently needed to control drug-resistant TB, especially the more dangerous MDR- and XDR- TB epidemics. Molecular diagnostics could potentially fill this need but require comprehensive information on the type and frequency of specific drug resistance-associated mutations. A recent study by Zhang et al. reported that through sequencing 161 Mtb isolates with a range of drug resistance profiles, they discovered 72 new genes, 28 intergenic regions, 11 non-synonymous SNPs and 10 intergenic region SNPs with strong and consistent associations with drug resistance in Mtb [11]. In this study, we further identified 36 unreported mutations in drug resistance-associated genes in clinical Mtb isolates. Our study adds to the growing body of knowledge on potential drug resistance-associated mutations in Mtb, which could lead to the development of rapid and more comprehensive molecular diagnostic methods for drug-resistant TB.

Some mutations we identified were seen in both susceptible and resistant isolates. For example, Rv1592c T70 (del) (INH), *gidB* S100F (SM), *embC* R927 (EMB), *ethA* Q360 H (ETH), *gidB* E92D (SM), gyrA (S95T) and *gyrA* E21Q (OFX, LVX) etc. were seen in identical or similar prevalence between susceptible and resistant isolates, thus those mutations should not be associated with drug resistance. Instead, they could be considered as reference specific mutations or phylogenetic informative mutations. The relatively low correlation between the phenotypic and genotypic drug resistance profiles for some drugs were consistent with our previous study and also some other studies [3,12]. Consistent with several previous studies, our observations showed that certain mutations were more frequently identified in resistant isolates vs. susceptible isolates and thus could serve as useful marker for rapid detection of resistance in the clinical Mtb isolates. For example, S315T and R463L in *katG* (for isoniazid resistance); S531L in *rpoB*; K43R and K88R in *rpsL* (for streptomycin resistance); M306V in *embB* (for ethambutol resistance); D94G and A90V in *gyrA* (for fluoroquinolone resistance); and G1332A and A1401G in *rrs* (for kanamycin resistance), etc. [13–15]. Except for those known drug resistance mutations, our studies also revealed a few more newly identified potential drug resistance mutations. For example, A1081 (del) in *rpoB*, A908C in *rrs*, A119D, A193 and A205 in *gidB*, A28 (del) in *pncA*, and A1789 (del) in *rpoB*. We then performed genetic studies by creating point mutation in the susceptible reference strain H37Rv using the pJV53K system [16] for those newly identified potential drug resistance-associated mutations, but failed to confirm their roles in causing drug resistance (data not shown). Our study further confirmed the notion that the genetic basis of drug resistance is more complex than previously anticipated, and a single point mutation could possibly contribute to but is not efficient for drug resistance [17,18].

Analysis of VNTR profiles specific to endemic strains in certain settings may be useful to identify the sources of outbreaks and transmission pathways. For example, genotyping analysis of MDR-TB isolates revealed that active transmission of MDR- and XDR-TB is taking place in Portugal, and that the high prevalence of observed XDR-TB is due to the continued transmission of particular genetic clusters [19,20]. But another recent study from Mexico showed that almost all MDR cases studied were epidemiologically unrelated, indicating that the genetic variations observed among those strains are suggestive of emergence of acquired drug-resistance during the course of treatment [21]. A large proportion of the analyzed cases in our study were primary MDR- or XDR-TB. In addition, the clustered MDR-, and XDR-TB cases among the new cases strongly indicated that those patients had been primarily infected with those highly drug-resistant Mtb strains. The observation that clustered isolates contained both susceptible and drug-resistant isolates in the UPGMA tree suggested that the drug resistance traits of Mtb strains are not conserved within defined strain clusters but rather that individual isolates within each cluster have evolved unique characteristics during long-term antibiotic treatment and complicated microbe-host interaction process. This observation further confirmed the previous notion that drug-resistant Mtb isolates have evolved and acquired different mutations independently [3]. We also observed higher proportion of re-treatment cases among patients with clustered isolates than those with unique MIRU-VNTR genotypes. This indicates that those re-treatment cases could be linked with reactivation or recent transmission of new strains. We did notice that several groups of isolates among clustered isolates had clinical epidemiological links, but since those epidemiologically linked clustered isolates had similar but not identical phenotypic and genotypic drug resistance profiles, thus those data could not support the incidence of patient-to-patient transmission during their hospitalization. A previous study indicated the limitations of using MIRU-VNTR typing and epidemiological investigations to detect transmission of Mtb in high burden settings and suggested that whole-genome sequencing (WGS) could provide better solution for tracing the transmission of Mtb in such settings [22]. Thus WGS could be used in the future to better discriminate the possibility of direct transmission events defined by the MIRU-VNTR typing in this study. Nevertheless, since the clustered isolates have a potential of transmission among patients as suggested by several previous studies [19,20,23], thus early detection of those isolates and strengthening of control measures to prevent transmission of them are warranted.

A limitation of this study is a possible sampling bias, which could be caused by the following factors. First, the sampling of the participants was not completely randomized but was based on voluntary participation of the patients and availability of the pure cultures of the isolates. Second, since the sample size is relatively small and some mutations were detected at a relatively low frequency, thus the statistical analysis used to identify potential drug resistance mutations in this study might not be suitable for some mutations. For example, D516V in *rpoB* was previously considered to be a drug resistance mutation [24], while the *P* value for this mutation was higher than 0.05. By contrast, a couple of mutations that were known not associated with drug resistance, such as *gidB* E92D [25] and *gyrA* S95T [26], were found with a *P* value lower than 0.05. Third, the sampling was not continuously done, which could lead to reduced clustering of cases. Nevertheless, this study provided important baseline data that may be used to understand and monitor the molecular epidemiology of Mtb isolates in China. Further studies are warranted to better understand the detailed molecular mechanisms in MDR- and XDR-Mtb isolates.

Supporting Information

Table S1 Demographic and clinical characteristics of the patients.

Table S2 Primers used in this study.

Table S3 Summarization of mutations identified in drug resistance-associated loci in *M. tuberculosis* isolates.

Table S4 Correlation between genotypic and phenotypic drug susceptibility testing results based on identification of reported drug resistance associated gene mutations.

Table S5 Drug resistance profiles and 28 loci MIRU-VNTR results of *M. tuberculosis* isolates.

Table S6 MIRU-VNTR fingerprinting results for 20 clustered *Mycobacterium tuberculosis* isolates.

Table S7 Drug resistance profiles and epidemiological information of the clustered isolates.

Author Contributions

Conceived and designed the experiments: CHL. Performed the experiments: QW CHL. Analyzed the data: QW SKPL FL PCYW CHL. Contributed reagents/materials/analysis tools: YZ HML PCYW CHL. Contributed to the writing of the manuscript: CHL QW SKPL BXL FL YZ HML YLH PCYW.

References

1. WHO (2013) Global tuberculosis report 2013.
2. Liu CH, Li L, Chen Z, Wang Q, Hu YL, et al. (2011) Characteristics and treatment outcomes of patients with MDR and XDR tuberculosis in a TB referral hospital in Beijing: a 13-year experience. PLoS One 6: e19399.
3. Liu CH, Li HM, Lu N, Wang Q, Hu YL, et al. (2012) Genomic sequence based scanning for drug resistance-associated mutations and evolutionary analysis of multidrug-resistant and extensively drug-resistant *Mycobacterium tuberculosis*. J Infect 65: 412–422.
4. Oelemann MC, Diel R, Vatin V, Haas W, Rusch-Gerdes S, et al. (2007) Assessment of an optimized mycobacterial interspersed repetitive- unit-variable-number tandem-repeat typing system combined with spoligotyping for population-based molecular epidemiology studies of tuberculosis. J Clin Microbiol 45: 691–697.
5. de Beer JL, Akkerman OW, Schurch AC, Mulder A, van der Werf TS, et al. (2014) Optimization of standard in-house 24-locus Variable Number of Tandem Repeat typing for *Mycobacterium tuberculosis* and its direct application to clinical material. J Clin Microbiol 52: 1338–1342.
6. Chen J, Tsolaki AG, Shen X, Jiang X, Mei J, et al. (2007) Deletion-targeted multiplex PCR (DTM-PCR) for identification of Beijing/W genotypes of *Mycobacterium tuberculosis*. Tuberculosis (Edinb) 87: 446–449.
7. Supply P, Allix C, Lesjean S, Cardoso-Oelemann M, Rusch-Gerdes S, et al. (2006) Proposal for standardization of optimized mycobacterial interspersed repetitive unit-variable-number tandem repeat typing of *Mycobacterium tuberculosis*. J Clin Microbiol 44: 4498–4510.
8. Luo T, Yang CG, Pang Y, Zhao YL, Mei J, et al. (2014) Development of a Hierarchical Variable-Number Tandem Repeat Typing Scheme for *Mycobacterium tuberculosis* in China. Plos One 9: e89726.
9. Allix-Beguec C, Wahl C, Hanekom M, Nikolayevskyy V, Drobniewski F, et al. (2014) Proposal of a Consensus Set of Hypervariable Mycobacterial Interspersed Repetitive-Unit-Variable-Number Tandem-Repeat Loci for Subtyping of *Mycobacterium tuberculosis* Beijing Isolates. Journal of Clinical Microbiology 52: 164–172.
10. Swofford DL (1993) Paup - a Computer-Program for Phylogenetic Inference Using Maximum Parsimony. Journal of General Physiology 102: A9.
11. Zhang HT, Li DF, Zhao LL, Fleming J, Lin N, et al. (2013) Genome sequencing of 161 *Mycobacterium tuberculosis* isolates from China identifies genes and intergenic regions associated with drug resistance. Nature Genetics 45: 1255–U1217.
12. Boonaiam S, Chaiprasert A, Prammananan T, Leechawengwongs M (2010) Genotypic analysis of genes associated with isoniazid and ethionamide resistance in MDR-TB isolates from Thailand. Clinical Microbiology and Infection 16: 397–399.
13. Poudel A, Maharjan B, Nakajima C, Fukushima Y, Pandey BD, et al. (2013) Characterization of extensively drug-resistant *Mycobacterium tuberculosis* in Nepal. Tuberculosis (Edinb) 93: 84–88.
14. Jnawali HN, Hwang SC, Park YK, Kim H, Lee YS, et al. (2013) Characterization of mutations in multi- and extensive drug resistance among strains of *Mycobacterium tuberculosis* clinical isolates in Republic of Korea. Diagn Microbiol Infect Dis 76: 187–196.
15. Imperiale BR, Zumarraga MJ, Di Giulio AB, Cataldi AA, Morcillo NS (2013) Molecular and phenotypic characterisation of *Mycobacterium tuberculosis* resistant to anti-tuberculosis drugs. Int J Tuberc Lung Dis 17: 1088–1093.
16. van Kessel JC, Hatfull GF (2007) Recombineering in *Mycobacterium tuberculosis*. Nature Methods 4: 147–152.
17. Wu W, Zheng H, Zhang L, Wen Z, Zhang S, et al. (2013) A genome-wide analysis of multidrug-resistant and extensively drug-resistant strains of *Mycobacterium tuberculosis* Beijing genotype. Mol Genet Genomics 288: 425–436.
18. Warner DF, Mizrahi V (2013) Complex genetics of drug resistance in *Mycobacterium tuberculosis*. Nat Genet 45: 1107–1108.
19. Perdigao J, Macedo R, Joao I, Fernandes E, Brum L, et al. (2008) Multidrug-resistant tuberculosis in Lisbon, Portugal: a molecular epidemiological perspective. Microb Drug Resist 14: 133–143.
20. Perdigao J, Macedo R, Silva C, Pinto C, Furtado C, et al. (2011) Tuberculosis drug-resistance in Lisbon, Portugal: a 6-year overview. Clin Microbiol Infect 17: 1397–1402.
21. Martinez-Guarneros A, Rastogi N, Couvin D, Escobar-Gutierrez A, Rossi LMG, et al. (2013) Genetic diversity among multidrug-resistant *Mycobacterium tuberculosis* strains in Mexico. Infection Genetics and Evolution 14: 434–443.

22. Luo T, Yang CG, Peng Y, Lu LP, Sun GM, et al. (2014) Whole-genome sequencing to detect recent transmission of *Mycobacterium tuberculosis* in settings with a high burden of tuberculosis. Tuberculosis 94: 434–440.
23. Zhang D, An J, Wang JM, Hu C, Wang ZG, et al. (2013) Molecular typing and drug susceptibility of *Mycobacterium tuberculosis* isolates from Chongqing Municipality, China. Infection Genetics and Evolution 13: 310–316.
24. Luo T, Zhao M, Li X, Xu P, Gui XH, et al. (2010) Selection of Mutations To Detect Multidrug-Resistant *Mycobacterium tuberculosis* Strains in Shanghai, China. Antimicrobial Agents and Chemotherapy 54: 1075–1081.
25. Spies FS, Ribeiro AW, Ramos DF, Ribeiro MO, Martin A, et al. (2011) Streptomycin Resistance and Lineage-Specific Polymorphisms in *Mycobacterium tuberculosis* gidB Gene. Journal of Clinical Microbiology 49: 2625–2630.
26. Giannoni F, Iona E, Sementilli F, Brunori L, Pardini M, et al. (2005) Evaluation of a new line probe assay for rapid identification of gyrA mutations in *Mycobacterium tuberculosis*. Antimicrobial Agents and Chemotherapy 49: 2928–2933.

4

Genotypes and Transmitted Drug Resistance among Treatment-Naive HIV-1-Infected Patients in a Northwestern Province, China: Trends from 2003 to 2013

Ke Zhao[1], Wenzhen Kang[1], Qingquan Liu[2], Yuan Li[1], Qing Liu[1], Wei Jiang[1], Yan Zhuang[1], Zisheng Guo[1], Zhuoran Yu[1], Xinhong Li[1], Chunfu Wang[1], Na Yao[1], Yongtao Sun[1]*

1 Department of Infectious Diseases, Tangdu Hospital, Fourth Military Medical University, Xi'an, Shaanxi Province, China, **2** Department of Endocrinology, Tangdu Hospital, Fourth Military Medical University, Xi'an, Shaanxi Province, China

Abstract

Background: Transmitted drug resistance (TDR) reduces the efficacy of initial antiretroviral treatment and has become a public health concern. Little information is available regarding the genetic diversity of HIV-1 and the prevalence of TDR among treatment-naïve patients in a northwestern province of China since the implementation of national free antiretroviral therapy (ART).

Methods: Blood samples from 372 HIV-1 treatment-naive patients were collected between 2003 and 2013 in Shaanxi province. Viral RNA was extracted for nested PCR, and phylogenetic reconstruction and recombination analyses were performed to characterize patterns of the HIV-1 subtypes. Genotypic drug resistance testing was performed using an in-house assay to determine trends in the prevalence of HIV-1 transmitted drug resistance.

Results: Multiple genotypes were identified among the patients in Shaanxi, including B (25.0%), C (0.3%), G (0.3%), and CRF01_AE (39.2%), CRF07_BC (32.7%), CRF08_BC (0.8%), CRF55_01B (1.1%), and URFs (0.6%). The subtypes were associated with the transmission routes ($\chi^2 = 77.113$, $p < 0.01$). In this study, a low baseline $CD4^+$ T cell count and a high viral load were found among CRF01_AE-infected patients compared with patients who were infected with non-CRF01_AE ($p < 0.01$) through sexual transmission; however, the CRF01_AE subtype was not associated with a low baseline $CD4^+$ T cell count or a high viral load in Chinese patients infected through blood transmission ($p = 0.249$). The overall TDR rate in this population was 4.4% between 2003 and 2013. A univariate logistic regression model revealed that a low CD4 T cell count (≤100 cells/μL) was associated with the development of drug-resistant strains.

Conclusion: Our work revealed diverse HIV-1 subtype distributions in Shaanxi province. We identified a low and stable TDR time trend among ART-naive patients. These findings enhance our understanding of HIV-1 genetic diversity and provide some guidelines for the improvement and implementation of a comprehensive public health strategy of HIV-1 TDR prevention.

Editor: Dimitrios Paraskevis, University of Athens, Medical School, Greece

Funding: This work was supported by the National Grand Program on Key Infectious Disease Control (No. 2014ZX10001002, 2013ZX10001002-002-003), Tangdu Technology Innovation and Development Program (No. 2012LCYJ002). The funders had no role in study design, data collection and analysis, decision to publish, or preparation of the manuscript.

Competing Interests: The authors have declared that no competing interests exist.

* Email: yongtaos@hotmail.com

These authors contributed equally to this work.

Introduction

Highly active antiretroviral therapy (HAART) has dramatically decreased the morbidity and mortality caused by HIV-1. According to the 2013 UNAIDS report, an estimated 35.3 million (range, 32.2–38.8 million) people were infected with HIV globally in 2012. The number of newly acquired HIV infections was 2.3 million (range, 1.9–2.7 million), showing a 33% decline in the number of new infections from 3.4 (3.1–3.7) million in 2001. The number of AIDS deaths is also declining, with 1.6 (1.4–1.9) million deaths in 2012, down from 2.3 (2.1–2.6) million deaths in 2005 [1]. Despite these achievements, HAART is not able to eliminate infected cells, and plasma viremia generally rebounds quickly after treatment is discontinued [2–3]. A safe and efficacious HIV-1 vaccine is essential for controlling the pandemic and eradicating HIV-1 infection [4]. However, the vaccine development attempts have been disappointing so far, and there is a lack of broad and potent protective neutralizing antibodies [5–6]. One of the greatest challenges in vaccine design is the rapidly evolving genetic

Table 1. Primers used in the optimized in-house methods.

Primer name	Sequence (5′-3′)	Location	Purpose
MAW 26[a]	TTGGAAATGTGGAAAGGAAGGAC	2028-2050	one-step RT-PCR
RT21[a]	CTGTATTTCTGCTATTAAGTCTTTTGATGGG	3539-3509	one-step RT-PCR
PRO-1[a]	CAGAGCCAACAGCCCCACCA	2147-2166	Second-round PCR
RT20[a]	CTGCCAGTTCTAGCTCTGCTTC	3462-3441	Second-round PCR
MAW26-07BC[b]	TGGAAATGTGGAAAAGAAGGAC	2029-2050	one-step RT-PCR
RT21-07BC[b]	CTGTATTTCAGCTATCAAGTCTTTTGATGGG	3539-3509	one-step RT-PCR
PRO1-07BC[b]	CAGAGCCAACAGCCCCACCA	2147-2166	Second-round PCR
RT20-07BC[b]	CTGCCAATTCTAATTCTGCTTC	3462-3441	Second-round PCR
MAW26-01AE[c]	TGGAAATGTGGRAARGAAGGAC	2029-2050	one-step RT-PCR
RT21-01AE[c]	GTAYTTCTGCYAYTAAGTCTTTTGATGGG	3537-3509	one-step RT-PCR
PRO1-01AE[c]	CAGAGCCAWCAGCCCCACCA	2147-2166	Second-round PCR and Forward sequencing
RT20-01AE[c]	CTGCCAAYTCTAATTCTGCTTC	3462-3441	Second-round PCR and Reverse sequencing
S1	GGACCTACACCTGTCAAC	2484-2501	Forward sequencing
S2	GCTGGGTGTGGTATTCC	2842-2826	Reverse sequencing
S3	CCTAGTATAAACAATGAGACAC	2946-2967	Forward sequencing

Primer c was preferred, Primer a or b was alternative.

diversity of HIV-1. As a result, monitoring the genetic diversity of HIV-1 is very important for understanding the molecular epidemiology and controlling the spread of the HIV-1 epidemic.

Different HIV-1 subtypes have distinct regional distribution patterns in China. CRF08_BC was found to be dominant in Yunnan province, CRF01_AE was dominant in Guangxi province, and subtype B (especially Thai-B) was dominant in Henan province [7]. Shaanxi province is located in the northwestern region of China, and it has a low prevalence of HIV-1 compared with the other provinces; However, the proportion of patients infected via MSM has been increasing rapidly each year. The HIV/AIDS epidemic has presented novel characteristics; However, comprehensive analyses on the viral genotypes in Shaanxi remain markedly limited.

Drug-resistant HIV-1 strains could be transmitted from one individual to another. Due to such transmitted drug resistance (TDR), a newly infected patient might carry a drug-resistant virus even if he or she has not yet used antiretroviral drugs. The expansion of antiretroviral treatment programs have led to increasing concern about the development of TDR. Available data revealed that between 10% and 17% of ART-naïve patients in Europe, the United States, Japan, and Australia have drug resistance to at least one antiretroviral drug [8–10]. A previous survey performed between 2004 and 2005 demonstrated that the rate of TDR was relatively low (3.8%) in China compared with the rate in developed countries [11]. However, the effect of TDR might be higher in patients in China than in patients in developed countries because in China, viral genotyping is typically unavailable, and transmitted resistance is rarely detected. Insufficiently strong drugs would be less effective in reducing the viral load, which could lead to the development of multi-class drug resistance. In addition, fewer first- or second-line treatment options are available for patients in China. Thus, surveillance data about the prevalence of TDR based on a large population over time are necessary to develop a rational public health strategy. HIV-1 TDR data from China are increasingly available. For example, a recent study revealed that the prevalence of TDR from 2007 to 2010 was 1.3% in Chengdu [12], and an overall 4.9% TDR rate was found among ART-naive MSM in 19 provinces/cities in 2012 [13].

Table 2. The List of primers used in Bootscanning analysis.

Primer name	Sequence (5′-3′)	Location	Purpose
PG3	AGGGACTTGAAAGCGAAAG	648-666	one-step RT-PCR
RT21	CTGTATTTCTGCTATTAAGTCTTTTGATGGG	3539-3509	one-step RT-PCR
gag-F2	ATGGGTGCGAGAGCGTCARTATTAA	790-814	Second-round PCR
RT20	CTGCCAGTTCTAGCTCTGCTTC	3462-3441	Second-round PCR
POL-1	TGGAAATGTGGRAARGAAGGAC	2029-2050	one-step RT-PCR
POL-2	CCTGTCTGYAGRCCCCAATATGTT	5264-5241	one-step RT-PCR
POL-3	ACTGARAGACAGGCKAATTTTTTAGGGA	2068-2095	Second-round PCR
POL-4	CTCCTAKTGGGATRTGTACTTCYGARCTTA	5221-5192	Second-round PCR

Table 3. The baseline characteristics of the study population.

Total	**364**	**Percentage**
Gender		
Male	304	83.5%
Female	60	16.5%
Age		
≤25	75	20.7%
26–35	130	35.7%
36–45	93	25.5%
>45	66	18.1%
Transmission routes		
Heterosexual contact	104	28.6%
Homosexual contact	177	48.6%
Blood transmission	49	13.5%
Intravenous drug injection	14	3.8%
Unknown	20	5.5%
$CD4^+$ T Count(cells/ul)		
≤100	109	29.9%
101–200	44	12.1%
201–350	92	25.3%
351–500	69	19.0%
>500	50	13.7%
HIV viral load(copies/ml)	364	
≤1000	34	9.3%
1001–10000	54	14.8%
10001–100000	105	28.9%
100001–1000000	132	36.3%
>1000000	39	10.7%
Year of HIV diagnosis		
2003–2005	41	11.3%
2006–2008	36	9.9%
2009–2011	122	33.5%
2012–2013	165	45.3%
HIV-1 subtype		
CRF01_AE	143	39.2%
CRF07_BC	119	32.7%
B	91	25.0%
CRF55_01B	4	1.1%
CRF08_BC	3	0.8%
URF(CRF01_AE/07_BC)	2	0.6%
C	1	0.3%
G	1	0.3%

URFs, unique recombinant forms; CRFs, circulating recombinant forms.

However, very few genetic studies have been performed on HIV-1 TDR in Shaanxi province.

In the present study, to assess the prevalence of transmitted drug resistance and identify the pattern of circulating subtypes among HAART-naive patients, plasma specimens from 372 HIV-1 treatment-naive patients were collected between 2003 and 2013. Viral RNA was extracted for nested PCR, and phylogenetic reconstruction and recombination analyses were performed to characterize the patterns of the HIV-1 subtypes. Genotypic drug resistance testing was conducted using an in-house assay to determine trends in the prevalence of HIV-1 transmitted drug resistance among HIV-1 infected individuals in Shaanxi province.

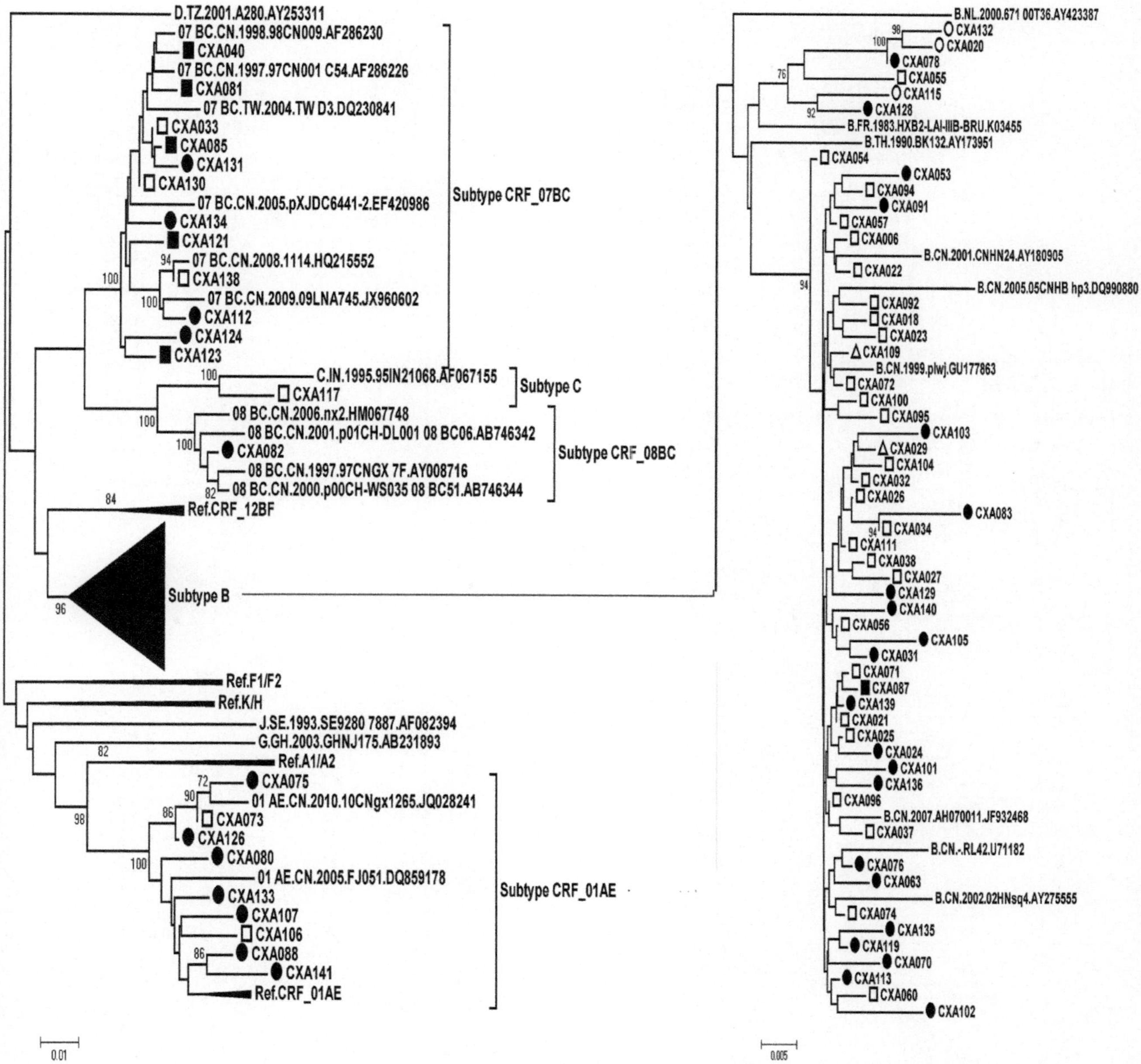

Figure 1. Phylogenetic tree analysis from ART-naïve patients diagnosed between 2003–2008. 77 sequences of HIV-1 gag-pol (1057 bp) were used and each reference sequence is labeled with the HIV-1 subtype, followed by country, year, sampling name, and accession number. The trees were constructed by the neighbor-joining method. Bootstrap values were calculated from 1,000 analyses, and bootstrap probabilities greater than 70% are indicated at the corresponding nodes of the tree. Open circle: homosexual contact, solid circle: heterosexual contact, open square: blood transmission, solid square: IDU, open triangle: unknown.

Materials and Methods

Study Subjects

A total of 372 newly diagnosed HIV-1-positive patients who visited the infectious diseases clinic in Tangdu hospital from January 2003 to December 2013 were enrolled in the study. The department of infectious diseases in Tangdu hospital is the main HIV/AIDS clinic in Shaanxi province. This clinic is responsible for HIV intervention and prevention programs, and provides free ART with care in Shaanxi province. HIV-1 infection status was determined by an Enzyme-Linked Immunosorbent Assay (ELISA, Livzon, China) and confirmed by Western blot assay (HIV BLOT 2.2, MP Diagnostics, Singapore). Inclusion criteria for this study were age 18–75 years, both genders, antiretroviral-naive, newly diagnosed and willingness to sign informed consent. The main exclusion criteria were pregnancy or breastfeeding, anticipated poor adherence.

Ethics Statement

This study was reviewed and approved by the institutional review board of the Fourth Military Medical University in Xi'an, Shaanxi, China. Written informed consent was obtained from every participant before blood sample donation.

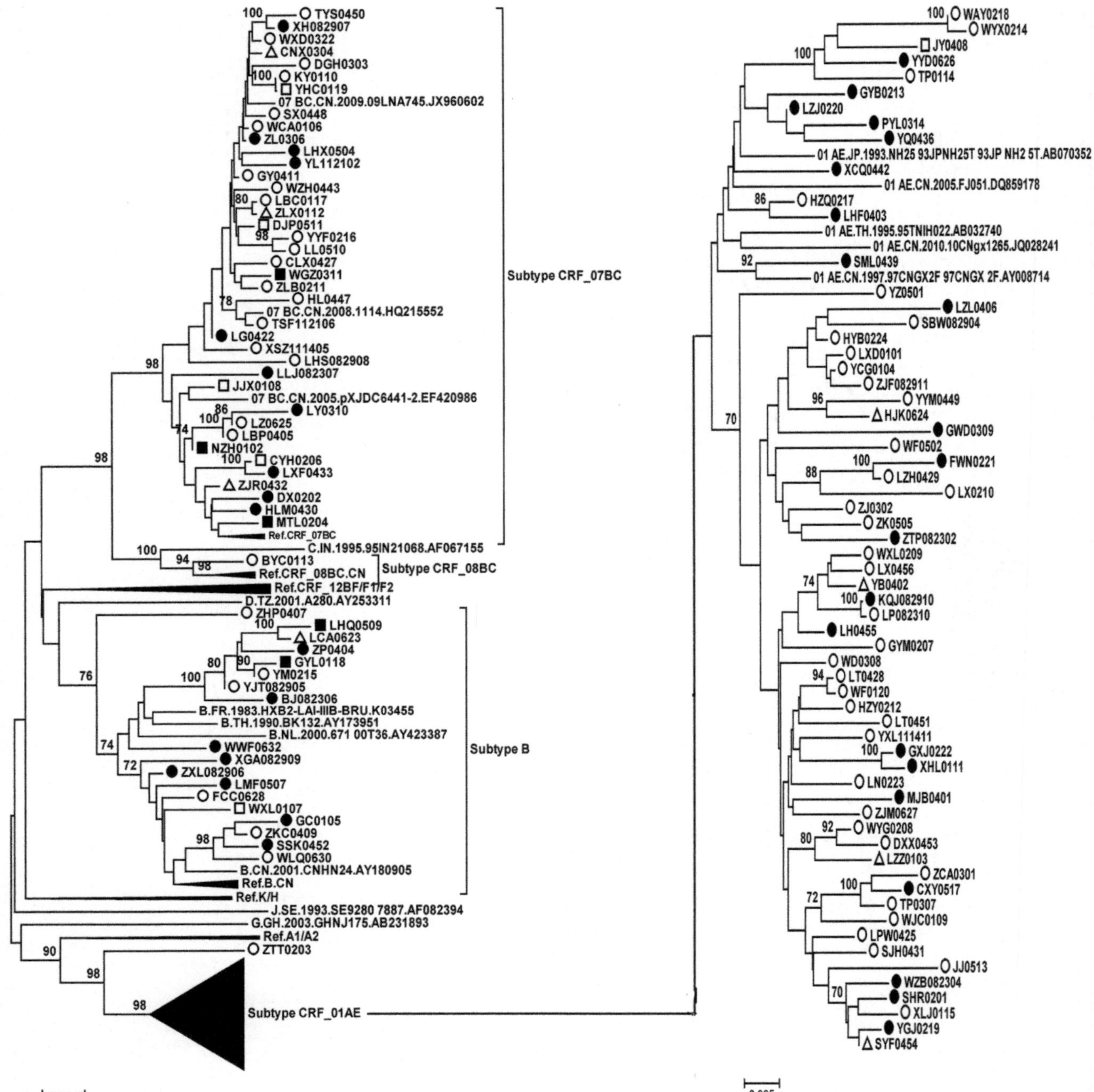

Figure 2. Phylogenetic tree analysis from ART-naïve patients diagnosed between 2009–2011. 122 sequences of HIV-1 gag-pol were used and the lengths were 1057 bp by using HXB2 as the reference genomic. Each reference sequence is labeled with the HIV-1 subtype, followed by country, year, sampling name, and accession number. Isolate ZTT0203 was not clustered with the basic reference subtypes/CRFs. The trees were constructed by the neighbor-joining method. Bootstrap values were calculated from 1,000 analyses, and bootstrap probabilities greater than 70% are indicated at the corresponding nodes of the tree. Open circle: homosexual contact, solid circle: heterosexual contact, open square: blood transmission, solid square: IDU, open triangle: unknown.

Plasma HIV-1 RNA assay and $CD4^+$ T-cell count

The HIV-1 viral load was determined using real-time reverse-transcription and polymerase chain reaction (RT-PCR) using a Cobas TaqMan 48 HIV-1 test with the high pure system (COBAS TaqMan48, Amplink version 3.2; Roche Molecular Systems, Branchburg, NJ), as previously described [14]. $CD4^+$ T cell counts were performed on a FACS Calibur (Becton Dickinson, San Jose, CA) using the TRITEST three-color CD4/CD8/CD3 reagent and TRUCOUNT tubes (Becton Dickinson, San Jose, CA) according to the manufacturer's instructions. The cells were analyzed with a multiset automatic analysis software program.

kang170
kang231
kang042
kang064
kang024
kang209
kang219
kang059
kang001a
kang092
kang029
kang094
kang126
kang008a
kang013
kang017
kang205
kang206
07 BC.CN.2009.09LNA745.JX960602
kang154
kang005a
kang069
kang127
kang194
kang138
kang088
kang109
kang098
kang025a
kang139
kang009a
kang215
kang020as
kang014
kang135
kang125
kang217
kang111
kang009
kang080
kang056
kang130
kang004
kang037
kang230
kang171
kang116
kang128
kang157
kang083
kang136
kang011
kang103
kang240
kang133
kang210
kang105
kang015as
kang107
kang095
07 BC.CN.2008.1114.HQ215552
kang158
kang220
kang038
kang120
Ref.CRF_07BC
kang144
kang241
kang077
kang022
kang203
kang108
Subtype CRF_07BC
C.IN.1995.95IN21068.AF067155
kang115
Ref.CRF_08BC.CN
Subtype CRF_08BC
kang074
kang142
kang003a
kang102
kang178
kang132
B.FR.1983.HXB2-LAI-IIIB-BRU.K03455
B.TH.1990.BK132.AY173951
B.NL.2000.671 00T36.AY423387
kang193
kang208
B.CN.-.RL42.U71182
kang030
kang112
kang113
B.CN.1999.plwj.GU177863
kang114
B.CN.2005.05CNHB hp3.DQ990880
B.CN.2007.AH070011.JF932468
kang081
B.CN.2002.02HNsq4.AY275555
B.CN.2001.CNHN24.AY180905
kang012
kang018
kang156
kang008
kang046
Subtype B
Ref.K/H/J/D/F1/F2/CRF_12BF
kang019as
kang140
kang124s
kang007a
kang190
kang253
kang001
G.GH.2003.GHNJ175.AB231893
Subtype G
A2.CD.1997.97CDKTB48.AF286238
A1.KE.1994.Q23 17.AF004885
Subtype CRF_01AE
0.01
kang002a
kang192
kang010a
kang202
kang117
kang012as
kang015
kang141
kang039
kang143
kang052
01 AE.CN.1997.97CNGX2F 97CNGX 2F.AY008714
01 AE.TH.1995.95TNIH022.AB032740
01 AE.JP.1993.NH25 93JPNH25T 93JP NH2 5T.AB070352
kang013as
kang155
kang174
01 AE.CN.2010.10CNgx1265.JQ028241
01 AE.CN.2005.FJ051.DQ859178
kang014as
kang131s
kang065
kang067
kang106
kang134
kang060
kang016as
kang165
kang087
kang072
kang062
kang145
kang034
kang035
kang251
kang011as
kang089
kang146
kang148
kang019
kang054
kang118s
kang100
kang176
kang254
kang099
kang191
kang055
kang032
kang010
kang104
kang007
kang026
kang252
kang084
kang051
kang079
kang024a
kang121
kang232
kang082
kang004a
kang153
kang119
kang137
kang031
kang159
kang224
kang101
kang221
kang097
kang218
kang025
kang093
kang096
0.005

Figure 3. Phylogenetic tree analysis from ART-naïve patients diagnosed between 2012–2013. 165 sequences of HIV-1 gag-pol were used and The lengths were 1057 bp by using HXB2 as the reference genomic. Each reference sequence is labeled with the HIV-1 subtype, followed by country, year, sampling name, and accession number. Isolates: kang007a, kang190, kang253, kang019as, kang140 were not clustered with the basic reference subtypes/CRFs. The trees were constructed by the neighbor-joining method. Bootstrap values were calculated from 1,000 analyses, and bootstrap probabilities greater than 70% are indicated at the corresponding nodes of the tree. Open circle: homosexual contact, solid circle: heterosexual contact, open square: blood transmission, solid square: IDU, open triangle: unknown.

Amplification and sequencing of HIV-1 pol

The viral RNA was extracted from 140 μL of plasma with the QIAamp Viral RNA Mini Kit (Qiagen, Hilden, Germany). Reverse transcription and first-round PCR were performed by one-step RT-PCR (Prime Script One Step RT-PCR Kit, Takara, Dalian, China). The RT-PCR and sequencing primers that were optimized based on previous reports [15] are shown in Table 1. The thermal cycling consisted of 50°C for 30 min and 94°C for 5 min, followed by 30 cycles of 94°C for 30 s, 55°C for 30 s, and 72°C for 2 min. The final incubation was at 72°C for 10 min. The nested PCR was performed in a 50-μL reaction mixture; the cycling conditions were 94°C for 5 min, followed by 30 cycles of PCR at 94°C for 30 s, 63°C for 30 s, 72°C for 2.5 min, and an extension at 72°C for 10 min. The PCR product of 1315 bp in length contained a full-length protease (PR) gene of 99 amino acid codons and the first 299-codon segment of the reverse transcriptase (RT) gene. The PCR products were purified using the QIAquick Gel Extraction Kit (Qiagen, Hilden, Germany) and sequenced using an ABI 3730 autosequencer followed by editing with SeqScape software v2.5 (Applied Biosystems, Foster City, CA).

Sequence analysis

The sequence contig assembly was performed using the analysis software, Sequencher 4.8b1 (Gene Codes Corporation, Ann Arbor, MI). Transmitted HIV-1 drug resistance was defined as one or more mutations from the Surveillance Drug Resistance Mutations (SDRM) list recommended by the World Health Organization [16]. The mutations were interpreted by the Stanford HIVdb Program Genotypic Resistance Interpretation Algorithm (http://sierra2.stanford.edu/sierra/servlet/JSierra) to predict the susceptibility of the viruses to antiretroviral drugs, and low level resistance or greater was reported.

HIV-1 Subtyping and Bootscanning analysis

The HIV-1 subtyping was performed by phylogenetic analyses of the pol sequences. The ClustalW Multiple Alignment [17] and manual editing were performed using BioEdit Sequence Alignment Editor (Ibis Biosciences, Carlsbad, CA) with HIV-1 reference sequences from the Los Alamos database (http://www.hiv.lanl.gov). The phylogenetic trees were generated by the neighbor-joining method under the Kimura two-parameter mode (1,000 replicates) with MEGA5.05 [18].

To demonstrate possible intersubtype mosaicism, sequences that were not grouped with any regular HIV-1 subtype were again amplified for entire full-length gag-pol gene sequencing (790 bp-5221 bp on HXB2). Two sets of primers designed for the determination of gag-pol (790 bp-3462 bp on HXB2) and pol (2068 bp-5221 bp on HXB2) sequences are shown in Table 2. For the gag-pol gene amplification, the cycling conditions of one step RT-PCR in a 25 μl reaction were 50°C for 30 min, 94°C for 5 min, 32 cycles of 94°C for 45 s, 52°C for 45 s, and 72°C for 5 min, followed by an extension at 72°C for 10 min. The nested PCR in a 50-μl reaction had the same conditions as above. For the pol amplification, the one Step RT-PCR cycling conditions were 50°C for 30 min, 94°C for 2 min, followed by 3 cycles of 94°C for 30 s, 55°C for 1 min, and 72°C for 4 min, 32 cycles of 94°C for 15 s, 55°C for 30 s, and 72°C for 3 min 30 s, and the final incubation was at 72°C for 10 min. The nested PCR cycling conditions were 94°C for 2 min, followed by 3 cycles of 94°C for 30 s, 62°C for 1 min, and 72°C for 4 min, 32 cycles of 94°C for 15 s, 62°C for 30 s, and 72°C for 3 min 30 s, and the final incubation was at 72°C for 10 min. The amplicons were directly sequenced using internal walking primers. The sequences were analyzed using the Recombination Identification Program (RIP, version 3.0; http://hiv-web.lanl.gov) [19], and the bootscanning analysis was performed using Simplot (SimPlot version 3.5.1; S. Ray, Johns Hopkins University, Baltimore, MD; http://sray.med.som.jhmi.edu/RaySoft/SimPlot/) [20]. The following parameters were used: window: 200 bp, step: 20 bp, gapstrip: on, replicates: 100, Kimura 2-parameter, T/t: 2.0, Neighbor-Joining.

Statistical analysis

The statistical analysis was conducted using the SPSS 17.0 package (SPSS Inc. Chicago, IL, USA). The figures were drawn using GraphPad Prism 5 (GraphPad Software, SanDiego CA) and Origin 7.5 (OriginLab Corporation, Northampton, MA). The categorical variables were compared using the chi-squared test or Fisher's exact test when appropriate, and two independent samples nonparametric tests (Mann-Whitney U) were used to compare the continuous data. A logistic regression analysis model was used to identify the factors associated with drug resistance. The factors with significant univariate P-values <0.1 were selected for inclusion in the multivariate model, which used a forward stepwise method. All of the statistical tests were 2-tailed ($P<0.05$).

Gene accession numbers

All of the sequences obtained in this study were submitted to GenBank with accession numbers KJ401414 to KJ401768, KJ 364641 and KJ364642.

Results

Three hundred seventy-two HIV-1 confirmatory samples were collected in this study; all of the samples were successfully amplified. Eight (2.15%, 8/372) PCR products could not be sequenced, resulting in a final analysis of 364 samples from treatment-naive, HIV-1 positive individuals in Shaanxi who newly diagnosed between 2003 and 2013. The baseline characteristics of the study population were shown in Table 3: 83.5% (304/364) of the patients were male, and the median age of the study group was 34 (IQR: 27–42) years. The median nadir $CD4^+$ T cell count was 257 (IQR: 70–386) cells per microliter, and the median viral load was 4.94 log10 (IQR: 4.03–5.56) copies/ml. Although an HIV viral load of greater than 3 log copies/ml was traditionally presumed to be necessary for genotypic resistance testing [21], the samples with an HIV viral load of 1.94–7.88 log10 copies/ml were successfully amplified using optimizing primers. All of the patients were recruited within 6 months of diagnosis.

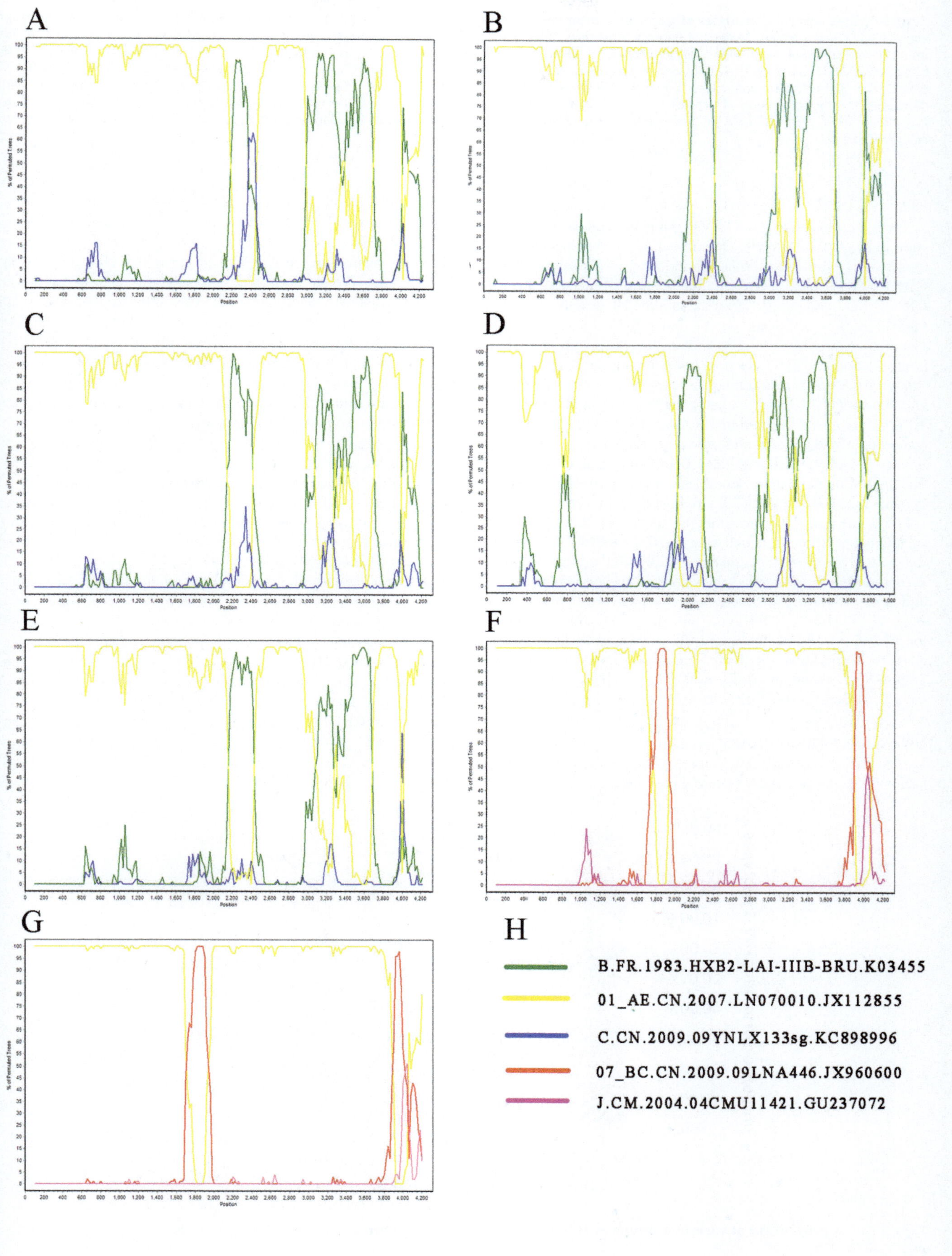
A
B
C
D
E
F
G
H
B.FR.1983.HXB2-LAI-IIIB-BRU.K03455
01_AE.CN.2007.LN070010.JX112855
C.CN.2009.09YNLX133sg.KC898996
07_BC.CN.2009.09LNA446.JX960600
J.CM.2004.04CMU11421.GU237072

Figure 4. Bootscanning analysis of gag-pol sequences (4300 bp) using SimPlot 3.5.1. A, Bootscanning analysis of gag-pol sequences of kang 007a. B, Bootscanning analysis of gag-pol sequences of kang 190. C, Bootscanning analysis of gag-pol sequences of kang 253. D, Bootscanning analysis of gag-pol sequences of ztt 0203. E. Bootscanning analysis of gag-pol sequences of the representative CRF55_01B (CN.2011.GDDG318.JX574662). F, Bootscanning analysis of gag-pol sequences of kang 019as. G, Bootscanning analysis of gag-pol sequences of kang 140. H, The reference sequences (subtype B, CRF01_AE, CRF07_BC) were shown and subtype C, J was used as an outlier. Bootscanning analysis illustrated the recombination structures for CRF55_01B and the newly identified URFs. The closely related parental strains for the URFs were identified: CRF01_AE and CRF07_BC. The condition used for C: Window: 200 bp, Step: 20 bp, GapStrip: on, Reps: 100, Kimura (2-parameter), T/t: 2.0. Neighbor-joining.

HIV-1 Variants

Three hundred sixty-four HIV-1 pol gene sequences (1057 bp) were subjected to subtype determination by phylogenetic tree analysis (Figure 1–3). One hundred forty-three (39.2%) CRF01_AE strains, one hundred and nineteen (32.7%) CRF07_BC strains, 91 (25.0%) subtype B strains, 3 (1.0%) CRF08_BC, 1 (0.27%) subtype G, and 1 (0.27%) subtype C were identified. Six isolates (kang007a, kang190, kang253, kang019as, kang140, and ztt0203) were not clustered with the basic reference subtypes/CRFs. Bootscanning analysis revealed that isolates ztt0203, kang007a, kang190, and kang253 were similar to a novel HIV-1 circulating recombinant form CRF55_01B, which is composed of CRF01_AE and subtype B, with four recombination breakpoints in the pol gene [22]. Bootscanning analysis revealed that two other isolates (kang019as and kang140) showed a recombinant CRF01_AE/07BC pol gene (Figure 4).

Over the past 10 years, the prevalence of the CRF01_AE subtype in patients increased from 9.76% to 43.03%. A similar trend was observed for subtype CRF07_BC infections, which accounted for 12.20% of all infections in the period 2003-2005 and increased during the last decade, reaching 41.21% in the years 2012–2013. During 2003–2005, subtype B infections accounted for 73.17% of all infections, and this decreased to 11.52% from 2012 to 2013 (Figure 5). Increasing HIV subtype diversity was found. Novel recombinant forms: CRF55_01B and URFs, were identified in the period 2012–2013.

HIV-1 transmission routes

The major transmission routes in the 364 patients were homosexual contact (48.6%) and heterosexual contact (28.6%), followed by blood transmission (13.5%), injection drug use (3.8%), and unknown (5.5%). Substantial changes in the transmission route occurred in the population of HIV-1 infected patients over time in Shaanxi province. The most important changes over the last decade were observed in the proportion of men who have sex with men (MSM). Before 2006, HIV-1 was spread mainly through blood and heterosexual contact transmission (56.10% and 29.27%, respectively). Between 2006 and 2008, the proportions of HIV-1 transmission through blood and homosexual contact were 25.56% and 27.78%, respectively. From the beginning of 2009, homosexual transmission of HIV displayed explosive growth and became the dominant route of HIV-1 transmission, accounting for 52.46% of the spread. In last time period (2012–13), homosexual transmission exceeded the other routes of HIV-1 transmission and further increased to 66.67% of the spread. The distribution of the routes of HIV-1 transmission in this region is shown in Figure 6. Subtypes were associated with the transmission route ($\chi^2 = 77.113$, $p<0.01$). IDU was the main risk group for the spread of CRF07_BC ($\chi^2 = 116.42$, $p<0.01$). Clade CRF01_AE was found significantly more frequently in homosexual patients than in the other risk groups ($\chi^2 = 169.552$, $P<0.01$). Subtype B was found significantly more frequently in blood transmission patients than in the other risk groups ($\chi^2 = 31.247$, $P<0.01$)

We classified HIV-1 infected individuals as CRF01_AE and non-CRF01_AE groups according to the subtypes. As shown in Figure 7-A, a low level baseline $CD4^+$ T cell count was found among the CRF01_AE-infected persons compared with the non-CRF01_AE patients infected through sexual transmission (homosexual contact and heterosexual contact) ($p = 0.027$ and $p = 0.005$, respectively); however, the $CD4^+$ T cell count was not lower

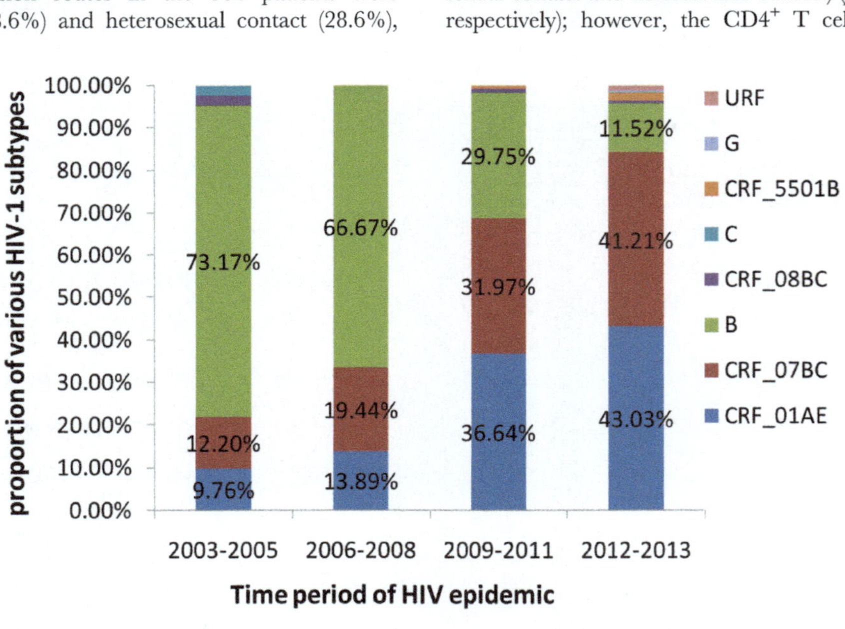

Figure 5. The proportion of various subtype of HIV-1 in Shaanxi province during four time periods: 2003–2005; 2006–2008; 2009–2011; 2012–2013.

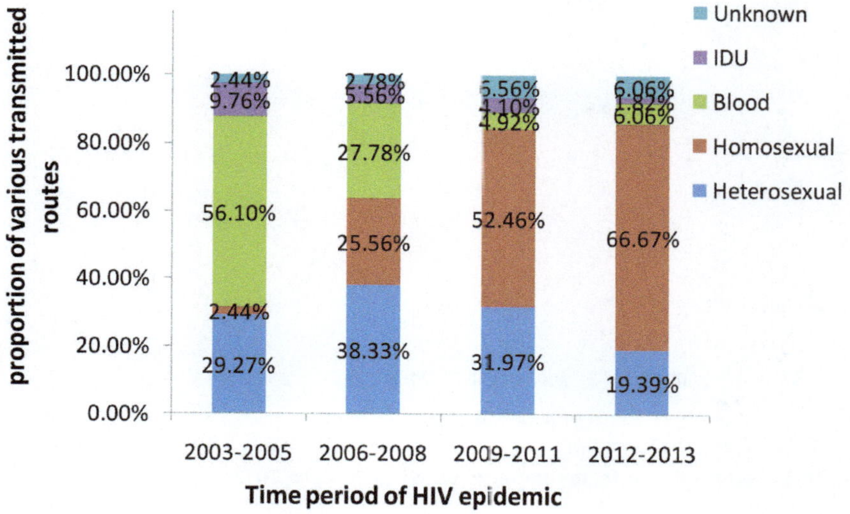

Figure 6. The proportion of various routes of HIV transmission in Shaanxi province during four time periods: 2003–2005; 2006–2008; 2009–2011; 2012–2013.

compared with the non-CRF01_AE patients who were infected through blood transmission ($p = 0.249$). The CRF01_AE group was associated with a higher viral load in patients who were infected through sexual transmission (homosexual contact and heterosexual contact) ($p = 0.0007$ and $p = 0.0044$, respectively); however, the viral load was not higher in the patients infected through blood transmission ($p = 0.66$) (Figure 7-B).

Time trends in TDR

The overall TDR rate in this population was 4.4% (16 of 364 patients) between 2003 and 2013. To assess the changes of the

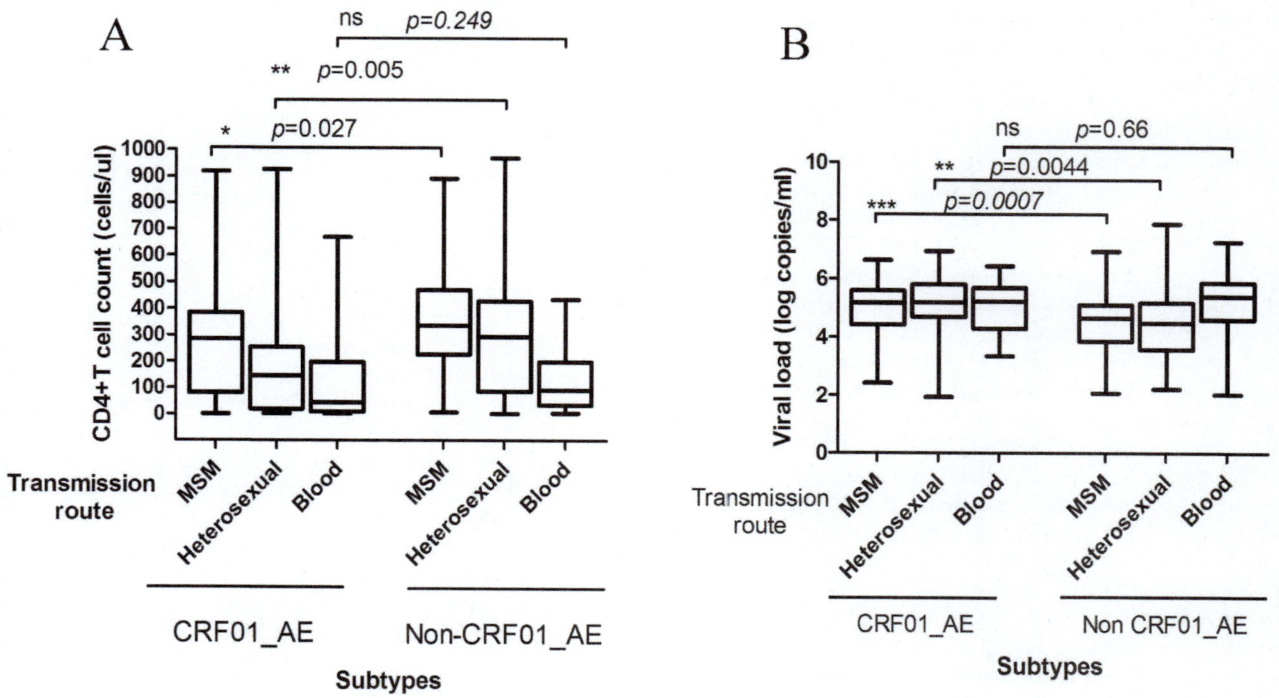

Figure 7. The levels of CD4$^+$ T cell count and viral load in CRF01_AE, non-CRF01_AE infected patients through different transmission routes. A, The levels of CD4$^+$ T cell count in CRF01_AE and non-CRFC1_AE infected patients through different transmission routes. A low level baseline CD4$^+$ T cell count was found among the CRF01_AE-infected persons compared with the non-CRF01_AE patients infected through sexual transmission; B, The levels of HIV viral load in CRF01_AE and non-CRF01_AE infected patients through different transmission routes. the CRF01_AE group was associated with a higher viral load in patients who were infected through sexual transmission (homosexual contact and heterosexual contact); however, the viral load was not higher in the patients infected through blood transmission. The statistical significance was calculated using the two independent samples nonparametric tests (Man-Whitney U). *, $p<0.05$; **, $p<0.01$; ***, $p<0.001$; ns, no significant.

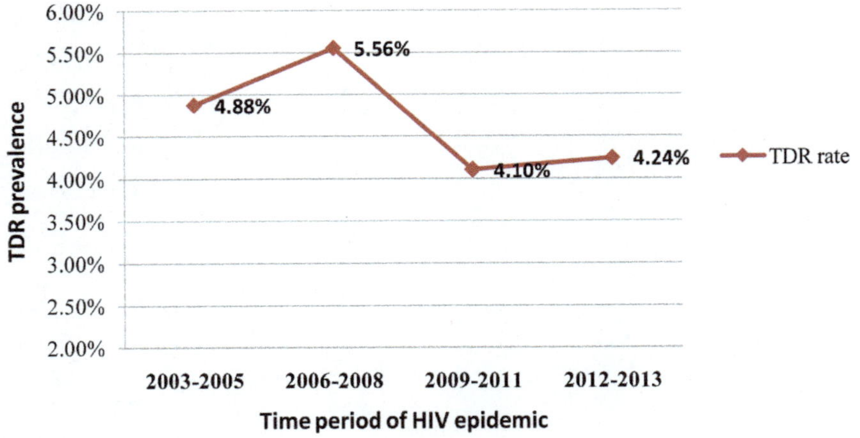

Figure 8. Time trends in TDR among HIV-1 infected patients between 2003 to 2013.

TDR prevalence over time, the sequences were grouped into four sampling time periods (2003–05, 2006–08, 2009–11, and 2012–13), according to the year of HIV-1 diagnosis. The TDR prevalence to any antiretroviral class slightly increased from the first period, peaked in 2006–08 (5.56%), decreased afterwards, and became stable between 2009 and 2013 ($P = 0.982$). (Figure 8).

Genotypic Analysis of DR

The proportion of sequences with resistance to NRTIs, NNRTIs, and PIs was 2.2% (8/364), and 1.9% (7/364), and 0.5% (2/364), respectively. When comparing resistance by drug class, dual class resistance was observed in only 1 of the 364 cases (0.27%), and none of the patients presented triple-class resistance. The PI-related mutations included I85V (0.27%, n = 1) and M46I (0.27%, n = 1). Eight NRTI-related mutations were found, including T69D (0.33%, n = 2), L74I (0.27%, n = 1), M184V (0.33%, n = 2), L210W (0.27%, n = 1), T215S (0.27%, n = 1), and K219N (0.27%, n = 1). The 7 NNRTIs-related mutations were K101E (0.27%, n = 1), K103N (0.33%, n = 2), V106M (0.33%, n = 1), G190A (0.27%, n = 2). and G190E (0.27%, n = 1). The demographic characteristics of these 16 treatment-naive individuals are shown in Table 4. According to the interpretation of the Stanford HIV Drug Resistance Database, the 16 drug resistant viruses showed different levels of resistance to 11 antiretroviral drugs (Table 5). The transmitted drug resistance to EFV and NVP was significantly higher than that to the other drugs during the study period. Although 2 NNRTIs (ETR and RPV) and 1 NRTI (FTC) were not used in local patients, resistance to these agents is present, and this resistance might be because of cross-resistance.

The baseline characteristics, including age, gender, transmission route, $CD4^+$ T-cell count, HIV-1 RNA viral load, and HIV-1 subtype, were included in the univariate logistic regression model to determine whether these factors were significantly associated with the presence of TDR (Table 6). The univariate model indicated that the only factor associated with the presence of TDR was the $CD4^+$ T cell level ($p = 0.007$). The patients classified as "≤ 100 cells/μL" were more likely to harbor TDR than were the patients in the "101–350 cells/μL" (OR = 0.195, 95% CI: 0.053 to 0.719) and the ">350 cells/μL" CD4 (OR = 0.148, 95% CI: 0.032 to 0.686) groups. Because no other factors were found to be significantly associated with TDR, the planned multivariate regression model was not attempted.

Discussion

In the present study, we performed the first systematic study of HIV-1 molecular epidemiology and transmitted drug resistance in Shaanxi province. Using the phylogenetic reconstruction and recombination analysis, a total of eight HIV-1 subtypes, including B, C, G, and CRF01_AE, CRF07_BC, CRF08_BC, CRF55_01B, and URFs, were detected. A low level prevalence of HIV-1 TDR from 2003 to 2013 was identified. The TDR rate to any antiretroviral class slightly increased from the first period (2003–2005), peaked in 2006–08 (5.56%), then decreased afterwards and became stable between 2009 and 2013.

Shaanxi province had a population of 38.3 million in 2010 according to results of the sixth nationwide population census [23]. From 2011 to 2013, the cumulative number of reported HIV-1 and AIDS cases increased each year, with the figures for each year standing at 2098, 3094 and 4332, respectively [24]. A previous study demonstrated that between 2002 and 2003, Thai-B was the most common genotype, followed by CRF07_BC and CRF01_AE [25]. Our results are in agreement with these findings. In our study, the predominance of subtype B decreased significantly from 2003 to 2013, and the predominance of CRF01_AE increased from 9.76% to 43.03% in the last decade, which showed a distinct genotypic profile compared with other regions in China. For example, in the neighboring Henan province, up to 92% of HIV-1 subtypes were Thai-B [26], CRF08_BC was predominant and accounted for 59.8% of cases in Yunnan province [7], and CRF07_BC infection remained at a high prevalence in Xinjiang province [27].

The transmission routes of HIV-1 have changed dynamically over time. While illegal and unsanitary blood collection practices have been significantly reduced in Shaanxi, the number of persons infected with HIV through MSM dramatically increased, suggesting that the trend of HIV transmission is shifting from high-risk groups to the general population. MSM was the most prevalent mode of transmission in Shaanxi between 2012 and 2013. Nationwide epidemiologic surveys from the National Center for AIDS/STD Control and Prevention have revealed that among the newly diagnosed HIV-1 infections in 2013, 69.1% of cases could be attributed to heterosexual transmission, and 20.8% could be attributed to homosexual transmission [28]. Our finding is in accordance with Beijing [29], Shijiazhuang [30], and Harbin [31] surveillance, suggesting that MSM have become the most

Table 4. Demographic characteristics of individuals infected by viruses containing TDR mutations.

ID	Gender	Age	Year Of Diagnosis	HIV-1 Subtype	Transmission route	CD4+ Count	Viral load	TDR mutation		
								PIs	NRTIs	NNRTIs
CXA-018	Male	25	2003	B	blood	46	194			**K103N**
CXA-119	Male	39	2005	B	heterosexual	8	5.82E+03	A71V		V106I **G190A**
CXA-095	Male	56	2006	B	blood	34	1.41E+07	A71T	**T69D**	
CXA-140	Female	47	2007	B	heterosexual	13	2.25E+05	A71V		**V106M**
JY102108	Male	45	2009	CRF01_AE	blood	27	9.88E+04		V75L **M184V** T215A	V106I
LPW103104	Male	28	2009	CRF01_AE	homosexual	2	6.26E+05		**T69ADN**	
CXY101017	Male	46	2010	CRF01_AE	heterosexual	31	9.89E+04			**G190E**
JJ101013	Male	31	2011	CRF01_AE	homosexual	390	1.01E+05		**M184V**	**K103N** V108I
WJC	Male	42	2011	CRF01_AE	homosexual	35	3.70E+06	L10V	**T215S**	
kang-007	Male	24	2012	CRF01_AE	homosexual	106	1.65E+05		**L74I**	
kang-035	Male	59	2012	B	heterosexual	17	1.44E+05			**K101E** E138K V179IT
kang-003a	Male	54	2013	B	heterosexual	29	5.63E+04		T69N	**G190A**
kang-016a	Male	34	2013	CRF01_AE	homosexual	83	3.48E+05	**M46I**		V179D
kang-119	Male	22	2013	CRF01_AE	homosexual	175	1.20E+03		**K219N**	
kang-128	Male	23	2013	CRF07_BC	homosexual	362	4.23E+03		**L210W**	
kang-218	Male	28	2013	CRF01_AE	unknown	146	3.57E+04	**I85V**		

Bold letters: TDR mutations; PIs: protease inhibitors; NRTIs: nucleoside reverse transcriptase inhibitors; NNRTIs: non-nucleoside reverse transcriptase inhibitors;

Table 5. The levels of resistance to antiretroviral drugs from 16 individuals with TDR strains.

NO.	ID	Low-level resistance	Intermediate resistance	High-level resistance
1	CXA-018			EFV NVP
2	CXA-119		EFV	NVP
3	CXA-095		DDI	
4	CXA-140			EFV NVP
5	JY102108	AZT TDF	ABC D4T	3TC FTC
6	LPW103104		DDI	
7	CXY101017		ETR RPV	EFV NVP
8	JJ101013	ABC		3TC FTC EFV NVP
9	WJC	AZT D4T		
10	kang-007		ABC	DDI
11	kang-035	EFV ETR	NVP	RPV
12	kang-003a		EFV	NVP
13	kang-016a	NFV		
14	kang-119			
15	kang-128	AZT D4T		
16	kang-218			

vulnerable population to HIV-1 infection in some northern metropolitan areas of China.

In this study, subtype B predominated among blood transmission, and CRF07_BC predominated among IDUs. CRF01_AE was dominant in the MSM group. These results are in agreement with other epidemiological findings from different studies [32–33] and suggest that the HIV-1 subtype was associated with the transmission route. We observed the highest viral diversity in MSM: five genotypes, consisting of four recombinants and subtype B, were detected in this population. Subtypes CRF55_01B and URFs (07BC/01AE) were first identified in Shaanxi. These results suggest that the coexistence of multiple HIV-1 genotypes may facilitate the generation of new recombination forms in MSM and make HIV-1 genetics intricate.

A recent study suggested that the CRF01_AE subtype is associated with higher viral load and faster progression from the estimated time of seroconversion to AIDS compared with non-CRF01_AE subtypes in patients infected through sexual transmission [34]. Another retrospective study revealed that a low baseline $CD4^+$ T cell count was found among the CRF01_AE subtype compared with the CRF07_BC subtype in MSM [35]. These findings suggest that the CRF01_AE subtype is associated with progression to AIDS and advanced immunosuppression. However, they did not address whether different risk populations are associated with different results. Our observation in patients infected through sexual transmission was in agreement with previous results; however, we did not observe this trend in patients infected by blood transmission, which suggests that the correlation between CRF01_AE and disease progression could be different in various risk populations. Because the number of IDUs is in our cohort was small, further studies will be performed to evaluate the influence of the HIV subtypes on HIV progression in IDU groups. Based on our results, we cannot get the conclusion that a lower CD4 cell count has a causal relationship with CRF01_AE as patients infected with a CRF01_AE subtype could have been diagnosed during a later stage of their infection than those infected with non-CRF01_AE subtype.

The overall TDR rate in this population remained low (<5%) despite 10 years of the active introduction of ART in Shaanxi province, reflecting the high efficacy of HAART in the last decade. The TDR rate between 2012 and 2013 was 4.24%, which was similar to the results from another study in Yunnan (4.3%) [15]; however, this was lower than the results from studies in Henan (14.3%) [26] and Guangdong (6.7%) [36]. Similar studies performed in the United States and in European countries reported significantly higher prevalence (10.9% and 14.2%, respectively) of TDR [8–9]. Several factors might explain the observed discrepancy. Shaanxi has a shorter history of using ART compared with the developed areas or countries, where combination therapy has been common since the late 1990s (North America, Europe). Additionally, the high rates of acquired drug resistance could affect the prevalence of TDR. In addition, the transmitted drug resistance could be potentially underestimated due to the limited sample size.

A logistic regression analysis revealed that low $CD4^+$ T cell counts (below 100 cells/μL) were closely correlated with the development of TDR. The median CD4 cell count was significantly lower among the patients with TDR compared with the patients without TDR. This finding is consistent with the results from the TREAT Asia and Hong Kong Studies [37–38], supporting the hypothesis that patients with advanced HIV-1 disease might be at greater risk of having HIV-1 TDR. Indeed, our study is a cross-sectional observation, it has several limits to identify the causal relationship between CD4 cell count and TDR. A further follow-up study will be needed to elaborate this relationship.

There are some limitations to the present study. First, HIV-1 TDR surveillance should be implemented in recently infected individuals, because over time, the mutations might revert to wild-type transmitted resistance variants (that might have reduced fitness capacity) or to the nonresistant genotypes, which could affect the accuracy of transmitted resistance surveillance [39]. Because recently infected persons are difficult to recruit and late diagnosis of HIV infection is common in china [40], our studies included individuals who might have been infected for a

Table 6. Univariate analysis of factors associated with the presence of HIV-1 TDR.

Factors	OR	95% CI	*P*	*Global P*
Gender				
Male	1			0.275
Female	0.321	0.042-2.476	0.275	
Age years				0.495
≤25	1			
26–35	0.553	0.134-2.281	0.412	
36–45	0.593	0.128-2.740	0.503	
>45	1.417	0.364-5.519	0.616	
Transmission route				0.980
Heterosexual contact	1			
Homosexual contact	0.828	0.256-2.681	0.753	
Blood transfusion	1.293	0.296-5.650	0.732	
IDU	0.000	0.000	0.999	
Unknown	1.141	0.125-10.382	0.907	
$CD4^+$ T cell (cells/ul)				0.007
≤100	1			
101–350	0.195	0.053-0.719	0.014	
>350	0.148	0.032-0.686	0.015	
HIV RNA (copies/ml)				0.974
≤1000	1			
1001–10000	2.020	0.201-20.269	0.550	
10001–100000	1.361	0.147-12.613	0.786	
100001–1000000	1.636	0.190-14.072	0.654	
>1000000	1.784	0.155-20.586	0.643	
HIV-1 subtype				0.118
CRF01_AE	1			
CRF07_BC	8.872	1.119-70.347	0.039	
B	6.860	0.787-59.783	0.081	

CI, confidence interval; OR, odds ratio.

considerably longer time and are considered to be "chronically infected". The results might be biased by the high proportion of patients with low $CD4^+$ T cell (≤200 cells/μL). The degree of this bias will be affected by the time between HIV-1 infection and diagnosis, which may have changed over time. Second, our testing was based on population sequencing and might have missed minority variants at levels <20–25% of the virus population [41].

Our study provided a timely update on the molecular epidemiology of HIV-1 among the various risk populations in Shaanxi province. The characteristics of transmitted drug resistance were described here for the first time. These findings provide important information for improving clinical practices and for policy-making.

Acknowledgments

The authors thank their colleagues in the Department of Infectious Diseases of Tang du hospital for their efforts in patient recruitment and for providing blood samples. The authors appreciate the kind assistance provided by Bing Chen (Beijing HUIHUACHEN science & technology Co., Ltd).

Author Contributions

Conceived and designed the experiments: KZ WK YS. Performed the experiments: KZ YL Qing Liu WJ CW ZY NY. Analyzed the data: KZ Qingquan Liu ZG XL. Contributed reagents/materials/analysis tools: YL ZG Qing Liu YZ. Wrote the paper: KZ. Reviewed the manuscript: WK YS.

References

1. UNAIDS website. UNAIDS report on the global AIDS epidemic. Available: http://www.unaids.org/. Accessed 2013 Dec.
2. Hamimi C, Pancino G, Barre-Sinoussi F, Saez-Cirion A (2013) Will it be possible to live without antiretroviral therapy? Curr Opin HIV AIDS 8: 196–203.
3. Chun TW, Fauci AS (2012) HIV reservoirs: pathogenesis and obstacles to viral eradication and cure. AIDS 26: 1261–1268.
4. Baden LR, Dolin R (2012) The road to an effective HIV vaccine. N Engl J Med 366: 1343–1344.
5. Buchbinder SP, Mehrotra DV, Duerr A, Fitzgerald DW, Mogg R, et al. (2008) Efficacy assessment of a cell-mediated immunity HIV-1 vaccine (the Step Study): a double-blind, randomised, placebo-controlled, test-of-concept trial. Lancet 372: 1881–1893.
6. Rerks-Ngarm S, Pitisuttithum P, Nitayaphan S, Kaewkungwal J, Chiu J, et al. (2009) Vaccination with ALVAC and AIDSVAX to prevent HIV-1 infection in Thailand. N Engl J Med 361: 2209–2220.

7. Li L, Sun G, Liang S, Li J, Li T, et al. (2013) Different distribution of HIV-1 subtype and drug resistance were found among treatment naive individuals in Henan, Guangxi, and Yunnan province of China. PLoS One 8: e75777.
8. Frentz D, Boucher CA, van de Vijver DA (2012) Temporal changes in the epidemiology of transmission of drug-resistant HIV-1 across the world. AIDS Rev 14: 17–27.
9. Poon AF, Aldous JL, Mathews WC, Kitahata M, Kahn JS, et al. (2011) Transmitted drug resistance in the CFAR network of integrated clinical systems cohort: prevalence and effects on pre-therapy CD4 and viral load. PLoS One 6: e21189.
10. World Health Organization website. WHO Drug Resistance Report. Available: http://www.who.int/hiv/pub/drugresistance/report2012/en/. Accessed 2012.
11. Liao L, Xing H, Shang H, Li J, Zhong P, et al. (2010) The prevalence of transmitted antiretroviral drug resistance in treatment-naive HIV-infected individuals in China. J Acquir Immune Defic Syndr 53 Suppl 1: S10–14.
12. Zeng P, Liu Y, He M, Gao Z, Zhou Y, et al. (2013) HIV-1 genotypic diversity and prevalence of drug resistance among treatment naive HIV-infected individuals in Chengdu of China. Virus Genes 47: 408–413.
13. Yang J, Xing H, Niu J, Liao L, Ruan Y, et al. (2013) The emergence of HIV-1 primary drug resistance genotypes among treatment-naive men who have sex with men in high-prevalence areas in China. Arch Virol 158: 839–844.
14. Huang D, Zhuang Y, Zhai S, Song Y, Liu Q, et al. (2010) HIV reverse transcriptase activity assay: a feasible surrogate for HIV viral load measurement in China. Diagn Microbiol Infect Dis 68: 208–213.
15. Chen M, Ma Y, Duan S, Xing H, Yao S, et al. (2012) Genetic diversity and drug resistance among newly diagnosed and antiretroviral treatment-naive HIV-infected individuals in western Yunnan: a hot area of viral recombination in China. BMC Infect Dis 12: 382.
16. Bennett DE, Camacho RJ, Otelea D, Kuritzkes DR, Fleury H, et al. (2009) Drug resistance mutations for surveillance of transmitted HIV-1 drug-resistance: 2009 update. PLoS One 4: e4724.
17. Larkin MA, Blackshields G, Brown NP, Chenna R, McGettigan PA, et al. (2007) Clustal W and Clustal X version 2.0. Bioinformatics 23: 2947–2948.
18. Hall BG (2013) Building phylogenetic trees from molecular data with MEGA. Mol Biol Evol 30: 1229–1235.
19. Siepel AC, Halpern AL, Macken C, Korber BT (1995) A computer program designed to screen rapidly for HIV type 1 intersubtype recombinant sequences. AIDS Res Hum Retroviruses 11: 1413–1416.
20. Lole KS, Bollinger RC, Paranjape RS, Gadkari D, Kulkarni SS, et al. (1999) Full-length human immunodeficiency virus type 1 genomes from subtype C-infected seroconverters in India, with evidence of intersubtype recombination. J Virol 73: 152–160.
21. Saravanan S, Vidya M, Balakrishnan P, Kumarasamy N, Solomon SS, et al. (2009) Evaluation of two human immunodeficiency virus-1 genotyping systems: ViroSeq 2.0 and an in-house method. J Virol Methods 159: 211–216.
22. Han X, An M, Zhang W, Cai W, Chen X, et al. (2013) Genome Sequences of a Novel HIV-1 Circulating Recombinant Form, CRF55_01B, Identified in China. Genome Announc 1.
23. Shaanxi Provincial Bureau of Statistics website. Bulletin of the sixth national population census in Shaanxi province 2010. Available: http://www.sn.stats.gov.cn/news. Accessed 2010 Dec.
24. Shaanxi Health and Family planning Commission website. The HIV/AIDS situation in Shaanxi. 2013. Available: http://www.sxhealth.gov.cn/newstyle/pub_newsshow.asp?id=1035215&chid=100235. Acessed 2013 Nov.
25. XING Ai-hua, XING Hui, Xiang L (2007) Analysis on subtype of epidemic HIV-1strains in Shaanxi province. Chin J Public Health 23.
26. Li L, Sun B, Zeng H, Sun Z, Sun G, et al. (2014) Relatively high prevalence of drug resistance among antiretroviral-naive patients from Henan, Central China. AIDS Res Hum Retroviruses 30: 160–164.
27. Xin R, He X, Xing H, Sun F, Ni M, et al. (2009) Genetic and temporal dynamics of human immunodeficiency virus type 1 CRF07_BC in Xinjiang, China. J Gen Virol 90: 1757–1761.
28. Chinese Centers for Disease Control Prevention website. 2013 China AIDS Response Progress Report.Available: http://www.chinacdc.cn/. Accessed 2013 Dec.
29. Beijing Health and Family planning Commission website. The HIV/AIDS situation in Beijing. 2013. Available: http://health.sina.com.cn/news/2013-11-29/0924114945.shtml. Accessed 2013 Nov.
30. Li L, Lu X, Li H, Chen L, Wang Z, et al. (2011) High genetic diversity of HIV-1 was found in men who have sex with men in Shijiazhuang, China. Infect Genet Evol 11: 1487–1492.
31. Shao B, Li Y, Yu L, Wang K, Chang M, et al. (2014) The HIV/AIDS epidemic characteristics in a northeast province of China–men who have sex with men have made a tremendous contribution to the growth of the HIV epidemic. J Infect 68: 273–280.
32. Ruan Y, Qin G, Liu S, Qian H, Zhang L, et al. (2005) HIV incidence and factors contributed to retention in a 12-month follow-up study of injection drug users in Sichuan Province, China. J Acquir Immune Defic Syndr 39: 459–463.
33. Pan XH, Zhang JF, Yao YP, Xu Y, Yang JZ, et al. (2010) [Subtype and transmission of HIV strain in both HIV infected spouses in Zhejiang province]. Zhonghua Liu Xing Bing Xue Za Zhi 31: 638–641.
34. Li Y, Han Y, Xie J, Gu L, Li W, et al. (2014) CRF01_AE subtype is associated with X4 tropism and fast HIV progression in Chinese patients infected through sexual transmission. AIDS 28: 521–530.
35. Li X, Xue Y, Zhou L, Lin Y, Yu X, et al. (2014) Evidence That HIV-1 CRF01_AE Is Associated with Low CD4+T Cell Count and CXCR4 Co-Receptor Usage in Recently Infected Young Men Who Have Sex with Men (MSM) in Shanghai, China. PLoS One 9: e89462.
36. Yang C, Liu S, Zhang T, Hou Y, Liu X, et al. (2012) Transmitted antiretroviral drug resistance and thumb subdomain polymorphisms among newly HIV type 1 diagnosed patients infected with CRF01_AE and CRF07_BC virus in Guangdong Province, China. AIDS Res Hum Retroviruses 28: 1723–1728.
37. Sungkanuparph S, Oyomopito R, Sirivichayakul S, Sirisanthana T, Li PC, et al. (2011) HIV-1 drug resistance mutations among antiretroviral-naive HIV-1-infected patients in Asia: results from the TREAT Asia Studies to Evaluate Resistance-Monitoring Study. Clin Infect Dis 52: 1053–1057.
38. Wong KH, Chan WK, Yam WC, Chen JH, Alvarez-Bognar FR, et al. (2010) Stable and low prevalence of transmitted HIV type 1 drug resistance despite two decades of antiretroviral therapy in Hong Kong. AIDS Res Hum Retroviruses 26: 1079–1085.
39. Pingen M, Nijhuis M, de Bruijn JA, Boucher CA, Wensing AM (2011) Evolutionary pathways of transmitted drug-resistant HIV-1. J Antimicrob Chemother 66: 1467–1480.
40. Zhang F, Dou Z, Ma Y, Zhang Y, Zhao Y, et al. (2011) Effect of earlier initiation of antiretroviral treatment and increased treatment coverage on HIV-related mortality in China: a national observational cohort study. Lancet Infect Dis 11: 516–524.
41. Li JZ, Kuritzkes DR (2013) Clinical implications of HIV-1 minority variants. Clin Infect Dis 56: 1667–1674.

5

An Automated High-Throughput Cell-Based Multiplexed Flow Cytometry Assay to Identify Novel Compounds to Target *Candida albicans* Virulence-Related Proteins

Stella M. Bernardo[1,2], Christopher P. Allen[3¤], Anna Waller[3], Susan M. Young[3], Tudor Oprea[3], Larry A. Sklar[3], Samuel A. Lee[1,2*]

1 Section of Infectious Diseases, New Mexico Veterans Healthcare System, Albuquerque, NM, United States of America, **2** Division of Infectious Diseases, University of New Mexico Health Science Center, Albuquerque, NM, United States of America, **3** Center for Molecular Discovery, University of New Mexico, Albuquerque, NM, United States of America

Abstract

Although three major classes of systemic antifungal agents are clinically available, each is characterized by important limitations. Thus, there has been considerable ongoing effort to develop novel and repurposed agents for the therapy of invasive fungal infections. In an effort to address these needs, we developed a novel high-throughput, multiplexed screening method that utilizes small molecules to probe candidate drug targets in the opportunistic fungal pathogen *Candida albicans*. This method is amenable to high-throughput automated screening and is based upon detection of changes in GFP levels of individually tagged target proteins. We first selected four GFP-tagged membrane-bound proteins associated with virulence or antifungal drug resistance in *C. albicans*. We demonstrated proof-of-principle that modulation of fluorescence intensity can be used to assay the expression of specific GFP-tagged target proteins to inhibitors (and inducers), and this change is measurable within the HyperCyt automated flow cytometry sampling system. Next, we generated a multiplex of differentially color-coded *C. albicans* strains bearing C-terminal GFP-tags of each gene encoding candidate drug targets incubated in the presence of small molecules from the Prestwick Chemical Library in 384-well microtiter plate format. Following incubation, cells were sampled through the HyperCyt system and modulation of protein levels, as indicated by changes in GFP-levels of each strain, was used to identify compounds of interest. The hit rate for both inducers and inhibitors identified in the primary screen did not exceed 1% of the total number of compounds in the small-molecule library that was probed, as would be expected from a robust target-specific, high-throughput screening campaign. Secondary assays for virulence characteristics based on null mutant strains were then used to further validate specificity. In all, this study presents a method for the identification and verification of new antifungal drugs targeted to fungal virulence proteins using *C. albicans* as a model fungal pathogen.

Editor: David R. Andes, University of Wisconsin Medical School, United States of America

Funding: This work was supported in part by funding from the Department of Veterans' Affairs (Merit Award 5I01BX001130), Biomedical Research Institute of New Mexico (SAL), and the National Institutes of Health (1U54MH084690-01) (LAS). The funders had no role in study design, data collection and analysis, decision to publish, or preparation of the manuscript.

Competing Interests: The authors of this manuscript have read the journal's policy and one author has the following competing interest: LAS is cofounder of IntelliCyt, which markets the HyperCyt high throughput flow cytometry platform.

* Email: SamALee@salud.unm.edu

¤ Current address: Department of Environmental and Radiological Health Sciences, Colorado State University, Fort Collins, CO, United States of America

Introduction

Candida infections are a major cause of morbidity and mortality in a wide range of patient populations. *Candida albicans* is a normal human commensal organism, although if host defense mechanisms are compromised, invasive candidiasis can occur in virtually any host. *Candida* species are now the fourth most common cause of hospital-acquired bloodstream infections, and the third most common cause of nosocomial urinary tract infections [1–3]. Despite improvements in antifungal therapy, the high attributable mortality rate due to *Candida* infections has improved little from two decades ago. Current antifungal agents are limited in the following aspects: (i) available formulations, (ii) accessibility to the site of infection and ability to achieve therapeutic dose, (iii) hepato- and nephro-toxicity, and/or (iv) emergence of antifungal drug resistant strains [4,5]. Azole antifungals, although usually well-tolerated, are limited by substantial resistance in several important non-albicans *Candida* species and within *C. albicans* biofilms. Amphotericin B, including lipid-formulations, while broadly effective against many fungal pathogens, is limited by substantial toxicities [6]. The echinocandins, are costly, and while highly effective against most *Candida* species, can be ineffective in the presence of point mutations in the *FKS1* gene which encodes the glucan synthase drug target [7,8]. Although the overall clinical prevalence for echinocandin-resistant *Candida* species remains low to date, there are increasing reports of echinocandin-resistant clinical isolates, particularly with the revision of Clinical Laboratory and Standards Institute (CLSI) clinical interpretative breakpoints for echinocandin susceptibilities [9–12]. Furthermore, echinocandins have limited central nervous system and urinary penetration, and are of limited to no utility against *Cryptococcus* species, some other rare opportunistic yeast species, and many invasive mold infections [13,14].

As a result of the continually recurring problem of drug resistance and antimicrobial therapy, efforts have turned towards developing therapeutic methods and preventative measures that target virulence-specific aspects of a microorganism rather than its growth. This approach is attractive because it can potentially circumvent the development of drug resistant isolates, as it targets the microorganism's potential to become virulent rather than directly targeting its growth. More importantly for antifungal drug development, there is currently a lack of fungal-specific targets, and virulence factors would thus serve as unique targets [15]. Research groups in the field have identified a growing number of *Candida albicans* virulence-related proteins, and many of these are cell surface, cell membrane-associated, or secreted proteins, including the secreted aspartyl proteases (Saps), which are degradative enzymes thought to assist in adherence and tissue invasion [16,17]. The ability to switch between yeast and hyphal morphology and/or to form biofilms has also received considerable attention [15,18,19] and there has been substantial recent effort to discover drugs against *Candida* biofilms including the development of a high-throughput *Candida* biofilm chip assay [20]. However, despite the identification and characterization of a number of potential drug targets, development of novel antifungal drugs has been lacking. Targeting of virulence factors has been an area of intensive study, particularly against bacterial pathogens. There are a number of promising anti-virulence compounds in pre-clinical development as anti-bacterial agents, although this strategy is still in its early stages [21]. This approach requires a detailed understanding of molecular pathogenesis. Importantly, significant resources are required for the pursuit of this strategy, with relatively fewer resources devoted to antifungal drug discovery. Thus, a concerted and collaborative effort between academic investigators, government and other funding sources, biotechnology companies, big pharma, and other stakeholders will be required to make substantial progress in the clinical development of anti-virulence compounds and vaccines for antifungal therapeutics [15]. Furthermore, new antifungal agents take a considerable amount of time in the developmental pipeline, including specific tests for safety-related issues. Thus, our objective was to design a method that would allow us to screen a small-compound drug library comprised of FDA-approved agents against specific proteins known to be required for antifungal resistance or virulence, in order to identify agents not previously known to have antifungal activity. The fact that the agents in this library are already FDA-approved, with long-term safety information already at hand, can potentially reduce the time and resources required to conduct clinical trials of the repurposed drug.

We first selected several membrane proteins associated with antifungal drug resistance or virulence in the opportunistic fungal pathogen *Candida albicans* as potential drug targets. *CDR1* encodes a multi-drug efflux pump that has been shown to be overexpressed in *C. albicans* laboratory and clinical isolates that are phenotypically resistant to fluconazole and other azole antifungal drugs [5,22–24]. In the clinically important biofilm lifestyle of *C. albicans*, which leads to intrinsic fluconazole resistance, overexpression of *CDR1* occurs in the early stages of biofilm formation [25,26]. *C. albicans FTR1* encodes a membrane protein required for iron transport [27,28]. Deletion of this gene results in a *C. albicans* mutant strain that is defective in virulence in a standard mouse model of disseminated candidiasis. *MLT1* encodes a vacuolar ABC-type transporter, which is also required for wild-type virulence [29]. *C. albicans SUR7* encodes a membrane-related protein that marks sites of endocytosis in the plasma membrane [30,31]. Absence of this gene results in defective biofilm formation, and in vitro macrophage killing [29], and attenuates virulence in the mouse model for disseminated candidiasis [32]. Thus, the membrane proteins encoded by these genes represent potential novel drug targets.

Next, we took advantage of the recent progress in flow cytometry and its applications for high-throughput screening to design a high-throughput screening approach [33,34]. Specifically, we employed the techniques of multiplexing [35] and the HyperCyt sampling system and its associated software programs [36,37], to increase the speed and efficiency by which data acquisition and analysis is performed. The technique of fluorescent cell barcoding, previously described by Krutzik and Nolan [35] increases the number of cell types that can be assayed in a single well by differentially staining each cell type with distinct fluorophores. Its use in a cell-based high-throughput screening assay was successfully demonstrated in the non-pathogenic yeast, *Saccharomyces cerevisiae*, in a study designed to identify novel inhibitors of the yeast TORC1 complex [38]. Further, we utilized the flow cytometry aspect of the screen to monitor protein levels of select *C. albicans* target proteins fused with the green fluorescent protein (GFP) at the C-terminal end, which maintains expression of each protein under control of the native promoter [39]. We reasoned that the fluorescence signals can be detected using a flow cytometer and act as a direct measure of protein levels within the live *C. albicans* cell. We thus developed a cell-based

Table 1. *Candida albicans* strains used in this study.

Strain Name	Parent	Genotype	Source
SC5314		wild-type	
DAY185		*ura3Δ::λimm434/ura3Δ::λimm434 his1::hisG/HIS1::his1::hisG arg4::hisG/ARG4::URA3::arg4::hisG*	[47]
BWP17		*ura3Δ/ura3Δ arg4Δ/arg4Δ his1Δ/his1Δ*	[48]
CDR1-GFP	BWP17	*ura3Δ/ura3Δ arg4Δ/arg4Δ his1Δ/his1Δ CDR1/CDR1-GFP-HIS1*	[42]
FTR1-GFP	BWP17	*ura3Δ/ura3Δ arg4Δ/arg4Δ his1Δ/his1Δ FTR1/FTR1-GFP-HIS1*	[42]
MLT1-GFP	BWP17	*ura3Δ/ura3Δ arg4Δ/arg4Δ his1Δ/his1Δ MLT1/MLT1-GFP-HIS1*	[42]
SUR7-GFP	BWP17	*ura3Δ/ura3Δ arg4Δ/arg4Δ his1Δ/his1Δ SUR7/SUR7-GFP-HIS1*	[30]

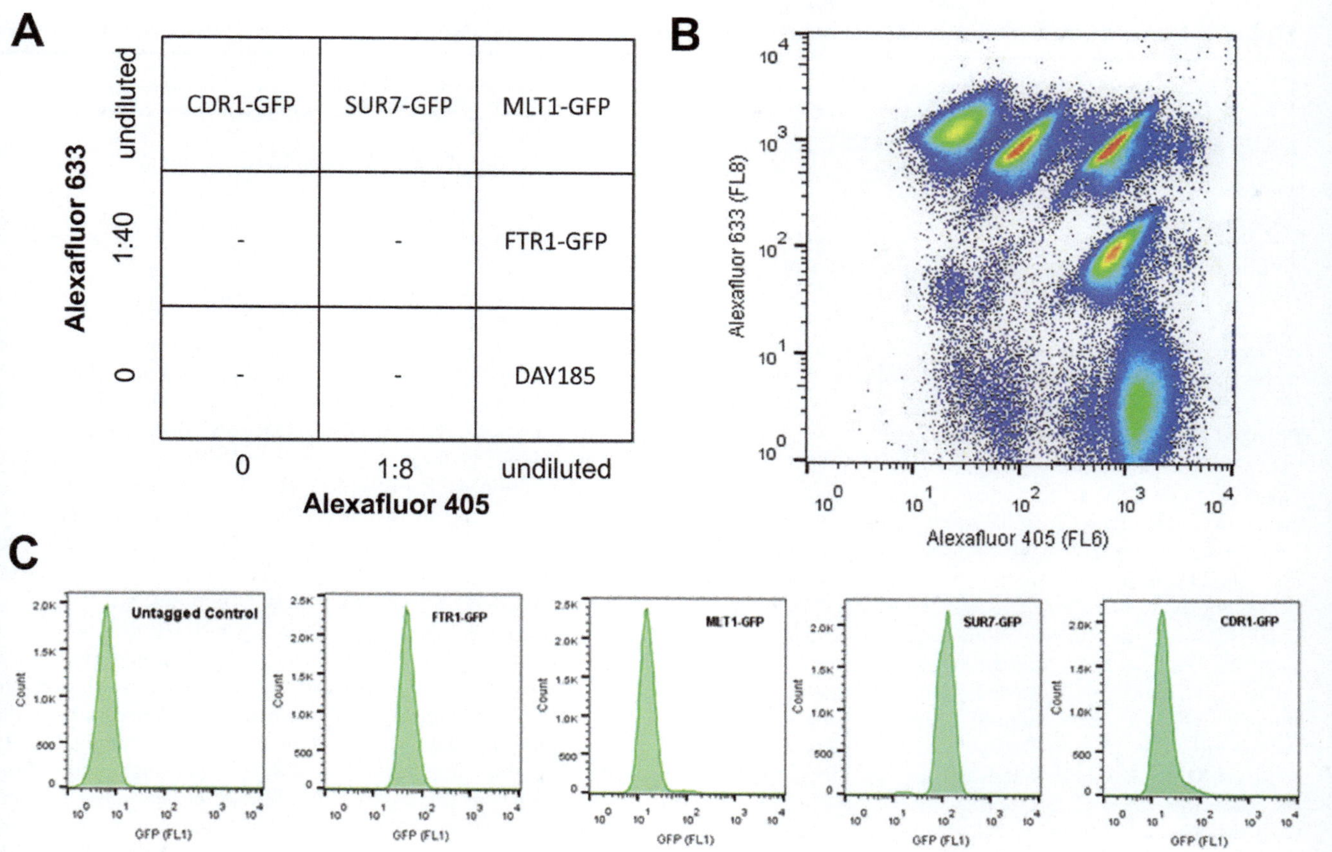

Figure 1. Multiplexing strategy to increase assay throughput. A schematic illustrating the amount of red and/or violet dyes used to differentially stain up to 9 different strains is shown in (A). Each sector of the 3×3 block represents one strain to be stained and details the specific concentration and combination of Alexafluor 633 and Alexafluor 405 dyes to be used for multiplexing. A typical sampling from one well carrying equal volumes of each stained cell is shown in a density plot in (B). Individual strains are visualized in a density plot of the two fluorescent channels (FL8 vs FL6), from which data on GFP levels can be extracted. Daughter cells that have lost some or all of the stain migrate to the lower left corner of the histogram. Gating on individual cell populations in the density plots depicted in (B) demonstrate that each population of cells is specific for one strain, with negligible contamination from migrating cells (C).

high-throughput multiplex screening assay to simultaneously probe several candidate drug targets in a single chemical library screen.

Materials and Methods

Strains and media

All strains are listed in Table 1. The strains carrying GFP-tagged virulence-related proteins are heterozygous for the *YFG*-GFP allele. Strains were routinely grown at 30°C in YPD [1% (w/v) yeast extract, 2% (w/v) peptone, 2% (w/v) glucose] or Complete Synthetic Medium (CSM) without uridine [0.67% (w/v) yeast nitrogen base without amino acids, 2% (w/v) glucose, 0.68% (w/v) complete synthetic mixture without uridine (MP Biochemicals, Solon, OH)]. Uridine was added at a final concentration of 80 µg ml^{-1} where required. Solid media was prepared by adding 2% (w/v) agar. For all high-throughput screening (HTS) assays, CSM was used as the base growth medium because YPD has fluorescence in the violet channel and confounds multiplex discrimination (data not shown). The final growth medium used for the HTS assays contained 0.015% (v/v) Pluronic F68 (Gibco, Life Technologies, Grand Island, NY). Iron-rich medium consisted of CSM supplemented with additional $FeCl_3$ at a final concentration of 1000 µM. Iron-deplete medium was CSM depleted of Fe^{3+} ions through the addition of 200 µM bathophenanthroline-disulfonic acid (Sigma, St. Louis, MO) [40].

Creating a multiplex

To increase the throughput of the HTS assay, each strain of interest was differentially color-coded with fluorescent dyes so that they can be combined and simultaneously interrogated in a single assay well (multiplex) [35]. The cell-based multiplex used in this study was generated using the five *C. albicans* strains of interest: CDR1-GFP, FTR1-GFP, MLT1-GFP, SUR7-GFP, and the untagged control strain, DAY185 (gift of A.P. Mitchell, Carnegie Mellon University). Differential staining was used to discriminate the different strains in the multiplex and was achieved by the addition of Alexafluor dye mixtures according to the diagram illustrated in Fig. 1. Thus, one milliliter cultures of each strain, grown as described in the Primary High-Throughput Screen (see below), were washed once with HHB buffer (30 mM HEPES, 110 mM NaCl, 10 mM KCl, 1 mM $MgCl_2 \cdot 6H_2O$, 10 mM glucose, pH 7.4) and resuspended in HHB buffer containing 0.015% (v/v) Pluronic F68 (Gibco, Life Technologies, Grand Island, NY). Next, 10 microliters of dye mixture was added and the cells were incubated at 30°C for 30 minutes. Following this incubation, the cells were then washed in HHB buffer and resuspended in CSM. Equal volumes of each strain were

Table 2. Measured green fluorescence (median channel fluorescence, MCF) of each strain used in the Primary Screen.

	CDR1-GFP	*SUR7-GFP*	*MLT1-GFP*	*FTR1-GFP*	**DAY185 (untagged control)**
MCF ± StdDev.	15.2±4.9	98.3±8.9	13.4±0.7	39.6±2.5	5.6±0.6
Relative MCF to control, DAY185	2.7x	17.5x	2.4x	7.0x	1x
Potential Z′-factor*	−0.73	0.69	0.50	0.72	

*Calculated with untagged control (DAY185) as the potential positive control for Inhibitor screens.

combined and sampled through the HyperCyt system to check for successful dye incorporation and good separation (Fig. 1).

Primary high-throughput screen

Overnight cultures of each strain, grown in YPD, were used to inoculate fresh CSM at a starting $OD_{600\ nm}$ of 0.2 which was grown to a final $OD_{600\ nm}$ of 0.5 at 30°C, followed by staining to create the multiplex. Following staining, the strains were combined and diluted into fresh media at 0.2 $OD_{600\ nm}$. 20 μL aliquots of the multiplex were transferred into 384-well plate format pre-populated with one library compound per well (columns 3 to 22) such that the final concentration of test compound was approximately 10 μM in growth medium using both Beckman Coulter Biomek NX MC and Span-8 liquid handling workstations (Beckman Coulter, Inc., Indianapolis, Indiana). The final concentration of DMSO in these wells was 1% (v/v). The first column of the plate was reserved for the negative or solvent-only control wells, wherein DMSO was added at a final concentration of 1% (v/v) in place of a test compound. The last two columns contained only growth medium and served as wash wells between row-to-row cell sampling to eliminate sample carry over. The second column would traditionally have been reserved for the positive control wells which would contain a known inhibitor of any one of the target proteins to be assayed. Currently, there are no known compounds that would inhibit expression of our proposed drug targets and thus the initial HTS assay was performed in the absence of a positive control compound, with the aim of identifying virulence-specific inhibitors. The assay plates were then incubated at 30°C for 2 hours using end-over-end rotation on a Hematology/Chemistry Mixer 346 (Fisher Scientific, Pittsburgh, PA). Following incubation at 30°C, the cells were diluted with Peptide Dilution Buffer (PDB) [0.1% (w/v) BSA, 1 mM magnesium chloride, 10 mM potassium chloride, 110 mM sodium chloride, 30 mM HEPES, pH 7.4] to achieve a final $OD_{600\ nm}$ of 0.2 prior to automated sampling of cells using the HyperCyt system [36,37] connected to a Beckman Coulter Cyan ADP flow cytometer (Beckman Coulter, Inc., Indianapolis, Indiana), with light scatter and emission of violet ($\lambda_{Ex} = 405$ nm, $\lambda_{Em} = 450$ nm), green ($\lambda_{Ex} = 488$ nm, $\lambda_{Em} = 530$ nm), and red ($\lambda_{Ex} = 635$ nm, $\lambda_{Em} = 665$ nm) fluorescence data analyzed using HyperView software. Compounds that had innate violet, green, or red fluorescence were excluded by mathematical filters based on the compound's 'shifting' impact on the lowest stained population. In total there were 5 green, 22 red, and 60 violet compounds, in the Prestwick Chemical Library. Identification of hits, either inducers or inhibitors, was based on constraints formulated from the Fold Response (the ratio between fluorescence measurement of sample well and fluorescence measurement of DMSO control for each strain). Compounds were considered hits according to the following criteria: A compound was classified as an inducer (inhibitor) if Fold Response was greater (lesser) than average plus (minus) three times the standard deviation of the particular strains' fold response across entire compound library. Compounds that fell under the definition of a "hit" but elicited a response in the untagged control strain DAY185 were classified as non-specific and were thus excluded from analysis in secondary assays.

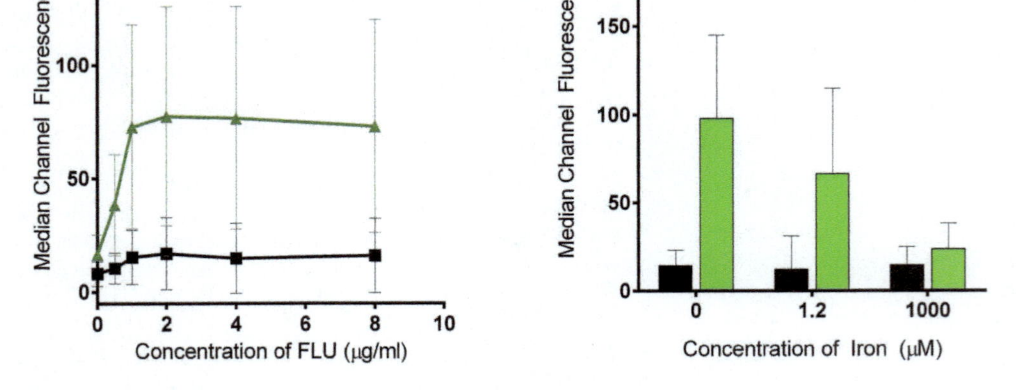

Figure 2. Changes in levels of GFP-tagged protein are readily detected. (A) Cells were grown in varying concentrations of fluconazole (FLU) and the response was measured using the HyperCyt system. Compared to the untagged control strain, a significant increase in fluorescence levels was detected for CDR1-GFP. Similarly, we queried the response of the FTR1-GFP strain to changes in iron content of the growth medium (B). These results demonstrate that we can detect changes to protein levels of our target proteins. Error bars indicate standard deviation values of the gated population of cells sampled and not of replicates of biological samples.

Table 3. Number of initial hits identified in the PCL screen.

	CDR1-GFP	SUR7-GFP	MLT1-GFP	FTR1-GFP
Inducers	22	1	0	0
Inhibitors	0	10	5	9
Confirmed Hits	12 (out of 22 tested)	4 (out of 11 tested)	3 (out of 5 tested)	6 (out of 9 tested)

Dose-response assay

Following the primary, high-throughput library screen, compounds that were identified as "hits" were assessed in a multiplexed, dose-response assay to verify target specificity. The cells and assay plates were prepared as for a primary screen but with drugs being serially diluted 3-fold across each row, starting at a concentration of 0.02 mg ml^{-1}. The final concentration of DMSO in all wells, including those that did not receive a test compound, was 1% (v/v). Following a two-hour incubation at 30°C with end-over-end rotation, the cells were diluted with PDB and sampled through the HyperCyt system. Data were collected and analyzed as described for the Primary High-Throughput Screen.

Statistical analysis

Zhang et al. [41] describe a statistical calculation, termed Z'-factor, that is used to determine if the difference observed between the positive and negative control is significant thus making a screen feasible. Z'-factor is calculated per the following equation:

$$Z'-factor = 1 - \frac{(3 \times Std_{+control} + 3 \times Std_{-control})}{|Mean_{+control} - Mean_{-control}|}$$

where Std is the standard deviation of the population, either positive control (+control) or negative (−control) and Mean is the average for those populations. Note the use of the absolute value of the differences between each Mean value. To compare the effect of a compound of interest (i.e., a "hit") on a target, a similar equation can be applied:

$$Z-factor = 1 - \frac{(3 \times Std_{+compound} + 3 \times Std_{-compound})}{|Mean_{+compound} - Mean_{-compound}|}$$

where data from when the compound of interest is present (+ compound) or absent (−compound) are used.

Secondary virulence-related assays

In order to validate specificity, virulence assays were performed to evaluate the effect of candidate compounds on the wild-type *C. albicans* strains SC5314 and DAY185. We had previously characterized the role of *C. albicans SUR7* in *Candida* virulence, specifically the result of loss-of-function of Sur7p [30], and thus selected a few potential inhibitors identified in our screen against strain SUR7-GFP. We performed characterization of wild-type strains in terms of lipase secretion, filamentation, biofilm formation, and virulence using an in vitro murine macrophage killing assay, as described previously [30], in the presence or absence of a potential inhibitor identified in the Primary High-Throughput Screen and Dose-response assays. The final concentration of DMSO, in which the compounds tested were dissolved, was 1% (w/v) in all assays performed.

Results and Discussion

Preliminary characterization of C. albicans strains encoding fluorescently-tagged membrane proteins

We developed a screening approach to assay the levels of *C. albicans* candidate drug targets by conjugating the protein of interest with green fluorescent protein (GFP), which can then be detected and quantified as the live cell carrying this fusion protein passes through the flow cytometer. In this study, we used a series of *C. albicans* strains in which GFP was fused to the C-termini of the plasma membrane transporters Cdr1p and Ftr1p, the vacuolar membrane ABC transporter Mlt1p, and the plasma membrane protein Sur7p [30,42]. The fluorescence level of each GFP-tagged strain (indicated by the median channel fluorescence, MCF) was compared to the untagged control strain, DAY185, using the HyperCyt system [36,37]. With respect to DAY185, the MCF of strains CDR1-GFP, FTR1-GFP, MLT1-GFP, and SUR7-GFP were on average 3-, 7-, 2-, and 18-fold higher, respectively (Table 2). These results show that GFP-tagged proteins of interest can readily be detected using the HyperCyt system and that the fluorescence of each strain is significantly higher than baseline fluorescence to allow for the detection of compounds that result in down-regulation of the GFP-tagged proteins. Due to the lack of known compounds that could act as positive controls for any of the strains, potential Z'-factors were calculated for each strain. The untagged control served as the positive control with the supposition that a compound that inhibits expression of each protein would result in fluorescence similar to the untagged control stain. The resulting calculations indicated that for three of the strains (SUR7-GFP, MLT1-GFP, and FTR1-GFP), high-throughput screens for inhibitors would be feasible (i.e., Z'-factor >0.5) (Table 2). The potential Z'-factor for strain CDR1-GFP suggests that this strain is not suitable for high-throughput screening, due to the relatively low MCF in combination with the resulting variability (standard deviation) in data generated from this strain (Table 2). This is in contrast to MLT1-GFP which yielded an acceptable potential Z'-factor of 0.5; despite having a relatively low MCF, the standard deviation of the data was very low.

Modulation and detection of the GFP-tagged protein target

Next, in order to provide proof-of-principle that modulation of GFP fluorescent intensity could be used to assay response to a drug, we grew the strains CDR1-GFP and FTR1-GFP under conditions which have been previously shown to alter expression of the target protein. Expression of the multidrug efflux pump, Cdr1p, has been shown to increase in response to azoles [23,43]. Thus, we queried the resulting GFP levels of the CDR1-GFP

Table 4. List of PCL compounds confirmed in dose-response assays.

	CDR1-GFP		SUR7-GFP		MLT1-GFP		FTR1-GFP	
Compound	(↑, ↓)*	EC_{50}	(↑, ↓)*	EC_{50}	(↑, ↓)*	EC_{50}	(↑, ↓)*	EC_{50}
3-alpha-hydroxy-5-beta-androstan-17-one	↑							
Alexidine dihydrochloride			↓	12.1	↓		↓	7.4
Amiodarone hydrochloride	↑	n.d.					↓	n.d.
Androsterone	↑	26.3						
Beclomethasone dipropionate	↑	5.6						
Benzamil hydrochloride			↓					
Benzbromarone			↓	18.4	↓	16.0	↓	
Berlambine	↑	2.4						
Butamben	↑							
Capsaicin	↑							
Clopamide			↓					
Daunorubicin hydrochloride	↑	13.0						
Dyclonine hydrochloride	↑							
Epiandrosterone	↑							
Equilin	↑	6.8						
Estradiol-17 beta	↑	17.8						
Flufenamic acid			↓					
Haloprogin			↓	13.0	↓	9.1	↓	4.7
Hexamethonium dibromide dihydrate			↑					
Isoconazole	↑							
Lasalocid sodium salt	↑	8.6						
Mechlorethamine hydrochloride			↓				↓	6.8
Menadione							↓	
Methyl benzethonium chloride			↓		↓		↓	3.7
Miconazole	↑	23.9						
Mometasone furoate	↑	1.0						
Norethynodrel	↑							
Nystatine			↓				↓	24
Piperine	↑	27.7						
Sulconazole nitrate	↑	23.9						
Suloctidil	↑							
Testosterone propionate	↑	6.4						

Table 4. Cont.

	CDR1-GFP		SUR7-GFP		MLT1-GFP		FTR1-GFP	
Thimerosal			↓	6.9	↓	1.4	↓	0.5
Tomatidine	↑							

*The arrow indicates whether a compound elicited an increase (inducer) or decrease (inhibitor) in expression levels, as measured by fold-response of the GFP-tagged protein.
EC_{50} (in µM) values are reported for each respective compound confirmed as a hit, as determined in subsequent dose-response assays.
"n.d." signifies that the hit was not confirmed in dose-response assays and therefore, no EC_{50} values have been determined.

strain grown in varying concentrations of fluconazole, over time. In response to sub-inhibitory concentrations of fluconazole, a large increase in fluorescence occurred between 5 and 6 hours, and increased over time (with a maximal effect noted at 24 hours). A modest baseline increase in fluorescence was seen in the untagged control strain incubated with fluconazole, even after multiple washes, possibly due to intrinsic fluorescence of fluconazole. However, there is a much greater relative induction of CDR1-GFP expression, in identical conditions. At a fluconazole concentration of 0.002 mg ml^{-1}, fluorescence is increased by 8.9 MCF (2.1-fold) in the untagged control strain, and by 60.8 MCF (4.9-fold) in the CDR1-GFP strain (Fig. 2A). Expression of *FTR1* mRNA is regulated by the amount of iron in the growth medium [27]. We thus grew strains FTR1-GFP and DAY185 in different media containing varying amounts of iron and measured the resulting changes in GFP levels using the HyperCyt system. We demonstrated that we could clearly detect compensatory increases (in low iron, or Fe-limiting conditions) and decreases (in high iron, or Fe-sufficient conditions) in Ftr1p-GFP expression, with the latter GFP-levels measuring close to the unlabeled control strain, DAY185, strongly suggesting complete depletion of the protein (Fig. 2B). The resulting changes in GFP levels are in agreement with the expected changes in mRNA levels in response to iron levels in the medium [27]. These results strongly suggest that modulation of *FTR1* by iron extends to the protein level, which to our knowledge has not been previously demonstrated. Taken together, these preliminary experiments demonstrate proof-of-principle that expression of GFP-tagged membrane proteins can be modulated by exposure to agents that can regulate protein expression and that these changes are readily detected and quantified using the HyperCyt system.

In this study, we expect to identify compounds that are able to modulate the levels of specific (target) proteins which have been previously been shown to be involved in *C. albicans* virulence. It should be noted that there could be many underlying mechanisms by which these compounds result in decreased levels of the proteins of interest, and that these compounds are not necessarily expected to interact with nor inhibit the target proteins directly. These specific mechanisms would be of great interest and would further our understanding of fungal virulence and pathogenesis. Whatever the specific mechanism may be, the objective is to identify compounds which decrease the levels of virulence-related proteins, thereby reducing virulence.

Differential staining of the *C. albicans* strains to increase high-throughput readiness

To markedly increase the throughput efficiency of the assay, each strain of interest was differentially color-coded with fluorescent dyes so that they could be simultaneously be queried with a compound in a single well [35]. Each strain, including the untagged control strain, was labeled with different amounts of Alexafluor dyes, as indicated in Fig. 1A. All five distinct populations of cells were readily distinguishable (Fig. 1B) from which MCF data for green fluorescence could be extracted. Furthermore, the GFP histogram from each population of cells exhibited the expected distribution of GFP levels from a single-strain population indicating that unrelated signals from other strains were not captured within the gated dataset (Fig. 1C). The multiplex also eliminated any potential confounding effect of unstained daughter cells as they simply migrated to the lower left-hand corner of the density plot and were excluded from measurement by gating. After a two-hour incubation, the number of mother (stained) cells left was more than sufficient for data analyses (data not shown).

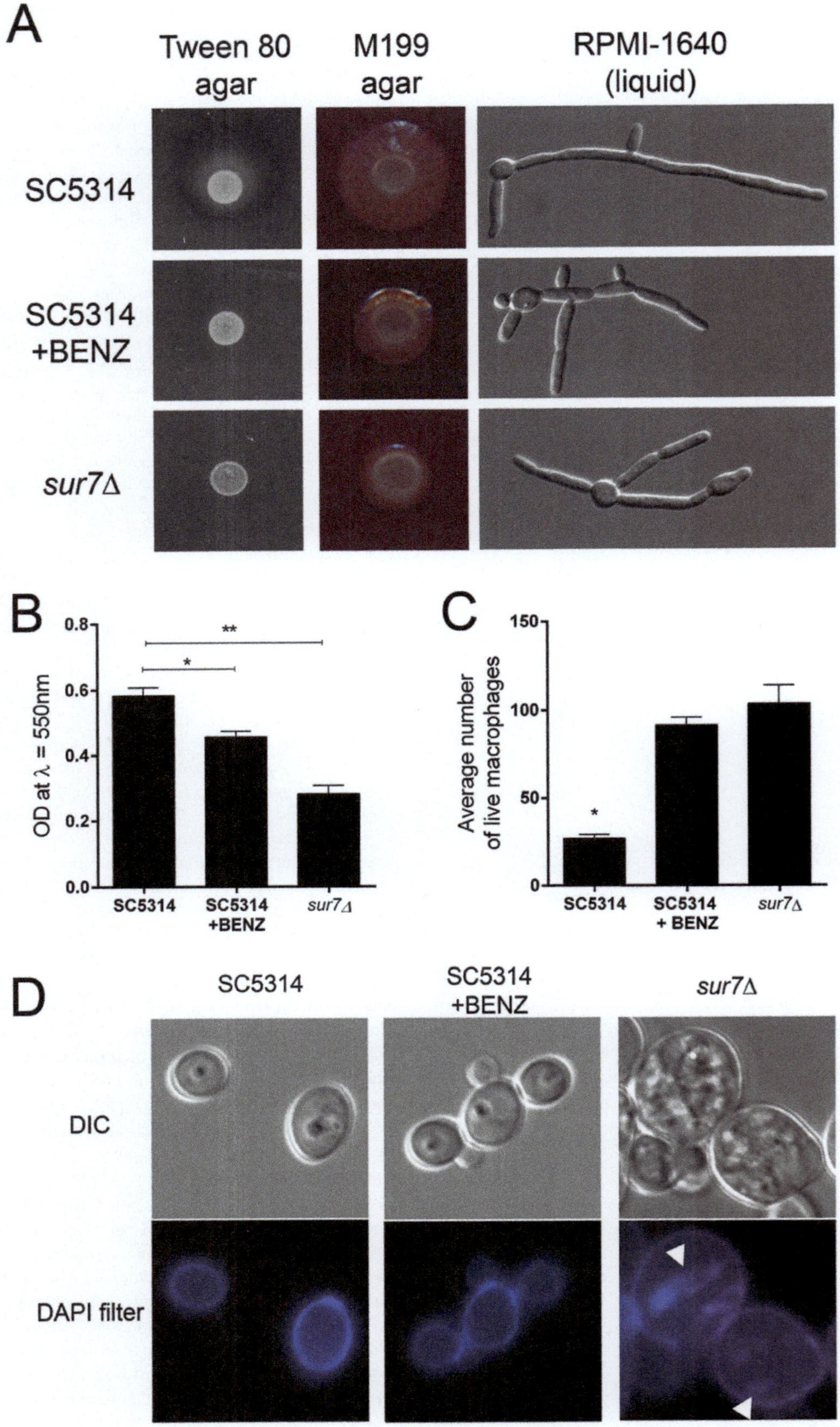

Figure 3. Secondary microbiologic and virulence assays further eliminate false-positive hits. To further eliminate false-positive hits, the effects of benzbromarone (BENZ, at a concentration of 18.4 μM) on wild-type *C. albicans* were tested and the results were compared to the phenotypic characteristics of the *sur7*Δ null mutant strain. In most cases, in the presence of benzbromarone, wild-type strains SC5314 and DAY185

mimicked the phenotypic profile of the *sur7Δ* mutant strain. Only results for SC5314 are shown in each panel. (A) The defects in secretion of degradative lipases, as indicated by the lack of the zone of precipitation surrounding the colony grown on Tween 80 agar, and in the inability to form mature hyphae on filamentation-inducing media (M199 agar and RPMI-1640) suggest that benzbromarone may act to reduce levels of Sur7p. (B) The ability of benzbromarone to inhibit biofilm formation of wild-type *C. albicans* was also tested and compared to the defective biofilm formed by the *sur7Δ* null mutant. Since metabolic activity can vary between strains of *C. albicans*, the resulting biofilm mass was measured using the crystal violet staining method [49]. When grown in the presence of benzbromarone, biofilm formation was reduced compared to wild-type growth. However, the reduction in biofilm formation of the benzbromarone-treated cells was not comparable to the biofilm formed by the *sur7Δ* strain. Statistical significance is indicated with asterisks, with $p<0.0001$. (C) The ability to kill macrophages by *C. albicans* was assessed by co-incubation of wild-type yeast cells with murine macrophage cells, in the presence or absence of benzbromarone. Virulent, wild-type *C. albicans* kill a significant proportion of macrophage cells when co-incubated for a period of 24 hours. When benzbromarone is present, a significant number of macrophage cells survive the co-incubation with wild-type *C. albicans*, similar to the *sur7Δ* null mutant strain. Statistical significance is indicated with an asterisk, with $p<0.0001$. (D) Cell wall morphology of the yeast form was characterized by staining yeast cells grown in the absence or presence of benzbromarone. Loss of function of *SUR7* results in plasma membrane invaginations and distinct cell wall-derived structures within the cell, which stain with calcofluor white as highlighted with arrows [30,31]. At the concentration of benzbromarone tested, the distribution and staining of cell wall material in wild-type cells grown in the presence of benzbromarone was indistinguishable from wild-type grown cells.

We also conducted single-plex experiments for each strain to examine changes in fluorescence between stained and unstained cells. There was no statistically significant difference in the ratio of median channel fluorescence to forward scatter (MCF/FSC ratio) (data not shown). Thus, although the median fluorescence was lower in the unstained population (likely due to the fact that they are new daughter cells with less surface area and therefore fewer membrane proteins), the relative fold change in GFP expression between daughter and parental cells was similar. Thus, any potential issue with unstained daughter cells during high-throughput screening is unlikely to be significant. Overall, these results demonstrate that multiplexing, by increasing the number of targets assayed per well, can greatly increase assay throughput.

High-throughput screening of the Prestwick Chemical Library (PCL)

Having demonstrated that we can detect modulation of the expression of our target proteins (both up- and down-regulation) and further increase high-throughput readiness of the assay by multiplexing, we next screened the Prestwick Chemical Library (PCL) for compounds that could alter expression of our target proteins. We probed the PCL, which consisted of 880 off-patent drugs at the time of screening, in a 384-well multiplex format. In this initial screen, we identified agonists and/or antagonists of GFP expression in the CDR1-GFP, FTR1-GFP, MLT1-GFP, and SUR7-GFP strains, respectively (Table 3). Of note, this screening approach identified a number of compounds in the PCL that have been known to induce expression of Cdr1p, such as azoles and estrogen [44,45] (Table 4). The relatively high hit rate for the number of inducers identified against Cdr1p is not surprising considering that the protein functions as a general efflux pump which protects against cell toxicity. Overall, the hit rate of the other target proteins did not exceed 1% of the total number of compounds tested, thus suggesting utility in screening much larger compound libraries.

We then cherry-picked 32 of the compounds for validation in dose-response multiplex assays utilizing the HyperCyt where cells were incubated with each compound of interest at varying concentrations. The number of compounds that confirmed activity in dose response assay with EC_{50} less than 30 μM are listed in Table 4, showing that across the strains the average percentage of false positives was 50%. Compounds that did not elicit a response were removed as false-positives. Moreover, using the calculation of Z-factors for each confirmed hit compound, we showed that the difference in GFP levels effected by each compound was statistically significant from baseline GFP levels of the target protein, thereby confirming these compounds as "hits" in the primary screen.

There are no known inhibitors of the candidate drug targets we pursued, and we anticipate that most, if not all, subsequent campaigns to identify targets of virulence will *a priori* lack appropriate inhibitors for use as a positive control in the screen. While the initial screen of the PCL was performed without a positive control, initial hits are verified using dose-response assays performed in exactly the same manner as the HTS library screen, and Z-factor calculations from the highest, triplicate concentration data points indicate that the effect of each compound tested is statistically significant (Z-factor >0.5). Thus, following assessment of specificity, any one of these hits can be used as positive controls for inhibitors (or inducers) of protein expression in subsequent screens of larger chemical libraries.

Secondary phenotypic assays to evaluate virulence specific to the target

The primary goal of the screen was to identify agents that could be used to target virulence as opposed to those that directly affect cell growth. The cell-based assay that we describe identifies potential compounds that specifically alter the expression of proteins involved in virulence, without necessarily targeting the protein directly. The proteins we present here are candidate drug targets based on previous phenotypic characterizations undertaken with classical, reverse-genetic approaches wherein modulation of protein levels have resulted in altered virulence or virulence-related characteristics of the organism. Thus we used secondary assays designed to validate the hits that have passed verification in dose-response assays. In theory, if the compound of interest decreases the levels of the protein of interest, addition of the antagonistic compound will result in a wild-type *C. albicans* strain that will exhibit the characteristics of a heterozygous (if haploinsufficiency exists) or homozygous null mutant strain.

Here, we present our analysis of a potential inhibitor affecting expression of *C. albicans* Sur7p. Benzbromarone (BENZ), a known uricosuric drug used to treat gout, which was one of 10 compounds that elicited a decrease in GFP levels from the SUR7-GFP strain. The Z-factor calculated from the dose-response assay was 0.91, indicating a statistically significant response of Sur7p to benzbromarone. Previous characterization of the role of *C. albicans SUR7* in pathogenesis demonstrated that the *C. albicans* null mutant strain is defective in a number of virulence-related phenotypes such as lipase secretion, filamentation, and biofilm formation [30]. Hence we tested wild-type strains DAY185 and SC5314 in these phenotypic assays, in the presence and absence of benzbromarone. Benzbromarone inhibited lipase secretion and filamentation of the wild-type strains to the same degree as that exhibited by the homozygous null mutant strain, *sur7Δ* (Fig. 3A). Likewise, biofilm formation is also inhibited in the presence of benzbromarone (Fig. 3B). We next assayed the effect of benzbromarone on the ability of wild-type *C. albicans* to kill macrophage in an in vitro assay for virulence utilizing murine macrophage. In this assay, we have previously shown that the *sur7*Δ homozygous null mutant is hypovirulent compared to the wild-type control strain, DAY185 [30]. In the presence of benzbromarone, both wild-type strains tested were hypovirulent compared to controls wherein benzbromarone was omitted (Fig. 3C). The resulting attenuation in virulence in the presence of benzbromarone was similar to levels exhibited by the *sur7*Δ mutant strain. ANOVA indicate that the difference between these conditions are statistically significant ($p<0.0001$). These preliminary results would seem to suggest that benzbromarone may affect Sur7p levels, thus resulting in homozygous null-like phenotypes. However, upon closer investigation of the cellular morphology of wild-type strains grown in the presence of benzbromarone, we did not observe the expected morphological defects in the plasma membrane that is characteristic of the *sur7*Δ homozygous null mutant (Fig. 3D). The latter phenotypic characterization suggests that the effect of benzbromarone is not limited to the reduced levels of Sur7p.

It is clear that the decrease in GFP levels detected by the flow cytometer do not necessarily translate to a wild-type strain that, when assayed in the presence of the inhibitory compound, exhibits reduced virulence similar to that of the corresponding homozygous null mutant. This is due to the many possible mechanisms by which a compound may effect a reduction in target protein levels. This caveat further stresses the importance of the secondary phenotypic virulence-related characterizations that follow verification of each identified hit, as part of the workflow in the identification of lead compounds that target virulence.

Identification of compounds with general antifungal activity

While the immediate goal of the method described here is to identify novel compounds that can be directed against proteins involved in virulence, either when used alone or in conjunction with standard therapy, we realize that it can also identify novel and re-purposed compounds with general, non-specific antifungal properties as well. Rather than be discarded, further detailed characterization of compounds with general antifungal properties, but not virulence-specific activity, would still be undertaken as part of a high-throughput screening workflow. To demonstrate the potential of non-specific compounds identified in our screen of the PCL, we recently evaluated flufenamic acid for its potential use as an antifungal agent [46]. High doses of flufenamic acid were effective for both the prevention of biofilm formation, and inhibition of mature *C. albicans* biofilms. More importantly, low dose flufenamic acid (8 μg ml^{-1}), when used in combination with standard antifungal agents, increased the potency of the latter in preventing the formation of *C. albicans* biofilms [46]. These results stress the importance of further studies on compounds that, although having been set aside in the initial high-throughput screen, may possess general antifungal activity.

Conclusions

We developed a novel, cell-based, high-throughput screening method that can be used for the identification of compounds that alter protein expression in the model opportunistic fungal pathogen *C. albicans*. In this method, the screen starts with the thoughtful identification of potential targets based on previous characterization of loss-of-function and/or gain-of-function mutations in the genes (encoding the target proteins) of the microorganism of interest based on detailed studies of microbial pathogenesis. With the completion of genome sequences of important and clinically relevant microorganisms, we have seen a dramatic increase of information on protein function, among which are potential virulence-specific drug targets. It would appear that a major gap remains in the ability to translate this information to clinically relevant applications. With the growing interest in targeting virulence in light of the clinical failures of current antimicrobial agents and well-publicized development of multiple drug-resistant "super-bugs", the method presented here represents a potential means to further such translational studies. This highly efficient, automated high-throughput screening method serves as a means to identify compounds that may be (i) general antifungal agents, (ii) used in virulence-targeted therapies, and/or (iii) inhibiting cellular pathways which, following mechanistic studies, can aid in the further understanding of pathogenesis of clinically relevant fungal pathogens.

Acknowledgments

We gratefully acknowledge Dr. Jun Chen for helpful suggestions on assay development.

Author Contributions

Conceived and designed the experiments: SMB CPA AW TO LAS SAL. Performed the experiments: SMB CPA SMY. Analyzed the data: SMB CPA AW LAS SAL. Contributed reagents/materials/analysis tools: AW TO LAS SAL. Contributed to the writing of the manuscript: SMB CPA AW LAS SAL.

References

1. Emori TG, Gaynes RP (1993) An overview of nosocomial infections, including the role of the microbiology laboratory. Clin Microbiol Rev 6: 428–442.
2. Jarvis WR (1995) Epidemiology of nosocomial fungal infections, with emphasis on *Candida* species. Clin Infect Dis 20: 1526–1530.
3. Banerjee SN, Emori TG, Culver DH, Gaynes RP, Jarvis WR, et al. (1991) Secular trends in nosocomial primary bloodstream infections in the United States, 1980–1989. National Nosocomial Infections Surveillance System. Am J Med 91: 86S–89S.
4. Kanafani ZA, Perfect JR (2008) Antimicrobial resistance: resistance to antifungal agents: mechanisms and clinical impact. Clin Infect Dis 46: 120–128.
5. Sanglard D, Coste A, Ferrari S (2009) Antifungal drug resistance mechanisms in fungal pathogens from the perspective of transcriptional gene regulation. FEMS Yeast Res 9: 1029–1050.
6. Fanos V, Cataldi L (2000) Amphotericin B-induced nephrotoxicity: a review. J Chemother 12: 463–470.
7. Wiederhold NP, Grabinski JL, Garcia-Effron G, Perlin DS, Lee SA (2008) Pyrosequencing to detect mutations in *FKS1* that confer reduced echinocandin susceptibility in *Candida albicans*. Antimicrob Agents Chemother 52: 4145–4148.
8. Park S, Kelly R, Kahn JN, Robles J, Hsu MJ, et al. (2005) Specific substitutions in the echinocandin target Fks1p account for reduced susceptibility of rare laboratory and clinical *Candida* sp. isolates. Antimicrob Agents Chemother 49: 3264–3273.
9. Beyda ND, Lewis RE, Garey KW (2012) Echinocandin resistance in *Candida* species: mechanisms of reduced susceptibility and therapeutic approaches. Ann Pharmacother 46: 1086–1096.
10. Fothergill AW, Sutton DA, McCarthy DI, Wiederhold NP (2014) Impact of new antifungal breakpoints on antifungal resistance in *Candida* species. J Clin Microbiol 52: 994–997.
11. Pfaller MA, Messer SA, Woosley LN, Jones RN, Castanheira M (2013) Echinocandin and triazole antifungal susceptibility profiles for clinical opportunistic yeast and mold isolates collected from 2010 to 2011: application of new CLSI clinical breakpoints and epidemiological cutoff values for characterization of geographic and temporal trends of antifungal resistance. J Clin Microbiol 51: 2571–2581.
12. Orasch C, Marchetti O, Garbino J, Schrenzel J, Zimmerli S, et al. (2013) *Candida* Species Distribution and Antifungal Susceptibility Testing According to EUCAST and New vs. Old CLSI Clinical Breakpoints: a Six-Year Prospective

Candidemia Survey from the Fungal Infection Network of Switzerland (FUNGINOS). Clin Microbiol Infect.
13. Miceli MH, Lee SA (2011) Emerging moulds: epidemiological trends and antifungal resistance. Mycoses 54: e666–678.
14. Miceli MH, Diaz JA, Lee SA (2011) Emerging opportunistic yeast infections. Lancet Infect Dis 11: 142–151.
15. Pierce CG, Lopez-Ribot JL (2013) Candidiasis drug discovery and development: new approaches targeting virulence for discovering and identifying new drugs. Expert Opin Drug Discov 8: 1117–1126.
16. Eder J, Hommel U, Cumin F, Martoglio B, Gerhartz B (2007) Aspartic proteases in drug discovery. Curr Pharm Des 13: 271–285.
17. Naglik JR, Challacombe SJ, Hube B (2003) *Candida albicans* secreted aspartyl proteinases in virulence and pathogenesis. Microbiol Mol Biol Rev 67: 400–428.
18. Shareck J, Belhumeur P (2011) Modulation of morphogenesis in *Candida albicans* by various small molecules. Eukaryot Cell 10: 1004–1012.
19. Saville SP, Lazzell AL, Bryant AP, Fretzen A, Monreal A, et al. (2006) Inhibition of filamentation can be used to treat disseminated candidiasis. Antimicrob Agents Chemother 50: 3312–3316.
20. Srinivasan A, Uppuluri P, Lopez-Ribot J, Ramasubramanian AK (2011) Development of a high-throughput *Candida albicans* biofilm chip. PLoS One 6: e19036.
21. Allen RC, Popat R, Diggle SP, Brown SP (2014) Targeting virulence: can we make evolution-proof drugs? Nat Rev Microbiol 12: 300–308.
22. Holmes AR, Lin YH, Niimi K, Lamping E, Keniya M, et al. (2008) ABC transporter Cdr1p contributes more than Cdr2p does to fluconazole efflux in fluconazole-resistant *Candida albicans* clinical isolates. Antimicrob Agents Chemother 52: 3851–3862.
23. White TC (1997) Increased mRNA levels of *ERG16*, *CDR*, and *MDR1* correlate with increases in azole resistance in *Candida albicans* isolates from a patient infected with human immunodeficiency virus. Antimicrob Agents Chemother 41: 1482–1487.
24. White TC, Marr KA, Bowden RA (1998) Clinical, cellular, and molecular factors that contribute to antifungal drug resistance. Clin Microbiol Rev 11: 382–402.
25. Mateus C, Crow SA Jr, Ahearn DG (2004) Adherence of *Candida albicans* to silicone induces immediate enhanced tolerance to fluconazole. Antimicrob Agents Chemother 48: 3358–3366.
26. Ramage G, Bachmann S, Patterson TF, Wickes BL, Lopez-Ribot JL (2002) Investigation of multidrug efflux pumps in relation to fluconazole resistance in *Candida albicans* biofilms. J Antimicrob Chemother 49: 973–980.
27. Ramanan N, Wang Y (2000) A high-affinity iron permease essential for *Candida albicans* virulence. Science 288: 1062–1064.
28. Knight SA, Vilaire G, Lesuisse E, Dancis A (2005) Iron acquisition from transferrin by *Candida albicans* depends on the reductive pathway. Infect Immun 73: 5482–5492.
29. Theiss S, Kretschmar M, Nichterlein T, Hof H, Agabian N, et al. (2002) Functional analysis of a vacuolar ABC transporter in wild-type *Candida albicans* reveals its involvement in virulence. Mol Microbiol 43: 571–584.
30. Bernardo SM, Lee SA (2010) *Candida albicans SUR7* contributes to secretion, biofilm formation, and macrophage killing. BMC Microbiol 10: 133.
31. Alvarez FJ, Douglas LM, Rosebrock A, Konopka JB (2008) The Sur7 protein regulates plasma membrane organization and prevents intracellular cell wall growth in *Candida albicans*. Mol Biol Cell 19: 5214–5225.
32. Douglas LM, Wang HX, Keppler-Ross S, Dean N, Konopka JB (2012) Sur7 promotes plasma membrane organization and is needed for resistance to stressful conditions and to the invasive growth and virulence of *Candida albicans*. MBio 3.
33. Robinson JP, Rajwa B, Patsekin V, Davisson VJ (2012) Computational analysis of high-throughput flow cytometry data. Expert Opin Drug Discov 7: 679–693.
34. Black CB, Duensing TD, Trinkle LS, Dunlay RT (2011) Cell-based screening using high-throughput flow cytometry. Assay Drug Dev Technol 9: 13–20.
35. Krutzik PO, Nolan GP (2006) Fluorescent cell barcoding in flow cytometry allows high-throughput drug screening and signaling profiling. Nat Methods 3: 361–368.
36. Kuckuck FW, Edwards BS, Sklar LA (2001) High throughput flow cytometry. Cytometry 44: 83–90.
37. Edwards BS, Young SM, Saunders MJ, Bologa C, Oprea TI, et al. (2007) High-throughput flow cytometry for drug discovery. Expert Opin Drug Discov 2: 685–696.
38. Chen J, Young SM, Allen C, Seeber A, Peli-Gulli MP, et al. (2012) Identification of a small molecule yeast *TORC1* inhibitor with a multiplex screen based on flow cytometry. ACS Chem Biol 7: 715–722.
39. Gerami-Nejad M, Berman J, Gale CA (2001) Cassettes for PCR-mediated construction of green, yellow, and cyan fluorescent protein fusions in *Candida albicans*. Yeast 18: 859–864.
40. Liang Y, Gui L, Wei DS, Zheng W, Xing LJ, et al. (2009) *Candida albicans* ferric reductase *FRP1* is regulated by direct interaction with Rim101p transcription factor. FEMS Yeast Res 9: 270–277.
41. Zhang JH, Chung TD, Oldenburg KR (1999) A Simple Statistical Parameter for Use in Evaluation and Validation of High Throughput Screening Assays. J Biomol Screen 4: 67–73.
42. Lee SA, Khalique Z, Gale CA, Wong B (2005) Intracellular trafficking of fluorescently tagged proteins associated with pathogenesis in *Candida albicans*. Med Mycol 43: 423–430.
43. Sanglard D, Kuchler K, Ischer F, Pagani JL, Monod M, et al. (1995) Mechanisms of resistance to azole antifungal agents in *Candida albicans* isolates from AIDS patients involve specific multidrug transporters. Antimicrob Agents Chemother 39: 2378–2386.
44. Karnani N, Gaur NA, Jha S, Puri N, Krishnamurthy S, et al. (2004) *SRE1* and *SRE2* are two specific steroid-responsive modules of *Candida* drug resistance gene 1 (*CDR1*) promoter. Yeast 21: 219–239.
45. Cheng G, Yeater KM, Hoyer LL (2006) Cellular and molecular biology of *Candida albicans* estrogen response. Eukaryot Cell 5: 180–191.
46. Chavez-Dozal AA, Jahng M, Rane HS, Asare K, Kulkarny VV, et al. (2014) In vitro analysis of flufenamic acid activity against *Candida albicans* biofilms. Int J Antimicrob Agents 43: 86–91.
47. Davis D, Edwards JE Jr, Mitchell AP, Ibrahim AS (2000) *Candida albicans RIM101* pH response pathway is required for host-pathogen interactions. Infect Immun 68: 5953–5959.
48. Wilson RB, Davis D, Mitchell AP (1999) Rapid hypothesis testing with *Candida albicans* through gene disruption with short homology regions. J Bacteriol 181: 1868–1874.
49. Ramage G, Lopez-Ribot JL (2005) Techniques for antifungal susceptibility testing of *Candida albicans* biofilms. Methods Mol Med 118: 71–79.

Dissemination of the Transmissible Quinolone-Resistance Gene *qnrS1* by IncX Plasmids in Nigeria

Eric T. Sumrall[1], Elizabeth B. Gallo[1], Aaron Oladipo Aboderin[2], Adebayo Lamikanra[3], Iruka N. Okeke[1]*

1 Department of Biology, Haverford College, Haverford, Pennsylvania, United States of America, **2** Department of Medical Microbiology and Parasitology, Obafemi Awolowo University, Ile-Ife, Osun State, Nigeria, **3** Department of Pharmaceutics, Obafemi Awolowo University, Ile-Ife, Osun State, Nigeria

Abstract

The plasmid-encoded quinolone resistance gene *qnrS1* was recently found to be commonly associated with ciprofloxacin resistance in Nigeria. We mapped the *qnrS1* gene from an *Escherichia coli* isolate obtained in Nigeria to a 43.5 Kb IncX2 plasmid. The plasmid, pEBG1, was sufficient to confer ciprofloxacin non-susceptibility, as well as tetracycline and trimethoprim resistance, on *E. coli* K-12. Deletion analysis confirmed that *qnrS1* accounted for all the ciprofloxacin non-suceptibility conferred by pEBG1 and tetracycline and trimethoprim resistance could be attributed to *tetAR* and *dfrA14* genes respectively. While it contained a complete IncX conjugation system, pEBG1 was not self-transmissible likely due to an IS*3* element inserted between the *pilX5* and *pilX6* genes. The plasmid was however efficiently mobilizable. pEBG1 was most similar to another *qnrS1*-bearing IncX2 plasmid from Nigeria, but both plasmids acquired *qnrS1* independently and differ in their content of other resistance genes. Screening *qnrS1*–positive isolates from other individuals in Nigeria revealed that they carried neither pEBG1 nor pNGX2-QnrS1 but that IncX plasmids were prevalent. This study demonstrates that the IncX backbone is a flexible platform that has contributed to *qnrS1* dissemination in Nigeria.

Editor: A. Mark Ibekwe, U. S. Salinity Lab, United States of America

Funding: This work was supported by a Branco Weiss Fellowship to INO from the Society-in-Science, ETHZ, Zürich and intramural funding from Haverford College. The funders had no role in study design, data collection and analysis, decision to publish, or preparation of the manuscript.

Competing Interests: The authors declare that no competing interests exist.

* Email: iokeke@haverford.edu

Introduction

When antibacterial quinolones were first introduced into clinical practice, it was thought that resistance would be slow to appear and that transmissible resistance was improbable [1]. Initial reports of quinolone resistance were due to point mutations in the genes encoding their gyrase and topoisomerase targets that made them less sensitive to the drug. In 1998, Martinez-Martinez et al [2] described a plasmid-borne gene, now termed *qnrA*, which conferred four-to-sixteen-fold resistance to quinolones on Enterobacteriaceae. *qnrA* is a pentapeptide repeat protein that protects DNA gyrase from quinolone binding and inhibition [3]. Other *qnr* genes have since been reported and many can be transmitted horizontally. Transmissible quinolone resistance is also attributable to genes encoding plasmid-encoded efflux pumps, such as *qepA* and *oqx* and, in the case of ciprofloxacin, the acetylating enzyme *aac(6')-Ib-cr*. While plasmid-encoded quinolone-resistance genes generally confer low-level resistance, their overall impact is great because they shield otherwise susceptible bacteria from the lethal effects of the quinolones, allowing them greater time and opportunity to evolve higher-level resistance.

Until recently, reports of quinolone resistance were almost non-existent from West Africa [4]. However, in the past decade Nigeria has seen a very rapid increase in fluoroquinolone use, due to the recent expiration of patents protecting ciprofloxacin and perfloxacin. Introduction of ciprofloxacin into Nigerian clinics was temporally associated with a significant rise in resistance among gut commensals. Five years after fluoroquinolones were introduced in a community in Western Nigeria, *Escherichia coli* strains showing quinolone-specific resistance mechanisms were isolated [5]. Although the majority of these isolates carried point-mutations in the quinolone-resistance determining regions of *gyrA* and *parC*, six strains bore the plasmid-encoded resistance gene *qnrS1*. In this study, we characterized a mobile element from one of these isolates in order to understand the mode of *qnrS1* dissemination in Western Nigeria.

Methods

Strains

Strain 09/22a and other *qnrS1*-bearing *E. coli* strains used in this study were isolated in 2009 during an earlier survey of quinolone resistance [5]. Other strains used in this study are listed in Table 1. Strains were maintained at −70°C in Luria broth: glycerol 1:1.

General molecular biology procedures

Genomic DNA was extracted using the Promega Wizard kit. Small-scale extractions of large, naturally occurring plasmids were

Table 1. Strains used in this study.

	Genotype and description	Reference or Source
09/22a	*qnrS1*-positive *E. coli* isolate from Nigeria	[5]
DH5α	F^- ø80d*lacZ*ΔM15 Δ(*lacZYA-argF*)U169 *deo*R *recA*1 *endA*1 *hsdR*17(rK^- mK^+) *phoA supE*44 λ– *thi*-1 *gyrA*96 *relA*1	Invitrogen
TOP10	F- *mcr*A (*mrr-hsd*RMS-*mcr*BC) ö80*lacZ*ΔM15.*lac*X74 *deo*R *rec*A1 *ara*D139 *Δ*(*ara-leu*)7697 *gal*U *gal*K *rps*L (StrR) *end*A1 *nup*G	Invitrogen
C600 Nal^R	Nalidixic acid-resistant derivative of C600.	[47]
SM10 λpir	λpir *E. coli* strain harboring an inc-P conjugation system in the chromosome	[35]
UMN026	Uropathogenic *E. coli* strain bearing IncP and IncX1 plasmids	[36]
NCTC 10418	Susceptibility testing control	
ATCC 35218	Susceptibility testing control	
LMG194	F-.*lac*X74 *gal* E *thi rps*L.*pho*A (*Pvu* II).*ara*714 *leu*::*Tn10*	Invitrogen [48]
EC1502	Rifampicin resistant *E. coli* strain	University of Bradford
J53	F^- *met pro* Azi^r	[49]

carried out by a modified boiling protocol as described previously [6]. A Qiagen miniprep kit was used to extract smaller recombinant plasmids. Large scale preparations of plasmids over 20 Kb in size were prepared after growth in terrific broth and induction with chloramphenicol [7] using the Qiagen large construct kit. Plasmids were electroporated into *E. coli* host strains using a Biorad micropulser according to manufacturers' instructions. DNA amplification was performed using Platinum PCR Supermix (Invitrogen) and 1 μM oligonucleotide primer in each reaction. Oligonucleotide primer sequences are listed in Table 2. All amplifications began with a two-minute hot start at 94°C followed by 30 cycles of denaturing at 94°C for 30 s, annealing at 5°C below primer annealing temperature for 30 s, and extending at 72°C for one minute per kilobase of DNA. When the target PCR product was over 3 Kb, we used *Pfx* polymerase (Invitrogen) in accordance with manufacturer's instructions, with annealing at 5°C below the temperature used for *Taq* PCRs. Where necessary, for sequencing, PCR amplicons were TA cloned into the pGEM-T Easy vector (Promega) according to manufacturer's directions and plasmids were transformed into chemically competent *E. coli* K-12 TOP10 cells. For expression, amplicons were cloned into pBAD/Thio-TOPO (Invitrogen) and induced with 0.2% arabinose. Expression was verified by western blotting. All plasmids used or constructed in the course of this study are listed in Table 3. Other molecular biology operations were performed using standard procedures [8].

Shot-gun sequencing and sequence analysis

Whole-replicon shotgun library preparation, Sanger sequencing and assembly of a large plasmid were performed by SeqWright DNA Technology Services (Houston, TX). Sequence analyses and annotation were performed in Artemis [9]. Open reading frames were initially defined by Glimmer. Annotations were made where BLAST e-values equaled or approached zero and there was 98% or greater identity at the nucleotide and amino acid levels. Direct and inverted repeats were identified by dot-plot analysis of pairwise FASTA alignments made using the BLAST suite. Open reading frame (Orf) and feature plots were prepared using Artemis and DNAPlotter [10]. Orf identify was determined using BLAST and Pfam [11,12]. Multiple sequence alignments were performed using Clustal and dendograms were computed using a Jones-Taylor-Thornton (JTT) model in MEGA5 [13].

Antimicrobial susceptibility testing

Antimicrobial susceptibility testing by disc diffusion was performed using the Clinical and Laboratory Standards Institute (CLSI, formerly NCCLS) [14]. Antimicrobial discs and control strain *E. coli* ATCC 35218 were obtained from Remel. The antimicrobial discs used contained ampicillin (10 μg), streptomycin (10 μg), trimethoprim (5 μg), tetracycline (30 μg), nalidixic acid (30 μg), chloramphenicol (30 μg), ciprofloxacin (5 μg) and sulphonamide (300 μg). Inhibition zone diameters were interpreted in accordance with CLSI guidelines with WHONET software version 5.3 [15]. Minimum inhibitory concentrations (MICs) to nalidixic acid were measured by the E-test (Biomerieux), in accordance with manufacturer's instructions and by the broth dilution technique in Mueller-Hinton broth as recommended by the CLSI and using *E. coli* ATCC 35218 as control [16].

In vitro conjugation

In vitro conjugation experiments were performed by solid and liquid mating. Donors and recipients were cultured in LB with appropriate selective antimicrobials. For liquid mating, 50 μl and 200 μl of mid-logarithmic phase LB cultures from donor and recipient respectively were mixed in 1 mL of LB, pre-warmed to 37°C. After one hour mating at 37°C, the cells were placed on ice to terminate the reaction. For solid mating, 0.5 mL of donor and recipient culture, grown overnight with selection, was spun down gently and resuspended in 20 μl of LB without antibiotics. The suspension was spotted onto dried LB plates, allowed to dry at room temperature for 15 minutes and then incubated at 37°C for three hours or overnight. The mating reaction was resuspended in 1 ml of LB with vortexing and placed on ice to terminate conjugation. After mating, serial ten-fold dilutions of each terminated reaction was made in cold phosphate buffered saline and plated onto plates containing tetracycline (or other appropriate antimicrobials for controls) to select for the plasmid and nalidixic acid, resistance to which is conferred chromosomally in the recipient. Transconjugant colonies were counted after overnight incubation at 37°C and verified by plasmid profiling, phenotype on MacConkey and Eosin methylene blue agars, PCR-RFLP for the *fliC* allele [17], and PCR for donor- and recipient-specific markers. Viable counts of donors and recipient were also performed. The number of transconjugant colonies per donor

Table 2. Oligonucleotide primers used for PCR.

Target gene	Sequences	Application	Reference
topB	F 5′ AGACTCTTTATTGCTGAAAAACCATC 3′; R 5′ TTTTTTAGCGAATTC-AAAACCTATTTTT 3′	Amplification of *topB* from pEBG1 for cloning into pBAD/Thio-TOPO	This study
fliC	F 5′ ATGGCACAAGTCATTAATACCCAAC 3′; R 5′ CTAACCCTGCAGCAGAGACA 3′	*fliC* PCR-RFLP	[17]
qnrS	F 5′ CAATCATACATATCGGCACC 3′; R 5′ TCAGGATAAACAACAATACCC 3′	Amplification of *qnrS1*	[28]
IncX1 (taxC)	F 5′ GCTTAGACTTTGTTTTATCGTT 3′; R 5′ TAATGATCCTCAGCATGTGAT 3′	Amplification of *taxC* from IncX1 plasmids	[20]
IncX2 (taxC)	F 5′ GCGAAGAAATCAAAGAAGCTA 3′; R 5′ TGTTGAATGCCGTTCTTGTCCAG 3′	Amplification of *taxC* from IncX1 plasmids	[20]
qnrS1comp	F 5′ TGAGGGGTTGTAATGTGTTGAT 3′; R 5′ TGCAAGGTTGACAATATTATTCGTTTT 3′	Cloning *qnrS1*	This study
BlplFrag	F 5′ AGTTGTTCCTCATGAG 3′; R 5′ ACTGACTGTCTTTGTATCA 3′	Verifying *qnrS1* deletion	This study
Blpl-2	F 5′ ATTCGCAGCGACTTTCGACGT 3′; R 5′TTCCGGTATCCGGGCTTCGGGTG 3′	Verifying *qnrS1* deletion	This study
pilX2	F 5′ ACCCTGTCTGTCTTCATGGC 3′; R 5′ ACCAGTTCGGCAGATAAGGC 3′	Amplification of *pilX2* from IncX plasmids	This study
pilX8	F 5′ GCACTTAATGCCGCTCTTCC 3′; F 5′ CCTGTACTGTTGCCGTTTGC 3′	Amplification of *pilX8* from IncX plasmids	This study

colony-forming units was computed as the plasmid transfer efficiency [18].

Plasmid stability

We serially passaged pEBG1 in its source strain, 09/22a and in DH5αE essentially as described by Sandegren et al [19]. Triplicate starting cultures were grown overnight at 37°C in 1 mL of LB supplemented with tetracycline (25 mg/L). Bacterial cells were washed by spinning down overnight cultures and resuspending the pellet in 1 mL of LB without antibiotics. An aliquot of 100 μL of washed cells was transferred into 1 mL of LB and incubated overnight at 37°C. Serial passaging of 100 μL of overnight culture to 1 mL of LB was performed daily, approximating 10 generations of growth per passage. Every 20 generations (48 hours), samples were diluted and plated on MacConkey and tetracycline plates. Once counts began to drop off, colonies were screened through replica plating on plates containing tetracycline, trimethoprim and ciprofloxacin to confirm that cells lost the plasmid.

Results

Quinolone-non-susceptible strain 09/22a from Nigeria carries its *qnrS1* gene on an IncX plasmid

Strain 09/22a is a quinolone-non-susceptible *E. coli* isolate from Nigeria that tested positive for *qnrS1* by PCR in a recent study [5]. The strain was unable to conjugate quinolone resistance to *E. coli* C600, TOP10 or EC1502. We therefore extracted plasmids from 09/22a by boiling and electroporated the extract into electro-competent *E. coli* DH5αE (Invitrogen). Selection on nalidixic acid plates (100 μg/mL) yielded no transformants. Selection on plates containing 1 mg/mL of ciprofloxacin produced transformants resistant to ciprofloxacin, trimethoprim and tetracycline, all of which carried a single large plasmid. The plasmid, termed pEBG1,

Table 3. Plasmids used in this study.

Plasmid	Description	Size	Reference or source
pEBG1	WT *qnrS1*-containing plasmid derived from *E. coli* strain 09/22a, confers resistance to ciprofloxacin, tetracycline and trimethoprim	43,530bp	This study
pETS4	pEBG1 plasmid with excised region containing *qnrS1*	41,387	This study
pINK2002	*qnrS1* cloned into the Tet gene of pACYC184	5,038 bp	This study
pETStop3	Expression vector (pBAD/Thio-TOPO vector) containing a *topB* insert	6,773 bp	This study
pBAD/Thio-TOPO	Arabinose inducible expression vector	4,454 bp	Invitrogen
p1ECUMN	IncP plasmid derived from *E. coli* strain UMN026	122,302 bp	[36]
p2ECUMN	IncX1 plasmid derived from *E. coli* strain UMN026	33,809 bp	[36]
pMG306	Naturally occurring *qnrS1*-bearing plasmid from a *Salmonella* isolate		[32]
pMB2	Naturally occurring *aac(6′)-Ib-cr*-bearing plasmid	125 Kb	This study
pMB80-2	Naturally occurring conjugative plasmid from enteropathogenic *E. coli* strain	>100 Kb	[34]
pGEM-T	Amp^R; TA-cloning vector	3,000 bp	Promega
pACYC184	Tet and Cm resistant cloning vector	4,244 bp	NEB

was purified and shot-gun sequenced. pEBG1 was found to be to be 43,534 bp, with an overall G+C content of 46.72%. Its sequence has been deposited in Genbank (Accession number KF738053). pEBG1 is most similar to IncX2 plasmid pNGX2-QnrS1(Genbank accession no. JQ269335) [20] and the two plasmids have 100% identical *taxC* genes. (IncX plasmids are sub-classified based on TaxC, which agrees with subclassifications based on polymorphic sites across whole plasmid sequences [20]). All other *taxC* genes are less than 90% identical to those of pEBG1 and pNGX2-QnrS1 and therefore pEBG1 is an IncX2 plasmid (Figure 1). The IncX2 plasmids demonstrate significant homology to other IncX plasmids, including the prototype R6K, with the most similarity occurring in the conjugative pilus operon [21].

As was pEBG1, pNGX2-QnrS1 was isolated in western Nigeria (Genbank accession no. JQ269335) [20] and the two plasmids currently represent the only completely sequenced IncX2 plasmids in the database. 71% of pEBG1 is 99% identical to 90% of pNGX2-QnrS1 and both plasmids carry the *qnrS1* gene (Figure 2). The backbones, including the region encoding the conjugation system, are nearly identical between the two plasmids, strongly suggesting that they share a recent ancestor. Their resistance modules are integrated at the same site butas shown in Figure 2, the plasmids diverge in the predicted mobile elements and resistance genes they carry and their resistance modules, based on the orientation of the *qnrS1* gene, are inserted in opposite orientations.

Like other IncX plasmids, pEBG1 contains alpha and beta origins of replication, and carries the necessary genes to facilitate plasmid replication and partition: *p1*, a replication initiation protein, *ddp3*, a DNA distortion polypeptide which alters the topology of the DNA to allow for transcription initiation (similar to *taxA* and *taxC*), and *parA* and *parG* which are plasmid partition proteins involved in segregation and stability [21]. As shown in Figure 3, the pEBG1 plasmid backbone contains all other core genes common to IncX plasmids, in the following order: *taxA*, encoding a DNA-distortion polypeptide, which is necessary for altering the DNA helix of plasmids that contain two replication origins [22]; *taxC*, encoding a relaxase required for replication and conjugative transfer of the plasmid [21]; *actX*, containing a NusG transcription termination motif and probably the last gene in that operon; the IncX-type pilus synthesis operon (*pilX1-pilX11*); putative relaxase, *taxB*; and putative nuclease *parB*. pEBG1 contains an IS*3* element in its pilus synthesis operon between *pilX5* and *pilX6*. This insertion sequence is not present in pNGX2-QnrS1 (Figures 2a and 3).

pEBG1 also carries open reading frames (orfs) not found in all IncX plasmids, including *orfEBG036*, predicted to encode a H-NS histone family protein, EBG071c, predicted to encode a PecM-like membrane protein associated with the *tetAR* genes, and a putative DnaJ-domain protein that is 39% similar to the N-terminal J-Domain of the *E. coli* heat-shock chaperone DnaJ/Hsp40. DnaJ-domain proteins are often encoded on plasmids, including pNGX2-QnrS1, but their function in this context is unknown. The function of the putative ABC transporter permease (PecM)-like protein is also unknown. Proteins belonging to this family are involved in small molecule efflux. Plasmid-borne H-NS family proteins are known to compete with chromosomal H-NS, reducing repression of A-T rich sequences present on the plasmid [23]. (The region containing the 657 bp *qnrS1* orf and 570 bp of intergenic sequence upstream, conserved among *qnrS1*-bearing plasmids, has a GC% of 44.42%, compared to 46.72% for the whole plasmid and 50.8% for the *E. coli* chromosome [24]). Another gene found on pEBG1, which is also found in many, but not all, other IncX-type plasmids, is the 2.3 Kb *topB*, coding for bacterial type III topoisomerase. *topB* is truncated at the 5′ end in pNGX2-QnrS1 by an antimicrobial resistance gene module but pEBG1 contains a full-length gene. Type III topoisomerases decatenate DNA during replication [25] but the function of plasmid-encoded *topB* genes, which are only partially conserved, has not been elucidated.

pEBG1 confers resistance to trimethoprim, tetracycline and the quinolones. The plasmid's trimethoprim resistance gene, *dfrA14*, probably originated in a class 1 integron. The 5′ end of this integron, including the *intI*1 gene is intact. The integron also includes a truncated *aadA* gene. The 3′ end of the integron is truncated by an IS*26* element that is 99% identical to the element downstream of *qnrS1*. Although both carry *qnrS1*, the region containing the *dfrA14* gene is not present on the similar plasmid

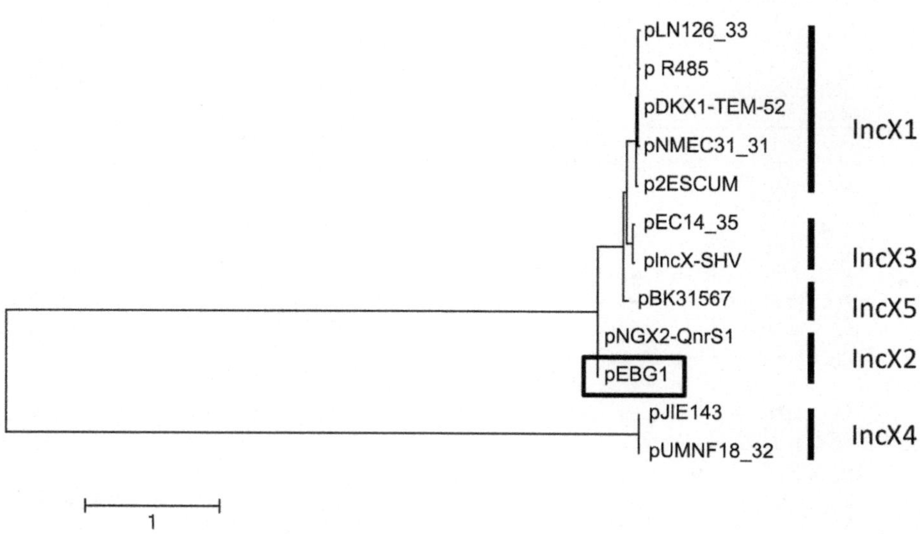

Figure 1. a. Maximum likelihood dendogram depicting evolutionary relationships among TaxC sequences from 12 IncX plasmids, including representatives from groups defined from whole plasmid polymorphisms [20] and pEBG1(boxed), the plasmid sequenced in this study.

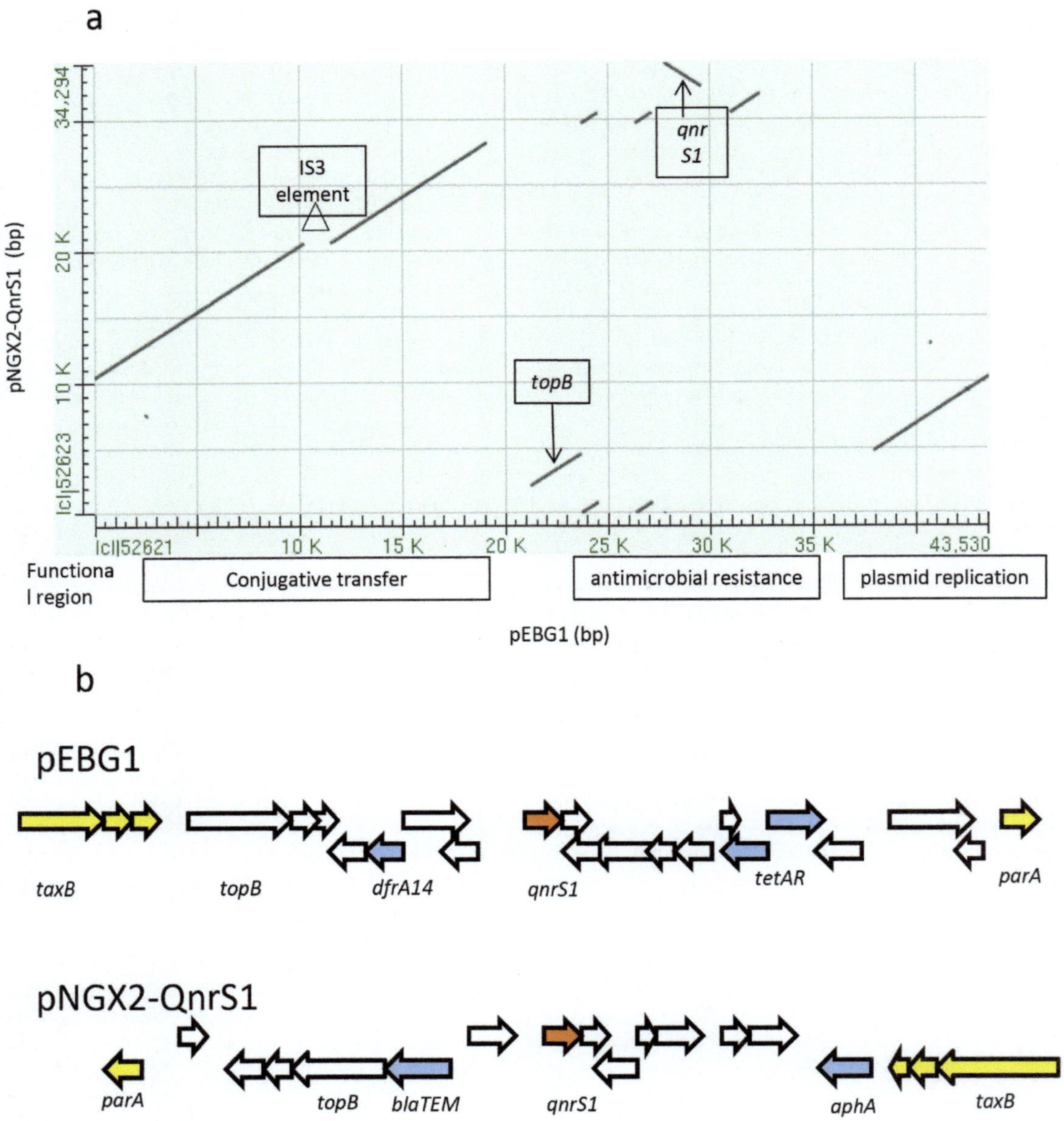

Figure 2. a. Dot plot of pairwise alignment between incX2 plasmids pEBG1 on the x-axis and pNGX2-QnrS1 on the y-axis. The plasmids are highly conserved overall but differ in the location and content of their antimicrobial resistance modules. The pEBG1 conjugation system is interrupted by an IS3 element not present in pNGX2-QnrS1. b. schematic of the resistance regiosn of plsdmifd pEBG1 and pNGX2-QnrS1. *qnrS1* is colored brown and other antimicrobial resistance genes are shaded blue. The flanking core plasmid regions are colored yellow. Note that the insertions are in diffirent orientations.

pNGX2-QnrS1, suggesting that it has been recently mobilized, perhaps via the IS*26* element. The resistance module encompassing *qnrS1* and *its* flanking resistance and transposase genes is nearly identical to that of other plasmids, including *E. coli* plasmid pT078 and pKOX105, from a *Klebsiella oxytoca* strain, with the important exception that beta-lactamase genes present on these plasmids are absent in pEBG1 [26,27]. Although the *qnrS1* modules of pEBG1 and pNGX2-QnrS1 are identical, the plasmid backbones are also highly similar, and the two plasmids were recovered from geographically proximal areas, it is clear that *qnrS1* was acquired independently by each plasmid. The location of *qnrS1* and its flanking IS*26* element varies between the plasmids, being at the 5′ end of *topB* in pNGX2-QnrS1 but with the resistance region downstream of *topB* in pEBG1. *qnrS1* is often adjacent to a *tnpA* transposase in other plasmids, and it has been hypothesized that the *qnrS1* gene is usually part of an IS*26* element [28]. IS*26* contains two 14 bp perfect terminal repeats. We identified four IS*26*-associated inverted repeats [29] adjacent to the *qnrS1* gene. These 14 bp sequences begin at coordinates 23,698, 24,504, 27,142 and 29,074, marking out a putative transposable element of 5,376 bp encompassing *dfrA14* and *qnrS1*.

In addition to *qnrS1* and *dfrA14*, pEBG1 also contains the genes *tetR* and *tetA*, which lie adjacent to the *pecM*-like gene EBG71 and another *tnpA* gene in a high G+C content region, whose collective sequence was also not found on pNGX2-QnrS1 (Figure 2). These genes are bordered by a putative relaxase and Tn*3* transposase, which outline the boundaries of the elevated GC

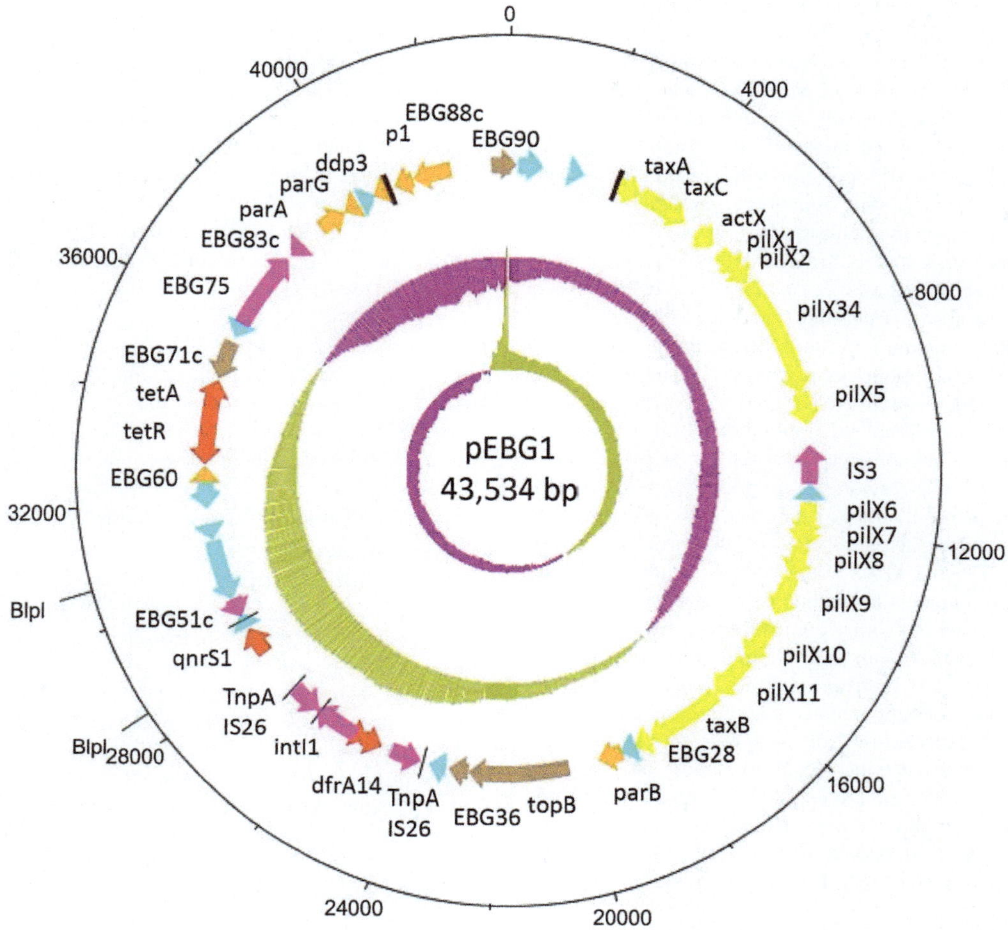

Figure 3. Circular map of pEBG1. and BlpI restriction sites conjugation genes are yellow and those that are predicted to encode proteins involved in DNA mobility or transposition are pink. Orange colored genes are involved in plasmid maintenance; other predicted open reading frames colored brown and hypothetical open-reading frames are colored light blue. Predicted OriT sites are marked as heavy dark brown lines. Thin black lines indicate the positions of *IS26* repeats.

content area, suggesting that the plasmid's tetracycline resistance also recently came into the plasmid via a transposable element. The existence of this module on other resistance plasmids [30], and its absence from homologous plasmid pNGX2-QnrS1 support this idea. No other areas of the plasmid show a significant difference in G+C content, save for the fact that the backbone of the plasmid has an overall lower G+C content than the rest of the plasmid (Figure 3).

qnrS1 is the sole quinolone resistance-conferring gene on pEBG1

pEBG1 encodes a number of hypothetical ORFs as well as ORFs that could contribute to topoisomerase activity (TopB) or antimicrobial efflux (PecM-like EBG071c). To verify that *qnrS1* is the only quinolone resistance gene on pEBG1, we deleted the gene by restricting the plasmid with *Blp*I (which cuts at positions 28,514 and 30,657 in our annotation effectively removing 2,143 bp, which includes the 3′ half of *qnrS1* as well as two other hypothetical orfs, Figure 3). The resulting plasmid, pETS4 was confirmed by PCR analysis with the *qnrS1*F and R primer pair and restriction analysis with *Nru*I, which has 10 sites in pEBG1 including a site at position 30,368, within the deleted region. pETS4 conferred a six-fold lower MIC to ciprofloxacin than pEBG1 on DH5α and a greater than 32-fold lower MIC on TOP10. Resistance was restored upon trans-complementation with *qnrS1* clone pINK2002. Deleting *qnrS1* from pEBG1 however produced little effect (about 2–3 fold) on MICs to nalidixic acid. We then determined the level of resistance conferred by *aac(6′)-Ib-cr*, a gene that confers ciprofloxacin resistance but not resistance to other quinolones [31], using plasmid pMB2, which carries this gene and no other known plasmid-mediated quinolone resistance gene. We found that pMB2 provided no significant alteration to the nalidixic acid MIC of DH5α, but increased resistance to ciprofloxacin three-fold. As *qnrS1* has previously been reported to confer significant resistance to multiple quinolones, including nalidixic acid, we additionally tested pMG306, a *qnrS1*-bearing plasmid isolated from non-typhoidal *Salmonella* [32]. As in the original report describing the plasmid [32], we confirmed that this plasmid conferred only a slightly greater level of resistance on TOP10, compared to pEBG1, and that ciprofloxacin resistance was increased over 10-fold in contrast to four-fold increase in nalidixic acid resistance. Similar results were seen with this plasmid in the DH5α andJ53 backgrounds. Thus *qnrS1* principally confers ciprofloxacin resistance. In addition to demonstrating the pivotal role that *qnrS1* plays in ciprofloxacin resistance, these data demonstrate that the host background is an important determinant of the effect of *qnrS1* (Table 4).

topB is a large, partially conserved gene that is present in most IncX plasmids, but the function of plasmid-encoded topoisomerases is unknown. Chromosomally encoded DNA gyrase and topoisomerase are the normal targets of the quinolones. While *topB* genes, encoding type III topoisomerases, should not normally be inhibited by antibacterial quinolones, which are better inhibitors of type II topoisomerases, bacterial *topB* genes have not previously been evaluated in this regard. Since over-expression of chromosomal topoisomerases can compensate for absence of gyrase [25], it is possible that plasmid-encoded TopB could confer resistance to the quinolones by supplementing the activity of inhibited chromosomal topoisomerases. As it is possible that *topB* is not expressed under conditions we used to test for quinolone resistance *in vitro*, we cloned the gene and placed it under the control of the arabinose promoter, using the vector pBAD/Thio-TOPO (Invitrogen). The resulting clone, pETS-Topo3 produced copious amounts of the TopB protein when induced with 0.2% arabinose. Expression was abrogated in the presence of 2% glucose (data not shown). MICs of the clone to nalidixic acid and ciprofloxacin were identical in arabinose and glucose, confirming the *topB* is not sufficient to confer quinolone resistance. Because it is known some *topB* alleles can promote homologous recombination by stabilizing and resolving Holliday junctions [33], we hypothesized that *topB* might enhance mutation rates and that this might serve as a mechanism of evolution to antimicrobial resistance by strains carrying IncX plasmids like pEBG1. However, tests with the pETS-Top3 clone in a TOP10 and LMG194 background did not reveal enhanced mutation rates leading to resistance to rifampicin determined by plating cultures on LB plates containing 50 µg/mL rifampicin (data not shown).

pEBG1 is mobilizable but not self-transmissible

We performed conjugation experiments on solid and in liquid media, using pMB80-2 as a control plasmid [34]. We were unable to conjugate pEBG1 into any of three recipient strains, C600, TOP10 or EC1502 even when mating was extended to 24 h. To determine whether pEBG1 could be mobilized *in vitro*, we electroporated it into strains carrying self-transmissible conjugation systems. Strain SM10, which contains conjugation genes from the broad host range IncP-type plasmid RP4 integrated into its chromosome [35] was unable to mobilize pEBG1. We then transferred two plasmids, IncP plasmid p1ECUMN and IncX1 plasmid p2ECUMN, from UMN026 [36], into C600 (Nal^R) by conjugation. The presence of both plasmids in strain C600 was confirmed by plasmid profiling and PCR. C600 (p2ECUMN, p1ECUMN, pEBG1) could mobilize pEBG1 into TOP10 ($Strep^R$). The conjugation rate for pEBG1 after a 3-hour mating was 1.7×10^{-3}. At 24 hours it was 3.38×10^{-7}.

Other *qnrS1*-bearing isolates from Western Nigeria carry non-identical plasmids

We screened genomic DNA from other *qnrS1*-bearing isolates recovered in the same study for pEBG1 plasmids. All these strains show lowered susceptibility to ciprofloxacin but only one meets the CLSI breakpoint criteria for resistance. The strains were screened using *taxC*-based primers designed by Johnson et al (2012) [20], as well as primers specific for *qnrS1, topB, pilX2* and *pilX8*. As shown in Table 5, all five pEBG1 markers were detected in another strain isolated from the same individual but not in the other isolates. As we were able to obtain two or more of the pEBG1 marker amplicons from three other strains, we screened them for the IncX1 *taxC* gene and found that only two strains, including 09/22a carried IncX1 plasmids.

Table 4. MICs to quinolones conferred by pEBG1, its derivatives and other Qnr-encoding plasmids.

Strain	Resistance to non-quinolone antimicrobials	Nalidixic acid MIC	Ciprofloxacin MIC
09/22a	Amp Chl Str Sul Tet Tmp	12	0.38
C600	-	>256	0.5
DH5α	-	24	0.032
DH5α (pEBG1)	Tet Tmp	12	0.125
DH5α (pETS4)	Tet Tmp	6	0.032
DH5α (pETS4, pINK2002)	Chl Tet Tmp	8	0.016
DH5α (pETS4, pACYC184)	Chl Tet Tmp	6	0.008
DH5α (pMG306)	Chl	24	0.25
DH5α (pMB2)	Tet Tmp	24	0.094
TOP10	Str	0.75	<0.002
TOP10 (pEBG1)	Str Tet Tmp	2	0.064
TOP10 (pETS4)	Str Tet Tmp	0.75	<0.002
TOP10 (pETS4, pINK2002)	Chl Str Tet Tmp	1	0.016
TOP10 (pETS4, pACYC184)	Chl Str Tet Tmp	0.75	<0.002
TOP10 (pMG306)	Chl	3	0.19
J53	Azi	4	0.008
J53 (pMG306)	Azi Chl	24	0.75
J53 (pMG282)	Azi Chl Str	16	0.38

pEBG1 is propagated in the medium term without selection

We studied the stability of pEBG1 following serial passage. pEBG1 was stably inherited by its original host strain 09/22a and by DH5α for over 200 generations without selection (Figure 4). However, extended subculture lead to a significant proportion of those lines loosing the plasmid.

Discussion

Plasmids harboring *qnrS* genes tend to be on average smaller than plasmids carrying other *qnr* subtypes, and are less likely to be self-transmissible [1,37]. They vary greatly, belonging to N, L, M, H, R and X incompatibility groups and residing in diverse enterobacterial genera [22,37]. Multiple studies have also noted that *qnrS1* tends not to be associated with integrons, and are instead associated with *tnpA* transposase genes, which likely contribute to intermolecular mobility [22,28,38].

In this study, we sequenced and functionally characterized a *qnrS1*-bearing plasmid obtained from an *E. coli* isolate in south-west Nigeria. At the time of isolation, *qnrS1* was the sole plasmid-encoded quinolone resistance gene identified in that locality and since the isolates were not clonal, we hypothesized that one or a few plasmids might account for the recently emerged quinolone resistance seen in the community. We identified and characterized pEBG1, a 43.5 Kb IncX2 plasmid. We find that other strains from Nigeria carry similar plasmids but that pEBG1 does not appear to have clonally expanded.

The IncX2 plasmid we characterized carries, in addition to *qnrS1*, genes conferring resistance to tetracycline and trimethoprim. The resistance gene repertoire is the most common variable among *qnrS1*-bearing IncX plasmids in and beyond Nigeria. This finding is worrisome, since it suggests that IncX backbones may be good scaffolds for incorporating new resistance genes. For example, in addition to plasmid-encoded quinolone resistance genes, extended-spectrum beta-lactamase genes have also been recently found on IncX plasmids [39,40]. Although the plasmid we identified in this study was from a human isolate, IncX plasmids, including one IncX2 plasmid, bearing *qnrS1* have also been found in isolates from migratory birds in Europe [41] and pNGX2-QnrS1 is a poultry isolate, pointing to possible animal reservoirs and global reach of these elements.

Qnr proteins confer low-level resistance by altering the interaction between the quinolones and their targets. What is known about Qnr function and spectrum has been gleaned from mechanistic studies with QnrA and QnrB. In this study, we found that *qnrS1* from pEBG1 conferred significant fluoroquinolone but only modest quinolone resistance on laboratory *E. coli* isolates. The finding that *qnrS1* confers ciprofloxacin non-susceptibility but very little nalidixic acid resistance on these strains is not wholly surprising since a previous study has indicated that *qnrS1* gene products sometimes confer no nalidixic acid resistance [42]. A separate study found that *qnrS* confers moderate nalidixic acid and ciprofloxacin resistance in *E. coli* strain HB101, but confers no resistance at all in *Shigella* strains [43]. The *qnrS1* gene was originally reported from a 47 Kb conjugative plasmid, pAH0376, from Japan in 2005 [43]. The plasmid was reported to confer on *E. coli* HB101 resistance to nalidixic acid (MIC, 16 µg/mL) and ciprofloxacin (MIC, 0.25 µg/mL). The following year, plasmid-borne *qnrS1* was reported from Germany and although the responsible plasmid pINF1 was not identical to pAH0376, the *qnrS1* genes on both plasmids were flanked by a Tn3 transposon carrying a bla_{TEM-1} gene. Other contexts for *qnrS1* have since been described [37,41,42]. Most of these reports demonstrate that

Table 5. IncX plasmid types in *qnrS1*-positive isolates from Nigeria.

Strain	Resistance pattern	*incX1*	*incX2*	*pilX2*	*pilX8*	*topB*	*qnrS1*
09/5d	Amp **Nal** Str Sul Tet Tmp	-	-	-	-	-	+
09/9a	Amp Chl **Nal** Str Sul Tet Tmp	-	-	-	-	-	+
09/16e	Amp Chl **Nal** Str Tet Tmp	-	-		+		+
09/22a	Amp Chl **Nal** Str Sul Tet Tmp	+	+	+	+	+	+
09/22d	Amp Chl **Nal** Str Tet Tmp	+	-	+	+	+	+
09/24c	Amp **Cip** Chl **Nal** Str Sul Tet Tmp	-	-	-	-	-	+
09/18d	Amp Chl **Nal** Str Sul Tet Tmp	-	+	-	-	-	+
UMN206		+	-	+	+	+	-
DH5α		-	-	-	-	-	-

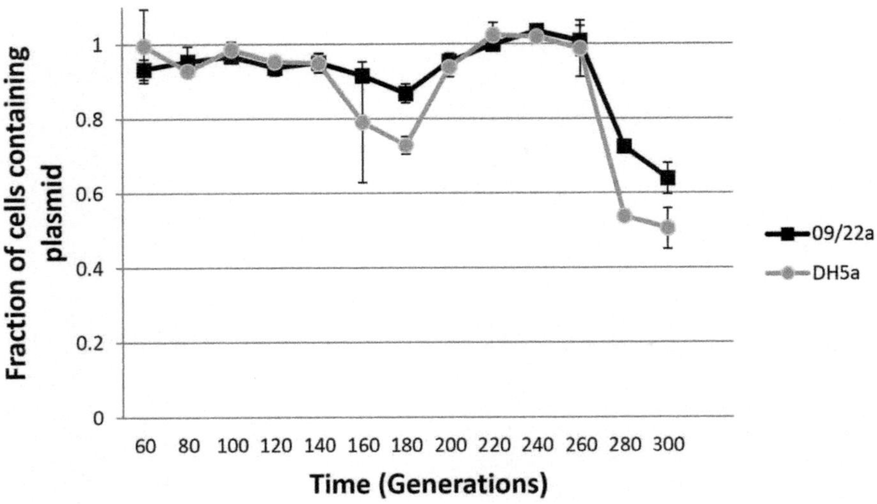

Figure 4. Stability of pEBG1 in the absence of selection in its natural host, 09/22a, and in laboratory strain DH5α.

qnrS1 confers low- to intermediate resistance to ciprofloxacin (0.25–0.5 μg/mL). Altogether, the findings of this and earlier studies indicate that the nature and degree of quinolone resistance conferred by *qnrS1* is dependent on host strain background [42]. In Nigeria, most of the quinolone resistance selective pressure comes from ciprofloxacin, as in Vietnam [37], where *qnrS1* is also highly prevalent, and it therefore makes sense that genes preferentially conferring resistance to this agent will be selected.

This research, along with work from other laboratories [20,44], points to multiple instances of *qnrS1*-bearing IncX plasmids in Nigeria. In this study, we determined that plasmid pEBG1 is highly mobilizable, though not self transmissible. The model IncX plasmid R6K is reported to conjugate at a rate of 1.7×10^{-3} into *E. coli* strain MC1061 [45], the same as the mobilization rate for pEBG1. The most likely explanation for the inability of pEBG1 to mediate its own conjugation is the insertion of an IS*3* element between *pilX5* and *pilX6*. Although this element does not disrupt a gene, it does interrupt the *pilX* operon and is likely polar on downstream genes. As the IS*3* element is absent from pNGX2-QnrS1, we anticipate that self transmissible IncX plasmids are also in circulation in Nigeria. Resistant strains in Nigeria frequently carry multiple plasmids [46]. Therefore the lack of a fully functional conjugative system is unlikely to greatly compromise the success of this *qnrS1*-bearing plasmid as long as it remains mobilizable. Overall, the high transmission rate and modular content observed by us and others [20,45] illustrates that IncX plasmids are good candidates for mediating the dissemination of resistance genes. pEBG1 was stably inherited without selection in the medium term but was lost after 16–18 passages. This suggests that IncX2 plasmids bearing *qnrS1* may be maintained by antibacterial drug selection and therefore their frequent occurrence in Nigeria may be linked to overuse of antibacterial drugs there, among which ciprofloxacin, trimethoprim and tetracycline are prominent.

Acknowledgments

We are grateful to Babatunde Odetoyin, Jose Otero-Vera and Erin Remaly for technical assistance. We thank the David C Hooper laboratory at Massachusetts General Hospital Boston and the University of Bradford for Strains.

Author Contributions

Conceived and designed the experiments: ETS EBG AL INO. Performed the experiments: ETS EBG AOA INO. Analyzed the data: ETS EBG AOA AL INO. Contributed reagents/materials/analysis tools: AL INO. Wrote the paper: ETS EBG AOA AL INO.

References

1. Strahilevitz J, Jacoby GA, Hooper DC, Robicsek A (2009) Plasmid-mediated quinolone resistance: a multifaceted threat. Clin Microbiol Rev 22: 664–689.
2. Martinez-Martinez L, Pascual A, Jacoby GA (1998) Quinolone resistance from a transferable plasmid. Lancet 351: 797–799.
3. Tran JH, Jacoby GA (2002) Mechanism of plasmid-mediated quinolone resistance. Proc Natl Acad Sci U S A 99: 5638–5642.
4. Robicsek A, Jacoby GA, Hooper DC (2006) The worldwide emergence of plasmid-mediated quinolone resistance. Lancet Infect Dis 6: 629–640.
5. Lamikanra A, Crowe JL, Lijek RS, Odetoyin BW, Wain J, et al. (2011) Rapid evolution of fluoroquinolone-resistant *Escherichia coli* in Nigeria is temporally associated with fluoroquinolone use. BMC Infect Dis 11: 312.
6. Kado CI, Liu ST (1981) Rapid procedure for detection and isolation of large and small plasmids. J Bacteriol 145: 1365–1373.
7. Sambrook J, Fritsch EF, Maniatis T (1989) Molecular cloning; a laboratory manual. New York: Coldspring Harbor Laboratory Press.
8. Sambrook J, Russell DW (2001) Molecular cloning: a laboratory manual. Cold Spring Harbor, N.Y.: Cold Spring Harbor Laboratory Press.
9. Rutherford K, Parkhill J, Crook J, Horsnell T, Rice P, et al. (2000) Artemis: sequence visualization and annotation. Bioinformatics 16: 944–945.
10. Carver T, Thomson N, Bleasby A, Berriman M, Parkhill J (2009) DNAPlotter: circular and linear interactive genome visualization. Bioinformatics 25: 119–120.
11. Altschul SF, Gish W, Miller W, Myers EW, Lipman DJ (1990) Basic local alignment search tool. J Mol Biol 215: 403–410.
12. Punta M, Coggill PC, Eberhardt RY, Mistry J, Tate J, et al. (2012) The Pfam protein families database. Nucleic Acids Res 40: D290–301.
13. Tamura K, Peterson D, Peterson N, Stecher G, Nei M, et al. (2011) MEGA5: molecular evolutionary genetics analysis using maximum likelihood, evolutionary distance, and maximum parsimony methods. Mol Biol Evol 28: 2731–2739.
14. NCCLS (2003) Performance standards for antimicrobial disk susceptibility tests, 8th Edition; Approved standard. Villanova, PA: National Committee for Clinical Laboratory Standards. NCCLS Document M2-A8 NCCLS Document M2-A8. 130 p.
15. O'Brien TF, Stelling JM (1995) WHONET: an information system for monitoring antimicrobial resistance. Emerg Infect Dis 1: 66.
16. CLSI (2006) Methods for dilution antimicrobial susceptiblity tests for bacteria that grow aerobically, 7th Edition; Approved standard. Wayne, PA: Clinical and Laboratory Standards Institute. CLSI Document M7-A7 CLSI Document M7-A7.

17. Fields P, Blom K, Hughes H, Helsel L, Feng P, et al. (1997) Molecular characterization of the gene encoding H antigen in *Escherichia coli* and development of a PCR-restriction fragment length polymorphism test for identification of *E. coli* O157:H7 and O157:NM. Journal of Clinical Microbiology 35: 1066–1070.
18. Lu J, Manchak J, Klimke W, Davidson C, Firth N, et al. (2002) Analysis and characterization of the IncFV plasmid pED208 transfer region. Plasmid 48: 24–37.
19. Sandegren L, Linkevicius M, Lytsy B, Melhus A, Andersson DI (2012) Transfer of an *Escherichia coli* ST131 multiresistance cassette has created a Klebsiella pneumoniae-specific plasmid associated with a major nosocomial outbreak. J Antimicrob Chemother 67: 74–83.
20. Johnson TJ, Bielak EM, Fortini D, Hansen LH, Hasman H, et al. (2012) Expansion of the IncX plasmid family for improved identification and typing of novel plasmids in drug-resistant Enterobacteriaceae. Plasmid 68: 43–50.
21. Nunez B, Avila P, de la Cruz F (1997) Genes involved in conjugative DNA processing of plasmid R6K. Mol Microbiol 24: 1157–1168.
22. Garcia-Fernandez A, Fortini D, Veldman K, Mevius D, Carattoli A (2009) Characterization of plasmids harbouring qnrS1, qnrB2 and qnrB19 genes in *Salmonella*. J Antimicrob Chemother 63: 274–281.
23. Doyle M, Fookes M, Ivens A, Mangan MW, Wain J, et al. (2007) An H-NS-like stealth protein aids horizontal DNA transmission in bacteria. Science 315: 251–252.
24. Blattner FR, Plunkett G 3rd, Bloch CA, Perna NT, Burland V, et al. (1997) The complete genome sequence of *Escherichia coli* K-12. Science 277: 1453–1474.
25. Usongo V, Tanguay C, Nolent F, Bessong JE, Drolet M (2013) Interplay between Type 1A Topoisomerases and Gyrase in Chromosome Segregation in *Escherichia coli*. J Bacteriol 195: 1758–1768.
26. Carattoli A, Aschbacher R, March A, Larcher C, Livermore DM, et al. (2010) Complete nucleotide sequence of the IncN plasmid pKOX105 encoding VIM-1, QnrS1 and SHV-12 proteins in Enterobacteriaceae from Bolzano, Italy compared with IncN plasmids encoding KPC enzymes in the USA. J Antimicrob Chemother 65: 2070–2075.
27. Huang SY, Zhu XQ, Wang Y, Liu HB, Dai L, et al. (2012) Co-carriage of qnrS1, floR, and bla(CTX-M-14) on a multidrug-resistant plasmid in Escherichia coli isolated from pigs. Foodborne Pathog Dis 9: 896–901.
28. Wu J-J, Ko W-C, Tsai S-H, Yan J-J (2007) Prevalence of plasmid-mediated quinolone resistance determinants QnrA, QnrB, and QnrS among clinical isolates of *Enterobacter cloacae* in a Taiwanese hospital. Antimicrob Agents Chemother 51: 1223–1227.
29. Mollet B, Iida S, Shepherd J, Arber W (1983) Nucleotide sequence of IS*26*, a new prokaryotic mobile genetic element. Nucleic Acids Res 11: 6319–6330.
30. Schlüter A, Heuer H, Szczepanowski R, Forney LJ, Thomas CM, et al. (2003) The 64 508 bp IncP-1β antibiotic multiresistance plasmid pB10 isolated from a waste-water treatment plant provides evidence for recombination between members of different branches of the IncP-1β group. Microbiology 149: 3139–3153.
31. Robicsek A, Strahilevitz J, Jacoby GA, Macielag M, Abbanat D, et al. (2006) Fluoroquinolone-modifying enzyme: a new adaptation of a common aminoglycoside acetyltransferase. Nat Med 12: 83–88.
32. Gay K, Robicsek A, Strahilevitz J, Park CH, Jacoby G, et al. (2006) Plasmid-mediated quinolone resistance in non-Typhi serotypes of *Salmonella enterica*. Clin Infect Dis 43: 297–304.
33. Wang JC (2002) Cellular roles of DNA topoisomerases: a molecular perspective. Nat Rev Mol Cell Biol 3: 430–440.
34. Nwaneshiudu AI, Mucci T, Pickard DJ, Okeke IN (2007) A second large plasmid encodes conjugative transfer and antimicrobial resistance in O119:H2 and some typical O111 enteropathogenic *Escherichia coli* strains. J Bacteriol 189: 6074–6079.
35. Simonsen C, Chen E, Levinson A (1983) Identification of the type I trimethoprim-resistant dihydrofolate reductase specified by the *Escherichia coli* R-plasmid R483: comparision with procaryotic and eucaryotic dihydrofolate reductases. Journal of Bacteriology 155: 1001–1008.
36. Lescat M, Calteau A, Hoede C, Barbe V, Touchon M, et al. (2009) A module located at a chromosomal integration hot spot is responsible for the multidrug resistance of a reference strain from *Escherichia coli* clonal group A. Antimicrob Agents Chemother 53: 2283–2288.
37. Vien LTM, AbuOun M, Morrison V, Thomson N, Campbell JI, et al. (2011) Differential phenotypic and genotypic characteristics of *qnrS1*-harboring plasmids carried by hospital and community commensal Enterobacteria. Antimicrobial Agents and Chemotherapy 55: 1798–1802.
38. Kehrenberg C, Hopkins KL, Threlfall EJ, Schwarz S (2007) Complete nucleotide sequence of a small qnrS1-carrying plasmid from *Salmonella enterica* subsp. enterica Typhimurium DT193. Journal of Antimicrobial Chemotherapy 50: 903–905.
39. Du X-X, Wang J-F, Fu Y, Zhao F, Chen Y, et al. (2013) Genetic characteristics of blaNDM-1-positive plasmid in Citrobacter freundii isolate separated from a clinical infectious patient. J Med Microbiol 62: 1332–1337.
40. Partridge SR, Ellem JA, Tetu SG, Zong Z, Paulsen IT, et al. (2011) Complete sequence of pJIE143, a pir-type plasmid carrying IS*Ecp1*-blaCTX-M-15 from an *Escherichia coli* ST131 isolate. Antimicrobial Ag Chemother 55: 5933–5935.
41. Literak I, Dolejska M, Janoszowska D, Hrusakova J, Meissner W, et al. (2010) Antibiotic-Resistant Escherichia coli Bacteria, Including Strains with Genes Encoding the Extended-Spectrum Beta-Lactamase and QnrS, in Waterbirds on the Baltic Sea Coast of Poland. Appl Environ Microbiol 76: 8126–8134.
42. Gunell M, Webber MA, Kotilainen P, Lilly AJ, Caddick JM, et al. (2009) Mechanisms of resistance in nontyphoidal Salmonella enterica strains exhibiting a nonclassical quinolone resistance phenotype. Antimicrob Agents Chemother 53: 3832–3836.
43. Hata M, Suzuki M, Matsumoto M, Takahashi M, Sato K, et al. (2005) Cloning of a novel gene for quinolone resistance from a transferable plasmid in *Shigella flexneri* 2b. Antimicrob Agents Chemother 49: 801–803.
44. Chah KF, Agbo IC, Eze DC, Somalo S, Estepa V, et al. (2010) Antimicrobial resistance, integrons and plasmid replicon typing in multiresistant clinical *Escherichia coli* strains from Enugu State, Nigeria. J Basic Microbiol 50: S18–S24.
45. Grudniak AM, Kraczkiewicz-Dowjat A, Wolska KI, Wild J (2007) Conjugal transfer of plasmid R6K gamma ori minireplicon derivatives from Escherichia coli to various genera of pathogenic bacteria. Curr Microbiol 55: 549–553.
46. Labar AS, Millman JS, Ruebush E, Opintan JA, Bishar RA, et al. (2012) Regional dissemination of a trimethoprim-resistance gene cassette via a successful transposable element. PLoS ONE 7: e38142.
47. Appleyard RK (1954) Segregation of new lysogenic types during growth of a doubly lysogenic strain derived from *Escherichia coli* K12. Genetics 39: 440–452.
48. Guzman LM, Belin D, Carson MJ, Beckwith J (1995) Tight regulation, modulation, and high-level expression by vectors containing the arabinose pBAD promoter. J Bacteriol 177: 4121–4130.
49. Yi H, Cho Y-J, Yong D, Chun J (2012) Genome sequence of *Escherichia coli* J53 a reference strain for genetic studies. Journal of Bacteriology 194: 3742–3743.

7

Treatment Effect and Drug-Resistant Mutations in Chinese AIDS Patients Switching to Second-Line Antiretroviral Therapy

Min Zhang[1,2], **Mingquan Shang[1,2]**, **Weiwei Yang[1,2]**, **Junli Chen[1,2]**, **Zhe Wang[3]**, **Hong Shang[1,2*]**

1 Key Laboratory of AIDS Immunology of National Health and Family Planning Commission, Department of Laboratory Medicine, The First Affiliated Hospital, China Medical University, Shenyang, China, **2** Collaborative Innovation Center for Diagnosis and Treatment of Infectious Diseases, Hangzhou, China, **3** Disease Prevention and Control Center of Henan Province, Zhengzhou, China

Abstract

This study aimed to investigate treatment effect, drug resistance changes, and their influencing factors in Chinese AIDS patients after switching to second-line antiretroviral therapy, and thus provide important information for the scale-up of second-line antiretroviral treatment in China. In Weishi county of Henan province, where second-line antiretroviral therapy was introduced early in China, 195 AIDS patients were enrolled, of which 127 patients met the switching criterion and 68 patients volunteered to switch drugs without meeting the switching criterion. CD4 cell count, viral load and in-house PCR genotyping for drug resistance were measured for all 195 subjects before drug switch, as well as 6 and 12 months after drug switch. Extensive secondary mutations to the protease inhibitor were observed, which suggested that long-term drug resistance surveillance is necessary for patients switching to second-line antiretroviral therapy. Multidrug resistance and cross-resistance were extensive in Chinese patients that experienced first-line treatment failure. Patients need timely CD4 count, viral load, and drug resistance monitoring in order to switch to second-line therapy under conditions of relatively good immunity and low viral duplication levels.

Editor: Xia Jin, University of Rochester, United States of America

Funding: This study was supported by the Twelfth Five-Year Project on Tackling Key Problems of National Science and Technology (2012ZX10001002-002-013) and National Natural Science Foundation of China (81171619). The funders had no role in study design, data collection and analysis, decision to publish, or preparation of the manuscript.

Competing Interests: The authors have declared that no competing interests exist.

* Email: hongshang100@hotmail.com

These authors contributed equally to this work.

Introduction

Highly active antiretroviral therapy (HAART) is currently the most effective therapy for the treatment of acquired immune deficiency syndrome (AIDS) [1]. However, the characteristic high mutation level of human immunodeficiency virus (HIV), along with selection pressure of antiretroviral drugs, usually cause rapid occurrence of drug resistance of HIV, which has been the main reason for clinical failure of antiretroviral treatment. In America and Europe, drug resistance of HIV existed in more than 60% of patients who were antiretroviral treatment failure [2].

China initiated a large-scale program for free antiretroviral treatment in 2003, with zidovudine (AZT)/stavudine (D4T) + didanosine (DDI) + nevirapine (NVP) as the main first-line therapy regimen. By October 2011, more than 130,000 patients with AIDS had received HAART. The increase in availability and access to free antiretroviral treatment had resulted in considerably prolonged survival time, improved quality of life and decreased mortality rate in AIDS patients [3]. But with the increase of HAART treatment duration, treatment failure and drug resistance occurred in more and more patients. In a cross-sectional survey conducted in 2007, Min Zhang et al. [4] studied HIV infected patients living in 7 provinces in China and reported that the rate of drug resistance was significantly higher in patients receiving first-line treatment for more than 6 months than that in patients receiving treatment for less than 6 months. In 2008, Min Zhang [5] conducted a cohort study on HIV infected patients living in Weishi county of Hunan province of China. The results showed that, among 88 patients who had received the first-line antiretroviral treatment, only 1% of patients had drug resistance before treatment. The rate increased to 15% after the patients received treatment for 12 months, and increased to 17% after 36 months. As the increasing rate of drug resistance resulted in diminishing efficacy of the free antiretroviral treatment, in 2009 China initiated the switch to second-line antiretroviral therapy for patients with failure of first-line treatment.

Some foreign researchers reported that extensive drug-resistant mutations were observed before switching therapy regimens. Regarding treatment results after switching to second-line antiretroviral regimens, there is variance among reports from different regions and countries [6–8]. In China, with the expanding of the free antiretroviral treatment and the extending of treatment duration, the rate of drug resistance increases over time. What would be the treatment results after patients switch to second-line antiretroviral therapy? Would accumulation of drug-resistant mutations during the first-line treatment have an impact on the second-line treatment results? Would new drug-resistant mutations appear soon after the administration of second-line regimens? In China, no study has examined efficacy and drug resistance after patients switch to second-line regimens. In Weishi county of Henan province, where antiretroviral treatment started early, more than 2,000 HIV infected individuals had received antiretroviral treatment, and rates of treatment failure and drug resistance were increasing annually [5]. From August 2009, some patients began to switch to second-line therapy. We conducted a prospective observational cohort follow-up study on these AIDS patients in Weishi county who switched to second-line therapy.

Materials and Methods

Ethics statement

This study was approved by the Institutional Review Board of the First Affiliated Hospital of China Medical University (Shen Yang, China). All participants ≥18 years of age, written informed consent was obtained from all subjects.

Study subjects

Our research is an observational study; we did not determine or influence the course of treatment. We just collected the therapy information and blood samples of patients in the second-line antiretroviral therapy pilots in Henan province.

One hundred and ninety-five AIDS patients living in Weishi county of Henan province were enrolled in August 2009 and February 2010, respectively. In accordance with Chinese National Free AIDS Antiretroviral Therapy Manual issued in 2009 [9] and WHO antiretroviral therapy for HIV infection in adults and adolescents [10], virological failure defines as plasma HIV-1 RNA level >400 copies/mL after 6 months of treatment or HIV-1 RNA level >1000 copies/mL after initial virological suppression. Virological failure of first-line therapy was the main reason to switch to second-line treatment. 127 patients were first-line virological failure, after first-line virological failure happened, the patients continued the old treatment for one month with rectified treatment compliance. After that, if the viral load of the patient was higher than 1000 copies/mL, the patient switched to second-line therapy. As a result, viral loads of all the 127 subjects were higher than 1000 copies/mL. The remaining 68 patients who were virological suppression after first-line therapy, but they requested to switch to the second-line drugs due to significant toxicity and side effects of their first-line drugs.

The second-line antiretroviral regimen used was Lamivudine (3TC) + tenofovir (TDF) and lopinavir + ritonavir (LPV/r), as recommended by WHO [9]. Patients were followed up and their CD4, viral loads and drug resistance were measured 6 and 12 months after starting second-line treatment. Peripheral venous blood was obtained from each patient using an EDTA-3 K tube containing anticoagulants. CD4+T lymphocytes were counted within 24 hours. Blood for measurement of viral loads and drug resistance was centrifuged for 15 minutes, and plasma separated from blood was stored at −80°C.

CD4+T lymphocyte count

Absolute CD4+, CD8+, and CD3+T lymphocytes were counted with the FACSCalibur flow cytometer (Becton Dickinson, USA).

Viral load measurement

Viral loads were measured using COBAS AmpliPrep/COBAS TaqMan HIV-1 Test (Roche, Switzerland).

Drug-resistant mutation analysis

HIV-1 viral RNA was extracted from 140 ul plasma with the QIAamp Viral RNA Mini Kit (Qiagen Inc, USA). Reverse transcription-polymerase chain reaction (RT-PCR) was carried out with the One-step RNA PCR Kit (AMV) within 4 hours, and 1300 bp of reverse transcriptase (RT) and protease (PR) genes of HIV-1 were amplified with nested PCR, including the whole PR and the first 300 amino acids of RT [5]. The PCR product was then analyzed with ABI 3130 Genetic Analyzers (ABI, USA) using the chain termination method.

Sequence analysis and statistical analysis

Sequencher 4.10.1, NTI advance 9 and ContigExpress were used for sequence editing, assembly and rectification. The resulting sequences were compared with the Stanford University HIV drug resistance database (http://hivdb.stanford.edu) version 6.1.1. for analysis of HIV drug-resistant mutations and drug resistance. Statistical Package for Social Sciences (SPSS) version 15.0 was used for statistical analysis; data are shown as means ± standard deviation. Differences between groups were compared with chi-square tests and Fisher exact probability tests. Logistic regression analysis was used to identify the influencing factors of viral inhibition failure (viral load <400 copies/mL) 12 months after switching to second-line treatment, and odds ratios (OR) and confidence intervals (CI) were used to express the relative strength of association. A P value <0.05 was considered statistically significant. Viral loads <400 copies/mL after treatment were considered to have virological inhibition. The average viral load was expressed with the logarithmic ($\log_{10}$) form.

Results and Discussion

General information and drug resistance of patients before drug switches

Of 127 patients who met the second-line treatment criterion, 71 patients (61.5%) were male and 56 patients (38.5%) were female. The average age was 45 (32–60) years. Among these 127 patients, 88 patients (69.3%) were drug resistant, 108 patients (85%) had CD4 cell counts <350 cells /mm^3, and all patients had viral loads >1,000 copies/mL (Table 1). Before drug switches, the average first-line treatment duration of 127 patients was 54.8 (13–73) months The first-line regimens used included AZT+DDI+NVP (70.0%), AZT+DDI+3TC (4.3%), AZT+3TC+NVP (8.6%), D4 T+DDI+NVP (8.6%) and others (8.5%).

Of 63 patients who did not meet the second-line treatment criterion, 42 patients (61.8%) were male and 26 patients (38.2%) were female. The average age was 44 (33–59) years. Among these 68 patients, 7 patients (10.3%) were drug resistant, 42 patients (61.8%) had CD4 cell counts <350 cells /mm^3, and all patients had viral loads <1,000 copies/mL (Table 1). Before drug switches, the average first-line treatment duration of 68 patients was 50.8 (24–70) months. The first-line regimens used included AZT+DDI+NVP (67.6%), AZT+DDI+3TC (6.0%), AZT+3TC+NVP (8.8%), D4 T+DDI+NVP (8.8%) and others (8.8%).

Treatment results after drug switches

All patients were 95% or more adherent to therapy after switching to second-line therapy according to the questionnaire results. One hundred and fifty-six patients were followed up 6

Table 1. General information of subjects before drug switch.

		Patients meeting drug switching criterion (127 patients)	Percent	Patients failing to meet Drug switching criterion (68 patients)	Percent
Gender	Male	71	61.50%	42	61.80%
	Female	56	38.50%	26	38.20%
CD4 cell count (cells /μl)	<200	72	56.70%	22	32.40%
	200–350	36	28.30%	20	29.40%
	≥350	19	15.00%	26	37.20%
viral loads (copies/mL)	>1,000	127	100%	0	0
Drug resistance	NRTI	69	54.30%	5	7.40%
	NNRTI	88	69.30%	7	10.30%
No. of sequence		122	96.10%	8	11.80%

months after switching to second-line therapy and one hundred and sixty-seven patients were followed up at 12 months.

Of 127 patients who met the second-line treatment criterion, 98 and 113 patients were followed up at 6 and 12 months. At 6 months, 29 subjects were loss-of-follow-ups during the study, 15 of these patients were migrant workers, when we went to collect the blood samples and therapy information they were away from Weishi county of Henan province, but according to the drug receiving records in CDC, they continued second-line therapy; The remaining 14 patients, we totally can not contact with them. At 12 months, we got the 15 migrant workers' blood samples and therapy information, but the 14 patients who we did not contact with at 6 months were still lost contact. After switching to second-line treatment, the CD4 average increased to 220.0±137.2 at 6 months with an increase of 16.6 cells/μl, and 280.0±160.0 ($P< 0.01$) at 12 months with an increase of 76.6 cells/μl, showing a gradual increase trend. Percentage of patients with CD4 <350 cells/μl reduced to 70% (79/113) at 12 months. At 6 months, the percentage of patients with viral loads <400 copies/mL increased from 0% before drug switches to 60.8%, and the viral load average VL($\log_{10}$) decreased from 4.6±0.8 before drug switches to 2.3±1.7; at 12 months, the percentage of patients with viral loads <400 copies/mL increased to 67.3%, and the viral load average VL ($\log_{10}$) decreased to 2.0±1.7 (Table 2) ($P<0.01$).

Of 68 patients who volunteered for second-line treatment, 58 and 54 patients were followed up at 6 and 12 months. The 14 loss-of-follow-up patients were migrant workers, they were away from Weishi county of Henan province that we can not contact with them. After switching to second-line treatment, the CD4 average was 322.2±164.8 at 6 months with an increase of 61.2 cells/μl, and 329.4±134.0 ($P>0.05$) at 12 months with an increase of 68.4 cells/μl, showing insignificant difference compared to that before drug switches. The percentage of patients with viral loads <400 copies/mL decreased from 89.7% before drug switches to 83.1% at 6 months, and increased to 90.7% at 12 months ($P>0.05$); the viral load average VL ($\log_{10}$) increased from 1.2±0.6 before drug switches to 1.6±1.4 at 6 months, and then decreased to 1.3±1.1 at 12 months (Table 2) ($P>0.05$).

Drug resistance in patients after drug switches

Of 127 patients, the rate of drug resistance was 69.3% (88/127) before second-line treatment, 23.5% (23/98) and 8.8% (10/113) after 6 and 12 months after drug switches, showing a significant decrease ($P<0.001$) compared to that before drug switches.

Of 68 patients who volunteered to participate in second-line treatment, drug-resistant mutations existed in 7 patients before drug switches. At 6 months, 58 patients were followed-up and 4 patients were found to have drug resistance, of which 2 patients were new to carry drug resistance. At 12 months, 54 patients were followed-up and 1 new patient was found to have drug resistance. All drug resistance found were high drug resistance to nonnucleoside reverse transcriptase inhibitor (NNRTI). No significant

Table 2. CD4 count, viral load and viral inhibition rates before, 6 months after, and 12 months after drug switch.

	CD4 count		Viral load (lg)		Viral inhibition	
Treatment duration	Patients meeting switching criteria	Patients failing to meet switching criteria	Patients meeting switching criteria	Patients failing to meeting switching criteria	Patients meet switching criteria	Patients failing to meet switching criteria
Before drug switch	203.4±153.0	261.0±127.0	4.8±0.8	1.2±0.6	0	89.7%
At 6 months	220.0±137.2	322.2±164.8	2.3±1.7	1.6±1.4	60.8%	83.1%
At 12 months	280.0±160.0	329.4±134.0	2.0±1.7	1.3±1.1	67.3%	90.7%

difference was observed between drug resistance before and after second-line treatment (P>0.05).

Cross-resistance and multidrug resistance before and after drug switches

Before second-line treatment, rates of cross-resistance and multidrug resistance were high. The rates of drug resistance to NRTI and cross-resistance to AZT/D4 T/DDI were 54.3% (69/127) and 51.2% (65/127), respectively. Rates of drug resistance to NRTI significantly decreased to 17.3% (17/98) and 7.1% (8/113) at 6 and 12 months (P<0.001) after second-line drug switches; Rates of cross-resistance to AZT/D4 T/DDI significantly decreased to 16.3% (16/98) and 7.1% (8/113) at 6 and 12 months (P<0.001), respectively. Before drug switches, the rate of drug resistance to NNRTI was 69.3% (88/127), it decreased to 23.5% (23/98) and 11.5% (13/113) at 6 and 12 months after drug switches, respectively. The multidrug resistance rate to NRTI and NNRTI decreased from 54.3% (69/127) to 7.1% (8/113) after one year of second-line treatment (Table 3).

Drug resistance to 3TC and TDF before and after drug switches

3TC and TDF, which were included in second-line therapy regimens, are NRTI. Rates of drug resistance to 3TC and TDF before drug switches were 30.7% (39/127) and 45.7% (58/127), respectively. After 6 and 12 months of second-line treatment, rates of drug resistance to 3TC decreased to 11.2% (11/98) and 2.7% (3/113) respectively, and rates of drug resistance to TDF decreased to 15.3% (15/98) and 6.2% (7/113) respectively (Table 4). At 6 and 12 months, rates of cross-resistance to 3TC and TDF also decreased from 28.3% (36/128) before second-line treatment to 9.2% (9/98) and 1.8% (2/113), respectively (Table 4). In which, there were 3 cases of new cross-resistance at 6 months, and 1 new case at 12 months. At 12 months, 86.1% (31/36) of patients who showed cross-resistance to 3TC and TDF had a HIV RNA <400 copies/mL.

Analysis of associated factors of treatment results

All patients were 95% or more adherent to therapy 6 and 12 months after second-line treatment initiation. At 12 months, viral loads in 32.7% (37/113) of patients were not inhibited under 400 copies/mL. Results of univariate analysis showed that HIV-1 RNA >100,000 copies/mL before drug switches correlated with ineffective inhibition of viral loads after regimen change, the OR values was 3.95 (95% CI: 1.20, 13.0) (P<0.05). The OR of CD4< 100 cells/µl was 5.14 (95% CI: 0.99, 26.71), however, the P value were insignificance statistical (P>0.05) (Table 5). The OR values of virological inhibition failure in female patients and 45 years old patients were 1.19 (95% CI: 0.34, 4.18) (P>0.05) and 2.11 (95% CI: 0.64, 6.95) (P>0.05) respectively. The second-line regimen with only 1 NRTI drug changed compared to the first-line regimen, OR = 1.31 (95% CI: 0.13, 13.74) (P>0.05). The OR of more than 3 years of first-line antiretroviral treatment duration was 1.01 (95% CI: 0.46, 1.68) (P>0.05) (Table 5).

In this study, 127 patients with first-line antiretroviral treatment failure switched to the second-line regimen after an average of 54.8 months of first-line treatment. Before second-line treatment, viral loads in all patients were higher than 1000 copies/mL. The patients had low immunity, 85% of which had a CD4 count lower than 350 cells/µl, and 69.3% harbored drug resistance. After switching to second-line regimen (3TC+TDF+LPV/r) for 12 months, 67.3% of patients had effective viral inhibition; the average of CD4 cell count had an increase of 76.6 cells/µl; the rate of CD4<350 cells/µl decreased to 70%; and the rate of drug resistance decreased to 8.8%. The significant improvement indicated that switching to second-line antiretroviral drugs led to good treatment effect.

Chinese second-line antiretroviral therapy is the first treatment choice in Europe and America, which produced satisfactory results in antiretroviral-naïve patients with AIDS [8,11]. While in most developing countries, second-line antiretroviral therapy is a substitute when first-line treatment fails. Extensive drug-resistant mutations accumulate in bodies of patients who experienced first-line treatment failure. MC Hosseinipour et al. [7] reported that 85.2% of 88 patients had viral loads lower than 400 copied/mL after one year treatment with (AZT+3TC+TDF+LPV/r) second-line therapy regimen, despite the extensive drug resistance at baseline. In a large multicenter study of 632 patients receiving second-line therapy for 16.6 months in Africa and Asia, only 54% of patients had effective inhibition of virus loads [8]. Possible reasons for significantly different efficacy of second-line therapy are accumulation of drug resistance before drug change or poor adherence to therapy. In this study, 67.3% inhibition rate of viral load was achieved 12 months after drug change, despite the 69.3% in drug-resistant mutation before second-line therapy. As second-line antiretroviral drugs are relatively costly, in addition to the timely switch to second-line drugs after first-line treatment failure, it is essential to strengthen the education of treatment adherence for patients to fully utilize the limited drug sources and obtain good treatment effect.

In our study, before switching to second-line antiretroviral therapy, drug resistance existed in 69.3% of patients who met drug switching criteria. Cross-resistance and multidrug resistance also existed severely. The rates of cross-resistance and multidrug resistance to NRTI and NNRTI were 69.3% and 54.3%,

Table 3. Drug resistance analysis before and after second-line treatment.

	NRTI		NNRTI		NRTI and NNRTI	
	Drug resistance	**High drug resistance***	**Drug resistance**	**High drug resistance**	**Drug resistance**	**High drug resistance**
Before drug switch(N = 127)	54.3% (69)	42.5% (54)	69.3% (88)	66.1% (84)	54.3% (69)	42.5% (54)
At 6 months (N = 98)	17.3% (17)	14.3% (14)	23.5% (23)	23.5% (23)	17.3% (17)	14.3% (14)
At 12 months (N = 113)	7.1% (8)	4.4% (5)	11.5% (13)	11.5% (13)	7.1% (8)	4.4% (5)

*A score >60 get from Stanford University drug resistance database was regarded as high drug resistance.

Table 4. Resistance to 3TC and TDF before and after drug switch.

	3TC	TDF	3TC & TDF
Before drug switches (N = 127)	30.7% (39)	45.7% (58)	28.3% (36)
At 6 months (N = 98)	11.2% (11)	15.3% (15)	9.2% (9)
At 12 months(N = 113)	2.7% (3)	6.2% (7)	1.8% (2)

respectively. However, after one year of second-line treatment, the overall drug resistance rate decreased to 8.8%, and rates of cross-resistance and multidrug resistance also decreased to 11.5% and 7.1%. Our results showed that, although severe cross-resistance and multidrug resistance existed after first-line treatment failure, second-line therapy effectively decreased the rate of drug resistance, as well as the epidemic risk of drug-resistant bacterial strains. Analysis of cross-resistance to NRTI (3TC and TDF), the main drug in the second-line antiretroviral regimen, showed that before drug change, 28.3% of patients had cross-resistance to 3TC and TDF, along with viral loads >1000 copies/mL. After 12 months of second-line treatment, viral loads in 86.1% of patients were effectively inhibited. In these patients, although there was only LPV/r playing a role, satisfying effect were observed and the cross-resistance rate to the 3TC and TDF decreased to 1.8%. It was noticeable that in our study, although drug resistance to the protease inhibitor (PI) was not found, extensive PI secondary drug-resistant mutations were detected. In case PIs primary drug-resistant mutations occur, it will lead to high PI drug resistance. Considering that there were new patients found to have drug-resistant mutations at 6 and 12 months, drug resistance surveillance should be continued to observe the cross-resistance, multidrug resistance and PI drug-resistant mutations after long-term second-line treatment.

Among patients meeting drug switching criteria, we performed a regression analysis of those patients whose virus loads were not inhibited under 400 copies/mL after drug change. The results showed that viral load higher than 100,000 copies/mL before drug change was a risk factor of virological inhibition failure. Similarly, a European study demonstrated that after switching to second-line antiretroviral regimen (including the protease inhibitor); levels of viral loads were more easily inhibited in patients with lower viral loads [8]. Similar findings have been reported that factor associated with HIV-1 RNA >400 copies/mL at 12 months on univariate analysis included having a presenting CD4 count <50 cells/µl and HIV-1 RNA >100,000 copies/mL [7]. However, in our research, lower CD4 count was not a significant factor associated with virological failure. This may be affected by the loss-of-follow-up subjects. In our study, 68 patients did not meet drug switching criteria and volunteered to change drugs. Before second-line treatment, the CD4 count average was 261.0 cells/µl, the rate of viral loads <400 copies/mL was 89.7%, and there were 7 cases of drug resistance; after 12 months of second-line treatment, the CD4 count average was 329.4 cells/µl, the rate of viral load inhibition was 90.7%, and there was 1 case of drug resistance. Although after second-line treatment, no significant differences were observed in virological immunological performance compared to that before drug change, ability of immunity and viral inhibition were still maintained in relatively high levels. These results suggested that in order to fully and effectively utilize the limited second-line drug sources and obtain good clinical effect, it is important to increase the measuring viral loads of treated patients.

Conclusions

Multidrug resistance and cross-resistance existed severely in Chinese patients that experienced first-line treatment failure.

Table 5. Associated factors of virological inhibition failure 12 months after second-line treatment initiation with logistic regression analysis.

Associated factors		OR (95% CI)
Female		1.19 (0.34, 4.18)
>45 years old		2.11 (0.64, 6.95)
CD4 cell count before second-line treatment	<100 cells/mL	5.14 (0.99, 26.71)
	100–200 cells/mL	1.72 (0.35, 8.48)
	≥200 cells/mL	1.0
Viral loads before second-line treatment	1000–10000 copies/mL	1.0
	10000–100000 copies/mL	0.87 (0.24, 3.08)
	≥100000 copies/mL	3.95 (1.20, 13.0)*
No. of NRTI drugs changed compared to the first-line regimen	1 NRTI drug	1.31 (0.13, 13.74)
	2 NRTI drugs	0.76 (0.07, 7.98)
Treatment duration of first-line therapy	<3 years	1.0
	≥3 years	1.01 (0.46, 1.68)

*P<0.05, which were considered statistical significance.

These patients need to be monitored timely on viral loads and drug resistance and switch to second-line therapy under conditions of relatively low viral duplication levels. Satisfactory results were observed in patients treated with Chinese second-line antiretroviral therapy regimen (3TC+TDF+LPV/r) for 12 months, with no drug resistance to PI found. However, extensive secondary mutations to PI were observed, which suggested that the long-term drug resistance surveillance is necessary for patients switching to second-line antiretroviral therapy.

Acknowledgments

We thank Serena Fuller, Wen Xiaoning of USCDC GAP and professor Ping Zhong of Shanghai CDC for their comments and suggestions which help to improve the presentation of the paper.

Author Contributions

Conceived and designed the experiments: MZ HS. Performed the experiments: MZ MS WY. Analyzed the data: MZ MS JC. Contributed reagents/materials/analysis tools: MZ HS ZW. Contributed to the writing of the manuscript: MZ JC.

References

1. Galárraga O, O'Brien ME, Gutiérrez JP, Renaud-Théry F, Nguimfack BD, et al. (2007) Forecast of demand for antiretroviral drugs in low and middle-income countries: 2007–2008. AIDS 21: S97–S103.
2. Boyd MA, Cooper DA (2007) Second-line combination antiretroviral therapy in resource-limited settings: facing the challenges through clinical research. AIDS 21: S55–S63.
3. Zhang F, Dou Z, Ma Y, Zhang Y, Zhao Y, et al. (2011) Effect of earlier initiation of antiretroviral treatment and increased treatment coverage on HIV-related mortality in China: a national observational cohort study. Lancet Infect Dis 11: 516–524.
4. Zhang M, Han XX, Cui WG, Jia MH, Meng XD, et al. (2008) The impacts of current antiretroviral therapy regimens on Chinese AIDS patients and their implications for HIV-1 drug resistance mutation. JPN J INFECT DIS 61: 361–365.
5. Zhang M, Han X, Hu Q, Cui W, Zhao B, et al. (2008) Cohort study of highly active antiretroviral therapy and drug resistant mutation in Henan Province, China. Chineses Journal of Laboratory Medicine 31: 1101–1104.
6. Pujades-Rodríguez M, O'Brien D, Humblet P, Calmy A (2008) Second-line antiretroviral therapy in resource-limited settings: the experience of Médecins Sans Frontières. AIDS 22: 1305–1312.
7. Hosseinipour MC, Kumwenda JJ, Weigel R, Brown LB, Mzinganjira D, et al. (2010) Second-line treatment in the Malawi antiretroviral programme: high early mortality, but good effect in survivors, despite extensive drug resistance at baseline. HIV Medicine 11: 510–518.
8. Pujades-Rodríguez M, Balkan S, Arnould L, Brinkhof MA, Calmy A, et al. (2010) Treatment failure and mortality factors in patients receiving second-line HIV therapy in resource-limited countries. JAMA 304: 303–312.
9. (2009) Chinese National Free AIDS Antiretroviral Therapy Manual.
10. WHO (2009) Rapid advice: antiretroviral therapy for HIV infection in adults and Adolescents.
11. Kempf DJ, King MS, Bernstein B, Cernohous P, Bauer E, et al. (2004) Incidence of resistance in a double-blind study comparing lopinavir/ritonavir plus stavudine and lamivudine to nelfinavir plus stavudine and lamivudine. J Infect Dis 189: 51–60.

8

Global Transcriptome and Physiological Responses of *Acinetobacter oleivorans* DR1 Exposed to Distinct Classes of Antibiotics

Aram Heo[1], Hyun-Jin Jang[2], Jung-Suk Sung[2], Woojun Park[1]*

1 Laboratory of Molecular Environmental Microbiology, Department of Environmental Science and Ecological Engineering, Korea University, Seoul, Republic of Korea,
2 Department of Life Science, Dongguk University, Seoul, Republic of Korea

Abstract

The effects of antibiotics on environment-originated nonpathogenic *Acinetobacter* species have been poorly explored. To understand the antibiotic-resistance mechanisms that function in nonpathogenic *Acinetobacter* species, we used an RNA-sequencing (RNA-seq) technique to perform global gene-expression profiling of soil-borne *Acinetobacter oleivorans* DR1 after exposing the bacteria to 4 classes of antibiotics (ampicillin, Amp; kanamycin, Km; tetracycline, Tc; norfloxacin, Nor). Interestingly, the well-known two global regulators, the *soxR* and the *rpoE* genes are present among 41 commonly upregulated genes under all 4 antibiotic-treatment conditions. We speculate that these common genes are essential for antibiotic resistance in DR1. Treatment with the 4 antibiotics produced diverse physiological and phenotypic changes. Km treatment induced the most dramatic phenotypic changes. Examination of mutation frequency and DNA-repair capability demonstrated the induction of the SOS response in *Acinetobacter* especially under Nor treatment. Based on the RNA-seq analysis, the glyoxylate-bypass genes of the citrate cycle were specifically upregulated under Amp treatment. We also identified newly recognized non-coding small RNAs of the DR1 strain, which were also confirmed by Northern blot analysis. These results reveal that treatment with antibiotics of distinct classes differentially affected the gene expression and physiology of DR1 cells. This study expands our understanding of the molecular mechanisms of antibiotic-stress response of environment-originated bacteria and provides a basis for future investigations.

Editor: Nancy E. Freitag, University of Illinois at Chicago College of Medicine, United States of America

Funding: This work was supported by the Mid-career Researcher Program through an NRF grant (2014R1A2A2A05007010) funded by the Ministry of Science, ICT & Future Planning (MSIP). The funders had no role in study design, data collection and analysis, decision to publish, or preparation of the manuscript.

Competing Interests: The authors have declared that no competing interests exist.

* Email: wpark@korea.ac.kr

Introduction

Antibiotics are abundant in various environmental habitats such as seawater, plants, sludge, and soils [1–3]. Because antibiotics affect our ecosystem, which includes the microbial diversity and abundance in the environment, they are widely considered to act as key pollutants [4,5]. Although antibiotics contaminate the environment, how antibiotics affect environment-originated bacteria and their evolution remains poorly understood. Because most antibiotics used for treating infections are produced by environmental microorganisms, antibiotic resistance genes and mechanisms could exist in nonclinical habitats [6]. In natural environments, antibiotic production and resistance might be considered as biochemical warfare to eliminate competing organisms because antibiotics suppress bacterial growth and metabolism [7]. Antibiotics of distinct classes act on different targets through specific mechanisms: β-lactams lead to autolysis by interfering with cell-wall biosynthesis [8]; aminoglycosides cause mistranslation by targeting the 30S subunit of the ribosome [9,10]; tetracycline inhibits protein synthesis by disrupting the binding of aminoacyl-tRNA to the mRNA-ribosome complex [11]; and fluoroquinolones inhibit DNA replication by binding with DNA gyrase and topoisomerase [12]. Antibiotic resistance could be acquired through several ways: i) the action of antimicrobial-inactivating enzymes, ii) reduced access of antimicrobials to bacterial targets (decreased outer-membrane permeability and overexpression of multidrug efflux pumps), and iii) mutations that change targets or cellular functions [13]. Many clinical and environmental bacteria have multiple antibiotic-resistance mechanisms [13].

The diesel-degrading *A. oleivorans* DR1 was isolated from the rice paddy soil and its genome was completely sequenced [14]. Our previous studies demonstrated that quorum sensing and biofilm formation are important for diesel-degradation in DR1 cells [14]. Most antibiotic resistance studies of *Acinetobacter* species have largely focused on pathogenic *Acinetobacter* such as *Acinetobacter baumannii* owing to high level of multidrug resistance. Transcriptional responses to various antibiotics and their regulation have not been extensively defined with *Acinetobacter* species. Reducing access to bacterial targets by means of decreasing permeability and using strong efflux systems has been reported as a major cause of multidrug resistance in *Acinetobacter* species [15]. Because the genome of DR1 is similar to those of the human pathogens *A. calcoaceticus* and *A. baumannii* [16], the

DR1 strain is appropriate for studying antibiotic effects in evolutionary aspect. To identify key genes and their functions in the antibiotic resistance of environment-originated bacteria, we performed whole-transcriptome profiling of *Acinetobacter oleivorans* DR1 using RNA-Seq technique. with four representative antibiotics: ampicillin (Amp), kanamycin (Km), tetracycline (Tc), and norfloxacin (Nor).

Bacteria could exhibit physiological changes by changing global gene expression pattern in response to low concentration of antibiotics [17]. To promote understanding how antibiotic resistance develops in DR1, we also conducted several physiological tests on DR1 under distinct antibiotic stresses. Herein we provide both transcriptomic and experimental evidence of antibiotic-resistance mechanisms in DR1. Elucidating transcriptional and physiological responses to distinct antibiotics might establish novel molecular basis for antibiotic-resistance mechanisms of *Acinetobacter* species.

Results

Comparative transcriptome analysis of *A. oleivorans* DR1 exposed to sub-MICs of antibiotics of distinct classes

Antibiotics have been reported to affect bacterial transcription in a concentration-dependent manner, and using antibiotics at concentrations as high as the MIC can cause extensive cellular stress and death [17]. To determine the appropriate concentration of antibiotics for the antibiotics induced transcriptome, we measured MICs of 4 classes of antibiotics in various cell densities (10^5–10^8 CFU/mL). When cell density increased, the MIC of antibiotics was increased (Figure S1). This result demonstrates the relationship between the cell density of bacteria and the MIC of antibiotics. Because of transcript modulation decreases at high antibiotics concentration, DR1 cells were exposed to sub-MIC of distinct antibiotic classes. Sub-MIC of antibiotic allows susceptible strains to grow, but induces stress responses. The highest MICs measured were for Amp (100–200 μg/mL), and by comparison, DR1 cells were more susceptible to other antibiotics (MICs, 1–8 μg/mL). We speculate that high number of lactamases encoded by the DR1 genome confer high resistance to Amp (and thus the high MIC ranges). In this study, we selected the genes that showed a 1.5-fold change in expression after antibiotic treatment when compared with the expression in control cells that were not exposed to antibiotics. In response to Amp, Km, Tc, and Nor, the expression levels of 1054 (26.6%), 1497 (37.33%), 1170 (29.52%), and 208 (5.25%) genes were markedly upregulated, and the levels of 1738 (43.86%), 910 (22.96%), 1254 (31.64%), and 635 (16.02%) genes were downregulated, respectively (Figure 1A, Table S1). The change in the expression of the same genes in response to each antibiotic treatment suggested that common responses were elicited by the 4 classes of antibiotics: 41 and 14 genes were commonly upregulated and downregulated, respectively (Figure 1B, Table 1, Table S2). Several upregulated genes appear to encode hypothetical proteins, a redox-sensing regulatory protein (*soxR*), RNA polymerase sigma factor (*rpoE*), dehydrogenases, and numerous transporter proteins. The commonly downregulated genes encoded a glycosyltransferase (*wcaA*), a lipoprotein (*rlpA*), and 3-dehydroquinate dehydratase (*aroQ*) (Table S2). Our RNA-Seq results were confirmed with quantitative real-time PCR (qRT-PCR). Commonly up- and down- regulated genes (*soxR, rpoE, lysR, wcaA*) and specifically induced genes were selected based on expression vales in 4 antibiotics conditions (Figure S2).

Effects of antibiotics on the expression of specific genes

Clusters of orthologous groups (COGs) were analyzed to examine specific gene-expression changes (Figure S3). DR1 cells treated with Amp exhibited altered expression of several COG categories: translation (COG J), transcription (COG K), and inorganic-ion transport/metabolism (COG P) categories were mainly downregulated, whereas lipid metabolism (COG I) and amino-acid metabolism and transport (COG E) categories were upregulated. By contrast, Km and Tc treatments boosted the expression of gene clusters involved in transcription (COG K), amino-acid metabolism/transport (COG E), and inorganic-ion transport/metabolism (COG P) categories, whereas the treatments downregulated the expression of gene clusters involved in cell-wall/membrane/envelop biogenesis (COG M). Under Nor treatment, most COG categories were not changed to the same degree as they were changed in response to other antibiotics; certain genes involved in inorganic-ion transport/metabolism (COG P) were upregulated and transcription genes (COG K) were downregulated (Figure S3). COG analyses of the transcriptomes revealed that the genes associated with amino-acid metabolism and transport and inorganic-ion transport and metabolism are critical for cellular-stress and cell-death responses under all antibiotic-treatment conditions. Our data suggest that amino-acid metabolism and transporter systems might play key roles in antibiotic-resistance mechanisms in *Acinetobacter* species.

Distinct antibiotics possess specific cellular targets such as DNA, RNA polymerase, ribosomal proteins, and cell walls [18]. The overexpression of the antibiotic targets could enhance the survival of bacterial cells under antibiotic treatment [19]. Our data showed that specific antibiotic targets were strongly upregulated under distinct antibiotic conditions. The DR1 genome contains 9 putative lactamase genes. Interestingly, not all β-lactamases were upregulated by Amp, and class-C-type β-lactamases were primarily induced by Amp (Table 2). The expression of genes encoding penicillin-binding proteins was downregulated under Amp treatment (Table 2), which might be because of the high concentration of Amp used. Km treatment downregulated several ribosomal-protein genes (data not shown). However, Tc induced the expression of several ribosomal-protein genes, including the expression of genes encoding ribosomal proteins S13 and S7, which are recognized to interact directly with Tc [20]. Moreover, Tc treatment induced enzymes that modify ribosomal proteins, such as 50S ribosomal-protein methyltransferase and 30S ribosomal-protein methylthiotransferase. Thus, although Km and Tc inhibit translation by binding to ribosomes, their influences on cellular responses appeared to differ. Cells treated with Nor exhibited 2.92- and 2.8-fold increases in the expression of *gyrA* (AOLE_18380) and *gyrB* (AOLE_00595), respectively, which are recognized as targets of fluoroquinolone-class antibiotics (Table 2). Interestingly, Tc and Nor did not induce any lactamase genes, but Km induced class-C β-lactamase (AOLE_17635) and metallo-β-lactamases (AOLE_00775 and AOLE_03925). *Acinetobacter* species appear to express numerous efflux-pump genes that are critical for the multidrug resistance of *A. baumannii* [21]. However, based on our transcriptome data, it is unclear which efflux pumps are crucial for conferring resistance against the antibiotics tested (Table 2). Certain efflux pumps might be specific to each antibiotic.

Interestingly, our data showed that the expression of fimbriae/pili-related genes changed in response to treatment with Amp, Km, and Tc (Table S3). Previously, we reported that *A. oleivorans* DR1 possesses 2 major fimbrial appendages [22]. Expression of fimbrial and pilin proteins is consistent with the cell aggregation and biofilm formations [23], and flagellar/pili appendages can

Table 1. Genes in *A. oleivorans* DR1 commonly upregulated by Amp, Km, Tc, and Nor.

Locus_tag DR1	Product	Genes	Fold-change			
			Amp	Km	Tc	Nor
AOLE_02445	Enoyl-CoA hydratase	*caiD*	4.62	1.78	2.34	1.87
AOLE_04025	Metal-dependent hydrolase		3.84	2.97	4.43	2.45
AOLE_06735	Putative short-chain dehydrogenase		4.15	2.37	3.63	1.92
AOLE_06795	Alkylhydroperoxidase		2.94	2.23	2.92	1.58
AOLE_08565	AraC-type DNA-binding domain-containing protein	*araC*	15.35	3.55	3.11	2.10
AOLE_08595	3-Oxoadipate enol-lactonase	*mhpC*	3.08	1.78	2.73	1.57
AOLE_08710	3-Oxoacyl-(acyl-carrier-protein) reductase	*fabG*	4.18	4.13	3.10	2.09
AOLE_08725	NIPSNAP family protein		3.35	3.55	6.22	2.09
AOLE_08765	Shikimate dehydrogenase	*aroE*	15.61	7.14	12.13	5.26
AOLE_09075	Transcriptional regulator	*lysR*	5.71	4.27	2.03	1.68
AOLE_09435	DoxX family protein		2.24	7.09	5.44	1.57
AOLE_09590	Putative tonB-like protein	*tonB*	1.57	3.90	3.42	1.67
AOLE_10175	Putative aliphatic sulfonate-binding protein	*tauA*	2.69	2.04	5.02	1.65
AOLE_11820	Major facilitator superfamily transporter	*araJ*	144.14	3.30	5.23	2.84
AOLE_11830	Methyltransferase domain-containing protein	*ubiE*	157.57	3.56	4.10	1.84
AOLE_12115	DMT-family permease		3.07	2.36	3.88	2.44
AOLE_12135	Redox-sensitive transcriptional activator SoxR	*soxR*	2.52	4.00	2.53	2.88
AOLE_12655	ECF subfamily protein RNA polymerase sigma-24 subunit	*rpoE*	5.05	3.56	1.82	1.75
AOLE_12705	Glycine betaine ABC transporter substrate-binding protein	*tauA*	1.71	1.77	1.75	2.49
AOLE_12875	Phenylacetic acid degradation protein	*paaI*	17.54	3.56	4.68	2.10
AOLE_13495	Competence-damaged family protein	*cinA*	5.05	13.36	7.42	2.63
AOLE_14540	Peptide deformylase	*def*	2.87	4.45	1.85	2.75
AOLE_14590	3-Phenylpropionate dioxygenase ferredoxin	*nirB*	2.44	2.85	3.75	1.58
AOLE_14800	RNA polymerase sigma factor FecI	*rpoE*	27.98	17.79	1.87	2.94
AOLE_16560	Short-chain dehydrogenase		1.71	2.44	2.24	1.57
AOLE_18975	GNAT family acetyltransferase		5.39	6.56	2.78	1.70

function as transporting or adhering machines in gram-negative bacteria [24]. Other transcriptome data demonstrated that the flagellar/pili metabolism-related genes were induced under diverse stress conditions such as osmotic stress, oxidative stress [25,26]. Beyond affecting target-gene expression, antibiotics are considered to exert secondary effects that are a part of the adaptive response to antibiotic stress.

Effect of distinct classes of antibiotics on physiology: cell morphology and membrane permeability

We investigated the effects of antibiotics on physiological changes such as cell death and alterations in cell morphology and membrane permeability in an effort to link gene-expression to physiology. Examination of the morphology of cells stained with 4′,6-diamidino-2-phenylindole (DAPI) showed that cells treated with antibiotics were longer than wild-type cells (Figure 2A, Figure S4), and the cells treated with Amp and Nor were nearly 4-times longer than control cells (Figure 3A, B). Previously, β-lactams were reported to lengthen cells by inhibiting peptidoglycan biosynthesis [27]. Cell filamentation is often associated with the SOS response [28]. The product of the *sulA* gene, a key component of the SOS response that leads to cell elongation by binding to FtsZ or DpiAB in a two-component system, induces cell filamentation [29]. Interestingly, no *sulA* gene homolog is present in *Acinetobacter* species, and thus it is worth identifying the roles of other genes involved in cell elongation in *Acinetobacter* species. In the cell walls of most bacteria, peptidoglycans play an essential role in antimicrobial resistance; peptidoglycans determine cell shape, and their biosynthesis is critical for antibiotics resistance [30]. Peptidoglycan hydrolase is a widely conserved outer-membrane protein that modulates cell shape in *E. coli* and *Pseudomonas aeruginosa* [31]. The expression of AOLE_00215, which encodes peptidoglycan hydrolase, was increased 1.5- and 1.6-fold by Km and Tc, respectively, but was not markedly affected by Amp and Nor.

We measured the change in membrane permeability by using ANS, a neutral, hydrophobic fluorescent probe; in membrane-damaged cells, fluorescence is increased because the enhanced permeability leads to ANS uptake [32]. The fluorescence-intensity values measured were divided by the OD_{600} values for normalizing the measurements, and the results showed that distinct antibiotic treatments altered membrane permeability to different degrees (Figure 2B). The membrane-permeability properties have a major impact on the susceptibility of microorganisms to antibiotics [33]. Membrane permeability was increased substantially after Km treatment, whereas only a slight increase of membrane permeability was induced by Amp and Tc, which might explain the sensitive response of DR1 cells to Km. Porins

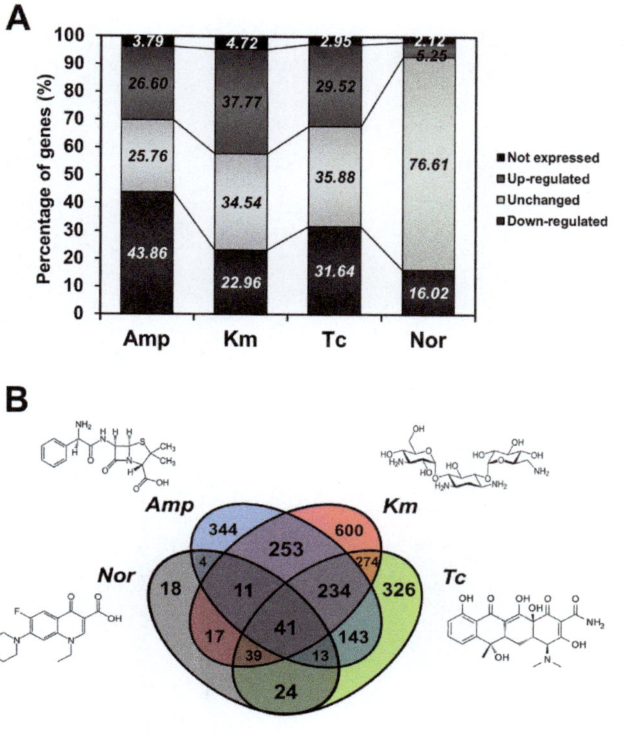

Figure 1. A summary of genes upregulated and downregulated by distinct classes of antibiotics. (A) The percentages of up- and down-regulated genes under treatment with 4 antibiotics. (B) Venn-diagram showing the number of overlapping genes upregulated by antibiotics of distinct classes. Fold-changes shown are a comparison of the RPKM values of exponentially growing control cells and of cells treated with each antibiotic. Upregulation of gene expression is >1.5-fold change in RPKM value, downregulation is <1.5-fold change.

are considered to be permanently open pores, and lowering porin expression reduces outer-membrane permeability [33]; thus, porin-mediated permeability is a critical aspect of antibiotic-resistance mechanisms. The DR1 genome contains several porin-encoding genes. The expression of *ompC* (AOLE_10405), which encodes an outer-membrane porin protein, increased 1.5-fold under Km treatment, but decreased in response to Amp (6.4-fold) and Nor (1.5-fold) and did not change after Tc treatment.

Oxidative stress, SOS response, and DNA repair in response to distinct antibiotics

Antibiotics have been widely reported to induce the production of reactive oxygen species (ROS), which causes oxidative stress damage [34]. We used the fluorescent probe DHR 123 and flow cytometry to monitor ROS generation following treatment with the 4 antibiotics (Figure 3): under the tested conditions, treatment with Amp, Km, and Nor, but not Tc, potently induced ROS generation. Interestingly, the expression profiles of oxidative stress-related genes were distinct following treatment with these antibiotics of different classes, based on which we suggest that distinct mechanisms exist that are used by bacteria for coping with disparate types and levels of oxidative stress induced by various antibiotics (Table 2). Peroxiredoxin (*ahpC*) and catalase (*katE1*) genes were induced by Amp and the thioredoxin (*trxA*) gene was highly upregulated by Km and Tc, whereas the redox-sensing regulatory gene *soxR* was induced by all antibiotics. Antibiotic-induced oxidative stress upregulated glyoxylate-bypass genes [35]. The expression levels of isocitrate lyase (*aceA*) and malate synthase (*aceB*) genes, which are link to glyoxylate bypass, were increased substantially in response to Amp and Nor, but not Tc and Km (Figure 4). These results suggest that distinct classes of antibiotics elicit different responses to oxidative stress by dissimilarly affecting the expression of genes associated with ROS defense and glyoxylate bypass.

Unexpectedly, only Nor treatment substantially upregulated the expression of these SOS response-related genes and DNA-repair genes: *recA, umuDC, dinP, uvrAC*, and *ssb* (Table 3). The SOS response is a global response to DNA damage in bacteria that is induced by a variety of environmental factors such as UV radiation, chemicals, and antimicrobial compounds [36]. The RecA protein and LexA repressor play central roles in SOS response [37,38], but a LexA-like transcriptional repressor has been studied only poorly in *Acinetobacter* species [39]. DNA damage increases the frequency of mutations when MMC is used, which indirectly confirms the presence of the SOS response [40]. Previously, MMC-induced mutation frequency was monitored by measuring the increase of colonies resistant to rifampicin [41]. MMC treatment increased the rifampicin-resistance mutation frequency 47-fold in DR1. When *E. coli* GC4468 and *A. baumannii* ATCC17978 were used as reference strains, the mutation frequency was determined to be increased 22- and 37-fold in *E. coli* and *A. baumannii*, respectively (Figure 5A). Our results reveal that crucial features of the canonical SOS response exist in the genome of DR1 cells. When we measured antibiotic-induced SOS response, we determined that rifampicin-resistance mutation frequency was strongly induced only by Nor (Figure 5B). Agreeing with these data, our reporter strains carrying GFP fused to the *recA* promoter region showed that Nor treatment induced the SOS response (Figure 5C). The fluorescence of these reporter cells depended on the concentration of Nor, although a high concentration of Amp increased *recA* expression. We could not rule out the possibility that *recA* transcription and GFP translation differ, because the RNA-seq results showed that *recA* expression increased under Km treatment. Antibiotic treatment can induce the SOS response, which can lead to the expression of *umuDC* [41]. Our transcriptome analysis revealed that the *umuDC* genes were induced only by Nor (Table 3). Thus, our results demonstrated that Nor, but not other antibiotics, strongly induced the SOS response in DR1 cells.

Loss of DNA-repair capability in response to Km and Tc treatment

The enzymes used in base excision repair (BER) are responsible for repairing endogenous DNA-damage lesions caused by ROS, environmental chemicals, and ionizing radiations [42,43]. BER is a highly conserved cellular mechanism in bacteria and humans [42], and the lesion in the damaged DNA is removed by a DNA glycosylase. Endonuclease IV, UDG, and Fpg are induced in response to oxidative stress and these molecules function in repairing DNA damage in *E. coli* [44]. We measured endonuclease activity after treatment with the 4 antibiotics and we used the DNA-excision assay and oligonucleotides including THF residues [44]. Unexpectedly, in response to Km and Tc, endonuclease IV did not exhibit BER activity that was distinct from the activity in control (Figure 6). We also tested the activities of the 2 other DNA-repair enzymes, UDG and Fpg (Figure S5). Fpg activity decreased under all antibiotic conditions, whereas UDG activity was not changed. In these assays, enzyme reactions performed using purified *E. coli* endonuclease IV, UDG, and Fpg served as positive controls. Our results showed that the DNA-repair capability of endonuclease IV was maintained only under Amp and Nor treatment, which suggests that each antibiotic distinctly

Table 2. Antimicrobial resistance-associated genes and target genes in *A. oleivorans* DR1.

Locus_tag DR1	Product	Description	Genes	Fold-change			
				Amp	Km	Tc	Nor
β-Lactamases							
AOLE_05220	β-Lactamase	Class C β-lactamase	*ampC*	1.05	−2.06	−3.45	−1.31
AOLE_06930	β-Lactamase class C	Class C β-lactamase	*ampC*	5.20	1.93	1.03	1.00
AOLE_12585	β-Lactamase	Class C β-lactamase	*ampC*	1.41	1.19	1.72	−1.09
AOLE_17635	β-Lactamase class C	Class C β-lactamase	*ampC*	4.20	3.44	1.53	1.25
AOLE_11070	β-Lactamase class D	Class D β-lactamase	$bla_{oxa\text{-}66}$	1.28	−2.14	−1.43	−1.65
AOLE_00775	Metallo-β-lactamase superfamily protein	Metallo-β-lactamase superfamily	*fpaA*	−2.31	2.51	1.51	−2.19
AOLE_03925	Putative metallo-β-lactamase	Metallo-β-lactamase superfamily	*gloB*	3.17	2.56	−2.55	−1.06
AOLE_10040	β-Lactamase	Metallo-β-lactamase superfamily	*gloB*	−1.32	1.72	−3.28	−1.11
AOLE_17515	Metallo-β-lactamase superfamily protein	Metallo-β-lactamase superfamily	*gloB*	−1.17	−1.05	−1.33	−1.05
AOLE_01440	Penicillin binding protein transpeptidase domain protein		*ftsI*	−1.03	−1.20	−1.62	−1.03
AOLE_01470	Putative penicillin-binding protein (PonA)		*mrcA*	−2.22	−4.35	−3.34	−1.11
AOLE_05610	Penicillin-binding protein 1B		*mrcB*	−1.80	−3.78	−3.30	−1.04
AOLE_14240	Penicillin-binding protein 2		*ftsI*	−4.09	−1.47	−1.68	−1.16
Aminoglycosides							
AOLE_08490	Predicted aminoglycoside phosphotransferase	aminoglycoside 6′-acetyltransferase	*aacA4*	3.93	1.30	−1.65	−1.07
AOLE_18475	Aminoglycoside2′-N-acetyltransferase(AAC(2′)-Ib)	aminoglycoside 2′-acetyltransferase	*aacB*	−8.13	2.15	1.99	−1.03
Fluoroquinolones							
AOLE_00020	DNA gyrase subunit B	GyrB mutation	*gyrB*	−1.82	−1.65	−1.48	1.24
AOLE_00595	DNA topoisomerase IV subunit B	GyrB mutation	*gyrB*	−1.18	1.61	−1.62	2.80
AOLE_04195	DNA gyrase subunit A	His-78→Asn	*gyrA*	−2.17	−1.51	−3.26	−1.01
AOLE_18380	DNA topoisomerase IV subunit A	GyrA mutation	*gyrA*	−1.22	−1.56	−2.00	2.92
Efflux pumps							
AOLE_00955	MFS transporter, metabolite:H+ symporter (MHS) family protein	MFS-family efflux pump	*uhpC*	−3.37	1.97	1.31	1.00
AOLE_00175	MFS-family transporter	MFS-family efflux pump	*araJ*	−1.63	4.02	2.99	1.38
AOLE_01040	MFS-family transporter	MFS-family efflux pump	*araJ*	−4.27	−3.45	1.14	−1.32
AOLE_12350	MFS-family transporter	MFS-family efflux pump	*araJ*	−4.03	−2.09	1.02	1.13
AOLE_00050	RND-type efflux pump	RND-family efflux pump	*dctP*	9.53	−1.03	1.52	1.02
AOLE_04230	Putative RND-family drug transporter	RND-family efflux pump	*emrA*	2.16	−15.53	−2.57	−1.21
AOLE_09410	RND-type efflux pump	RND-family efflux pump		6.59	1.19	3.25	1.58
AOLE_18750	RND-superfamily exporter	RND-family efflux pump		77.60	19.11	1.34	−1.13
AOLE_00035	ABC transporter ATP-binding protein	ABC-family efflux pump	*uup*	−3.32	−1.18	−2.10	1.22
AOLE_01345	Putative ABC transporter ATP-binding protein	ABC-family efflux pump	*uup*	−3.46	1.23	−3.34	1.06
AOLE_17260	ABC transporter ATP-binding protein	ABC-family efflux pump	*uup*	−3.46	−2.54	−3.51	−1.01

Table 2. Cont.

Locus_tag DR1	Product	Description	Genes	Fold-change			
				Amp	Km	Tc	Nor
AOLE_00290	Multidrug-resistance protein norM	MATE-family efflux pump	*norM*	−3.58	−1.34	1.92	−1.03
AOLE_00530	Na+-driven multidrug efflux pump	MATE-family efflux pump	*norM*	2.40	1.14	1.73	1.16
AOLE_05880	MATE efflux family protein	MATE-family efflux pump	*norM*	-2.29	−1.69	2.25	1.28
AOLE_17460	Multidrug ABC transporter	MATE-family efflux pump	*norM*	−3.07	1.57	−1.22	−1.20
AOLE_05535	Quaternary ammonium compound-resistance protein QacE	SMR-family efflux pump	*qacE*	−2.23	2.48	1.71	1.47
AOLE_16200	Quaternary ammonium compound-resistance protein SugE	SMR-family efflux pump	*sugE*	−1.59	4.62	1.97	1.15

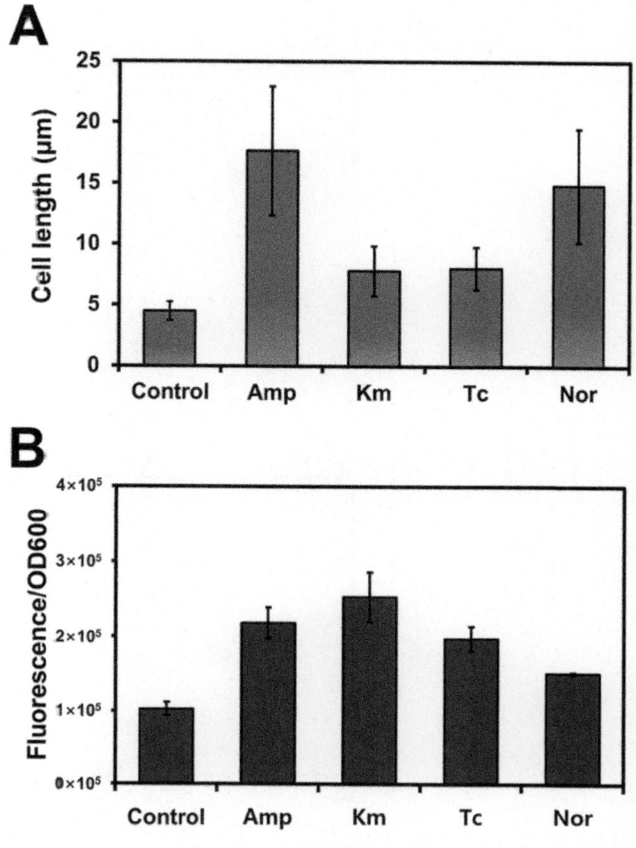

Figure 2. Influence of distinct classes of antibiotics on cell morphology, and membrane permeability in DR1. (A) The average cell size was measured from 50 cells treated with antibiotics. (B) Membrane permeability was measured using ANS. The error bars indicate standard deviation from triplicate experiments.

affects the genes encoding DNA-repair enzymes. The expression of endonuclease IV (AOLE_14840) was upregulated by Km but not the other 3 antibiotics, and the expression of Fpg (AOLE_03065) was decreased 2.3-fold and increased 1.7-fold in response to Amp and Km, respectively, but was unaffected by Tc and Nor. Our data reveal that the activity of DNA-repair enzymes was not correlated with the expression of the genes encoding these enzymes.

Discussion

In this study, we conducted a comparative transcriptome analysis and examined the physiological changes in soil-borne *A. oleivorans* DR1 exposed to antibiotics of distinct classes. Although the antibiotic resistance of *A. baumannii* has been widely studied [45], the transcriptional response elicited by various antibiotics in other *Acinetobacter* species remains poorly documented. The effects of antibiotics and the antibiotic-resistance mechanism in DR1 have been described previously [22,46,47], but this is first study in which the transcriptional changes induced in DR1 cells by 4 antibiotics have comparatively analyzed. Our results revealed that the MIC of Amp exhibited extremely high ranges, which could be due to high number of lactamases encoded by the DR1 genome. Amp was hydrolyzed by various β-lactamases present in the periplasm before Amp can reach its targets [48]. Moreover, Amp induced the genes involved in glyoxylate bypass (Figure 4). Glyoxylate bypass is induced in numerous bacteria when carbon

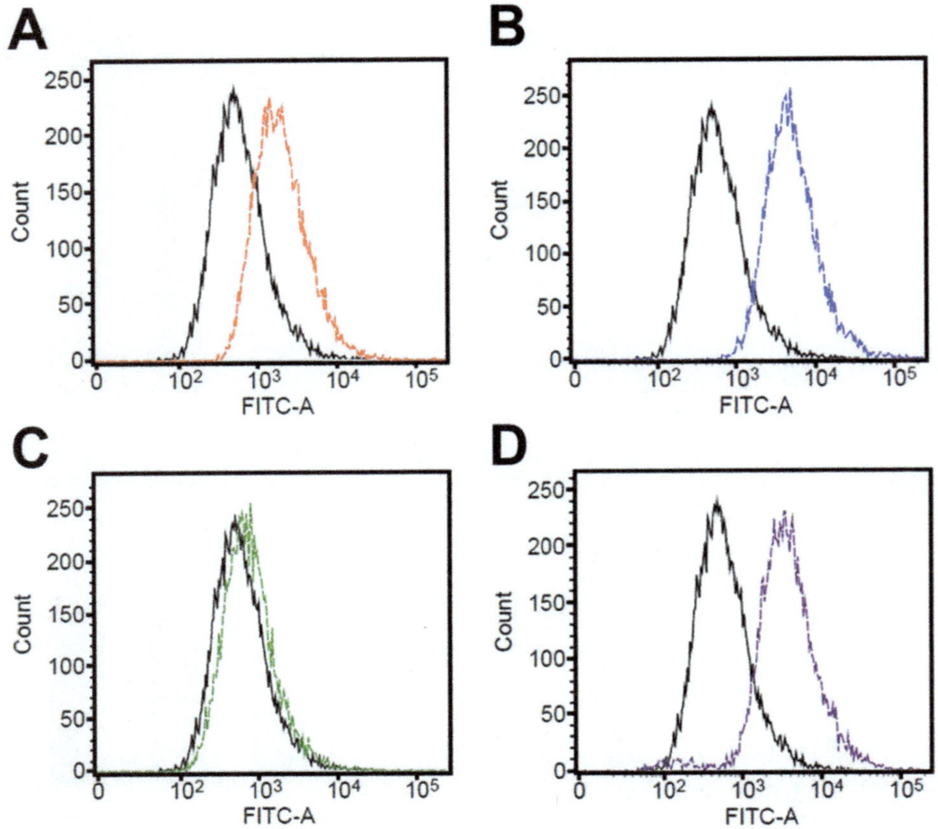

Figure 3. Measurement of oxidative stress induced by antibiotics. Intracellular superoxide-anion generation was measured using DHR 123. Fluorescence intensity was determined using flow cytometry and is represented as a histogram. FITC-A indicates the intensity of green fluorescence and the number of cells exhibiting the corresponding fluorescence intensity (amount of ROS production). The fluorescence histograms are of the samples before and after antibiotic treatment; solid and dotted lines are untreated cells and antibiotic-treated cells, respectively. (A) Amp, (B) Km, (C) Tc, (D) Nor. A shift to stronger fluorescence indicates a greater generation of oxidative stress.

and energy sources are scarce or when oxidative stress is generated [49,50]. Copper stress, which causes oxidative stress, induced glyoxylate bypass in *Pseudomonas* [51]. Glyoxylate bypass was particularly induced under Amp and Nor conditions (Figure 4). Km strongly induced oxidative stress and caused growth defects, but could not induce glyoxylate bypass. Therefore, we speculated

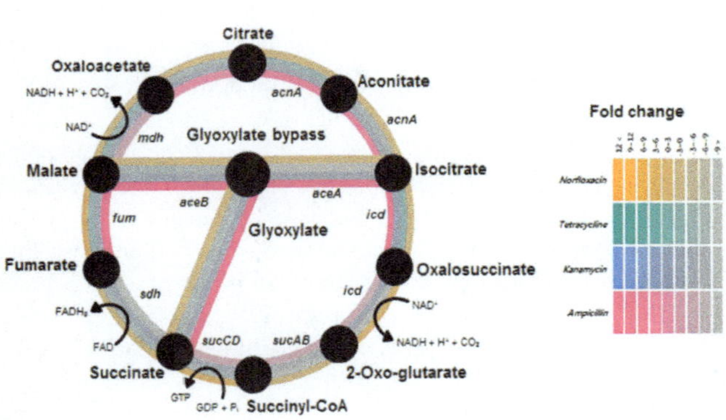

Figure 4. Expression of citrate-cycle genes in *A. oleivorans* DR1 treated with distinct antibiotics. Gene-expression changes are represented by a color gradient that is based on the fold-changes of gene expression in response to antibiotic treatments.

that there are other factors that induce glyoxylate bypass in DR1 under antibiotic conditions.

In *E. coli*, sublethal concentrations of aminoglycosides increased the expression of several genes involved in heat-shock response, such as *htpG*, *ibpA*, *groES*, and *asrA* [52]. Aminoglycosides also induced the Lon protease in *P. aeruginosa* [53]. Our data showed that genes encoding chaperones and proteases (DnaK, AOLE_19360; GroEL, AOLE_03915; GroES, AOLE_03910) exhibit high RPKM values under Km treatment. These results suggest that chaperones and proteases might play a key role in mistranslation under Km condition in DR1 cells. Our data showed that endonucleases did not exhibit DNA-repair capabilities in DR1 cells treated with Km and Tc. Intriguingly, only ribosome-targeting antibiotics caused a loss of DNA-repair capability; this is probably because of the long protein-maturation times required for DNA-repair enzymes. Antibiotics can interfere with the metabolic pathways of bacteria, and this can cause structural alterations in the bacterial cell wall and surface appendages including flagella, fimbriae, and pili [54]. Bacteria employ extracellular structures such as pili and fimbriae in attachment and invasion, biofilm formation, cell motility, and transport across membranes [55]. Km and Tc have similar target regions, and they inhibit protein synthesis by binding to the 30S subunit of the ribosome [11,13]. Our transcriptomic data showed that Km and Tc markedly induced fimbriae/pili-related genes. Interestingly, these antibiotics also upregulated the natural

Table 3. Expression change of functional gene clusters.

Locus_tag DR1	Product	Genes	Fold change			
			Amp	Km	Tc	Nor
SOS-response genes						
AOLE_07085	Nucleotidyltransferase/DNA polymerase	*dinP*	−2.67	1.48	2.34	1.57
AOLE_07375	Recombinase A	*recA*	2.05	3.04	1.02	5.17
AOLE_07965	DNA-directed DNA polymerase UmuC	*umuC*	−1.43	−2.26	2.24	4.20
AOLE_07970	DNA polymerase V component		−1.43	−1.69	−1.10	5.41
AOLE_11745	SOS-response transcriptional repressor (RecA-mediated autopeptidases)	*umuD*	−1.18	1.82	−1.57	3.15
AOLE_14875	DNA polymerase V component		1.18	3.20	2.18	6.51
AOLE_14880	DNA-directed DNA polymerase UmuC	*umuC*	−4.42	−2.49	1.51	2.13
AOLE_18420	DNA polymerase IV	*dinP*	−1.67	1.34	−1.01	−1.53
DNA repair-related genes						
AOLE_05830	Putative DNA-binding/iron metalloprotein/AP endonuclease		−8.18	2.00	−1.26	−1.33
AOLE_13505	Metalloendopeptidase-like membrane protein	*nlpD*	2.96	1.35	−2.53	1.06
AOLE_14215	Endonuclease III	*nth*	−3.09	1.81	−1.31	−1.19
AOLE_14840	HNH endonuclease		−1.04	5.00	−1.12	−1.43
AOLE_18425	Endoribonuclease L−PSP family protein	*tdcF*	−1.31	−1.42	−7.04	1.11
AOLE_18840	Endoribonuclease L-PSP family protein	*tdcF*	5.89	−2.73	1.24	−1.16
AOLE_03065	Formamidopyrimidine-DNA glycosylase	*mutM*	−2.32	1.72	−1.24	−1.03
AOLE_10805	Uracil-DNA glycosylase	*ung*	−2.90	1.89	−1.41	−1.02
Oxidative stress-related genes						
AOLE_01750	Cu/Zn superoxide dismutase	*sodC*	2.99	−1.24	1.10	−1.07
AOLE_02915	Peroxiredoxin	*ahpC*	2.72	−1.38	−3.47	−1.98
AOLE_05305	Superoxide dismutase	*sodA*	−1.57	−1.27	−1.54	−1.16
AOLE_07635	Thioredoxin	*trxA*	−1.77	5.36	8.61	2.10
AOLE_11770	Catalase	*katE*	3.98	1.22	1.56	−1.17
AOLE_12135	Redox-sensitive transcriptional activator SoxR	*soxR*	2.52	4.00	2.53	2.88
AOLE_12755	Catalase	*katE*	1.41	1.54	2.09	−1.75
AOLE_13380	Peroxiredoxin	*ahpC*	1.23	1.24	−1.87	−1.23
AOLE_14380	Hydrogen peroxide-inducible genes activator	*oxyR*	1.11	1.40	−1.79	1.08
AOLE_16430	Thioredoxin	*trxA*	1.09	−1.24	−2.07	−1.09
AOLE_17390	Catalase	*katG*	−1.19	−3.11	−2.88	1.34
AOLE_18445	SoxR-family transcriptional regulator	*soxR*	1.17	1.38	1.47	−1.32

competence-associated type-IV pilus-assembly proteins encoded by AOLE_15230 (3.5-fold) and AOLE_17785 (3.69-fold).

Fluoroquinolones can induce the SOS response [56], key regulators of which are the proteins LexA and RecA [35,36]. However, the lack of a LexA homolog indicates a critical role of other regulators for SOS response in *Acinetobacter* species [57]. Here, transcriptome analysis demonstrated that in DR1 cells, Nor strongly induced genes involved in the typical SOS response and DNA-repair genes. The relative amounts of SOS gene expression are determined primarily through by transcriptional regulation. Our previous study showed that, Nor treatment caused target-gene mutation in *gyrA* (AOLE_04195) and persister formation in DR1 [46]. Our data additionally validated the SOS response of *Acinetobacter* species by showing that DNA damage enhanced mutation frequency. This characteristic of DR1 might be helpful for having resistance to antibiotics stress.

Noncoding RNAs are commonly referred to as small RNAs because they are 50–500 nucleotides in size [58]. Small RNAs are potent regulatory molecules that function at the transcriptional or posttranscriptional level [59]. Interestingly, RNA-seq mapping data revealed that the noncoding regions of DR1 contain sequences of small-RNA candidates (Table S4). Three small-RNA candidates are conserved in certain *Acinetobacter* species. The Northern blot analysis confirmed the expression of small-RNA candidates (Figure S6).

In a recent study on *A. baumannii*, 31 putative small RNAs were identified using computational approaches [60]. Two of these small RNAs display sequence similarities with those of the DR1 strain and other *Acinetobacter* species. However, these 2 small RNAs were not induced under our tested conditions. Small RNAs play key roles in efflux-pump regulation and antimicrobial-agent resistance in *A. baumannii* [60], and efflux pumps are widely accepted to bestow clinically relevant resistance to antibiotics [61]. How small RNAs involved in expression of efflux pumps remains to be investigated in DR1 cells. Our study will serve as a baseline for understanding the effects of antibiotics on *Acinetobacter*

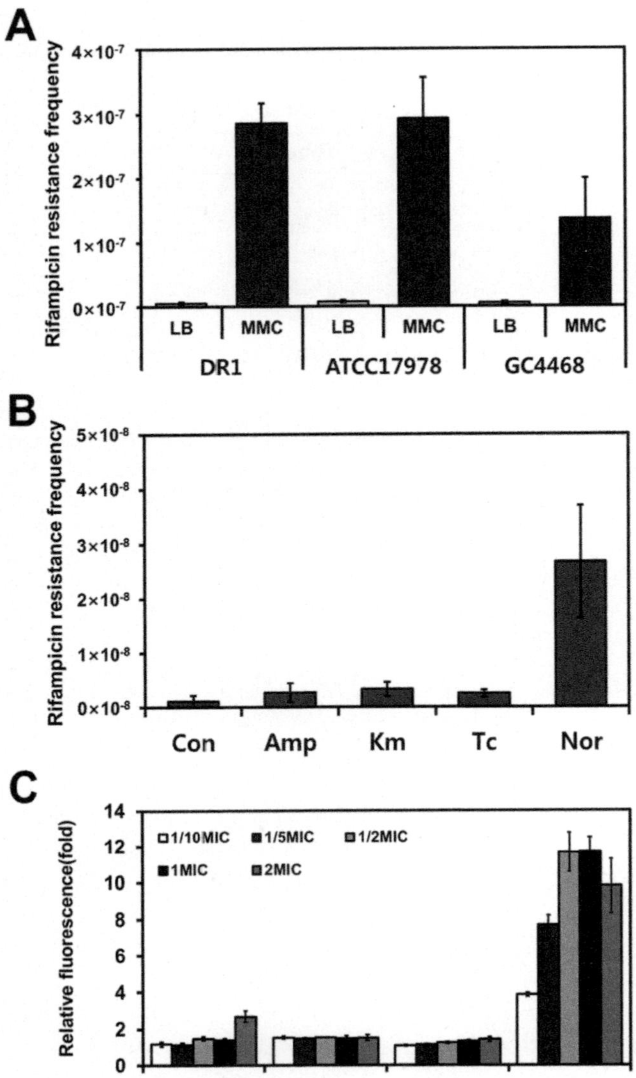

Figure 5. SOS-response induction in *Acinetobacter oleivorans* DR1. The mutation frequency, which corresponds to the rifampicin-resistance CFU count divided by the total CFU count, was measured and is represented on the Y-axis in the case of each antibiotic. (A) MMC-induced mutagenesis frequency. (B) Mutagenesis frequency induced by antibiotics of distinct classes. (C) Effect of antibiotics on *recA* expression was confirmed using a GFP fusion protein.

species, and it should help in developing a new strategy for predicting novel antibiotic-resistance mechanisms, as well as for preventing multidrug resistance across multiple species of bacteria by using this soil-borne bacterium.

Materials and Methods

Bacterial strains, growth conditions, and antibiotics

The bacterial strains used in this study are listed in Table S5. Environment-originated nonpathogenic *A. oleivorans* DR1 was grown in nutrient broth at 30°C with rotational shaking at 220 rpm. Bacteria harboring plasmids and wild-type bacteria were cultured under the same conditions. *Escherichia coli* GC 4468 and *A. baumannii* ATCC17978 were grown at 37°C in LB and aerated by means of shaking. In bacterial antibiotic-treatment

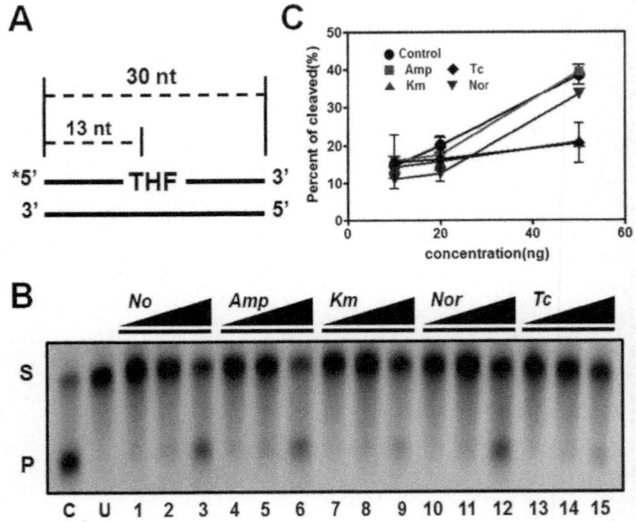

Figure 6. Verification of endonuclease IV activity by using the base-excision DNA-repair assay. DNA-repair capability of endonuclease IV was measured in DR1 exposed to distinct classes antibiotics. (A) Schematic representation of DNA substrate containing a site-specific THF residue. (B) A representative autoradiograph of gel electrophoresis to measure *in vitro* BER products. (C) Quantification of endonuclease IV BER activity. S, substrate; P, product; C, positive control; U, untreated negative control. Error bars indicate the S.D. calculated for each data point (n = 2).

experiments, we used commercially available Rifampicin (Sigma-Aldrich, USA), Amp (Bioshop, Canada), Km (Bioshop, Canada), Tc (Sigma-Aldrich, USA), and Nor (Sigma-Aldrich, USA).

Determination of antibiotic minimum inhibitory concentrations (MICs) of *A.* oleivorans DR1

MICs were determined in liquid nutrient medium by using 96-well polystyrene microtiter plates (Costar, USA). DR1 cells were grown overnight in nutrient broth at 30°C with shaking at 220 rpm. The cells were washed twice with phosphate-buffered saline (PBS) and inoculated at a cell density of $10^5 \sim 10^8$CFU/mL in 200 μL of nutrient broth containing 0–256 μg/mL of each antibiotic (Amp, Km, Tc, Nor), and then grown in 96-well polystyrene plates at 30°C for 24 h without shaking. MICs were determined by measuring the optical density at 600 nm (OD_{600}) by using a microtiter-plate reader (PowerWaveXS, Bio-Tek, USA); the MICs were the lowest concentrations of the 4 antibiotics at which OD_{600} was <0.04.

RNA extraction, sequencing, and analysis

Total RNA of DR1 cells grown in nutrient media was isolated from exponential-phase cells ($OD_{600} \sim 0.4$). Cells were grown at 30°C with shaking at 220 rpm and when they reached the exponential phase, they were treated without or with each antibiotic at the sub-MIC (Amp,100 g/mL, Km, 4 g/mL, Tc: 1 g/mL, Nor: 4 g/mL) for 15 min. Total RNA was extracted using RNeasy Mini kits (Qiagen, USA) by following the manufacturer's instructions. The isolated RNA was stored at −80°C until use. All RNA-sequencing and alignment procedures were conducted by Chunlab (Seoul, South Korea). The RNA was subjected to a subtractive Hyb-based rRNA-removal process by using the MICROBExpress Bacterial mRNA Enrichment Kit (Ambion, USA), and subsequent processes, including library construction, were performed as described previously (Table S1) [62]. RNA sequencing was performed using 2 runs of the Illumina

HiSeq to generate single-ended 100-bp reads. The genome sequence of *A. oleivorans* DR1 was retrieved from the NCBI database (accession number NC_014259.1). Quality-filtered reads were aligned to the reference-genome sequence by using the CLC Genomics Workbench 6.5.1 tool (CLC bio, Denmark). Mapping was based on a minimal length of 100 bp, with an allowance of up to 2 mismatches. The relative transcript abundance was measured in reads per kilobase of exon sequence per million mapped sequence rea20kds (RPKM) [63]. The mapping results were visualized using the CLRNAseq program (Chunlab, South Korea). The RNA-seq data were deposited in the National Center for Biotechnology Information (NCBI) GEO site under accession numbers GSE38340, GSE44428, GSE58166 and GSE58167.

Quantitative real-time PCR (qRT-PCR) analysis

cDNA was synthesized from 1 μg each RNA extract by using gene specific primers. (Table S5) and the primers for genes were used as templates for quantitative real-time PCR (qRT-PCR). The 25 μl PCR mixture included 12.5 μl iQ SYBR Green Supermix (Bio-Rad, USA), 1 μl of each primer (0.5 μM), 2 μl cDNA, and 8.5 μl distilled water. The PCR reactions were conducted at 95°C for 3 min, followed by 40 cycles consisting of 30 s at 95°C, 30 s at 60°C, and 30 s at 72°C. The expression level of each gene was normalized to the 16S rRNA expression level that was quantified with 16s rRNA-341F/16s rRNA-534R primers. Relative quantifications were performed in triplicate.

Cell membrane permeability assays

The fluorescent probe 8-anilino-1-naphthylenesulfonic acid (ANS; Sigma-Aldrich, USA) was used for assessing the integrity of bacterial cell membranes. Overnight cultures were diluted 100-fold in 5 mL of fresh medium and grown to the logarithmic-growth phase at 30°C and 220 rpm. After the cells were treated with or without each antibiotic at the exponential phase (OD_{600} ~0.4) for 15 min, 1 mL of the cell cultures was harvested by centrifugation (13,000× *g*, 1 min) and washed twice with PBS. The resuspended solutions were supplemented with ANS (1 μL, 3 mM) and maintained at room temperature for 10 min in the dark. The fluorescence intensity of cells was measured using a microplate reader. The filter set used for fluorescence measurements included a 555-nm excitation filter and 590-nm emission filter. The possibility that distinct growth rates were measured under various experimental conditions was excluded by normalizing protein amounts (in μg). Cell membrane permeability assays were performed 3 times independently.

Microscopic observation

Antibiotics (used at the sub-MICs) was added to the cells at the exponential phase (OD600nm = 0.4), and the cells were then incubated for 30 min at 30°C. 1 mL of the cell cultures was harvested by centrifugation (13,000× *g*, 1 min) and washed twice with PBS. The resuspended solutions were supplemented with 4′,6-diamidino- 2-phenylindole (DAPI) (1 μL, 2 μg/mL), and maintained at room temperature for 10 min in the dark. DAPI - treated cells was washed and resuspended using PBS. Then, 5 μL of cells was placed on a glass slide and observed. Bacteria treated with antibiotics were viewed with a Carl ZeissAxio Imager microscope (ZEISS, Germany).

Measurement of oxidative stress

Intracellular superoxide-anion generation was measured using dihydrorhodamine (DHR) 123 (Sigma-Aldrich, USA). The cells were grown to exponential phase (OD_{600}~0.4) and treated for 15 min with the antibiotics (used at the sub-MICs). The cells were washed twice and resuspended using PBS and then treated with DHR 123 (2.5 μg/mL) for 1 h in the dark at 30°C. DHR-123-treated cells were washed and resuspended using PBS, and the intracellular superoxide anion-mediated oxidation of DHR 123 was assayed be means of FACSverse flow cytometry (BD Biosciences, San Jose, CA, USA). The samples were analyzed by using a fluorescein isothiocyanate (FITC) argon-ion laser for excitation, and fluorescence intensity was determined and analyzed by measuring 10,000 cell counts. BD FACSuite software was used for data analysis.

DNA damage-induced mutagenesis frequency

Cells were grown to the exponential phase (OD_{600}~0.4) and treated without (control) or with 1 g/mL MMC (sub-MIC) and antibiotics (used at the sub-MICs) for 1 h. After the treatment, cells were washed twice and resuspended using PBS and inoculated at a cell density of 5×10^8 CFU/mL in 5 mL of fresh nutrient broth and grown with shaking at the appropriate temperature for 24 h. The cultures were collected and diluted in PBS and then plated on nutrient or LB agar media containing either 100 μg/mL rifampicin or no rifampicin to calculate the frequencies of rifampicin-resistance mutations. Colonies were counted after incubation for 24 h at the appropriate temperatures. Mutation frequency was determined from the relative percentage of CFU/mL ((CFU at 100 g/mL rifampicin/CFU at no rifampicin) × 100).

Construction of transcriptional-fusion green fluorescent protein (GFP) and quantification of GFP fluorescence

The broad-host-range expression vector pRK415 was used for constructing transcriptional-fusion GFP. A fragment of the *recA* promoter region was amplified by means of Polymerase chain reaction (PCR) performed using pRKprecA-gfp-F/pRKprecA-gfp-R primer pairs (Table S5). A 190-bp fragment of the promoter region of *recA* was cloned into the *Kpn*I/*Bam*HI cloning site of the multi cloning site of the pRK415 vector. The amplicon (715 bp) obtained using pRKgfp-F/pRKgfp-R was cloned into the *Bam*HI/*Eco*RI cloning site of the pRK415 vector to generate transcriptional-fusion GFP. The plasmid was extracted using a Dyne Plasmid Miniprep Kit (DYNEBIO, Korea). The constructed plasmid was then introduced into *E. coli* Top10 and A. *oleivorans* DR1 by electroporation. Competent cells (50 μl) were transformed with 2.5 μl plasmid DNA samples using a Micropulser (Bio-Rad, USA) with a time constant range of 3.0–3.5 ms and a constant voltage of 4.5–5 kV. PCR was conducted to confirm insertion of the gfp gene using the GFP-F/GFP-R primer set (Table S5). Overnight cultures of the DR1 harboring constructing transcriptional-fusion GFP grown in nutrient broth were diluted 100-fold in 5 mL of fresh medium and then incubated with shaking. At the exponential-growth phase (OD_{600}~0.4), the antibiotics were added and the cells were incubated for 1 h. A 1-mL aliquot of each GFP fusion-strain culture was harvested and centrifuged at 13,000×*g* for 1 min and then washed twice with PBS. The resuspended cells were transferred to polystyrene 48-well microtiter plates (BD Biosciences, USA) and the GFP fluorescence intensity of the cells was quantified using a Multi-Detection Microplate Reader (Sense, HIDEX, Finland). The GFP fusion-strain expressed a stable GFP variant that has an excitation wavelength of 488 nm and emission wavelengths of 507–510 nm. The OD_{600} of each culture was measured using a microtiter-plate reader (PowerWaveXS; Bio-Tek, USA). The possibility of distinct growth rates being measured under various experimental conditions was excluded by normalizing the measured fluorescence intensity relative to the OD_{600} value. One fluorescence unit was defined as [(fluorescence intensity of cells/fluorescence intensity of PBS buffer)/OD_{600} of cells], and a relative fluorescence unit (fold) was defined as [fluorescence unit of treated cells/fluorescence unit of control (untreated) cells].

In vitro base-excision repair (BER) assay

Overnight cultures were diluted 100-fold in 500 mL of fresh nutrient medium and grown to the exponential phase (OD_{600}~0.4) at 30°C and 180 rpm. After the cells were treated with or without antibiotics for 15 min, a cell cultures was harvested by centrifugation (10,000×*g*, 30 min) and washed twice with PBS. The cell pellet was resuspended in <5 ml of sonication buffer (50 mM Tris-HCl (pH 8.0), 1 mM EDTA, and 0.1 mM DTT), and cells were lysed by sonification. After cell debris was removed by centrifugation (13,000×*g*, 20 min) at 4°C, supernatant was collected and placed on ice. Cell-free extract was transferred to Eppendorf tubes (0.5 mL) and stored in 100 µl aliquots at −80°C. The radionucleotide [γ-^{32}P] ATP was obtained from PerkinElmer Life Sciences (Wellesley, USA). We purchased *E. coli* uracil-DNA glycosylase (UDG), formamidopyrimidine-DNA glycosylase (Fpg), endonuclease IV, and T4 polynucleotide kinase (New England Biolabs, UK). Micro Bio-Spin 30 Chromatography Columns were from Bio-Rad. DNA oligonucleotides containing uracil, tetrahydrofuran (THF), or 8-oxoguanine (8-oxoG) residues were provided by Dr. B. Demple, SUNY-Stony Brook (Stony Brook, USA), and these were amplified by means of PCR performed using 30F-F/30F-R, U30-F/U30-R, OxoG-F/OxoG-R primer pairs (Table S5). The endonuclease-IV-activity assay was performed in a reaction mixture 10 µL containing 50 mM Hepes-KOH (pH 7.5), 8 mM $MgCl_2$, 5% glycerol, 0.5 mM DTT, 0.1 mg/mL BSA, and 1 nM 5′-end-labeled duplex-DNA substrate containing THF residues. The reactions were initiated by adding 10, 20, or 50 ng of cell-free extracts and were incubated at 37°C. Aliquots of each reaction were withdrawn at 30 min, and the reactions were terminated by adding formamide loading buffer. The reaction products were separated by performing electrophoresis; we used 15% denaturing polyacrylamide gels containing 7 M urea in 90 mM Tris, 90 mM boric acid, and 2 mM EDTA. Gels were dried using a gel dryer (Model 583, Bio-Rad, USA), and products were visualized by means of autoradiography and quantified using ImageQuant software v5.2. The percentage of cleaved AP sites was calculated from amount of products divided by the sum of total products and substrates.

Northern blot assay

Total RNA (5 µg) were run on denaturing agarose gels containing 0.25 M formaldehyde, and the gels were stained with ethidium bromide (EtBr) to visualize 23S and 16S rRNA. The fractionated RNA was transferred to nylon membranes (Schleicher & Schuell, Germany) using a Turboblotter (Schleicher & Schuell, Germany). The mRNA levels were determined by hybridizing the membrane with a gene specific, ^{32}P-labeled probe (Takara, Japan) prepared by PCR amplification with their respective primer pair as indicated in Table S5. Autoradiography was conducted using an IP plate (Fujifilm, Japan) and a Multiplex Bio-Imaging system (FLA-7000; Fujifilm, Japan).

Supporting Information

Figure S1 Determination of MIC under different cell density in *A.* oleivorans DR1.

Figure S2 Confirmation of RNA-Seq results with qRT-PCR. (A) Commonly up- and down- regulated genes were confirmed the gene expression on 4 antibiotics conditions. (B) Three genes were selected based on expression value on each antibiotics condition.

Figure S3 COG assignments of differently expressed genes under distinct antibiotics conditions. The percentage of up-regulated and down-regulated genes was sorted by general COG categories. Colors of the bars indicate the changes of gene expression. Red, gene expression is >1.5-fold change in RPKM value, Brown, gene expression is <1.5-fold change in RPKM value, Gray, gene expression of between a −1.5 and 1.5-fold change in value. COG abbreviations for the functional categories: A, RNA processing and modification; B, chromatin structure and dynamics; C, energy production and conversion; D, cell cycle control and mitosis; E, amino acid metabolism and transport; F, nucleotide metabolism and transport; G, carbohydrate metabolism and transport; H, coenzyme metabolism; I, lipid metabolism; J, translation, including ribosome structure and biogenesis; K, transcription; L, replication, recombination, and repair; M, cell wall structure and biogenesis and outer membrane; N, secretion, motility and chemotaxis; O, molecular chaperones and related functions; P, inorganic ion transport and metabolism; Q, secondary metabolite biosynthesis, transport, and catabolism; T, signal transduction; U, intracellular trafficking, secretion, and vesicular transport; V, defense mechanisms.

Figure S4 Microscopic observation of antibiotics treated and untreated DR1 cells. Morphology observation of cell treated with antibiotics. Phage contrast and staining with DAPI are shown. The scale bar represents 20 µm.

Figure S5 Verification of UDG and Fpg activity by using the base-excision DNA-repair assay.

Figure S6 Confirmation of expression of small RNA candidates using Northern blot. The expression of small RNA candidate was determined under antibiotics conditions using Northern blot. The ethidium bromide (EtBr) staining demonstrated consistent loading in all lanes.

Table S1 Total number of reads aligning with the regions of interest (coverage) of the five libraries constructed from the RNA samples.

Table S2 Fimbriae/pili related gene expression profiles by different class antibiotics.

Table S3 Bacterial strains, plasmids, and primers used in this study.

Table S4 The feature of small RNA genes in *A. oleivorans* DR1.

Table S5 Bacterial strains, plasmid and oligonucleotides sequence used in this study.

Author Contributions

Conceived and designed the experiments: AH HJ JS WP. Performed the experiments: AH HJ. Analyzed the data: AH WP. Contributed reagents/materials/analysis tools: AH HJ JS WP. Contributed to the writing of the manuscript: HJ WP.

References

1. Hartzell JD, Kim AS, Kortepeter MG, Moran KA (2007) *Acinetobacter* pneumonia: a review. Med Gen Med 9: 4.
2. Forster DH, Daschner FD (1998) *Acinetobacter* species as nosocomial pathogens. Eur J Clin Microbiol Infect Dis 17: 73–77.
3. Towner KJ, Bergogne-Bérézin E, Fewson CA (1991) *Acinetobacter*: Portrait of a Genus. In Towner KJ, Bergogne-Bérézin E, Fewson CA, editors. The biology of *Acinetobacter*: taxonomy, clinical importance, molecular biology, physiology, industrial relevance. New York, NY: Plenum Press pp. 1–24.
4. Martinez JL (2009) Environmental pollution by antibiotics and by antibiotic resistance determinants. Environ Pollut 157: 2893–2902.
5. Kümmerer K (2003) Significance of antibiotics in the environment. J Antimicrob Chemother 52: 5–7.
6. Alonso A, Sánchez P, Martínez JL (2001) Environmental selection of antibiotic resistance genes. Environ Microbiol 3: 1–9.
7. Martínez JL (2008) Antibiotics and antibiotic resistance genes in natural environments. Science 321: 365–367.
8. Pitout JD, Sanders CC, Sanders WE Jr (1997) Antimicrobial resistance with focus on beta-lactam resistance in gram-negative bacilli. Am J Med 103: 51–59.
9. Davis BD (1987) Mechanism of bactericidal action of aminoglycosides. Microbiol Rev 51: 341–350.
10. Kohanski MA, Dwyer DJ, Wierzbowski J, Cottarel G, Collins JJ (2008) Mistranslation of membrane proteins and two-component system activation trigger antibiotic-mediated cell death. Cell 135: 679–690.
11. Chopra I, Roberts M (2001) Tetracycline antibiotics: mode of action, applications, molecular biology, and epidemiology of bacterial resistance. Microbiol Mol Biol Rev 65: 232–260.
12. Hooper DC (2001) Mechanisms of action of antimicrobials: focus on fluoroquinolones. Clin Infect Dis 32 Suppl 1: S9–S15.
13. Manchanda V, Sanchaita S, Singh N (2010) Multidrug resistant *Acinetobacter*. J Glob Infect Dis 2: 291–304.
14. Jung J, Baek JH, Park W (2010) Complete genome sequence of the diesel-degrading *Acinetobacter* sp. strain DR1. J Bacteriol 192: 4794–4795.
15. Poirel L, Bonnin RA, Nordmann P (2011) Genetic basis of antibiotic resistance in pathogenic *Acinetobacter* species. IUBMB Life 63: 1061–1067.
16. Jung J, Madsen EL, Jeon CO, Park W (2011) Comparative genomic analysis of *Acinetobacter oleivorans* DR1 to determine strain-specific genomic regions and gentisate biodegradation. Appl Environ Microbiol 77: 7418–7424.
17. Chopra I, Roberts M (2001) Tetracycline antibiotics: mode of action, applications, molecular biology, and epidemiology of bacterial resistance. Microbiol Mol Biol Rev 65: 232–260.
18. Fajardo A, Martínez JL (2008) Antibiotics as signals that trigger specific bacterial responses. Curr Opin Microbiol 112: 161–167.
19. Kohanski MA, Dwyer DJ, Collins JJ (2010) How antibiotics kill bacteria: from targets to networks. Nat Rev Microbiol 8: 423–435.
20. Tenover FC (2006) Mechanisms of antimicrobial resistance in bacteria. Am J Med 119: S3-S10.
21. Yonath A, Bashan A (2004) Ribosomal crystallography: initiation, peptide bond formation, and amino acid polymerization are hampered by antibiotics. Annu Rev Microbiol 58: 233–251.
22. Adams MD, Goglin K, Molyneaux N, Hujer KM, Lavender H (2008) Comparative genome sequence analysis of multidrug-resistant *Acinetobacter baumannii*. J Bacteriol 190: 8053–8064.
23. Kang YS, Park W (2010) Trade-off between antibiotic resistance and biological fitness in *Acinetobacter* sp. strain DR1. Environ Microbiol 12: 1304–1318.
24. Prigent-Combaret C, Sanguin H, Champier L, Bertrand C, Monnez C, et al. (2012) The bacterial thiopurine methyltransferase tellurite resistance process is highly dependent upon aggregation properties and oxidative stress response. Environ Microbiol 14: 2645–260.
25. Fronzes R, Remaut H, Waksman G (2008) Architectures and biogenesis of non-flagellar protein appendages in gram-negative bacteria. EMBO J 27: 2271–2280.
26. Chen SH, Chen RY, Xu XL, Xiao WB (2012) Microarray analysis and phenotypic response of *Pseudomonas aeruginosa* PAO1 under hyperbaric oxyhelium conditions. Can J Microbiol 58: 158–169.
27. Fléchard M, Cortes MA, Répérant M, Germon P (2012) New role for the ibeA gene in H2O2 stress resistance of *Escherichia coli*. J Bacteriol 194: 4550–4560.
28. Yao Z, Kahne D, Kishony R (2012) Distinct single-cell morphological dynamics under beta-lactam antibiotics. Mol Cell 48: 705–712.
29. Bi E, Lutkenhaus J (1993) Cell division inhibitors SulA and MinCD prevent formation of the FtsZ ring. J Bacteriol 175: 1118–1125.
30. Jones C, Holland IB (1985) Role of the SulB (FtsZ) protein in division inhibition during the SOS response in *Escherichia coli*: FtsZ stabilizes the inhibitor SulA in maxicells. Proc Natl Acad Sci U S A 82: 6045–6049.
31. Turner RD, Vollmer W, Foster SJ (2014) Different walls for rods and balls: the diversity of peptidoglycan. Mol Microbiol 91: 862–874.
32. Frirdich E, Gaynor EC (2013) Peptidoglycan hydrolases, bacterial shape, and pathogenesis. Curr Opin Microbiol 16: 767–778.
33. Braun V, Oldmixon E (1979) Relationship between steps in 8-anilino-1-naphthalene sulfonate (ANS) fluorescence and changes in the energized membrane state and in intracellular and extracellular adenosine 5′-triphosphate (ATP) levels following bacteriophage T5 infection of *Escherichia coli*. J Supramol Struct 10: 329–347.
34. Nikaido H, Vaara M (1985) Molecular basis of bacterial outer membrane permeability. Microbiol Rev 49: 1–32.
35. Kohanski MA, Dwyer DJ, Hayete B, Lawrence CA, Collins JJ (2007) A common mechanism of cellular death induced by bactericidal antibiotics. Cell 130: 797–810.
36. Rui B, Shen T, Zhou H, Liu J, Chen J, et al. (2010) A systematic investigation of *Escherichia coli* central carbon metabolism in response to superoxide stress. BMC Syst Biol 4: 122.
37. Eisenstadt E, Carlton BC, Brown BJ (1994) Gene mutation. In Gerhardt P, Murray RGE, Wood WA, and Krieg NR editors. Methods for general and molecular bacteriology. American Society for Microbiology. Washington, D.C: Plenum Press pp. 297–316.
38. Aranda J, Bardina C, Beceiro A, Rumbo S, Cabral MP, et al. (2010) *Acinetobacter baumannii* RecA protein in repair of DNA damage, antimicrobial resistance, general stress response, and virulence. J Bacteriol 193: 3740–3747.
39. Kelley WL (2006) Lex marks the spot: the virulent side of SOS and a closer look at the LexA regulon. Mol Microbiol 62: 1228–1238.
40. Hare JM, Bradley JA, Lin CL, Elam TJ (2012) Diverse responses to UV light exposure in *Acinetobacter* include the capacity for DNA damage-induced mutagenesis in the opportunistic pathogens Acinetobacter baumannii and Acinetobacter ursingii. Microbiology 158: 601–611.
41. Baharoglu Z, Mazel D (2011) *Vibrio cholerae* triggers SOS and mutagenesis in response to a wide range of antibiotics: a route towards multiresistance. Antimicrob Agents Chemother 55: 2438–2441.
42. Hare JM, Perkins SN, Gregg-Jolly LA (2006) A constitutively expressed, truncated *umuDC* operon regulates the *recA*-dependent DNA damage induction of a gene in *Acinetobacter baylyi* Strain ADP1. Appl Environ Microbiol 72: 4036–4043.
43. Zharkov DO (2008) Base excision DNA repair. Cell Mol Life Sci 65: 1544–1165.
44. Blainey PC, van Oijen AM, Banerjee A, Verdine GL, Xie XS (2006) A base-excision DNA-repair protein finds intra helical lesion bases by fast sliding in contact with DNA. Proc Natl Acad Sci U S A 103: 5752–5757.
45. Ide H, Tedzuka K, Shimzu H, Kimura Y, Purmal AA, et al. (1994) Alpha-deoxyadenosine, a major anoxic radiolysis product of adenine in DNA, is a substrate for *Escherichia coli* endonuclease IV. Biochemistry 33: 7842–7847.
46. Poirel L, Bonnin RA, Nordmann P (2011) Genetic basis of antibiotic resistance in pathogenic *Acinetobacter* species. IUBMB Life 63: 1061–1067.
47. Kim J, Noh J, Park W (2013) Insight into norfloxacin resistance of *Acinetobacter oleivorans* DR1: target gene mutation, persister, and RNA-Seq analyses. J Microbiol Biotechnol 23: 1293–1303.
48. Hong H, Ko HJ, Choi IG, Park W (2014) Previously undescribed plasmids recovered from activated sludge confer tetracycline resistance and phenotypic changes to *Acinetobacter oleivorans* DR1. Microb Ecol 67: 369–379.
49. Gupta V, Garg R, Garg S, Chander J, Attri AK (2010) Coexistence of extended spectrum beta-lactamases, AmpC beta-lactamases and metallo-beta-lactamases in *Acinetobacter baumannii* from burns patients: a report from a tertiary care centre of India. Ann Burns Fire Disasters 26: 189–192.
50. Lemuth K, Hardiman T, Winter S, Pfeiffer D, Keller MA (2008) Global transcription and metabolic flux analysis of *Escherichia coli* in glucose-limited fed-batch cultivations. Appl Environ Microbiol 74: 7002–7015.
51. Schroeter R, Voigt B, Jürgen B, Methling K, Pöther DC (2011) The peroxide stress response of *Bacillus licheniformis*. Proteomics 11: 2851–2866.
52. Li K, Pidatala RR, Ramakrishna W (2012) Mutational, proteomic and metabolomic analysis of a plant growth promoting copper-resistant *Pseudomonas* spp. FEMS Microbiol Lett 335: 140–148.
53. Kindrachuk K, Fernández L, Bains M, Hancock R (2011) Involvement of an ATP-dependent protease, PA0779/AsrA, in inducing heat shock in response to tobramycin in *Pseudomonas aeruginosa*. Antimicrob Agents Chemother 55: 1874–1882.
54. Marr AK, Overhage J, Bains M, Hancock RE (2007) The Lon protease of *Pseudomonas aeruginosa* is induced by aminoglycosides and is involved in biofilm formation and motility. Microbiology 153: 474–482.
55. Braga PC, Ricci D (2011) Imaging bacterial shape, surface, and appendages before and after treatment with antibiotics. Methods Mol Biol 736: 391–399.
56. Fronzes R, Remaut H, Waksman G (2008) Architectures and biogenesis of non-flagellar protein appendages in gram-negative bacteria. EMBO J 27: 2271–2280.
57. Phillips I, Culebras E, Moreno F, Baquero F (1987) Induction of the SOS response by new 4-quinolones. J Antimicrob Chemother 20: 631–638.

58. Aranda J, Bardina C, Beceiro A, Rumbo S, Cabral MP, et al. (2011) *Acinetobacter baumannii* RecA protein in repair of DNA damage, antimicrobial resistance, general stress response, and virulence. J Bacteriol 193: 3740–3747.
59. Waters LS, Storz G (2009) Regulatory RNAs in bacteria. Cell 136: 615–628.
60. Gottesman S (2005) Micros for microbes: non-coding regulatory RNAs in bacteria. Trends Genet 21: 399–404.
61. Sharma R, Arya S, Patil SD, Sharma A, Jain PK, et al. (2014) Identification of novel regulatory small RNAs in *Acinetobacter baumannii*. PLoS One 9: e93833.
62. Li XZ, Nikaido H (2004) Efflux-mediated drug resistance in bacteria. Drugs 64: 159–204.
63. Yi H, Cho YJ, Won S, Lee JE, Jin YH, et al. (2011) Duplex-specific nuclease efficiently removes rRNA for prokaryotic RNA-Seq. Nucleic Acids Res. 39: e140.

9

Alarming Levels of Drug-Resistant Tuberculosis in HIV-Infected Patients in Metropolitan Mumbai, India

Petros Isaakidis[1]*, Mrinalini Das[1], Ajay M V Kumar[2], Christopher Peskett[1], Minni Khetarpal[3], Arun Bamne[4], Balkrishna Adsul[5], Mamta Manglani[6], Kuldeep Singh Sachdeva[7], Malik Parmar[8], Avinash Kanchar[9], B.B. Rewari[9], Alaka Deshpande[10], Camilla Rodrigues[11], Anjali Shetty[11], Lorraine Rebello[1], Peter Saranchuk[12]

1 Médecins Sans Frontières, Mumbai, India, **2** International Union Against Tuberculosis and Lung Disease (The Union), South-East Asia Regional Office, New Delhi, India, **3** City TB Office, Bawala Wadi, Chinckpokli, Mumbai, India, **4** Brihanmumbai Municipal Corporation, Mumbai, India, **5** Mumbai District AIDS Control Society (MDACS), Mumbai, India, **6** Pediatric Centre of Excellence for HIV Care, L.T.M. Medical College & General Hospital, Sion, Mumbai, India, **7** Central TB Division, Directorate General of Health Services, Ministry of Health and Family Welfare, New Delhi, India, **8** World Health Organisation - Country Office for India, New Delhi, India, **9** Department of AIDS Control, National AIDS Control Organisation, New Delhi, India, **10** MGM Institute of Health Sciences, Navi Mumbai, India, **11** P.D. Hinduja National Hospital and Medical Research Centre, Mumbai, India, **12** Southern Africa Medical Unit (SAMU), Médecins Sans Frontières, Cape Town, South Africa

Abstract

Background: Drug-resistant tuberculosis (DR-TB) is a looming threat to tuberculosis control in India. However, no countrywide prevalence data are available. The burden of DR-TB in HIV-co-infected patients is likewise unknown. Undiagnosed and untreated DR-TB among HIV-infected patients is a major cause of mortality and morbidity. We aimed to assess the prevalence of DR-TB (defined as resistance to any anti-TB drug) in patients attending public antiretroviral treatment (ART) centers in greater metropolitan Mumbai, India.

Methods: A cross-sectional survey was conducted among adults and children ART-center attendees. Smear microscopy, culture and drug-susceptibility-testing (DST) against all first and second-line TB-drugs using phenotypic liquid culture (MGIT) were conducted on all presumptive tuberculosis patients. Analyses were performed to determine DR-TB prevalence and resistance patterns separately for new and previously treated, culture-positive TB-cases.

Results: Between March 2013 and January 2014, ART-center attendees were screened during 14135 visits, of whom 1724 had presumptive TB. Of 1724 attendees, 72 (4%) were smear-positive and 202 (12%) had a positive culture for *Mycobacterium tuberculosis*. Overall DR-TB was diagnosed in 68 (34%, 95% CI: 27%–40%) TB-patients. The proportions of DR-TB were 25% (29/114) and 44% (39/88) among new and previously treated cases respectively. The patterns of DR-TB were: 21% mono-resistant, 12% poly-resistant, 38% multidrug-resistant (MDR-TB), 21% pre-extensively-drug-resistant (MDR-TB plus resistance to either a fluoroquinolone or second-line injectable), 6% extensively drug-resistant (XDR-TB) and 2% extremely drug-resistant TB (XDR-TB plus resistance to any group-IV/V drug). Only previous history of TB was significantly associated with the diagnosis of DR-TB in multivariate models.

Conclusion: The burden of DR-TB among HIV-infected patients attending public ART-centers in Mumbai was alarmingly high, likely representing ongoing transmission in the community and health facilities. These data highlight the need to promptly diagnose drug-resistance among all HIV-infected patients by systematically offering access to first and second-line DST to all patients with 'presumptive TB' rather than 'presumptive DR-TB' and tailor the treatment regimen based on the resistance patterns.

Editor: Anil K. Tyagi, University of Delhi, India

Funding: This study was funded by Médecins Sans Frontières, Belgium. The funder was involved in study design, data collection and analysis. However, final preparation of the manuscript and the decision to publish rests with the first author.

Competing Interests: The authors have declared that no competing interests exist.

* Email: msfocb-asia-epidemio@brussels.msf.org

Introduction

India is a high burden country for tuberculosis (TB) and multidrug-resistant TB (MDR-TB). The World Health Organization has estimated that India accounted for 26% of the total number of TB cases worldwide in 2012, with 2.2% and 15% of the new and retreatment cases respectively being caused by multidrug-resistant strains [1]. Further, India is home to approximately 2.4 million people living with HIV [2] and considered to have a high burden on account of the large absolute numbers of people living with HIV in the country.

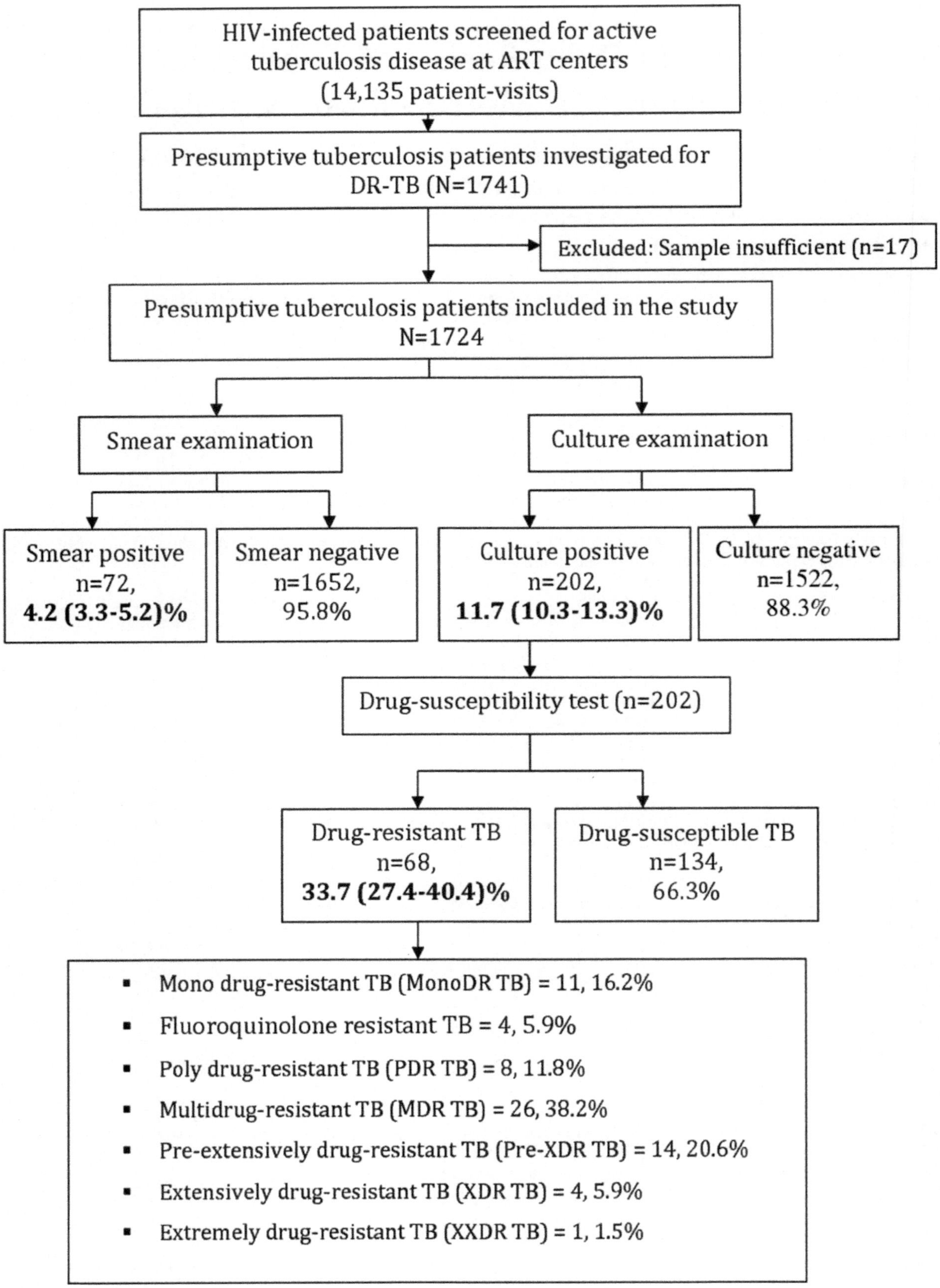

Figure 1. Drug-resistant tuberculosis among HIV-infected patients with presumptive tuberculosis, Mumbai, India.

The dual burden of HIV and TB/DR-TB in India is significantly high with a combined rate of 5.2%, ranging from 0.4% to 28.8% in various studies, with increasing trends noted in states having a higher burden of HIV infection [3–7]. However, nation-wide studies do not exist and previous studies have occurred mainly in hospitals and tertiary care centres [2,6–11]. A crude estimate from these studies suggests that 2500–3000 HIV-infected persons develop MDR-TB annually in India.

Country-wide or state-wide drug resistance surveys (DRS) aim to estimate the DR-TB burden at the country or state level. While this approach is scientifically and operationally acceptable, it may mask significant and important variance in the magnitude of the epidemic in different localities, communities or specific populations. For India, a vast country with an enormous burden of TB and a relatively large burden of HIV in absolute numbers, this statement seems to hold true; from an overcrowded impoverished slum in Mumbai to a small isolated village in the Northern Eastern Provinces of the country, one can assume that several different epidemics may exist. A description of such local epidemics is necessary so as to complement the country-wide prevalence estimate. While there is an urgent need for a nationally representative, country-wide DRS in India, specific studies to identify pockets of extremely high DR-TB prevalence or extensive drug resistance patterns are equally needed in order to advocate for and implement effective control strategies.

The overall aim of this study was to assess the burden of drug-susceptible and drug-resistant tuberculosis among HIV-infected patients attending antiretroviral treatment (ART) centers in the metropolitan area of Mumbai. The specific objectives were a) to determine the proportion of HIV-infected patients with DR-TB among those attending public ART centers, b) to describe drug susceptibility patterns among *Mycobacterium tuberculosis* isolates from this population, and c) to identify factors associated with TB and drug-resistant TB among HIV patients. We aimed to contribute to the evidence base that informs policies and practices and help to estimate the resources needed to control the epidemic in this specific group, as well as the community.

Methods

Ethics

The study was approved by the Institutional Ethics Committee of Grant Medical College and Sir J.J. Group of Hospitals (Mumbai, India), the Ethics Review Board of Médecins Sans Frontières (Geneva, Switzerland) and the Ethics Advisory Group of the International Union Against Tuberculosis and Lung Disease (Paris, France). The study protocol was approved by the Indian

Table 1. Demographic and clinical characteristics of HIV-infected patients with presumptive TB, Mumbai, India.

Characteristics	HIV-infected patients with presumptive TB (N = 1724)
	n (%)
Age [years, median (IQR)]	35.0 (24.3–44.0)
Sex of patients	
Male	1042 (60.4)
Female	671 (38.9)
Transgender	11 (0.6)
Family income per month (in Rupees)	
Less than 3500	88 (5.1)
3500–6999	910 (52.8)
7000 and above	454 (26.3)
Patient did not disclose	272 (15.6)
TB site	
Pulmonary	1688 (97.9)
Extra-pulmonary	36 (2.1)
ART status	
On ART	1386 (80.4)
Pre-ART	338 (19.6)
CD4 count, last visit (in cells/µl)	
Less than 200	258 (15.0)
200–349	351 (20.4)
350–499	289 (16.8)
500 and above	684 (39.7)
No information	142 (8.2)
ART duration* [months, median (IQR)]	26.0 (10.7–47.5)
Previous episode of TB	
Yes	933 (54.1)
No	791 (45.9)

ART: Antiretroviral treatment* Patients on ART with available information about ART initiation date, N = 1370.

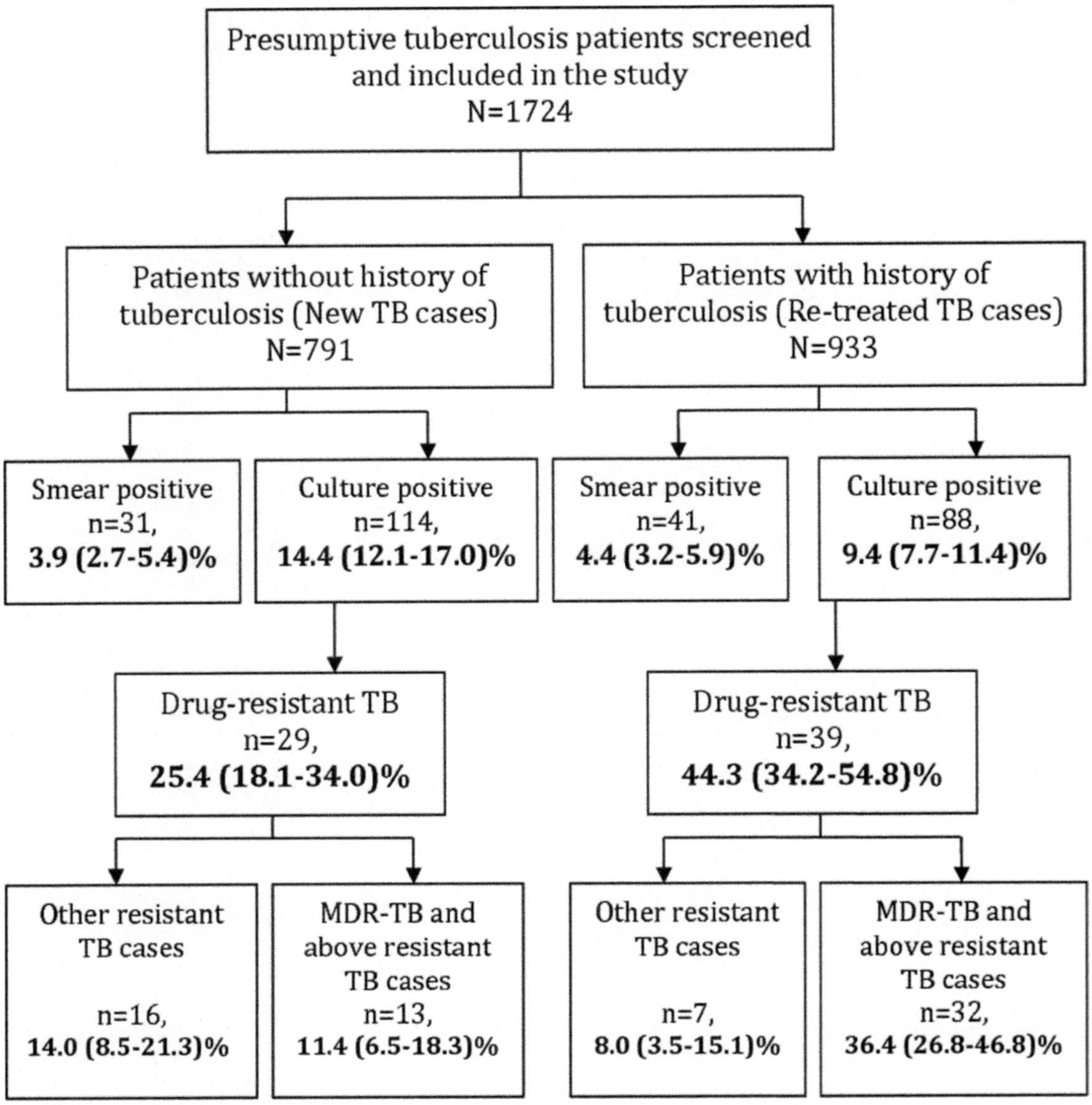

Figure 2. Distribution of Drug-resistant tuberculosis among HIV-infected (new and previously treated) with presumptive tuberculosis patients, Mumbai, India.

Council of Medical Research (ICMR), New Delhi, India. Informed consent was obtained from all study participants.

Study design

This was a cross-sectional survey among HIV-infected adult and paediatric patients attending public and public-private ART clinics in the greater metropolitan Mumbai area. All patients with presumptive pulmonary or extra-pulmonary TB were assessed with smear microscopy and conventional liquid culture. All M. tuberculosis isolates underwent drug susceptibility testing (DST) for first- and second-line anti-TB drugs.

Sample size

The desired sample size was determined separately for new and previously treated culture positive TB cases. Previous tuberculosis treatment was defined as any anti-tuberculosis treatment reported by the patient. Assuming a prevalence of MDR-TB of 3% among new cases and 17% among retreated cases, based on a DST survey conducted in Gujarat [12], a sample size of 123 confirmed new cases and 110 confirmed retreatment cases was sought in order to estimate the prevalence of MDR-TB, with 95% confidence intervals having a margin of error of 3% for new cases and 7% for retreated cases respectively.

Study setting and study population

The study was carried out in five Mumbai District AIDS Control Society (MDACS) ART Centres [1) KEM Hospital, 2) SION Hospital, 3) SION Centre of Excellence in Paediatric HIV Care, 4) Godrej Hospital, Vikhroli and 5) Larsen & Toubro Hospital, Andheri] as well as in two Maharashtra State AIDS Control Society (MSACS) ART Centres [1) Thane Civil Hospital and 2) Navi Mumbai (Vashi) Municipal Corporation Hospital].

All HIV-infected adult and paediatric patients enrolled in the ART centres were potentially eligible to be enrolled in the study, if they had presumptive pulmonary or extra-pulmonary TB based on symptom screening, regardless of the time they were enrolled in the centres or whether they were on ART or not at the time of the

Table 2. Resistance profile (first and second-line) for all drug-resistant tuberculosis patients, Mumbai, India.

Resistance profile (culture-based DST)	Total TB Patients (N = 68), n (%)	New TB patients (N = 29), n (%)	Previously treated TB patients (N = 39), n (%)
H-mono	11 (16.2)	7 (24.1)	4 (10.3)
R-mono	-	-	-
Ofx-mono	1 (1.5)	1 (3.4)	-
Ofx Mfx	3 (4.4)	2 (6.9)	1 (2.6)
HE	7 (10.3)	6 (20.7)	1 (2.6)
HE Eto Ofx Mfx	1 (1.5)	-	1 (2.6)
HR	10 (14.7)	4 (13.8)	6 (15.4)
HRE	6 (8.8)	2 (6.9)	4 (10.3)
HR Eto	5 (7.4)	-	5 (12.8)
HR E Eto	5 (7.4)	2 (6.9)	3 (7.7)
HR Ofx Mfx E	2 (2.9)	2 (6.9)	-
HR Ofx Mfx Eto	1 (1.5)	1 (3.4)	-
HR Ofx Mfx E Lin	1 (1.5)	-	1 (2.6)
HR Ofx Mfx E Eto	8 (11.8)	1 (3.4)	7 (17.9)
HR Ofx Mfx E Eto PAS	2 (2.9)	-	2 (5.1)
HR Ofx Mfx Km Eto	1 (1.5)	-	1 (2.6)
HR Ofx Mfx Km E Eto	2 (2.9)	1 (3.4)	1 (2.6)
HR Ofx Mfx Km Cm E Eto	1 (1.5)	-	1 (2.6)
HR Ofx Mfx Km Cm E Eto PAS	1 (1.5)	-	1 (2.6)

H-isoniazid, R-rifampicin, E-ethambutol, Eto-ethionamide, Km-kanamycin, Cm-capreomycin, Ofx-ofloxacin, Mfx-Moxifloxacin, Lin- Linezolid, PAS- para-aminosalicylic acid.

study. Patients on TB treatment at the time of the study were excluded.

Recruitment and sampling procedure

All HIV-infected ART center attendees were screened by an MSF-employed nurse during the study period. Patients with presumptive TB were investigated using a standard diagnostic algorithm recommended by the World Health Organization [13] that included TB culture and DST. The nurse explained in detail the objectives of the study to the patient and/or caregiver and obtained the signature or thumbprint of the patient if consent was given to participate. When pulmonary TB was presumed, two sputum specimens were collected on the same day, one hour apart, at each study site/hospital laboratory. When extra-pulmonary TB (EPTB) was presumed, biological specimens (fine needle aspirates, pleural fluid, cerebrospinal fluid, etc) were obtained from extra-pulmonary sites. All specimens were transferred to Hinduja Hospital Microbiology Laboratory in Mumbai for culture and first- and second-line DST.

Conventional microscopy with Ziehl-Neelsen (ZN) staining for acid-fast bacilli and further sputum decontamination was performed using the N-acetyl-L-cysteine and sodium hydroxide method. Concentrated sediment was inoculated in one liquid culture tube for testing using the Mycobacterial growth indicator tube (MGIT 960) method. Positive cultures underwent microscopy with ZN staining to confirm cord formation, and speciation with MPT 64 antigen detection by Immunochromatography was carried out to confirm *M. tuberculosis* complex. Specimens fulfilling the above criteria underwent further testing with phenotypic DST using the MGIT System for the following drugs: isoniazid, rifampicin, ethambutol, ofloxacin, moxifloxacin, kanamycin, capreomycin, PAS, ethionamide, clofazimine and linezolid. Non-tuberculous Mycobacteria (NTM) speciation was done by molecular methods using Reverse Line Blot Hybridisation. Hinduja laboratory is quality controlled and has been accredited for first-line DST by the WHO Supranational Reference Laboratory in Bangalore and the College of American Pathologists. The laboratory was also accredited by the TB programme for second-line DST in December 2013; prior to this date, if a strain was suspected to have resistance to one or more second-line anti-TB drugs, it was sent to the National Tuberculosis Institute Laboratory in Bangalore for confirmation.

Multidrug-resistant tuberculosis (MDR-TB) was defined as resistance to both isoniazid and rifampicin; pre-XDR-TB was defined as MDR-TB with additional resistance to either a fluoroquinolone or a second-line injectable agent; and extensively drug-resistant tuberculosis (XDR-TB) was defined as MDR-TB with additional resistance to both a fluoroquinolone and an injectable agent. Extremely drug-resistant tuberculosis (XXDR-TB) was defined as XDR-TB with additional resistance to any group IV and/or group V TB drugs (PAS, ethionamide, clofazimine, linezolid) [13].

Management of those diagnosed with DR-TB

All patients diagnosed with MDR- or XDR-TB were managed in accordance with the national DR-TB treatment guidelines [14], while those with pre-XDR-TB were offered individualized treatment with 4 drugs likely to be effective.

Data collection and analysis

Demographics, clinical and laboratory data, antiretroviral treatment (yes/no) and duration on ART, as well as data on

Table 3. Demographic and clinical factors associated with culture-positive tuberculosis in HIV-infected patients, Mumbai, India.

Explanatory Variable	Patients with tuberculosis (N = 202), n (%)	Patients without tuberculosis (N = 1522), n (%)	Chi-square/t-test (p-value)	aPR[a] (95% CI)
Age [years, median (IQR)]	38.0 (32.0–43.3)	35.0 (22.0–44.0)	8.9 (<0.01)	0.99 (0.99–1.00)
Sex of patients				
Male	138 (13.1)	915 (86.9)	5.0 (0.02)	1.01 (0.99–1.03)
Female	64 (9.5)	607 (90.5)		
Family income per month⁺ (in Rupees)				
Less than 5000	96 (10.9)	788 (89.1)	0.2 (0.65)	
5000 and above	66 (11.6)	502 (88.4)		
ART status				
Pre-ART	72 (21.3)	266 (78.7)	37.3 (<0.01)	1.07 (1.04–1.10)
On ART	130 (9.4)	1256 (90.6)		
CD4 count, last visit* (in cells/µl)				
Less than 200	58 (22.5)	200 (77.5)	34.7 (<0.01)	1.08 (1.05–1.11)
200 and above	127 (9.6)	1197 (90.4)		
ART duration** [months, median (IQR)]	19.3 (5.2–35.7)	27.1 (11.3–47.9)	9.9 (<0.01)	
Previous episode of TB				
Yes	88 (9.4)	845 (90.6)	10.3 (<0.01)	1.01 (0.99–1.03)
No	114 (14.4)	677 (85.6)		

ART: Antiretroviral treatment, IQR: Inter-quartile range, CI; Confidence Intervals ⁺Patients with recorded family income, N = 1452* Patients with available information on CD4, last visit, N = 1582** Patients on ART with available information about ART initiation date, N = 1370[a] aPR; adjusted Prevalence Ratios (calculated by Poisson regression using multiple imputation for CD4 missing data).

previous TB treatment were doubly-entered into an EpiData database (Version 3.1, EpiData Association, Odense, Denmark), validated and analyzed.

To identify factors associated with TB and DR-TB, univariate and multivariate analyses were performed using Poisson and binary logistic regression models. Factors significant ($p = 0.05$) on univariate analysis were entered into the multivariate logistic regression models. Factors were coded as categorical variables and missing values for CD4 cell counts were imputed using a multiple imputation method. Transgender individuals (all were male to female) were grouped with biological males in the models. All factors were entered as a block into multivariate logistic regression models. Data analysis, including multivariate logistic regression models, was conducted with SPSS Version 20.0. Armonk, NY: IBM Corp. Released 2011).

Results

Screening for presumptive TB was carried-out during 14,135 patient visits at seven ART centers in metropolitan Mumbai between March 2013 and January 2014 (Figure 1). Individual patients might have been screened more than once during the study period. A total of 1741 HIV-infected patients with presumptive tuberculosis (TB) were identified. All of them consented to participate in the study and were investigated for drug-resistant TB. The sputum specimens of 17 patients were found insufficient for laboratory investigations and had to be excluded. Thus, 1724 (99%) of the eligible patients were included in the study.

Patient characteristics

The median age of the 1724 patients was 35 (Inter-quartile range, IQR: 24–44) years (Table 1) and the majority (60%) were male. A large proportion (53%) of patients had an average family income between 3500 and 7000 Indian National Rupees (equivalent to 60–120 USD) per month. Most of the patients (98%) had pulmonary TB. Among the entire study cohort, 80% were on ART during the study period and the majority (52%) had CD4 cell counts lower than 500 cells/µL at their last visit to an ART center. The median duration of exposure to ART prior to enrollment in the study was 26 months (IQR: 10.7–47.5). More than half (933/1724) of the presumptive TB patients had had at least one episode of active TB disease in the past.

Culture-positive and drug-resistant tuberculosis

All of the 1724 patients with presumptive TB included in the study (Figure 1) underwent smear, culture and drug susceptibility testing (DST). Of these, 72 (4.2%; 95% Confidence Intervals (CI): 3.3–5.2) patients had smear-positive TB while 202 (11.7%; 95% CI: 10.3–13.3) patients had culture-positive TB. Eleven TB

Table 4. Demographic and clinical factors associated with drug-resistant tuberculosis in HIV-infected tuberculosis patients, Mumbai, India.

Explanatory Variable	Patients with drug-resistant tuberculosis (N = 68), n (%)	Patients without drug-resistant tuberculosis (N = 134), n (%)	Chi-square/t-test (p-value)	aOR[a] (95% CI)
Age [years, median (IQR)]	35.5 (28.5–42.8)	38.0 (33.8–44.0)	2.15 (0.14)	0.98 (0.96–1.01)
Sex of patients				
Male	46 (33.3)	92 (66.7)	0.02 (0.88)	0.95 (0.49–1.82)
Female	22 (34.4)	42 (65.6)		
Family income per month[+] (in Rupees)				
Less than 5000	36 (37.5)	60 (62.5)	1.8 (0.17)	
5000 and above	18 (27.3)	48 (72.7)		
ART status				
Pre-ART	22 (30.6)	50 (69.4)	0.48 (0.49)	0.96 (0.49–1.90)
On ART	46 (35.4)	84 (64.6)		
CD4 count, last visit* (in cells/µl)				
Less than 200	19 (32.8)	39 (67.2)	0.22 (0.88)	0.96 (0.48–1.93)
200 and above	43 (33.9)	84 (66.1)		
ART duration** [months, median (IQR)]	19.3 (5.7–31.3)	19.2 (3.2–43.7)		
Previous episode of TB				
Yes	39 (44.3)	49 (55.7)	7.93 (<0.01)	2.31 (1.24–4.30)
No	29 (25.4)	85 (74.6)		

ART: Antiretroviral treatment, CI; Confidence Intervals [+]Patients with recorded family income, N = 162* Patients with available information about CD4 count, last visit N = 185** Patients on ART with available information about ART initiation date, N = 126[a] aOR; adjusted Odds ratios (calculated by binary logistic regression using multiple imputation for CD4 missing data).

patients were smear-positive but culture negative and 141 patients were culture-positive but smear-negative. Those patients having a history of TB had a higher rate of smear-positivity (4.4% versus 3.9%), but lower culture-positivity rate (9.4% versus 14.4%) as compared to patients without TB history (Figure 2).

Among all culture-positive TB patients, 68 or 33.7% (95% CI: 27.4–40.4) had some form of drug-resistant TB. A high proportion of MDR-TB and pre-XDR-TB, 38% (26/68) and 21% (14/68) respectively, was observed amongst drug-resistant TB patients. Table 2 shows the detailed resistance patterns of all patients with DR-TB.

Of the newly diagnosed (114/791) and previously treated (88/933) culture-positive TB patients, 25.4% (95% CI: 18.1–34.0) and 44.3% (95% CI: 34.2–54.8) patients had drug-resistant TB. The proportion of patients with multidrug-resistant TB and more advanced TB resistance profiles was higher (36% versus 11%) in previously treated patients compared to newly diagnosed TB patients.

Children and extra-pulmonary tuberculosis patients

In the study, 283 children aged less than 15 years were investigated. The median (IQR) age of these children was 11 (8–13) years, just over half of them were male (56%), and sixty-eight percent were on ART. Of the 283 children investigated, 5% (15/283) had culture-positive TB, of whom seven (46.7%, 7/15) had drug-resistant TB; one had polydrug-resistant TB, two had MDR-TB and four had pre-XDR-TB.

Among 1724 HIV-infected patients investigated during the study period, 36 patients had presumptive extra-pulmonary TB (EPTB). The median (IQR) age of these patients was 45 (40–47) years and three-quarters of them were male (27/36, 75%). Among the 36 investigated presumptive EPTB patients, 14% (5/36) patients had culture-positive TB. Of these, 40% (2/5) patients had DR-TB: one had INH mono-resistant TB while another had pre-XDR-TB.

Factors associated with culture-confirmed TB, DR-TB and MDR-TB

The demographic and clinical factors were assessed for association with culture-confirmed TB, DR-TB and MDR-TB. The univariate and bivariate analyses found age, ART status, CD4 count at last visit and previous episode of TB significantly related to culture-positive TB (Table 3). A multivariate Poisson regression model showed that older age, pre-ART status (i.e. not yet on ART), CD4 count less than 200 cells/µL at the last visit and a previous episode of TB were associated with culture-positive TB. None of the factors other than previous history of TB were associated with drug-resistant TB (Table 4) and multi-drug resistant TB (Table 5) in bivariate and multivariate binary logistic regression models.

Table 5. Demographic and clinical factors associated with multidrug-resistant tuberculosis in HIV-infected tuberculosis patients, Mumbai, India.

Explanatory Variable	Patients with multidrug-resistant tuberculosis (N = 45), n (%)	Patients with drug-susceptible tuberculosis (N = 134), n (%)	Chi-square/t-test (p-value)	aOR[a] (95% CI)
Age [years, median (IQR)]	38.0 (30.0–42.5)	38.0 (33.8–44.0)	1.85 (0.18)	0.98 (0.95–1.01)
Sex of patients				
Male	30 (24.6)	92 (75.4)	0.06 (0.80)	0.83 (3.8–1.83)
Female	15 (26.3)	42 (73.7)		
Family income per month⁺ (in Rupees)				
Less than 5000	21 (25.9)	60 (74.1)	1.03 (0.31)	
5000 and above	11 (18.6)	48 (81.4)		
ART status				
Pre-ART	12 (19.4)	50 (80.6)	1.69 (0.19)	0.85 (0.37–1.96)
On ART	33 (28.2)	84 (71.8)		
CD4 count, last visit* (in cells/μl)				
Less than 200	13 (25.0)	39 (75.0)	0.00 (1.00)	1.02 (0.43–2.41)
200 and above	28 (25.0)	84 (75.0)		
ART duration** [months, median (IQR)]	19.7 (5.7–34.8)	19.2 (3.2–43.7)	0.02 (0.88)	
Previous episode of TB				
Yes	32 (39.5)	49 (60.5)	16.2 (<0.01)	4.16 (1.93–8.95)
No	13 (13.2)	85 (86.7)		

ART: Antiretroviral treatment, CI; Confidence Intervals ⁺Patients with recorded family income, N = 140* Patients with available information about CD4 count, last visit N = 164** Patients on ART with available information about ART initiation date, N = 126[a]aOR; adjusted Odds ratios (calculated by binary logistic regression using multiple imputation for CD4 missing data).

Discussion

To our knowledge this is the first DR-TB survey carried out among HIV clinic attendees in India. This study shows that, among HIV-infected children and adults in Mumbai, the burden of drug-resistant tuberculosis is extremely high: almost one in four new TB cases and one in two of those previously treated for TB have a drug-resistant strain. Of just as great concern, a large proportion of these strains was resistant to one or more second-line tuberculosis drugs, especially fluoroquinolones.

The overall rate of culture positivity amongst presumptive TB cases was surprisingly low (11.7%). We hypothesize that this was due neither to limitations in laboratory techniques nor the presence of NTM disease, but instead to the broad inclusion criteria that required a person attending a study site to have just one of four possible TB symptoms as recommended by WHO [13]; a person with 'current cough', for example, who was otherwise stable was eligible for enrolment. Another possible contributor to the low rate of TB culture positivity was the relatively large number of poor quality specimens (e.g. consisting of saliva) despite active instruction being given by a dedicated study nurse at each site. In any case, this finding warrants further investigation.

Even though the overall yield of TB was small in the pediatric cohort as well, it remains significant that almost half of the children with TB were infected with drug-resistant strains, most commonly pre-XDR-TB. Since bacteriological confirmation of DR-TB is more challenging in young children than in adults, as they cannot expectorate sputum and are more likely to have paucibacillary and extra-pulmonary TB, we hypothesize that the burden of TB and DR-TB is likely to be underestimated among children in this study, similar to what has been found in a recent meta-analysis [15]. With less than 2% of all study participants having specimens taken from extrapulmonary sites, it is almost certain that EPTB is being underdiagnosed as well in this cohort. A separate analysis found no significant association between EPTB and DR-TB in children or adults.

Our statistical models revealed no significant associations between most demographic and clinical factors and the risk of DR-TB and MDR-TB. We believe that these findings are important for their lack of associations; it seems that most TB/HIV co-infected patients attending ART centers in Mumbai are at risk for DR-TB. Although the relatively small sample size limits the power of our analyses and calls for cautious interpretation, the lack of associations suggests that all those infected with HIV and presumed to have active TB be tested for drug-resistant strains.

Given the high population density in Mumbai, in which a large proportion of the population lives in slums under extreme poverty, the very high TB prevalence and the relatively high HIV burden reported in greater metropolitan Mumbai, these data are unlikely to be representative of a country as vast and diverse as India. Nevertheless the living conditions in Mumbai and common practices in the public and private health sectors (as for example the prescribing of inappropriate regimens and over-the-counter availability of fluoroquinolones and other drugs with anti-TB properties) are similar to those of other large metropolitan centres in the country, so these data could very well represent the DR-TB situation in such cities as New Delhi, Kolkata and Chennai.

While it may not be possible to generalise our estimates for the entire country or even for HIV-uninfected populations, they serve to highlight the overall magnitude of the DR-TB epidemic in Mumbai, which is not unknown [16,17]. A high prevalence of MDR-TB strains (11–68%) was reported in tertiary health facilities as early as 1991, followed by further documentation in 2006 [18–20], including information on the magnitude of the epidemic in children [21]. A study by D'Souza et al in 2009 [18] documented high levels of multiple drug resistance (both MDR and poly-drug resistance) amongst previously untreated cases in urban parts of Mumbai. In 2011 Udwadia et al reported a cases-series of totally-drug resistant TB (a term that has not officially been endorsed by WHO) in Mumbai, which captured the attention of local and international media [22,23]. However to-date such findings are often overlooked and their importance minimized as representing only selected populations, laboratory or tertiary care settings and small case-series. Our study confirms that there is more than one epidemic ongoing in Mumbai and reinforces the urgent need to accurately measure the overall prevalence and incidence of DR-TB around the country in order to define appropriate interventions. Studies in selected populations such as this complement the overall estimates and can help in directing resources and prioritizing interventions targeted at the most vulnerable groups.

This survey is subject to the usual limitations in survey design and data collection. There is likely to be a tendency for patients to not report previous treatment either because they do not remember (recall limitation) or, on purpose, to avoid going through a long course of treatment that includes daily injectable medication and is known among patients for debilitating side effects [24]. Such bias could have led to an overestimate of DR-TB among new cases and an underestimate among retreatment cases. However, most HIV-infected patients attending ART clinics are usually aware of tuberculosis and have been counseled and screened for TB on several occasions, so recall limitation is rather unlikely.

The majority of HIV-infected patients attending public and public-private ART centers in the city are likely to access the public national TB programme for TB diagnosis and treatment. However many still seek care from private practitioners or may switch between the public and private sectors. The contribution to DR-TB levels from suboptimal treatment regimens prescribed in the unregulated Indian private health sector has been well documented [25–27]. Cox et al in 2007 have shown that even under well-established DOTS programmes in areas with high levels of drug resistance, high levels of amplification of drug resistance are to be expected [28].

The high level of resistance to three or more first-line anti-TB drugs and to fluoroquinolones has been previously described by others [29]. The proportion of previously untreated cases in our study that were resistant to more than three drugs, especially isoniazid, rifampicin and a fluoroquinolone, was particularly alarming and highlights two major issues in the management of TB in the setting of HIV/ART clinics. Firstly, it points to the scenario of nosocomial transmission of TB and DR-TB. Those attending an ART clinic at least once a month are more likely to be exposed to susceptible and resistant strains of *M. tuberculosis* than the general population. Given that the ART centers in Mumbai are usually extremely busy, constantly crowded and that they often lack adequate TB infection control interventions, this scenario is not unlikely. Instead of hypothesizing that most cases of DR-TB are due to non-adherence among patients on treatment, exogenous infection or re-infection should first be considered [30,31]. Secondly, considering the high levels of resistance to second-line TB drugs and especially fluoroquinolones in this population, it is reasonable to assume that patients with presumptive TB may actually have pre-XDR-TB or even XDR-TB. This statement implies a huge investment in laboratory capacity in an already constrained public sector in Mumbai in order to screen all TB patients at the outset for strains that are resistant to fluoroquinolones and anti-TB injectables Nevertheless, we believe that it is a reasonable investment to make if the epidemic of DR-TB is to be controlled in the city in the future. Conversely, if DST is only offered afterward to those failing their TB treatment regimen, a large proportion of DR-TB cases will be missed due to the high risk of mortality among HIV-positive patients with untreated DR-TB [32].

There is an ongoing plan to systematically offer molecular TB diagnosis (mainly using Xpert MTB/RIF, also known as GeneXpert) to all HIV-infected patients in Mumbai and elsewhere in the country. While this is a giant leap forward, since GeneXpert can rapidly detect MTB and rifampicin resistance within 2 hours, we are concerned that 'scale up' of DR-TB diagnosis using this particular diagnostic may lead to suboptimal practices, since a diagnosis of rifampicin resistance alone and/or assumption that it represents a diagnosis of MDR-TB, may mask a diagnosis of pre-XDR or XDR-TB (or worse); the risks then associated with giving a suboptimal treatment regimen are significant both in terms of morbidity and mortality for the patient, as well as amplification of resistance and subsequent community transmission of resistant strains. While GeneXpert is an excellent and efficient diagnostic tool for MTB and screening test for DR-TB, in settings like Mumbai it is essential that it be complemented by culture and DST involving first- and second-line anti-TB drugs. The national programme has recently changed the policy to account for this risk, starting with HIV-infected patients in Mumbai and Maharashtra.

Our initial study protocol included fingerprinting studies using spoligotyping, which we had to abandon due to the high cost. Cox et al have in the past found a strong association between the Beijing genotype and amplification in situations of preexisting resistance in a central Asian setting [33]. Similarly, the proportion of the Beijing genotype was reported to be 35% in the urban Mumbai population studied by Almeida et al [34]. We need fingerprinting studies to establish how often nosocomial transmission occurs and to guide TB infection control interventions. Another area of research that is urgently needed relates to chemoprophylaxis for child contacts of DR-TB cases in Mumbai; preventative regimens that have shown to be effective in other settings are unlikely to prevent development of active disease in many children in Mumbai due to the high baseline rate of fluoroquinolone resistance [35].

Conclusion

Our findings strongly suggest that there is an ongoing DR-TB epidemic among people living with HIV and attending ART centers in Mumbai, which requires urgent, innovative and feasible models of care that allow for rapid and accurate detection and treatment of as many DR-TB patients as possible. Ideally all patients with presumptive TB attending any ART center in Mumbai, or settings with similar drug resistance patterns, should be screened with a rapid molecular diagnostic followed by DST to first- and second-line anti-TB drugs, including for fluoroquinolones, so that the correct diagnosis is made as early as possible and followed by prompt treatment initiation with an appropriate individualized regimen. The high rate of DR-TB amongst new TB patients also highlights the need for better TB infection control measures in order to prevent ongoing transmission of DR-TB in the community and health facilities, especially those attended by vulnerable populations, such as those living with HIV.

Acknowledgments

The authors wish to acknowledge the strong collaboration of hospital responsibles, heads of medical departments and ART centres and all staff from the seven ART Centers involved in this study. We thank the MSF nurses working within the 7 ART centres: Ujwala Pongurlekar, Prachi Salunke, Sonakshi Jadhav, Mansi Kulkarni, Jobin Joseph, Albert Thomas, and Louis Fernandes. We also thank Meenakshi Iyer for laboratory analyses and data entry. Finally, we acknowledge the participation within this study of the patients suffering from HIV and tuberculosis in Mumbai.

Author Contributions

Conceived and designed the experiments: PI. Performed the experiments: CP CR MM AS. Analyzed the data: PI MD. Contributed reagents/materials/analysis tools: AMVK MK AB BA MM MP AK BBR AD LR PS KSS. Contributed to the writing of the manuscript: PI MD PS. Critically reviewed several versions of the manuscript: AMVK MK AB BA MM MP AK BBR AD LR PS KSS CP AS.

References

1. World Health Organization (WHO) (2013), Global tuberculosis report 2013. WHO Press, Geneva, WHO/HTM/TB/2013.11.
2. Department of AIDS Control (2013), National AIDS Control Organization, Annual Report 2012–2013, Ministry of Health & Family Welfare, Government of India.
3. Paramasivan CN, Venkataraman P (2004) Drug resistance in tuberculosis in India. Indian J Med Res; 120: 377–386.
4. Deivanayagam CN, Rajasekaran S, Venkatesan R, Mahilmaran A, Ahmed PR, et al. (2002) Prevalence of acquired MDR TB and HIV co-infection. Indian J Chest Dis Allied Sci 44: 237–242.
5. Williams BG, Granich R, Chauhan LS, Dharmshaktu NS, Dye C (2005) The impact of HIV/AIDS on the control of tuberculosis in India. Proc Natl Acad Sci U S A 102: 9619–9624.
6. Swaminathan S, Paramasivan CN, Ponnuraja C, Iliayas S, Rajasekeran S (2005) Anti-tuberculosis drug resistance in patients with HIV and tuberculosis in South India. Int J Tuberc Lung Dis 9: 896–900.
7. Maniar JK, Kanuth RR, Mandalia S, Shah K, Maniar A (2006) HIV and tuberculosis: partners in crime. Indian J Dermatol Venereol Leprol 72: 276–82.
8. Pereira M, Tripathy S, Inamdar V, Ramesh K, Bhavsar M, et al. (2005) Drug resistance pattern of Mycobacterium tuberculosis in seropositive and seronegative HIV-TB patients in Pune, India. Indian J Med Res 121: 235–239.
9. Sethi S, Mewara A, Dhatwalia SK, Singh H, Yadav R, et al. (2013) Prevalence of multidrug resistance in *Mycobacterium tuberculosis* isolates from HIV seropositive and seronegative patients with pulmonary tuberculosis in north India. BMC Infect Dis 1471–2334/13/137.
10. Menon S, Dharmshale S, Chande C, Gohil A, Lilani S, et al. (2012) Drug resistance profiles of *Mycobacterium tuberculosis* isolates to first line anti-tuberculous drugs: a five years study. Lung India 29: 227–231.
11. Kumar P, Balooni V, Sharma BK, Kapil V, Sachdeva KS, et al. (2014) High degree of multi-drug resistance and hetero-resistance in pulmonary TB patients from Punjab state of India. Tuberculosis (Edinb) 94(1): 73–80.
12. Ramachandran R, Nalini S, Chandrasekar V, Dave PV, Sanghvi AS, et al. (2009) Surveillance of drug-resistant tuberculosis in the state of Gujarat, India Int J Tuberc Lung Dis 13(9): 1154–1160.
13. World Health Organization (2011) Guidelines for the programmatic management of drug-resistant tuberculosis. 2011 update. Geneva, Switzerland: WHO.
14. Central TB Division (2012) Programmatic Management for Drug-resistant Tuberculosis guidelines-May version, Directorate General of Heath Services, Ministry of Health and Family Welfare. Available: http://www.tbcindia.nic.in/pdfs/Guidelines%20for%20PMDT%20in%20India%20-%20May%202012.pdf. Accessed 2014 May 5.
15. Jenkins HE, Tolman AW, Yuen CM, Parr JB, Keshavjee S, et al. (2014) Incidence of multidrug-resistant tuberculosis disease in children: systematic review and global estimates. Lancet 383(9928): 1572–1579.
16. Almeida D, Rodrigues C, Udwadia ZF, Lalvani A, Gothi GD, et al. (2003) Incidence of multidrug-resistant tuberculosis in urban and rural India and implications for prevention. Clin Infect Dis 36: e152–4.
17. Singh S, Sankar MM, Gopinath K (2007) High rate of extensively drug-resistant tuberculosis in Indian AIDS patients. AIDS 21(17): 2345–7.
18. D'souza DT, Mistry NF, Vira TS, Dholakia Y, Hoffner S, et al. (2009) High levels of multidrug resistant tuberculosis in new and treatment-failure patients from the Revised National Tuberculosis Control Programme in an urban metropolis (Mumbai) in Western India. BMC Public Health 29: 9–211.
19. Rodrigues C, Shenai S, Sadani M, Thakkar P, Sodha A, et al. (2006) Multi drug-resistant tuberculosis in Mumbai: it's only getting worse. Int J Tuberc Lung Dis 10(12): 1421–1422.
20. Chowgule RV, D Lina (1998) Pattern of secondary acquired drug resistance to antituberculosis drugs in Mumbai, India 1991–1995. Ind J Chest Dis Allied Sciences 40: 23–31.
21. Karande S, Bavdekar SB (2002) Children and multidrug-resistant tuberculosis in Mumbai (Bombay), India. Emerg Infect Dis 8(11): 1360–1361.
22. Udwadia ZF, Amale RA, Ajbani KK, Rodrigues C (2011) Totally Drug-Resistant Tuberculosis in India. Clin Infect Dis. doi:10.1093/cid/cir889.
23. TIME Magazine (2013). Contagion; Why drug-resistant tuberculosis threatens us all. March 4, 2013.
24. Isaakidis P, Rangan S, Pradhan A, Ladomirska J, Reid T, et al. (2013) 'I cry every day': experiences of patients co-infected with HIV and multidrug-resistant tuberculosis. Trop Med & Int Health 18(9): 1128–1133.
25. Uplekar M, Juvekar S, Morankar S, Rangan S, Nunn P (1998) Tuberculosis patients and practitioners in private clinics in India. Int J Tuberc Lung Dis 2(4): 324–329.
26. Bhargava A, Pinto L, Pai M (2011) Mismanagement of tuberculosis in India: Causes, consequences, and the way forward. Hypothesis 9(1): e7.
27. Udwadia ZF, Pinto LM, Uplekar MW (2010) Tuberculosis management by private practitioners in Mumbai, India: has anything changed in two decades? PLoS One 5: e12023.
28. Cox HS, Niemann S, Ismailov G, Doshetov D, Orozco JD, et al. (2007) Risk of acquired drug resistance during short-course directly observed treatment of tuberculosis in an area with high levels of drug resistance. Clin Infect Dis 441421–1427.
29. Agrawal D, Udwadia ZF, Rodriguez C, Mehta A (2009) Increasing incidence of fluoroquinolone-resistant Mycobacterium tuberculosis in Mumbai, India. Int J Tuberc Lung Dis 13(1): 79–83.
30. Andrews JR, Gandhi NR, Moodley P, Shah NS, Bohlken L, et al. (2008) Exogenous reinfection as a cause of multidrug-resistant and extensively drug-resistant tuberculosis in rural South Africa. JID 198: 1582–1589.
31. March F, Garriga X, Rodriguez P, Moreno C, Garrigo M, et al. (1997) Acquired drug resistance in Mycobacterium tuberculosis isolates recovered from compliant patients with human immunodeficiency virus-associated tuberculosis. Clin Infect Dis 25: 1044–1047.
32. Gandhi NR, Shah NS, Andrews JR, Vella V, Moll AP, et al. (2010) HIV coinfection in multidrug- and extensively drug-resistant tuberculosis results in high early mortality. Am J Respir Crit Care Med 181: 80–86.
33. Cox HS, Kubica T, Doshetov D, Kebede Y, Rüsch-Gerdess S, et al. (2005). The Beijing genotype and drug resistant tuberculosis in the Aral Sea region of Central Asia. Respiratory research, 6(1): 134.
34. Almeida D, Rodrigues C, Ashavaid TF, Lalvani A, Udwadia ZF, et al. (2005). High incidence of the Beijing genotype among multidrug-resistant isolates of Mycobacterium tuberculosis in a tertiary care center in Mumbai, India. CID, 40(6): 881–886.
35. Seddon JA, Hesseling AC, Finlayson H, Fielding K, Cox H, et al. (2013) Preventive therapy for child contacts of multidrug-resistant tuberculosis: a prospective cohort study. CIDcit655.

10

Acinetobacter Peritoneal Dialysis Peritonitis: A Changing Landscape over Time

Chia-Ter Chao[1,2], Szu-Ying Lee[3], Wei-Shun Yang[4], Huei-Wen Chen[5], Cheng-Chung Fang[6], Chung-Jen Yen[7], Chih-Kang Chiang[8], Kuan-Yu Hung[5], Jenq-Wen Huang[5]*

1 Division of Nephrology, Department of Internal Medicine, National Taiwan University Hospital Jin-Shan Branch, New Taipei City, Taiwan, **2** Graduate Institute of Toxicology, National Taiwan University, Taipei, Taiwan, **3** Division of Nephrology, Department of Internal Medicine, National Taiwan University Hospital Yun-Lin Branch, Yun-Lin County, Taiwan, **4** Division of Nephrology, Department of Internal Medicine, National Taiwan University Hospital Hsin-Chu Branch, Hsin-Chu County, Taiwan, **5** Division of Nephrology, Department of Internal Medicine, National Taiwan University Hospital, Taipei, Taiwan, **6** Department of Emergency Medicine, National Taiwan University Hospital, Taipei, Taiwan, **7** Department of Geriatrics and Gerontology, National Taiwan University Hospital, Taipei, Taiwan, **8** Department of Integrative diagnostics and Therapeutics, National Taiwan University Hospital, Taipei, Taiwan

Abstract

Background: *Acinetobacter* species are assuming an increasingly important role in modern medicine, with their persistent presence in health-care settings and antibiotic resistance. However, clinical reports addressing this issue in patients with peritoneal dialysis (PD) peritonitis are rare.

Methods: All PD peritonitis episodes caused by *Acinetobacter* that occurred between 1985 and 2012 at a single centre were retrospectively reviewed. Clinical features, microbiological data, and outcomes were analysed, with stratifications based upon temporal periods (before and after 2000).

Results: *Acinetobacter* species were responsible for 26 PD peritonitis episodes (3.5% of all episodes) in 25 patients. *A. baumannii* was the most common pathogen (54%), followed by *A. lwoffii* (35%), with the former being predominant after 2000. Significantly more episodes resulted from breaks in exchange sterility after 2000, while those from exit site infections decreased ($P = 0.01$). The interval between the last and current peritonitis episodes lengthened significantly after 2000 (5 vs. 13.6 months; $P = 0.05$). All the isolates were susceptible to cefepime, fluoroquinolone, and aminoglycosides, with a low ceftazidime resistance rate (16%). Nearly half of the patients (46%) required hospitalisation for their *Acinetobacter* PD-associated peritonitis, and 27% required an antibiotic switch. The overall outcome was fair, with no mortality and a 12% technique failure rate, without obvious interval differences.

Conclusions: The temporal change in the microbiology and origin of *Acinetobacter* PD-associated peritonitis in our cohort suggested an important evolutional trend. Appropriate measures, including technique re-education and sterility maintenance, should be taken to decrease the *Acinetobacter* peritonitis incidence in PD patients.

Editor: Shamala Devi Sekaran, University of Malaya, Malaysia

Funding: This study is financially supported by National Taiwan University Hospital intramural fund (project NO. 102-N2249). The funders had no role in study design, data collection and analysis, decision to publish, or preparation of the manuscript.

Competing Interests: The authors have declared that no competing interests exist.

* Email: 007378@ntuh.gov.tw

Introduction

The utilisation of peritoneal dialysis (PD) in end-stage renal disease (ESRD) has been long purported for its economical advantage and similar effects on patient outcomes compared with haemodialysis (HD) [1,2]. However, the underuse of PD in most countries could be attributable to several factors, including the infectious complications [1,3]. Among these complications, PD-associated peritonitis is the most dreadful, accounting for 8% to 21% of infection-related mortality and 30% of technique failures [4,5]. Gram-negative bacteria (GNB) cause 25% to 40% of PD peritonitis episodes worldwide, secondary to gram-positive bacteria, and the number is still increasing [6,7].

Among GNB peritonitis, the acronym SPICE (*Serratia*, *Providencia*, indole-positive *Proteus/Acinetobacter/Morganella*, *Citrobacter*, *Enterobacter*, or *Hafnia*) denotes a spectrum of organisms with an intrinsic tendency for antibiotic resistance [8]. These pathogens demonstrate inducible beta-lactamase during initial therapy, potentially leading to empirical regimen failure, prolonged peritoneal damage, and worse outcomes [9]. In the ANZDATA registry, the SPICE group of pathogens accounts for 26.9% of all non-*Pseudomonas* GNB PD-associated peritonitis, and *Acinetobacter* species are responsible for <20% of these cases [10,11]. Intuitively thinking, patients with *Acinetobacter* PD-associated peritonitis might have poorer prognoses compared with the other Enterobacteriaceae members, owing to their high antibiotic resistance rates. However, past reports suggested just the opposite, with technique failure rates of 9% [12] and 9.2% [13] in several studies. In addition, several researchers proposed that *Acinetobacter* peritonitis might occur with an immunocompromised status [12,14]. These reports were all published nearly 2 decades ago, and the temporal changes of clinical characteristics, microbiologic features, and outcomes of patients with *Acinetobacter* PD peritonitis are unknown. Utilising a longitudinal PD registry of more than 2 decades in a single centre, we conducted the present study to explore the transition of these parameters over different time periods.

Subjects and Methods

Ethical considerations

The ethics committee of the NTUH approved the current study (No. 201212165RINC). The local institutional review board did not mandate patient consent, as no interventions were performed and patient privacy was not breached. The local institutional review board waived the need for written informed consent from the participants.

Study design and setting

In 1985, the National Taiwan University Hospital (NTUH) PD program was established [9,15]. Patients with ESRD who underwent PD for >3 months were eligible for the present study. Patients with episodes of culture-confirmed *Acinetobacter* species PD peritonitis between 1985 and 2012 were selected. Peritonitis was diagnosed according to the presence of symptomatology and cloudy effluent, with a white blood cell count >100/μL and neutrophil percentage >50% [5].

Clinical data collection

All enrolees' demographic profiles, including age, sex, and comorbidities (diabetes mellitus [DM], heart failure [HF], and autoimmune diseases), were reviewed [16,17]. The aetiologies of their ESRD were also documented. Through a chart review, patients' medications, including steroid, immunosuppressant, and antibiotic use, within 1 month before the selected peritonitis episodes were recorded, along with laboratory data.

For each peritonitis episode caused by *Acinetobacter*, we documented the PD vintage and modality (continuous ambulatory peritoneal dialysis [CAPD] or automated peritoneal dialysis [APD]) and the initial symptoms. Laboratory data during the peritonitis episodes were obtained. Pathogen species and the corresponding antibiogram were identified. The presumptive origins of each peritonitis episode were divided into a break in sterility during the exchange procedure, gastrointestinal (GI) microflora transmural migration, exit site infection/tunnel infection, or undetermined, according to past reports [18,19]. Empirical antibiotic regimens for PD peritonitis in our institute were as follows: first-generation cephalosporin with aminoglycoside before 1998 and first-generation cephalosporin with third-generation cephalosporin after 1998, according to the recommendations of the International Society for Peritoneal Dialysis (ISPD), unless otherwise indicated [5,9,15,20,21].

Outcome variables

The outcome measures in the present study consisted of primary response, antibiotic switch (secondary response), relapse peritonitis, repeat peritonitis, Tenckhoff catheter removal, and patient death. The primary response was defined as symptomatic improvement of the patient within 3 days of receiving empirical antibiotics, accompanied by effluent leukocyte count <100/μL. The secondary response was defined as a response to the second-line antibiotics when effluent did not turn clear after first-line antibiotic use [9,15]. If an antibiotic switch occurred, the duration of empirical antibiotic use was also recorded. Repeat peritonitis was defined as peritonitis recurring 4 weeks after the previous episode, whereas relapse peritonitis was defined as peritonitis recurring within 4 weeks after the treatment of the previous episode, both involving the same pathogen [5,21]. Peritonitis-related death was considered if the patient's death occurred within 1 month of the peritonitis episode and was attributable to peritonitis [9,15].

Results

During the study period, 369 of 1278 chronic PD patients developed 744 episodes of PD-related peritonitis over 40,499 patient-months, with an overall peritonitis incidence of 1 per 54 patient-months (before 2000, 1 per 37 patient-months; after 2000, 1 per 64 patient-months). Among the 744 episodes of PD-related peritonitis, 25 patients developed 26 peritonitis episodes from *Acinetobacter* species (3.5% of all episodes), with an incidence of 1 per 1558 patient-months. One patient developed 2 separate episodes of *Acinetobacter* peritonitis, and the remaining patients each experienced 1 episode. To delineate the temporal changes of their clinical features, we divided the *Acinetobacter* peritonitis cohort into 2 phases, those that occurred before 2000 and those that occurred during or after 2000.

Clinical features of Acinetobacter PD-associated peritonitis

The demographic profiles and clinical characteristics of the patients who developed PD-related peritonitis caused by *Acinetobacter* are listed in Table 1, according to the time of onset. Overall, the mean age and PD vintage of the patients with *Acinetobacter* peritonitis were 52 years and 29 months, respectively, without differences between phases. Three-fourths of the patients chose CAPD as their dialysis modality. One-third of *Acinetobacter* peritonitis patients had DM, but only 12% of patients had HF.

The most common aetiologies of ESRD in the *Acinetobacter* peritonitis patients included DM (31%) and chronic glomerulonephritis (31%), followed by lupus (4%) and hypertension (4%). No significant difference was found between those that occurred before 2000 and those that occurred after 2000. Only 1 patient had undergone intra-abdominal surgery (cholecystitis with open cholecystectomy) before developing *Acinetobacter* peritonitis. Another patient received an antibiotic within the month before the current peritonitis episode (cephalexin for cellulitis for 1 week), and still another lupus patient was receiving prednisolone 10 mg daily.

Table 2 shows the relevant features of the current *Acinetobacter* peritonitis episodes. The current peritonitis episode represented the first peritonitis episode in 62% of the patients, and for those who had already developed peritonitis, there were on average 2 episodes before the *Acinetobacter* episodes (Table 2). No significant difference in past peritonitis episode counts was observed between those that occurred before 2000 and those that occurred after 2000, but it was interesting to note that the time from the last peritonitis episode was borderline significantly longer in the after-2000 group ($P = 0.07$; Table 2). To verify this result, we further constructed a diagram comparing latency between the last peritonitis and *Acinetobacter* episodes within the historic cohorts and ours. We observed a temporal evolution of this feature, with lengthening of such latency (Figure 1; $P<0.01$).

The presenting symptoms of PD-related peritonitis caused by *Acinetobacter* were abdominal pain (85%), followed by nausea/vomiting (31%), fever (27%), and diarrhoea (15%). All the patients exhibited a turbid dialysate. The mean effluent leukocyte count

Table 1. Baseline features of patients with *Acinetobacter* PD-associated peritonitis.

Characteristics	*Total (n = 26)*	*Before 2000 (n = 12)*	*After 2000 (n = 14)*	*P value**
Incidence (one episode per patient-months)	1558	805	2053	<0.01
Age (years)	52 (23–84)	48 (37–71)	55 (23–84)	0.11
Gender (male %)	11 (42)	4 (33)	7 (50)	0.32
Vintage (months)	29 (1–80)	27 (1–80)	31 (3–60)	0.45
Modality (CAPD %)	19 (73)	10 (83)	9 (64)	0.29
Comorbidities				
DM	8 (31)	2 (16)	6 (43)	0.12
Heart failure	3 (12)	3 (25)	0 (0)	0.08
Autoimmune disorders	2 (8)	1 (8)	1 (7)	0.96
ESRD aetiology				0.15
DM	8 (31)	2 (16)	6 (43)	
CGN	8 (31)	4 (33)	4 (29)	
Hypertension	1 (4)	0 (0)	1 (7)	
Lupus	1 (4)	1 (8)	0 (0)	
Congenital renal dysplasia	1 (4)	0 (0)	1 (7)	
Unknown	7 (23)	5 (42)	2 (7)	
Laboratory data				
Albumin (mg/dL)	3.7 (2.5–4.2)	3.6 (2.5–4.2)	3.7 (2.6–4.2)	0.36
Hemoglobin (g/dL)	9.6 (6.3–13.8)	8.4 (6.3–13.8)	10.6 (8.4–11.7)	0.01
Creatinine (mg/dL)	11 (7.1–16.1)	12.2 (8.8–16.1)	10 (7.1–14.4)	0.07
Total cholesterol (mg/dL)	187.8 (113–321)	198 (124–321)	179.1 (113–267)	0.38

*Comparison between before-2000 group and after-2000 group.
Continuous variables are expressed in mean (ranges), while categorical variables are expressed in number (percentage in parentheses).
Abbreviations: CAPD, continuous ambulatory peritoneal dialysis; CGN, chronic glomerulonephritis; DM, diabetes mellitus; ESRD, end-stage renal disease.

was 5296/μL (range, 148–13,300/μL), with 95% neutrophils (range, 90–98%). The initial blood leukocyte levels were 8094/μL (range, 3400–15,930/μL), with 82% neutrophils (range, 55–94%). The most common species from the effluent was *Acinetobacter baumannii* (54%), followed by *A. iwoffii* (35%), *A. ursingii* (4%), and *A. junii* (4%). *A. baumannii* was the predominant pathogen after 2000, while *A. iwoffii* outnumbered the others before 2000 ($P = 0.01$; Table 2). Two episodes (8%) showed polymicrobial growth (both involved *A. baumannii* and occurred after 2000). Coagulase-negative staphylococci and *Acremonium* species were each responsible for 1 episode of polymicrobial growth. None of the patients had concomitant *Acinetobacter* bacteremia.

The most common identifiable causes of PD-related peritonitis caused by *Acinetobacter* were a break in exchange sterility (19%) and translocation of GI microflora (19%), followed by exit site infection/tunnel infection (8%; Table 2). Approximately half of the episodes did not have an identifiable origin. It is interesting to speculate that an exchange sterility break emerged as the predominant origin of *Acinetobacter* peritonitis after 2000, while the proportions of causes related to exit site infection/tunnel infection and unidentified peritonitis declined significantly after 2000 ($P<0.01$).

The antibiotic susceptibilities of the *Acinetobacter* species are shown in Table 2. All the identified *Acinetobacter* species were susceptible to aminoglycosides (including gentamicin and amikacin), fluoroquinolones, and a fourth-generation cephalosporin (cefepime) after 1995. Approximately 16% of *Acinetobacter* species were non-susceptible to ceftazidime, among which three-fourths were resistant. In addition, 8% of isolates showed intermediate resistance to piperacillin, while 4% were resistant to ticarcillin. Intraperitoneal cefazolin/ceftazidime was initially used in 12 episodes (46%), and intraperitoneal cefazolin/aminoglycosides were used in another 12 episodes (46%; all before 1998). Two patients (8%) were initially treated with intravenous vancomycin/ceftazidime for septic shock. Antibiotics were shifted to intravenous ceftazidime and intraperitoneal gentamicin in 1 patient and intraperitoneal ceftazidime/gentamicin in the other after the patients stabilised with fair responses.

Outcomes of Acinetobacter PD-associated peritonitis

Table 3 lists the clinical outcomes of the patients with PD-related peritonitis caused by *Acinetobacter*. A primary response was achieved in 16 episodes (62%), and after antibiotics, a secondary response was obtained in another 7 episodes (27%). There was no difference between time groups regarding antibiotic response percentage. The average time to antibiotic switch was 4 days (range, 2–9 days), and episodes that occurred after 2000 presented a significantly shorter time to antibiotic switch ($P<0.01$). Those responding to antibiotics were maintained on susceptible antibiotics for an average of 17 days (range, 12–21 days). About half of the episodes (46%) required hospitalisation, with more patients in the after-2000 group admitted ($P = 0.01$) but no significant difference in hospitalisation duration. A total of 3 patients (12%) with PD-related peritonitis caused by *Acinetobacter* had technique failure with catheter removal. The overall outcome was fair. None of the patients with *Acinetobacter* PD-associated

Table 2. Features of *Acinetobacter* PD-associated peritonitis episodes.

Clinical features	*Total (n = 26)*	*Before 2000 (n = 12)*	*After 2000 (n = 14)*	*P value**
Parameters regarding previous episodes				
Previous peritonitis episodes numbers (if present)	2.1 (n = 10)	2 (n = 5)	2.2 (n = 5)	0.77
Time to last peritonitis (months)	9.3 (1–27)	5 (1–13)	13.6 (3–27)	0.05
Most recent PD peritonitis pathogen	Culture-negative (6) Coagulase-negative staphylococci (2) *Enterococcus fecalis* (1) MRSA (1)	Culture-negative (4) Coagulase-negative staphylococci (1)	Culture-negative (2) Coagulase-negative staphylococci (1) *Enterococcus fecalis* (1) MRSA (1)	
Parameters of current episode				
Pathogen				0.01
Acinetobacter baumannii	14 (54)	5 (42)	9 (64)	
Acinetobacter iwoffii	9 (35)	6 (50)	3 (21)	
Acinetobacter ursingii	1 (4)	0 (0)	1 (7)	
Acinetobacter junii	1 (4)	1 (8)	0 (0)	
Unspecified *Acinetobacter*	1 (4)	0 (0)	1 (7)	
Presumed peritonitis origin				<0.01
Break in exchange sterility	5 (19)	0 (0)	5 (36)	
Gastrointestinal bacterial translocation	5 (19)	1 (8)	4 (29)	
Exit site infection or tunnel infection	2 (8)	2 (16)	0 (0)	
Un-identified	14 (54)	9 (75)	5 (36)	
Antibiotic susceptibilities				
Aminoglycosides (gentamicin/amikacin)	**S** 26 (100)	**S** 26 (100)	**S** 26 (100)	
Ceftazidime	**S** 22 (84)	**S** 9 (76)	**S** 13 (93)	
	I 1 (4)	**I** 1 (8)	**I** 0 (0)	
	R 3 (12)	**R** 2 (16)	**R** 1 (7)	
Cefepime	**S** 26 (100)	**S** 26 (100)	**S** 26 (100)	
Ticarcillin	**S** 24 (92)	**S** 10 (84)	**S** 26 (100)	
	I 1 (4)	**I** 1 (8)		
	R 1 (4)	**R** 1 (8)		
Piperacillin	**S** 24 (92)	**S** 10 (84)	**S** 26 (100)	
	I 2 (8)	**I** 2 (16)		
Fluoroquinolone (ciprofloxacin, levofloxacin)	**S** 26 (100)	**S** 26 (100)	**S** 26 (100)	
Carbapenem	**S** 26 (100)	**S** 26 (100)	**S** 26 (100)	

*Comparison between before-2000 group and after-2000 group.
Continuous variables are expressed in mean (ranges), while categorical variables are expressed in number (percentage in parentheses).
Abbreviations: MRSA, methicillin-resistant *Staphylococcus aureus*; PD, peritoneal dialysis.

peritonitis developed relapse or repeat peritonitis, and there was no mortality from such peritonitis.

Discussion

In the present study of *Acinetobacter* peritonitis episodes over more than 20 years, we found that *A. baumannii* was the most common species identified from the effluent, and the leading causes of infection were sterility break and GI microflora translocation. Nearly half required hospitalisation, but the outcome was good, with only 12% of technique failure and no mortality cases. In addition, temporal differences in the clinical features, including the interval between the last and current peritonitis episodes, increased significantly after 2000, along with more episodes from sterility break and less from exit site infection/tunnel infection. However, the fair outcome of *Acinetobacter* peritonitis surprisingly did not change with time.

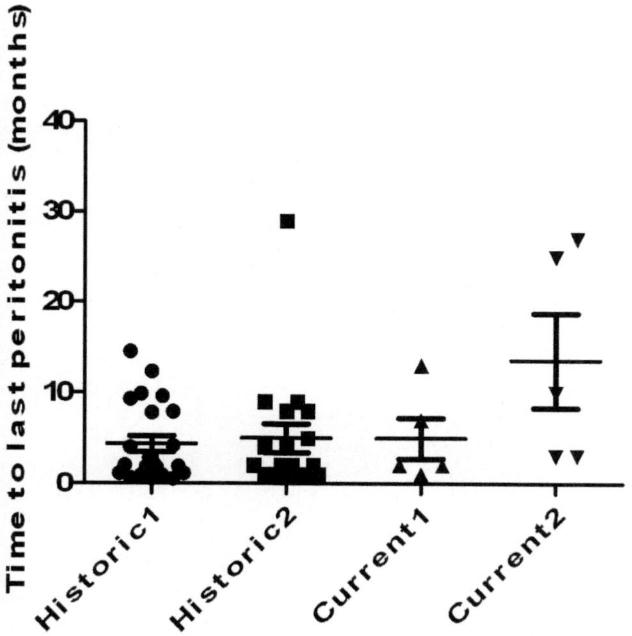

Figure 1. Diagram plotting the distribution of time between the last and current peritonitis episodes in different reports. Our contemporary cohort demonstrated a significantly longer latency compared with the others ($P<0.01$ between the after-2000 cohort and the historic 1 and 2 cohorts). Historic 1 cohort, data derived from *Am J Kidney Dis* 1989; 14(2): 101–4 Historic 2 cohort, data derived from *Perit Dial Int* 1994; 14(2): 174–7.

Compared with past reports of *Acinetobacter* peritonitis, our patients (before 2000) were of similar age and carried a similar percentage of comorbidities (DM and autoimmune disorders; Tables 1 and 4). However, the mean age of the patients with *Acinetobacter* peritonitis seemed to be younger than the others with peritonitis from different pathogens (mean age: historic cohort, 41 years; our cohort, 52 years; ANZDATA, 61–64 years) [7,10]. This might account for the fact that these patients manifested a milder comorbidity profile than the others with peritonitis (Table 1).

Acinetobacter species are strictly aerobic, pleomorphic, and non-motile GNB. To date, 26 named species and 9 genomic species have been identified, among which *A. baumannii* and *Acinetobacter* genomic species 3 and 13TU are collectively referred to as the *A. baumannii* complex [22,23]. Towner and colleagues identified 3 major overlapping populations of *Acinetobacter* species, each with different epidemiologic and microbiologic features [24]. First, *A. baumannii* complex species are prone to develop antibiotic resistance and frequently cause nosocomial outbreaks through their persistent presence in health-care facilities. Second, *A. iwoffii* and *A. johnsonii* are antibiotic-susceptible skin commensal flora of animals and humans, mostly resulting in food spoilage. Finally, *A. calcoaceticus* and other *Acinetobacter* members are ubiquitous colonisers in soil and wastewater, with low virulence and high antibiotic sensitivity. In our PD peritonitis series, the predominant pathogen differed with time; *A. iwoffii* was the predominant offender before 2000 but was replaced by *A. baumannii* after 2000 (Table 2). This seems to suggest that the ecology of *Acinetobacter* breaching the peritoneum has changed temporally, from common skin flora (*A. iwoffii*) to colonisers of health-care settings (*A. baumannii*). This transition might reflect the improvement of exchange systems and increased utilisation of health-care resources as time passed.

Galvao *et al.* proposed that *Acinetobacter* PD peritonitis often occurred during a "vulnerable period, when hosts were immunocompromised," especially in the months immediately following another PD peritonitis episode [12]. This theory was subsequently embraced by other researchers, and the average period between the last peritonitis and *Acinetobacter* episodes was 4 to 5 months at that time (Table 4) [13,14]. However, from our findings (Figure 1), the formerly recognised susceptible period of developing *Acinetobacter* peritonitis seemed vanishing in the modern PD era, and now most episodes did not follow the previous peritonitis as closely as they did in the past. Furthermore, the proportion of PD peritonitis cases from sterility breaks in our cohort also increased significantly after 2000, with a decrease in peritonitis originating

Table 3. Outcomes of *Acinetobacter* PD-associated peritonitis.

Variables	*Total (n = 26)*	*Before 2000 (n = 12)*	*After 2000 (n = 14)*	*P value**
Antibiotics				
Primary response	16 (62)	7 (58)	9 (64)	0.76
Secondary response	7 (27)	3 (25)	4 (29)	0.62
Time to antibiotic switch (days)	4 (2–9)	6 (2–14)	3 (3–5)	<0.01
Total antibiotic duration (days)	17 (12–21)	16 (12–21)	18 (14–21)	0.28
Hospitalization	12 (46)	3 (25)	9 (64)	0.01
Length of stay	13 (5–23)	13 (10–14)	14 (5–23)	0.77
Tenckhoff catheter removal	3 (12)	2 (17)	1 (7)	0.47
Relapse peritonitis	0 (0)	0 (0)	0 (0)	
Repeat peritonitis	0 (0)	0 (0)	0 (0)	
Mortality	0 (0)	0 (0)	0 (0)	

*Comparison between before-2000 group and after-2000 group.
Continuous variables are expressed in mean (ranges), while categorical variables are expressed in number (percentage in parentheses).
Abbreviations: PD, peritoneal dialysis.

from unidentified sources (Tables 2 and 4). We proposed that the development of *Acinetobacter* peritonitis in the modern era might not result from a dysregulated host immune status in PD patients (as before), but more from the hygiene breaks and failure to perform sterile exchange procedures. This phenomenon could be reminiscent of the routes of staphylococci infections.

The choice of treatment for *Acinetobacter* PD-associated peritonitis is contrary to what we might expect if extrapolating from the experiences of hospital-acquired *Acinetobacter* species infections [25]. In our cohort, all isolates were susceptible to aminoglycosides, fourth-generation cephalosporins (available after 1995), and fluoroquinolones, whilst only 10% to 20% were non-susceptible to third-generation cephalosporins or anti-pseudomonal penicillins (Table 2). However, strains with resistance to aminoglycosides or fluoroquinolones did exist (Table 4). Based on the findings of past reports and ours, we propose that fourth-generation cephalosporins might be a more suitable choice for *Acinetobacter* PD peritonitis, followed by aminoglycosides or fluoroquinolones. Empirical use of ceftazidime might carry a 10% to 30% risk for treatment failure.

An interesting issue among our findings is that the outcomes of *Acinetobacter* peritonitis remained fair throughout the follow-up period, despite the increase in *A. baumannii* percentage. *A. baumannii* infections are reportedly associated with 10% to 20% higher attributable mortality and longer length of hospital stay [2,26,27], and inappropriate antibiotic regimens further increase the risk [28]. Thus, the observed fair outcome in this series of predominantly *A. baumannii* PD-associated peritonitis is unusual. We proposed several reasons for this finding. First, the overall antibiotic resistance rate in our *Acinetobacter* effluent isolates is quite low (including *A. baumannii*), and this phenomenon holds true in other major PD peritonitis series as well (Table 4) [8,12,14]. Second, the empirical antibiotic regimens recommended by the ISPD for peritonitis invariably contain aminoglycosides or ceftazidime, both with fair efficacy against *Acinetobacter* isolates [5,20,21]. Consequently, the chance of treatment failure would be expectedly low; therefore, clinical outcomes would be better in this specific type of *Acinetobacter* infection than in the other types. Finally, as we explained, the origin of *Acinetobacter* PD peritonitis has gradually shifted from poor host immunity to

Table 4. Comparison between the current (after-2000) and the historic cohort.

Characteristics	*Historic 1 (n = 23, 1989)*[#]	*Historic 2 (n = 28, 1994)*[&]	*Current (after 2000) (n = 14)*
Age (years)	41 (9–66)	NA	55 (23–84)
Gender (male %)	16 (70)	NA	7 (50)
Comorbidities			
DM	5 (22)	NA	6 (43)
Autoimmune disorders	2 (9)	NA	1 (7)
Time to last peritonitis (months)	4 (0.5–14.5)	5 (1–29)	13.6 (3–27)
Pathogen			
Acinetobacter baumannii	8 (35)	27 (96)	9 (64)
Acinetobacter iwoffii	3 (13)	1 (4)	3 (21)
Acinetobacter ursingii	0 (0)	0 (0)	1 (7)
Unspecified Acinetobacter	12 (52)	0 (0)	1 (7)
Presumed peritonitis origin			
Break in exchange sterility	4 (17)	NA	5 (36)
Gastrointestinal bacterial translocation	3 (13)		4 (29)
Exit site infection or tunnel infection	0 (0)		0 (0)
Un-identified	16 (70)		5 (36)
Antibiotic susceptibilities			
Aminoglycoside	**S** 20 (87)	NA	**S** 14 (100)
	I 3 (13)		
Ceftazidime	**S** 23 (100)	**S** 19 (69)	**S** 13 (93)
		R 9 (31)	**R** 1 (7)
Cefepime	NA	NA	**S** 14 (100)
Fluoroquinolone	**S** 23 (100)	**S** 22 (79)	**S** 14 (100)
		R 6 (21)	
Outcomes			
Tenckhoff catheter removal	2 (9)	10 (36)	1 (7)
Recurrent peritonitis	1 (4)	0 (0)	0 (0)
Mortality	0 (0)	1 (4)	0 (0)

NA, not available.
[#]Adapted from Galvao C et al. *Am J Kidney Dis* 1989; 14(2): 101–4.
[&]Adapted from Lye WC et al. *Perit Dial Int* 1994; 14(2): 174–7.

sterility break. This might also play a role in the relatively fair outcome of *Acinetobacter* peritonitis.

To our knowledge, this study is the most up-to-date and comprehensive report of PD-associated peritonitis caused by *Acinetobacter*, with focused descriptions of their clinical features. The temporal changes in the characteristics of patients with *Acinetobacter* peritonitis include an increasing proportion of *A. baumannii*, more cases from breaks in exchange sterility, fewer cases from catheter infections, and an increasing interval between the previous and the present peritonitis episodes. These features might have important epidemiologic meanings in the PD field and warrant our continuous attention.

Acknowledgments

We are grateful to the staffs of the Center of Peritoneal Dialysis in National Taiwan University Hospital (NTUH) for their contributions, the staffs of the Second core lab of the department of Medical Research of NTUH for their technical support.

Author Contributions

Conceived and designed the experiments: CTC JWH. Performed the experiments: CTC SYL WSY HWC CCF CJY CKC KYH JWH. Analyzed the data: CTC SYL WSY HWC CCF CJY CKC KYH JWH. Contributed reagents/materials/analysis tools: CTC JWH. Wrote the paper: CTC SYL WSY HWC CCF CJY CKC KYH JWH.

References

1. Li PK-T, Chow KM (2013) Peritoneal Dialysis–First Policy Made Successful: Perspectives and Actions. Am J Kidney Dis 62: 993–1005.
2. Baboolal K, McEwan P, Sondhi S, Spiewanowski P, Wechowski J, et al (2008) The cost of renal dialysis in a UK setting–a multicentre study. Nephrol Dial Transplant 23: 1982–1989.
3. Lafrance JP, Rahme E, Iqbal S, Elftouh N, Vallee M, et al (2012) Association of Dialysis Modality with Risk for Infection-Related Hospitalization: A Propensity Score–Matched Cohort Analysis. Clin J Am Soc Nephrol 7: 1598–1605.
4. Najafi I, Alatab S, Atabak S, Majelan NN, Sanadgol H, et al (2013) Seventeen years' experience of peritoneal dialysis in Iran: first official report of the Iranian peritoneal dialysis registry. Perit Dial Int June 3 [Epub].
5. Li PK-T, Szeto CC, Piraino B, Bernadini J, Figueiredo AE, et al (2010) Peritoneal dialyusis-related infections recommendations: 2010 update. Perit Dial Int 30: 393–423.
6. Zelenitsky S, Barns L, Findlay I, Alfa M, Ariano R, et al (2000) Analysis of microbiological trends in peritoneal dialysis–related peritonitis from 1991 to 1998. Am J Kidney Dis 36: 1009–1013.
7. Barraclough K, Hawley CM, McDonald SP, Brown FG, Rosman JB, et al (2010) Polymicrobial Peritonitis in Peritoneal Dialysis Patients in Australia: Predictors, Treatment, and Outcomes. Am J Kidney Dis 55: 121–131.
8. Friedman O, Jassal SV, Bargman JM (2008) Acinetobacter Peritoneal Dialysis Peritonitis: Description and Relation to the SPICE Family of Organisms. Perit Dial Int 28: 195–197.
9. Chao CT, Lee SY, Yang WS, Chen HW, Fang CC, et al (2013) Citrobacter peritoneal dialysis peritonitis: rare occurrence with poor outcomes. Int J Med Sci 10: 1092–1098.
10. Jarvis EM, Hawley CM, McDonald SP, Brown FG, Rosman JB, et al (2010) Predictors, treatment, and outcomes of non-Pseudomonas Gram-negative peritonitis. Kidney Int 78: 408–414.
11. Szeto CC, Leung CB, Chow KM, Kwan BC, Law MC, et al (2005) Change in bacterial aetiology of peritoneal dialysis-related peritonitis over 10 years: experience from a centre in south-east Asia. Clin Microbiol Infect 11: 837–839.
12. Galvao C, Swartz R, Rocher L, Reynolds J, Starmann B, et al (1989) Acinetobacter peritonitis during chronic peritoneal dialysis. Am J Kidney Dis 14: 101–104.
13. Lye WC, Lee EJ, Ang KK (1991) Acinetobacter peritonitis in patients on CAPD: characteristics and outcome. Adv Perit Dial 7: 176–179.
14. Lye W, Lee EJ, Leong S, Kumarasinghe G (1994) Clinical characteristics and outcome of Acinetobacter infections in CAPD patients. Perit Dial Int 14: 174–177.
15. Chao CT, Lee SY, Yang WS, Chen HW, Fang CC, et al (2013) Peritoneal dialysis peritonitis by anaerobic pathogens: a retrospective case series. BMC Nephrol 14: 111.
16. Chao CT, Wu VC, Lai CF, Shiao CC, Huang TM, et al (2012) Advanced age affects the outcome-predictive power of RIFLE classification in geriatric patients with acute kidney injury. Kidney Int 82: 920–927.
17. Chao CT, Lin YF, Tsai HB, Wu VC, Ko WJ (2013) Acute kidney injury network staging in geriatric postoperative acute kidney injury patients: shortcomings and improvements. J Am Coll Surg 217: 240–250.
18. Prowant B, Nolph K, Ryan L, Twardowski Z, Khanna R (1986) Peritonitis in continuous ambulatory peritoneal dialysis: analysis of an 8-year experience. Nephron 43: 105–109.
19. Kern E, Newman L, Cacho C, Schulak JA, Weiss MF (2002) Abdominal catastrophe revisited: the risk and outcome of enteric peritoneal contamination. Perit Dial Int 22: 323–334.
20. Keane WF, Bailie GR, Boeschoten E, Gokal R, Golper TA, et al (2000) Adult peritoneal dialysis-related peritonitis treatment recommendations: 2000 update. Perit Dial Int 20: 396–411.
21. Piraino B, Bailie G, Bernardini J, Boeschoten E, Gupta A, et al (2005) Peritoneal dialysis-related infections recommendations: 2005 update. Perit Dial Int 25: 107–131.
22. Gerner-Smidt P, Tjernberg I, Ursing J (1991) Reliability of phenotypic tests for identification of Acinetobacter species. J Clin Microbiol 29: 277–282.
23. McConnell MJ, Actis L, Pachón J (2013) Acinetobacter baumannii: human infections, factors contributing to pathogenesis and animal models. FEMS Microbiol Rev 37: 130–155.
24. Towner KJ (2009) Acinetobacter: an old friend, but a new enemy. J Hosp Infect 73: 355–363.
25. Zarrilli R, Pournaras S, Giannouli M, Tsakris A (2013) Global evolution of multidrug-resistant Acinetobacter baumannii clonal lineages. Int J Antimicrob Agent 41: 11–19.
26. Falagas M, Bliziotis I, Siempos I (2006) Attributable mortality of Acinetobacter baumannii infections in critically ill patients: a systematic review of matched cohort and case-control studies. Crit Care 10: R48.
27. Abbo A, Carmeli Y, Navon-Venezia S, Siegman-Igra Y, Schwaber MJ (2007) Impact of multi-drug-resistant Acinetobacter baumannii on clinical outcomes. Eur J Clin Microbiol Infect Dis 26: 793–800.
28. Falagas ME, Kasiakou SK, Rafailidis PI, Zouglakis G, Morfou P (2006) Comparison of mortality of patients with Acinetobacter baumannii bacteraemia receiving appropriate and inappropriate empirical therapy. J Antimicrob Chemother 57: 1251–1254.

11

Characterization of Egg Laying Hen and Broiler Fecal Microbiota in Poultry Farms in Croatia, Czech Republic, Hungary and Slovenia

Petra Videnska[1], Md. Masudur Rahman[1], Marcela Faldynova[1], Vladimir Babak[1], Marta Elsheimer Matulova[1], Estella Prukner-Radovcic[2], Ivan Krizek[2], Sonja Smole-Mozina[3], Jasna Kovac[3], Ama Szmolka[4], Bela Nagy[4], Karel Sedlar[5], Darina Cejkova[1], Ivan Rychlik[1]*

1 Veterinary Research Institute, Brno, Czech Republic, **2** Faculty of Veterinary Medicine, University of Zagreb, Zagreb, Croatia, **3** Biotechnical Faculty, University of Ljubljana, Ljubljana, Slovenia, **4** Institute for Veterinary Medical Research, Hungarian Academy of Sciences, Budapest, Hungary, **5** Department of Biomedical Engineering, Brno University of Technology, Brno, Czech Republic

Abstract

Poultry meat is the most common protein source of animal origin for humans. However, intensive breeding of animals in confined spaces has led to poultry colonisation by microbiota with a zoonotic potential or encoding antibiotic resistances. In this study we were therefore interested in the prevalence of selected antibiotic resistance genes and microbiota composition in feces of egg laying hens and broilers originating from 4 different Central European countries determined by real-time PCR and 16S rRNA gene pyrosequencing, respectively. *strA* gene was present in 1 out of 10,000 bacteria. The prevalence of *sul1*, *sul2* and *tet(B)* in poultry microbiota was approx. 6 times lower than that of the *strA* gene. *tet(A)* and *cat* were the least prevalent being present in around 3 out of 10,000,000 bacteria forming fecal microbiome. The core chicken fecal microbiota was formed by 26 different families. Rather unexpectedly, representatives of *Desulfovibrionaceae* and *Campylobacteraceae*, both capable of hydrogen utilisation in complex microbial communities, belonged among core microbiota families. Understanding the roles of individual population members in the total metabolism of the complex community may allow for interventions which might result in the replacement of *Campylobacteraceae* with *Desulfovibrionaceae* and a reduction of *Campylobacter* colonisation in broilers, carcasses, and consequently poultry meat products.

Editor: Jose Luis Balcazar, Catalan Institute for Water Research (ICRA), Spain

Funding: This study was supported by PROMISE project 7th Framework Programme and AdmireVet project CZ.1.05/2.1.00/01.0006–ED0006/01/01 from the Czech Ministry of Education. A. Szmolka is holder of a Bolyai Janos Stipend from the Hungarian Academy of Sciences. The funders had no role in study design, data collection and analysis, decision to publish, or preparation of the manuscript.

Competing Interests: The authors have declared that no competing interests exist.

* Email: rychlik@vri.cz

Introduction

Poultry meat is the most common protein source of animal origin for humans. However, extensive breeding and selection for increased meat or egg production together with animals living in confined spaces have led to an increased susceptibility of chickens to infections with pathogens, some of them with zoonotic potential. Moreover, if an infected flock is treated with antibiotics, then the antibiotic resistant bacteria are positively selected and the poultry products become contaminated with such bacterial clones.

The composition of gut microbiota is known to affect many host functions including nutrient utilization, gut epithelium nourishment and the development and activity of the gut immune system [1–4]. One of the possibilities how to reduce pathogen colonization as well as antibiotic usage in poultry production is therefore to maintain a normal gut microbiota composition. This can be achieved either by using an appropriate feed formula or by providing chickens with live beneficial bacterial cultures. However, most of the experiments in this area were based on an empirical basis as it is quite difficult to determine the effect of any such preparations on gut microbiota composition when culture conditions for the majority of gut microbiota are not known.

Since the introduction of next generation sequencing, culture independent characterization of gut microbiota has become possible and has allowed for the characterization of complex microbial communities including those in the intestinal tract of chickens. We and others have already characterized changes in chicken gut microbiota after antibiotic therapy, pathogen infection or throughout the chicken's life [5–8]. Besides understanding the behavior of gut microbiota under extreme conditions such as the administration of antibiotics, it is also important to understand what is normal and common to at least the majority of poultry flocks. Accumulation of such knowledge will then allow for a gradual definition of an aberrant microbiota composition and setting up of experiments towards the selection of microbiota with a positive effect on host performance.

In this study we therefore tested to what extent egg or meat production systems in different Central European countries may affect the composition of chicken fecal microbiota and the prevalence of antibiotic resistance genes in whole bacterial population. We considered 2 alternative hypotheses. First, that the microbiota composition would be highly diverse among the countries in which case we would continue with a detailed epidemiological investigation. The alternative hypothesis was that the microbiota composition would not be extremely different, under which circumstance we should be able to define core chicken microbiome. To test both hypotheses we collected over 100 fecal samples from both broilers and egg layers in 4 different countries in which we compared the prevalence of antibiotic resistance genes as well as the composition of fecal microbiota. As the fecal microbiota composition of broilers and hens across four EU countries was quite similar, we finally defined bacterial families representing core chicken fecal microbiome.

Results

Antibiotic resistance gene prevalence in fecal microbiota of egg layers and broilers

The *strA* was the most prevalent gene in all the samples. The median prevalence of this gene in poultry microbiota was around 0.0001 which means that it was present in 1 copy per 10,000 copies of 16S rRNA genes. Since copies of 16S rRNA genes approximately correlate with a number of bacteria, *strA* gene was present in 1 out of 10,000 bacteria forming fecal microbial population. The prevalence of *sul1*, *sul2* and *tet(B)* in poultry microbiota was approx. 6 times lower than that of the *strA* gene and *tet(A)* and *cat* genes were the least prevalent, being present in around 3 out of 10,000,000 bacteria. When the samples were compared according to country of origin, those of Czech or Hungarian origin usually had the lowest antibiotic resistance gene prevalence. On the other hand, *strA* and *sul1* genes were significantly increased in broiler and egg layer microbiota from Croatia whilst an increased prevalence of *tet(B)* or *cat* was characteristic of the microbiota of egg laying hens from Slovenia. When the microbiota of broilers was compared with the microbiota of egg layers, no significant differences were observed in the prevalence of antibiotic resistance genes (Fig. 1 and Table S1).

Microbiota composition determined by 16S rRNA pyrosequencing

Because of the differences in the antibiotic resistance gene prevalence, next we were interested whether these differences were reflected also in microbiota composition. Pyrosequencing of 16S rRNA amplification products in 45 selected samples showed that the microbiota of egg layers was usually more complex than the microbiota of broilers (Table S2), except for the microbiota of hens from Slovenia which was of low complexity, similar to that of broilers (Fig. 2). In both egg layers and broilers, *Firmicutes*, *Bacteroidetes* and *Proteobacteria* represented the major phyla in fecal samples. Broiler microbiota was dominated by *Firmicutes* (76.2%) followed by *Proteobacteria* (14%). *Bacteroidetes* formed only 6.5% of the total fecal microbiota of broilers and *Actinobacteria* were present at 3.8% in broiler microbiota. The microbiota of egg layers was also dominated by *Firmicutes* (58.8%). However, representatives of *Bacteroidetes* formed 22.1% of the total microbiota and *Proteobacteria* formed 16.9%. *Actinobacteria* were present at 0.6% and, instead, *Fusobacteria* were detected as a minority subpopulation reaching 1.4% in egg layer fecal microbiota (for more details see Table S3).

Search for the most common OTUs (Operational Taxonomic Unit, bacterial species with at least 97% similarity in 16S rRNA sequences) showed that 3 OTUs, all belonging to the genus *Lactobacillus* were found in all 45 analysed samples. In addition, other 2 OTUs detected in 43 and 41 samples, respectively, also belonged to the genus *Lactobacillus*. The remaining common OTUs included an unclassified representative of *Peptostreptococcaceae*, and those of genera *Streptococcus* and *Escherichia* (Table 1). However, the more appropriate identification of core chicken microbiota was achieved after a combined view at family and class taxonomic levels. This showed that in addition to the above mentioned genera, representatives of families *Clostridiaceae*, *Ruminococcaceae*, *Lachnospiraceae*, *Veillonellaceae* and classes *Bacteroidia* and *Actinobacteria* were common to chicken microbiota (Table 1).

In the next step we analyzed whether there are any OTUs which would occur only among hen microbiota and were absent in broiler microbiota, and vice versa. Not a single OTU completely specific for either of the poultry categories was identified. However, the OTUs with the highest preference to either broiler or hen microbiota are listed in Table 2. Data in this table show that OTUs specific to young broilers usually belonged to the phylum *Firmicutes* while those characteristic of adult hens commonly belonged to the phylum *Bacteroidetes*.

To further identify mutual interconnections among chicken fecal microbiota, the presence of bacterial families was correlated with each other based on the microbiota composition of 45 samples. Although altogether representatives of 108 families were identified in this study in at least one sample, in correlation analysis we included only the families which were present in more than 15 out of 45 analyzed samples, *i.e.* we excluded bacterial families, whose representatives were recorded only sporadically as their inclusion resulted in distorted outputs. This analysis grouped bacterial families into three main clusters (Fig. 3). Cluster I included 21 different families, with equal representation of phyla *Bacteroidetes* and *Firmicutes,* each having 8 families in this cluster. Some of the families, *e.g.* *Lachnospiraceae*, *Ruminococcaceae*, *Bacteroidaceae*, *Rikenellaceae*, *Veilonellaceae* and *Prevotellaceae*, were frequently reported to be present in the intestinal microbiota of chickens or hens [9–11]. The families of Cluster I are therefore characteristic of developed and established microbiota of adult hens. Interestingly, representatives of *Helicobacteraceae* and *Campylobacteraceae* also clustered with these families, although as the most distant members.

Cluster II included 6 bacterial families such as *Pseudomonadaceae*, *Xanthomonadaceae*, *Moraxellaceae* or *Flavobacteriaceae* which we recently detected in the water used for transporting ornamental fish [12] and can be considered as of environmental origin.

The last cluster III included 14 bacterial families, eight of them belonging to the phylum *Firmicutes*, five of them to the phylum *Actinobacteria* and the last one to the phylum *Proteobacteria*. The phylum *Bacteroidetes* was not represented in this group. Families such as *Lactobacillaceae*, *Turicibacteraceaea*, *Corynebacteriaceae*, *Brevibacteriaceae*, *Staphylococcaceae*, *Streptococcaceae*, *Peptostreptococcaceae*, *Enterococcaceae* or *Enterobacteriaceae* were also included in this cluster. These families overlapped with those covering the most common microbiota members listed in Table 1 or those covering microbiota characteristic of broilers (Table 2).

Discussion

In this study we first analyzed the prevalence of selected antibiotic resistance genes followed by the composition of fecal

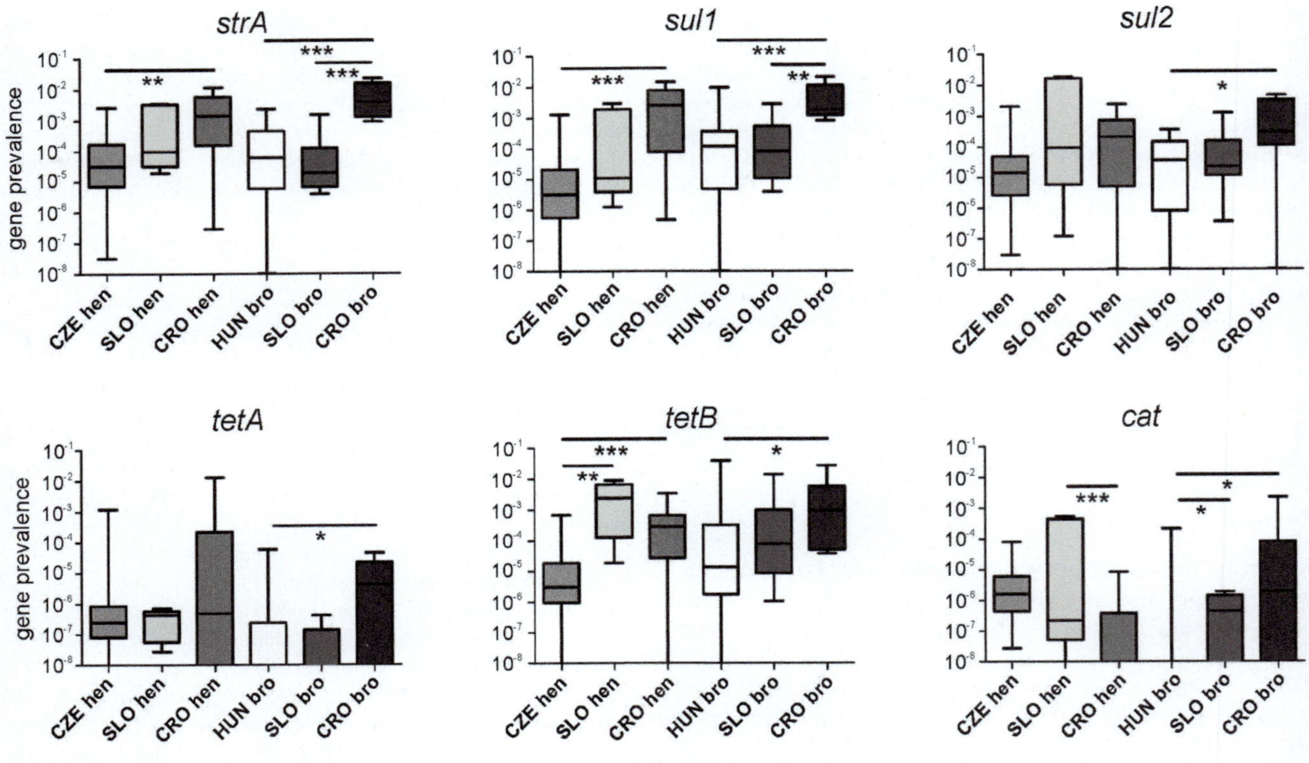

Figure 1. Antibiotic resistance gene prevalence in poultry fecal microbiota. Antibiotic resistance gene prevalence is presented as the median with 25th and 75th percentile (box) and the whiskers indicating minimum and maximum values recorded. Mind the logarithmic scaling of Y axis.

microbiota in broilers and egg laying hens. The prevalence of antibiotic resistance genes was quite low, similar to our previous report [13]. These results are consistent with the absence of recent therapy in the flocks since the frequencies in antibiotic resistance genes increase during such therapies but decrease very quickly after antibiotic withdrawal [13]. The frequencies of antibiotic resistance genes in different microbial populations vary around similar prevalencies such as 10^{-3}–10^{-4}, with *sul1*, *sul2* and *strB* genes being usually more frequent than *tet(A)* or *tet(B)* genes [14–16]. However, a high prevalence can occasionally be detected as was the case of *sul1* and *sul2* gene prevalence in a report from

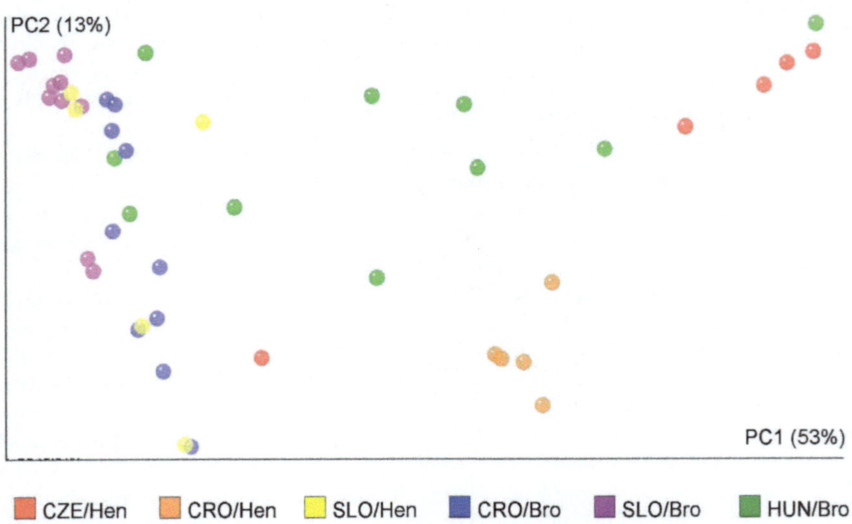

Figure 2. Microbiota diversity in fecal samples of broilers and hens. UniFrac analysis followed by PCoA showed that the two main factors explained 69% of the variability observed in the microbiota composition in poultry fecal samples. Slovenian egg layer microbiota were similar to samples originating from broilers and Hungarian samples of broiler origin clustered between broiler samples from Slovenia and Croatia, and egg laying hen samples from Croatia and the Czech Republic. SLO – Slovenia, CRO – Croatia, CZE – Czech Republic, HUN – Hungary, Hen - samples from egg laying hens, Bro – samples from broilers.

Table 1. List of OTUs and their taxonomic classification found in at least 40 out of 45 samples tested by pyrosequencing 16S rRNA genes.

OTU	Genus	Family	Class	Phylum
Lactobacillus (45)	***Lactobacillus (45)***	***Lactobacillaceae (45)***	***Clostridia (45)***	***Firmicutes (45)***
Lactobacillus (45)	*Streptococcus (42)*	*Peptostreptococcaceae (44)*	***Bacilli (45)***	***Bacteroidetes (45)***
Lactobacillus (45)	*Escherichia (41)*	*Enterobacteriaceae (44)*	***Bacteroidia (45)***	*Proteobacteria (44)*
Lactobacillus (43)		*Clostridiaceae (43)*	*Gammaproteobacteria (44)*	*Actinobacteria (42)*
Peptostreptococcaceae (42)		*Streptococcaceae (42)*	*Actinobacteria (42)*	
Streptococcus (41)		*Ruminococcaceae (42)*		
Lactobacillus (41)		*Lachnospiraceae (40)*		
Escherichia (40)		*Veillonellaceae (40)*		

China reaching prevalence of 10^{-2} in porcine manure and even around 10^{-1} in the manure of chicken origin [14].

Microbiota composition was characterized by pyrosequencing the V3 and V4 variable region of 16S rRNA genes. Though the results can be affected by the efficiency of cell lysis of different bacterial species, efficiency of primer annealing and extension as well as the different rRNA gene copy number in different species [17,18], pyrosequencing 16S rRNA genes is currently the most

Table 2. List of OTUs present 10 or more times in either hen or broiler fecal microbiota.

	Reads		Classification			
OTU	**Hens***	**Broilers***	**Phylum**	**Order**	**Family**	**Genus**
784	11	0	*Firmicutes*	*Clostridiales*	*Veillonellaceae*	*Megasphaera*
3113	11	0	*Bacteroidetes*	*Bacteroidales*		
4272	11	0	*Bacteroidetes*	*Bacteroidales*		
5509	11	0	*Bacteroidetes*	*Bacteroidales*	*Porphyromonadaceae*	*Parabacteroides*
5896	11	0	*Firmicutes*	*Coriobacteriales*	*Coriobacteriaceae*	*Atopobium*
519	11	1	*Bacteroidetes*	*Bacteroidales*	*Bacteroidaceae*	*Bacteroides*
776	10	0	*Firmicutes*	*Lactobacillales*	*Lactobacillaceae*	*Lactobacillus*
1210	10	0	*Bacteroidetes*	*Bacteroidales*	*Prevotellaceae*	*Prevotella*
1622	10	0	*Bacteroidetes*	*Bacteroidales*	*Bacteroidaceae*	*Bacteroides*
2571	10	0	*Bacteroidetes*	*Bacteroidales*		
4806	10	0	*Bacteroidetes*	*Bacteroidales*	*Paraprevotellaceae*	
7750	10	0	*Firmicutes*	*Clostridiales*	*Veillonellaceae*	*Megamonas*
9129	10	0	*Bacteroidetes*	*Bacteroidales*	*Bacteroidaceae*	*Bacteroides*
3822	10	1	*Bacteroidetes*	*Bacteroidales*	*Bacteroidaceae*	*Bacteroides*
8364	10	1	*Bacteroidetes*	*Bacteroidales*	*Paraprevotellaceae*	*Prevotella*
8059	1	18	*Firmicutes*	*Lactobacillales*	*Lactobacillaceae*	
3467	0	15	*Bacteroidetes*	*Bacteroidales*	*Bacteroidaceae*	*Bacteroides*
2592	0	14	*Bacteroidetes*	*Bacteroidales*	*Rikenellaceae*	
2652	0	13	*Actinobacteria*	*Actinomycetales*	*Corynebacteriaceae*	*Corynebacterium*
4021	1	13	*Actinobacteria*	*Actinomycetales*	*Brevibacteriaceae*	*Brevibacterium*
6880	1	13	*Firmicutes*	*Lactobacillales*	*Lactobacillaceae*	*Lactobacillus*
1226	1	12	*Firmicutes*	*Lactobacillales*	*Lactobacillaceae*	
8226	1	12	*Actinobacteria*	*Actinomycetales*	*Corynebacteriaceae*	*Corynebacterium*
962	0	11	*Firmicutes*	*Lactobacillales*	*Lactobacillaceae*	*Lactobacillus*
2087	1	11	*Bacteroidetes*	*Bacteroidales*	*Rikenellaceae*	
20	0	10	*Firmicutes*	*Clostridiales*	*Ruminococcaceae*	*Ruminococcus*
1373	0	10	*Firmicutes*	*Clostridiales*	*Ruminococcaceae*	
7469	0	10	*Firmicutes*	*Bacillales*	*Staphylococcaceae*	

* numbers show the number of samples in which particular OTU was recorded, out of 15 hen and 30 broiler fecal microbiota samples.

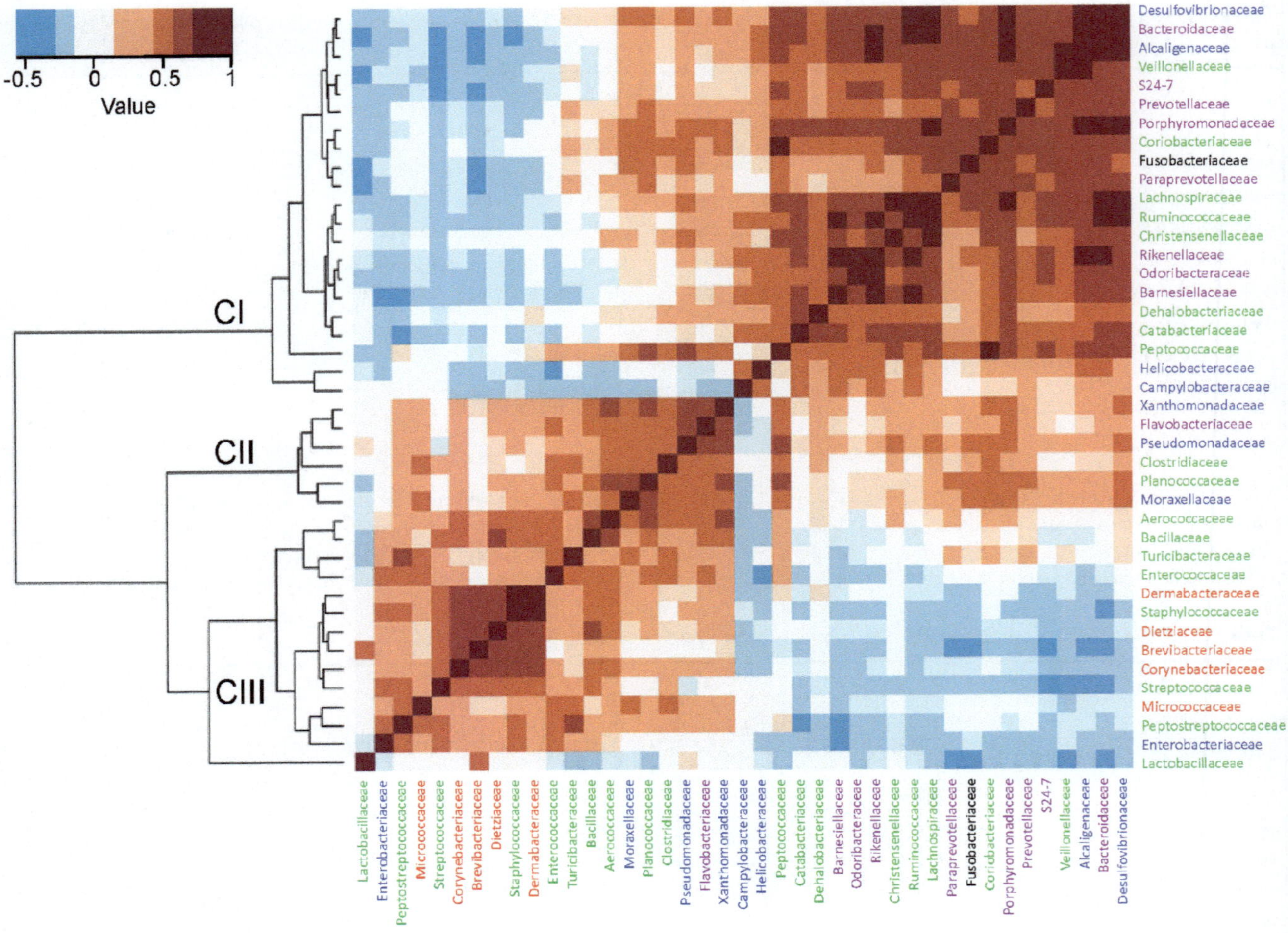

Figure 3. Identification of core chicken fecal microbiota. Cluster I (CI) of positively correlating bacterial families represents families common to established chicken fecal microbiota of adult hens. Interestingly, representatives of *Helicobacteraceae* and *Campylobacteraceae* were clustered with these families, although as the most distant members. Cluster II (CII) was formed mainly by the families which are characteristic of aquatic environments. Cluster III (CIII) of positively correlating bacterial families included bacterial families that were mainly characteristic of gut microbiota of young broilers. Families in green belong to the phylum *Firmicutes*, purple to *Bacteroidetes*, blue to *Proteobacteria* and red to *Actinobacteria*. Shades of brown – positively correlating families; shades of blue – negatively correlating families.

frequently used tool for the characterization of complex microbial communities. The microbiota of hens originating from Slovenia differed from the microbiota of hens from the Czech Republic and Croatia (Fig. 2 and Figs. S1 and S2). Although we are unable to explain this observation exactly, we noticed that hens from the Czech Republic and Croatia analyzed in this study were 30 weeks old whilst hens from Slovenia were 61 weeks old. Interestingly, in our previous study, the feces of young hens were quite complex whilst fecal microbiota from older hens consisted mainly of *Lactobacillaceae* [8]. However, in another study where we followed the development of cecal, *i.e.* not fecal, microbiota in hens over their whole life, no such reduction in microbiota composition in the cecum was recorded (unpublished observations). It is possible that the frequency of cecal excretion decreases with increasing age which results in a lower presence of cecal microbiota in the feces. And since *Lactobacillaceae* are present in the small intestine in large quantities [8,19], these may become enriched also in the feces [20].

In an attempt to identify the core microbiota, all 21 families forming cluster I and selected families from cluster III, namely *Lactobacillaceae*, *Peptostreptococcaceae*, *Streptococcaceae*, *Clostridiaceae* and *Enterobacteriaceae* have to be considered. Although it is likely that additional studies may increase or decrease the number of families defined as the core chicken microbiome, families such as *Lachnospiraceae*, *Ruminococcaceae*, *Clostridiaceae*, *Lactobacillaceae*, *Bacteroidaceae*, *Rickenellaceae*, *Veilonellaceae*, *Prevotellaceae* or *Enterobacteriaceae* are repeatedly reported as common members of not only chicken [9–11] but also human gut microbiota [21].

Representatives of *Desulfovibrionaceae* or *Campylobacteraceae*, which were detected among the core microbiota, are not frequently mentioned as common microbiota members. However, these may play important roles in the overall microbiota metabolism. Molecular hydrogen is released during fermentation by strict anaerobes of phyla *Firmicutes* and *Bacteroidetes* [22]. Since metabolism of such anaerobes is inhibited by the end product of their fermentation, *i.e.* by hydrogen, the hydrogen has to be removed. Both *Desulfovibrionaceae* and *Campylobacteraceae*, as well as *Megamonas* from the family *Veillonelaceae*, are capable of hydrogen utilization thus serving as a hydrogen sink in complex microbial communities [23,24] which may explain the presence of *Desulfovibrionaceae*, *Campylobacteraceae* and *Megamonas* as common members of chicken microbiota. In addition, we observed a high correlation between the occurrence of *Bacteroi-*

Table 3. List of primers used in this study.

Primer	Target	Primer sequence 5′ - 3′	Reference
strA_F	aminoglycoside phosphotransferase	ACCCTAAAACTCTTCAATGC	[13]
strA_R	aminoglycoside phosphotransferase	TCCCCAATACATTGAATAGG	[13]
sul1_F	dihydropteroate synthase	GTCTAAGAGCGGCGCAATAC	[13]
sul1_R	dihydropteroate synthase	GGATCAGACGTCGTGGATGT	[13]
sul2_F	dihydropteroate synthase	CGCAATGTGATCCATGATGT	[13]
sul2_R	dihydropteroate synthase	GCGAAATCATCTGCCAAACT	[13]
tetB_F	tetracycline efflux protein	TACAGGGATTATTGGTGAGC	[13]
tetB_R	tetracycline efflux protein	ACATGAAGGTCATCGATAGC	[13]
tetA_F	tetracycline efflux protein	CGATCTTCCAAGCGTTTGTT	[13]
tetA_R	tetracycline efflux protein	CCAGAAGAACGAAGCCAGTC	[13]
cat_F	chloramphenicol acetyl transferase	GGGAAATAGGCCAGGTTTTC	[13]
cat_R	chloramphenicol acetyl transferase	TCCATGAGCAAACTGAAACG	[13]
16S_univ_F	all bacteria	GAGGAAGGIGIGGAIGACGT	[30]
16S_univ_R	all bacteria	AGICCCGIGAACGTATTCAC	[30]
16S_univ-1F	all bacteria	GTGSTGCAYGGYTGTCGTCA	[29]
16S_univ-1R	all bacteria	ACGTCRTCCMCACCTTCCTC	[29]

daceae and *Desulfovibrionaceae* (Fig. 3). It has been described that the representatives of *Bacteroidaceae* are capable of sulphate release from sulphated chondroitin or mucin produced by host cells [25]. The released sulphate can in turn be respired to H_2S by *Desulfovibrio* sp. [26] explaining the high correlation of *Desulfovibrionaceae* and *Bacteroidaceae*. Such ecological and metabolic bacterial mutualisms in complex microbiota systems may be used to prevent pathogen colonization. *E.g.* administering bacterial species from families *Desulfovibrionaceae* or *Veillonelaceae* with a similar metabolism and role as *Campylobacteraceae* to chickens may result in the replacement of *Campylobacteraceae* with *Desulfovibrionaceae* or *Veillonelaceae* and the reduction of *Campylobacter* sp. colonization in broilers, carcasses, and consequently poultry meat products.

Material and Methods

Ethics Statement

The owners of the farms were aware that the droppings were being collected for this study and gave their permission for doing so. Such permissions can be obtained from EPR for Croatian samples, IR for the Czech samples and SSM for Slovenian samples. Sampling of Hungarian broilers was performed on freshly slaughtered animals at two slaughter houses under the supervision and with the permission of the district veterinary officer and poultry owners. BN can be contacted for further information on Hungarian samples.

Sample characterization

Altogether 137 fresh fecal samples originating from healthy flocks of both egg laying hens and broilers from 4 countries were analyzed in this study. Sixty-seven samples of fresh fecal droppings from 3 egg laying hen farms in the Czech Republic were taken, with 30, 28 and 9 samples collected at each farm. Fifteen fecal samples originated from 3 farms in Slovenia (five samples per farm). Ten of them originated from 3 and 4 week old broilers and the remaining 5 samples were collected from 62 week old laying hens. Thirty samples originated from 6 different farms in Croatia (five samples per farm). Twenty of these originated from 4 egg laying hen farms and the remaining 10 samples were collected at 2 different broiler farms. Thirty Hungarian samples originated from 3 different broiler farms, collecting 10 samples at each farm. Upon collection, the samples were kept frozen at −20°C preceding DNA purification. All the samples originated from farms following basically the same intensive production system of raising broilers or eggs with no antimicrobials having been provided while in egg or meat production.

DNA purification and real-time PCR

After slowly defrosting at room temperature, approx. 250 mg of feces were homogenized for 1 min at 7000 RPM in a MagNALyzer (Roche Diagnostics) using zirconia silica beads (BioSpec Products). Following homogenization, the DNA was extracted using a QIAamp DNA Stool Mini Kit (Qiagen) according to the manufacturer's instructions and the purified DNA was stored at −20°C until use. The primer design for quantification of antibiotic resistance genes was described in our previous study [13] and all the primers are listed in Table 3. Target antibiotic resistance genes were selected to represent those commonly found in genomes of poultry microbiota [13,27]. In addition, two primer pairs specific for the conservative regions of 16S rRNA genes (domain *Bacteria* universal primer pairs) were used to determine the total bacterial DNA present in the samples [28,29]. Real-time PCR was carried out using QuantiTect SYBR Green PCR Kit (Qiagen) in a LightCycler 480 thermocycler (Roche). PCR was initiated with a hot start for 15 min at 95°C followed by 45 cycles of 20 sec at 95°C, 30 sec at 60°C and 30 sec at 72°C. Melting temperatures were determined after PCR to verify the correctness of each PCR product. The Ct values of the genes of interest were subtracted from an average Ct value of amplifications performed with *Bacteria* universal primer pairs (ΔCt) and the relative amount of each antibiotic resistance gene in the total bacterial population, *i.e.* antibiotic resistance gene prevalence in a given bacterial population, was finally calculated as $2^{-\Delta Ct}$.

Pyrosequencing

Forty-five samples representing 5 egg laying hen samples from each country and 10 broiler samples per country were selected for pyrosequencing and microbiota characterization. The purified DNA was used as a template in PCR with the forward primer 5′ CGTATCGCCTCCCTCGCGCCATCAG – MID-*GGAGG-CAGCAGTRRGGAAT* 3′, and reverse primer 5′ CTATGC-GCCTTGCCAGCCCGCTCAG- MID- *CTACCRGGGTATC-TAATCC* 3′ using HotStarTaq Master Mix Kit following the manufacturer's instructions (Qiagen). The underlined sequences were required at different steps of pyrosequencing while those in italics are sequences complementary to the conserved parts of 16S rRNA genes flanking the V3/V4 hypervariable region [8,30]. Cycling conditions consisted of a hot start at 95°C for 15 min followed by 30 cycles of incubation at 94°C for 40 s, 55°C for 55 s and 72°C for 60 s. PCR ended with a final extension at 72°C for 5 min. After PCR, the amplification products, approx. 525 bp in size, were separated electrophoretically in a 1.2% agarose gel, gel-purified using a QIAquick Gel Extraction Kit (Qiagen) and subjected to pyrosequencing. Pyrosequencing was performed using GS Junior Titanium sequencing chemistry and a GS Junior 454 sequencer according to the manufacturer's instructions (Roche).

Sequence analysis

Fasta and qual files generated as an output of the pyrosequencing were uploaded into Qiime software [31]. Quality trimming criteria included no mismatch in MID sequences and a maximum of 1 mismatch in primer sequences. The obtained sequences with a qual score higher than 20 were shortened to the same length of 350 bp and classified with RDP Seqmatch with an OTU discrimination level set to 97%. In the next step, chimeric sequences were predicted and excluded from the analysis. Diversity analyses (rarefaction curves and Chao1 richness) on OTU clusters were performed using all sequences available for each sample. Finally, UniFrac analysis [32] followed by weighted principal coordinate analysis (PCoA) was used to characterize the diversity in the microbial populations tested. The raw sequence reads have been deposited in the NCBI Short Read Archive under the accession number SRP045877.

Statistics

Data from real-time PCR are presented as medians ± 25th and 75th percentile. The comparison of antibiotic resistance gene representation in particular samples was evaluated by Kruskal-Wallis test using GraphPad Prism 5 software. A heat map based on Spearman's correlation coefficients was constructed in R using gplots package. Bipartite graphs of microbiota composition in different groups of birds were calculated using Matlab 2013a and visualized in Gephi 0.8.2.

Supporting Information

Figure S1 Composition of chicken fecal microbiota at the family level in 45 selected samples.

Figure S2 Comparison of broiler and egg layer fecal microbiota.

Table S1 List of raw Ct data for the quantification of antibiotic gene prevalence.

Table S2 Dominant phyla found in the samples analyzed in this study.

Table S3 List of all OTUs identified in this study.

Acknowledgments

Authors would like to thank Peter Eggenhuizen and Lee Clayton Elsheimer for their English language corrections.

Author Contributions

Conceived and designed the experiments: PV IR BN SSM EPR. Performed the experiments: PV MMR MF MEM IK JK AS. Analyzed the data: VB KS DC. Contributed reagents/materials/analysis tools: DC. Wrote the paper: PV IR BN SSM EPR.

References

1. Lillehoj HS (1993) Avian gut-associated immune system: implication in coccidial vaccine development. Poult Sci 72: 1306–11.
2. Gaboriau-Routhiau V, Rakotobe S, Lecuyer E, Mulder I, Lan A, et al. (2009) The key role of segmented filamentous bacteria in the coordinated maturation of gut helper T cell responses. Immunity 31: 677–89.
3. Ismail AS, Behrendt CL, Hooper LV (2009) Reciprocal interactions between commensal bacteria and gamma delta intraepithelial lymphocytes during mucosal injury. J Immunol 182: 3047–54.
4. Hill DA, Hoffmann C, Abt MC, Du Y, Kobuley D, et al. (2010) Metagenomic analyses reveal antibiotic-induced temporal and spatial changes in intestinal microbiota with associated alterations in immune cell homeostasis. Mucosal Immunol 3: 148–58.
5. Lu J, Idris U, Harmon B, Hofacre C, Maurer JJ, et al. (2003) Diversity and succession of the intestinal bacterial community of the maturing broiler chicken. Appl Environ Microbiol 69: 6816–24.
6. Stanley D, Keyburn AL, Denman SE, Moore RJ (2012) Changes in the caecal microflora of chickens following *Clostridium perfringens* challenge to induce necrotic enteritis. Vet Microbiol 159: 155–62.
7. Videnska P, Sisak F, Havlickova H, Faldynova M, Rychlik I (2013) Influence of *Salmonella enterica* serovar Enteritidis infection on the composition of chicken cecal microbiota. BMC Vet Res 9: 140.
8. Videnska P, Faldynova M, Juricova H, Babak V, Sisak F, et al. (2013) Chicken faecal microbiota and disturbances induced by single or repeated therapy with tetracycline and streptomycin. BMC Vet Res 9: 30.
9. Callaway TR, Dowd SE, Wolcott RD, Sun Y, McReynolds JL, et al. (2009) Evaluation of the bacterial diversity in cecal contents of laying hens fed various molting diets by using bacterial tag-encoded FLX amplicon pyrosequencing. Poult Sci 88: 298–302.
10. Qu A, Brulc JM, Wilson MK, Law BF, Theoret JR, et al. (2008) Comparative metagenomics reveals host specific metavirulomes and horizontal gene transfer elements in the chicken cecum microbiome. PLoS One 3: e2945.
11. Nordentoft S, Molbak L, Bjerrum L, De Vylder J, Van Immerseel F, et al. (2011) The influence of the cage system and colonisation of *Salmonella* Enteritidis on the microbial gut flora of laying hens studied by T-RFLP and 454 pyrosequencing. BMC Microbiol 11: 187.
12. Gerzova L, Videnska P, Faldynova M, Sedlar K, Provaznik I, et al. (2014) Characterization of microbiota composition and presence of selected antibiotic resistance genes in carriage water of ornamental fish. PLoS One 9: e103865.
13. Faldynova M, Videnska P, Havlickova H, Sisak F, Juricova H, et al. (2013) Prevalence of antibiotic resistance genes in faecal samples from cattle, pigs and poultry. Vet Med Czech 58: 298–304.
14. Czekalski N, Berthold T, Caucci S, Egli A, Bürgmann H (2012) Increased levels of multiresistant bacteria and resistance genes after wastewater treatment and their dissemination into lake Geneva, Switzerland. Front Microbiol 3:106.
15. Walsh F, Ingenfeld A, Zampicolli M, Hilber-Bodmer M, Frey JE, et al. (2011) Real-time PCR methods for quantitative monitoring of streptomycin and tetracycline resistance genes in agricultural ecosystems. J Microbiol Methods 86:150–155.
16. Cheng W, Chen H, Su C, Yan S (2013) Abundance and persistence of antibiotic resistance genes in livestock farms: a comprehensive investigation in eastern China. Environ Int 61:1–7.
17. Kembel SW, Wu M, Eisen JA, Green JL (2012) Incorporating 16S gene copy number information improves estimates of microbial diversity and abundance. PLoS Comput Biol 8:e1002743.
18. Rastogi R, Wu M, Dasgupta I, Fox GE (2009) Visualization of ribosomal RNA operon copy number distribution. BMC Microbiol 9:208.

19. Cressman MD, Yu Z, Nelson MC, Moeller SJ, Lilburn MS, et al. (2010) Interrelations between the microbiotas in the litter and in the intestines of commercial broiler chickens. Appl Environ Microbiol 76: 6572–82.
20. Sekelja M, Rud I, Knutsen SH, Denstadli V, Westereng B, et al. (2012) Abrupt temporal fluctuations in the chicken fecal microbiota are explained by its gastrointestinal origin. Appl Environ Microbiol 78: 2941–8.
21. Qin J, Li R, Raes J, Arumugam M, Burgdorf KS, et al. (2010) A human gut microbial gene catalogue established by metagenomic sequencing. Nature 464: 59–65.
22. Vignais PM, Billoud B (2007) Occurrence, classification, and biological function of hydrogenases: an overview. Chem Rev 107: 4206–72.
23. Kim T, Mundt E (2011) Metagenomic analysis of intestinal microbiomes in chickens. Methods Mol Biol 733: 185–94.
24. Sergeant MJ, Constantinidou C, Cogan TA, Bedford MR, Penn CW, et al. (2014) Extensive microbial and functional diversity within the chicken cecal microbiome. PLoS One 9: e91941.
25. Benjdia A, Martens EC, Gordon JI, Berteau O (2011) Sulfatases and a radical S-adenosyl-L-methionine (AdoMet) enzyme are key for mucosal foraging and fitness of the prominent human gut symbiont, *Bacteroides thetaiotaomicron*. J Biol Chem 286: 25973–82.
26. Rey FE, Gonzalez MD, Cheng J, Wu M, Ahern PP, et al. (2013) Metabolic niche of a prominent sulfate-reducing human gut bacterium. Proc Natl Acad Sci U S A 110: 13582–7.
27. Szmolka A, Anjum MF, La Ragione RM, Kaszanyitzky EJ, Nagy B (2012) Microarray based comparative genotyping of gentamicin resistant *Escherichia coli* strains from food animals and humans. Vet Microbiol 156: 110–8.
28. Levesque C, Piche L, Larose C, Roy PH (1995) PCR mapping of integrons reveals several novel combinations of resistance genes. Antimicrob Agents Chemother 39: 185–91.
29. Tseng CP, Cheng JC, Tseng CC, Wang C, Chen YL, et al. (2003) Broad-range ribosomal RNA real-time PCR after removal of DNA from reagents: melting profiles for clinically important bacteria. Clin Chem 49: 306–9.
30. Nossa CW, Oberdorf WE, Yang L, Aas JA, Paster BJ, et al. (2010) Design of 16S rRNA gene primers for 454 pyrosequencing of the human foregut microbiome. World J Gastroenterol 16: 4135–44.
31. Caporaso JG, Kuczynski J, Stombaugh J, Bittinger K, Bushman FD, et al. (2010) QIIME allows analysis of high-throughput community sequencing data. Nat Methods 7: 335–6.
32. Lozupone C, Knight R (2005) UniFrac: a new phylogenetic method for comparing microbial communities. Appl Environ Microbiol 71: 8228–35.

12

Characterization of the Commercially-Available Fluorescent Chloroquine-BODIPY Conjugate, LynxTag-CQ_{GREEN}, as a Marker for Chloroquine Resistance and Uptake in a 96-Well Plate Assay

Cheryl C. Y. Loh[1]☯, Rossarin Suwanarusk[2]☯, Yan Quan Lee[1,3]☯, Kitti W. K. Chan[1], Kit-Ying Choy[4], Laurent Rénia[2], Bruce Russell[1], Martin J. Lear[5], François H. Nosten[6,7,8], Kevin S. W. Tan[1,3]*, Larry M. C. Chow[4]*

1 Department of Microbiology, Yong Loo Lin School of Medicine, National University of Singapore, Singapore, Singapore, **2** Singapore Immunology Network, Agency for Science Technology and Research, Biopolis, Singapore, Singapore, **3** NUS Graduate School for Integrative Sciences and Engineering, National University of Singapore, Singapore, Singapore, **4** Department of Applied Biology and Chemical Technology, The State Key Laboratory of Chirosciences, The Hong Kong Polytechnic University, Hung Hom, Kowloon, Hong Kong SAR, PR China, **5** Department of Chemistry, Graduate School of Science, Tohoku University, Aza Aramaki, Aoba-ku, Sendai, Japan, **6** Centre for Tropical Medicine, Nuffield Department of Medicine, University of Oxford, Oxford, United Kingdom, **7** Shoklo Malaria Research Unit, Mahidol-Oxford Tropical Medicine Research Unit, Faculty of Tropical Medicine, Mahidol University, Mae Sot, Thailand, **8** Centre for Tropical Medicine, University of Oxford, Churchill Hospital, Oxford, United Kingdom

Abstract

Chloroquine was a cheap, extremely effective drug against *Plasmodium falciparum* until resistance arose. One approach to reversing resistance is the inhibition of chloroquine efflux from its site of action, the parasite digestive vacuole. Chloroquine accumulation studies have traditionally relied on radiolabelled chloroquine, which poses several challenges. There is a need for development of a safe and biologically relevant substitute. We report here a commercially-available green fluorescent chloroquine-BODIPY conjugate, LynxTag-CQ_{GREEN}, as a proxy for chloroquine accumulation. This compound localized to the digestive vacuole of the parasite as observed under confocal microscopy, and inhibited growth of chloroquine-sensitive strain 3D7 more extensively than in the resistant strains 7G8 and K1. Microplate reader measurements indicated suppression of LynxTag-CQ_{GREEN} efflux after pretreatment of parasites with known reversal agents. Microsomes carrying either sensitive- or resistant-type PfCRT were assayed for uptake; resistant-type PfCRT exhibited increased accumulation of LynxTag-CQ_{GREEN}, which was suppressed by pretreatment with known chemosensitizers. Eight laboratory strains and twelve clinical isolates were sequenced for PfCRT and Pgh1 haplotypes previously reported to contribute to drug resistance, and *pfmdr1* copy number and chloroquine IC_{50}s were determined. These data were compared with LynxTag-CQ_{GREEN} uptake/fluorescence by multiple linear regression to identify genetic correlates of uptake. Uptake of the compound correlated with the $logIC_{50}$ of chloroquine and, more weakly, a mutation in Pgh1, F1226Y.

Editor: Georges Snounou, Université Pierre et Marie Curie, France

Funding: As part of the Oxford Tropical Medicine Research Program of Wellcome Trust–Mahidol University, Shoklo Malaria Research Unit (SMRU) is funded by the Wellcome Trust of Great Britain. The authors further thank the National Research Foundation (NRF2009NRF-POC002–102), the National Medical Research Council (NMRC/1310/2011; NMRC/EDG/1038/2011), and the Agency for Science, Technology and Research (A*STAR, Singapore) for their generous support. The funders had no role in study design, data collection and analysis, decision to publish, or preparation of the manuscript.

Competing Interests: KSWT and MJL are founding directors of BioLynx Technologies (Singapore), a private company that specializes in fluorophore-conjugated drug surrogates including LynxTag-CQGREEN. Other authors declare no competing interests. KSWT and MJL own minority shares in BioLynx Technologies (Singapore).

* Email: kevin_tan@nuhs.edu.sg (KSWT); larry.chow@polyu.edu.hk (LMCC)

☯ These authors contributed equally to this work.

Introduction

Despite years of intense global effort to eradicate it, malaria is still one of the deadliest infectious diseases, killing more than 600 000 people in 2010 alone [1,2]. The severest form of malaria is caused by the protozoan parasite *Plasmodium falciparum*. Chloroquine (CQ), once a spectacularly successful antimalarial drug, was first discovered by the German chemist Johann Andersag but was mistakenly thought to be too toxic for therapeutic purposes, an incident which became known as "the resochin error" (resochin being the name given to the compound by Andersag) [3,4]. CQ was so effective that it inspired optimism for the eradication of malaria. However, resistance soon arose, first appearing along the Thai-Cambodian border in the 1950s. By the 1970s, CQ resistance had spread throughout the world [5,6]. This resistance is generally attributed to mutations in the *pfcrt* (*P.*

falciparum chloroquine resistance transporter) gene, which codes for a transporter situated on the membrane of the parasite digestive vacuole (DV).

During parasite development in the intraerythrocytic cycle, haemoglobin is digested in the DV and the toxic heme moiety is released, which the parasite crystalizes into non-toxic hemozoin [7]. CQ is generally thought to kill the parasite by inhibiting the formation of hemozoin and thus preventing the detoxification of free heme [8–10]. In wild-type parasites CQ diffuses through the DV membrane and is diprotonated in the acidic environment of the DV, acquiring a net positive charge which prevents it from escaping the DV; however, mutant PfCRT found in CQ-resistant parasites effluxes this charged CQ out of the DV, removing it from its site of action [11]. Although the current first line artemisinin-combination therapies are effective in clearing parasitaemia, resistance against artemisinins has emerged [12–17]. There is therefore an urgent need to develop novel antimalarial strategies. Several research groups, including our own, have tried different approaches to tackle the problem of CQ resistance by either reversing CQ resistance with a PfCRT inhibitor or synthesizing "reversed" CQ analogues that cannot be effluxed by PfCRT [18–24]. The ultimate goal is to reintroduce CQ as a viable treatment for malaria. Both development of PfCRT inhibitors and synthesis of "reversed" CQ analogues require a sensitive assay for CQ uptake which is typically performed by the use of radiolabelled CQ [22,25–28]. Such methods are difficult to adopt in a high-throughput screen and may raise concerns of safety. To overcome this technical difficulty, fluorescent derivatives of chloroquine have recently been developed and used for this purpose; fluorophores used include 6-(N-(7-nitrobenz-2-oxa-1, 3-diazol-4-yl)amino)hexanoic acid (NBD) [29], coumarin [21,30], and 4, 4-difluoro-4-bora-3a, 4a-diaza-*s*-indacene (BODIPY) [31].

BODIPY derivatives typically exhibit strong fluorescence and are relatively inert in biological conditions [32]. Furthermore, their maximum emission wavelengths are in the green-red region [32], allowing them to be used with many DNA dyes that fluoresce blue, such as the DAPI and Hoechst stains. These properties make BODIPY a promising candidate as a marker for CQ uptake in *P. falciparum*. We therefore present here the characterization of a commercially-available BODIPY-CQ conjugate, LynxTag-CQ_{GREEN}, in several laboratory strains and clinical isolates.

Methods

Parasite culture and synchronization

P. falciparum laboratory strains 3D7 (MRA-102), K1 (MRA-159), 7G8 (MRA-154), HB3 (MRA-155), CS2 (MRA-96), T9-94 (MRA-153), and Dd2 (MRA-156) were obtained from MR4, ATCC Manassas Virginia. Strain T9/96 was obtained from The European Malaria Reagent Repository. A further twelve clinical isolates were collected from the Mae Sot district, Tak Province, in northwest Thailand at the Shoklo Malaria Research Unit; these isolates are prefixed 'SMRU'. Parasites were continuously cultured in complete malaria culture media (MCM) consisting of RPMI 1640 (Life Technologies) supplemented with 0.5% (w/v) Albumax I (Invitrogen), 0.005% (w/v) hypoxanthine, 0.03% (w/v) L-glutamate, 0.25% (w/v) gentamycin, with human erythrocytes at 2.5% haematocrit. Cultures were gassed with a mixture of 3% CO_2, 4% O_2 and 93% N_2 and incubated at 37°C. Synchronization of parasite cultures was performed by resuspending erythrocytes in 5% (w/v) D-sorbitol and incubating at 37°C for 10 min, after which the erythrocytes were washed twice, resuspended in MCM and returned to culture conditions. Thin Giemsa smears were made before each experiment to determine parasitemia and parasite stage.

Compound preparation

For work involving parasites, chlorpheniramine maleate salt, chlorpromazine hydrochloride, desipramine hydrochloride, promethazine hydrochloride, verapamil hydrochloride and CQ diphosphate (all from Sigma-Aldrich) were dissolved in PBS to a working concentration of 1 mM. LynxTag-CQ_{GREEN} (BioLynx Technologies, Singapore; hereafter abbreviated to 'CQ_{GREEN}') was dissolved in DMSO to the same concentration. All compounds were stored at −20°C and protected from light. For microsome uptake assays, methiothepin mesylate salt, metergoline, loperamide hydrochloride, octoclothepin maleate salt, mibefradil dihydrochloride hydrate, L703,606 oxalate salt hydrate, and chlorprothixene hydrochloride (all from Sigma-Aldrich) were dissolved in DMSO to 10 mM and stored at 4°C. Verapamil hydrochloride, adenosine triphosphate (ATP), and CQ diphosphate (all from Sigma-Aldrich) were dissolved in water to 7.5 mM, 50 mM and 0.1 M respectively and stored at −20°C. Tritiated CQ (^{3}H-CQ; from Moravek Biochemicals and Radiochemicals) was diluted in water to 5.32 μM and stored at −20°C; specific activity was 4.7 Ci/mmol.

Reinvasion half-maximal inhibitory concentration (IC_{50})

Synchronized ring-stage cultures at 1–2% parasitemia, 1.25% haematocrit were incubated with either CQ or CQ_{GREEN} at a range of concentrations for 48 h in 96-well flat-bottomed plates at culture conditions. Following this, cells were stained with 1 μg/ml of Hoechst 33342 (Invitrogen) for 30 min at 37°C, washed twice and resuspended in PBS. Parasitemia was then assessed with the CyAn ADP flow cytometer (Beckman Coulter). IC_{50}s were determined by plotting the measurements in Graphpad Prism 5 using a variable slope logistic curve.

Confocal imaging

200 μl cultures of 3D7 at 3% parasitemia, 1.25% haematocrit were incubated with CQ_{GREEN} for 2 h at 2 μM in a 96-well plate format. Erythrocytes were then washed twice and stained with Hoechst 33342 as previously. Wet mounts of stained parasites were visualized under ×100 magnification with the Fluoview FV1000 confocal microscope (Olympus). Hoechst and CQ_{GREEN} were excited at 405 nm and 488 nm with emissions captured at 430–470 nm and 505–525 nm respectively.

Parasite CQ_{GREEN} uptake assay

Synchronized trophozoite-stage cultures at 3–5% parasitemia were resuspended in 200 μl of MCM with 2 μM of CQ_{GREEN} to 2.5% haematocrit in a 96-well plate format. The parasites were then incubated for 2 h at culture conditions, after which they were washed twice and resuspended in PBS. Cells were allowed to settle in a Nunc F96 MicroWell black non-treated polystyrene plate (Thermo Scientific) for 1 h. Fluorescence was then measured with the Infinite M200 microplate reader (Tecan) with excitation and emission wavelengths of 488 nm and 520 nm respectively. K1 chemoreversal assays were performed by pretreatment with 10 μM of the reversal agents for 30 min prior to the addition of CQ_{GREEN}.

Preparation of microsomes carrying PfCRT

PfCRT originating from *P. falciparum* strains Dd2 or 3D7 were expressed in *Pichia pastoris* KM71 and microsomes harvested as described previously [33]. Microsomal levels of PfCRT were determined by western blot with standard curves generated from blots of purified PfCRT.

Uptake kinetics in microsomes

In order to assess the Michaelis-Menten kinetics of CQ_{GREEN} uptake by the microsomes, total or non-specific uptake was measured. Non-specific uptake was determined by pretreating microsomes with unlabelled CQ at 1000 times of the concentration of CQ_{GREEN} used, for 15 min at 37°C, before adding CQ_{GREEN}; total uptake was determined without the pretreatment. Reactions were carried out in accumulation buffer (0.25 M sucrose, 10 mM Tris–HCl, 5 mM $MgCl_2$, and 3 mM ATP, pH 7.5). Following this, microsomes were washed twice in accumulation buffer, then lysed in lysis buffer (0.75 M HCl, 1% Triton X-100, 77.5% isopropanol) on ice for at least one hour. Unlabelled CQ was excluded from the washing step as the addition of CQ was observed to cause displacement of CQ_{GREEN} from the microsomes, possibly due to the higher affinity of CQ for PfCRT. Measurement of the fluorescence intensity was performed using the FLUOstar Galaxy microplate reader (BMG Labtech) with excitation and emission wavelengths of 480 nm and 520 nm respectively. For each experiment, measurements were made in triplicate and the mean calculated. Specific uptake was then calculated by subtracting non-specific uptake from total uptake. Non-linear regression analysis with the Michaelis-Menten model (GraphPad Prism 5) was then applied to obtain the V_{max} and K_m of specific uptake.

CQ_{GREEN} uptake in PfCRT microsomes

Unless stated otherwise, microsomes were incubated with 15 μM CQ_{GREEN} at 37°C for 15 min. For uptake inhibition assays, microsomes were pre-incubated with chemoreversal compounds [21] for 15 min at 37°C prior to the addition of CQ_{GREEN}. All data presented are specific uptake based on the calculations stated above.

^{3}H-CQ uptake in PfCRT microsomes

^{3}H-CQ uptake was measured as described previously [33], with some modifications. Non-specific uptake was determined by pre-incubation with 200-fold unlabelled CQ. Incubation was performed with ^{3}H-CQ at 308 nM for 5 min, after which 200-fold unlabelled CQ was added to stop the reaction. Microsomes were then precipitated by the addition of polyethylene glycol (PEG) 8000 and washed twice with accumulation buffer containing 200-fold unlabelled CQ to remove excess ^{3}H-CQ. Microsomes were then resuspended in scintillation buffer and agitated overnight. Radioactivity was measured using the LS 5600 Scintillation Counter (Beckman). All data presented are specific uptake.

Genotyping of strains and isolates

To assess *pfmdr1* polymorphisms, parasite DNA from *in vitro* cultures was extracted with the QIAamp DNA Mini kit (Qiagen) as per the manufacturer's instructions. For *pfcrt* polymorphisms, total RNA was extracted with the RNeasy Mini Kit (Qiagen) and reverse transcription performed with SuperScript III (Invitrogen) as per manufacturers' instructions. Polymerase chain reaction (PCR) mixtures were made with 200 μM of each dNTP, 0.5 μM forward primer, 0.5 μM reverse primer, 0.02 U/μl Phusion DNA polymerase (Thermo Scientific), 6 μl of 5× Phusion HF buffer, and 1 μl of genomic DNA or cDNA to a total reaction volume of 30 μl. Thermocycler parameters were as follows: 98°C for 30 s, followed by 35 cycles of 98°C for 10 s, 60°C for 30 s, and 72°C for 1 min. Primers used for *pfmdr1* sequencing were 5′- ATGGG-TAAAGAGCAGAAAGA and 5′- TCCACAATAACTTGCAA-CAGT, or 5′- GTCAAGCGGAGTTTTTGC and 5′- TAT-TCTCTGTTTTTGTCCAC. *Pfcrt*-specific primers were 5′-GACGAGCGTTATAGAGAAT and 5′- CTTCGGAATCTT-CATTTTCT. PCR products were purified with the QIAquick PCR purification kit as per manufacturer's instructions. Purified PCR products were sequenced by a commercial vendor (AIT Biotech, Singapore). Copy number of *pfmdr1* was assessed by real-time PCR as previously reported [34]. Briefly, reaction mixtures were prepared with TaqMan universal PCR master mix (Applied Biosystems), 5.5 mM $MgCl_2$, 300 nM dNTPs, 300 nM each of forward and reverse primers, and 100 nM of the probe. Thermocycler parameters were 95°C for 10 min, then 40 cycles of 95°C for 15 s and 60°C for 1 min. Forward and reverse primers used were 5′- TGCATCTATAAAACGATCAGACAAA and 5′-TCGTGTGTTCCATGTGACTGT respectively, and TaqMan probe was 5′- 6FAM-TTTAATAACCCTGATCGAAATG-GAACCTTTG-TAMRA. A reference gene, β-tubulin, was also included; primers were 5′- TGATGTGCGCAAGTGATCC and 5′- TCCTTTGTGGACATTCTTCCTC, while the probe was 5′- VIC-TAGCACATGCCGTTAAATATCTTCCATGTCT-TAMRA. The threshold cycle (C_t) was analysed by the comparative C_t method, based on DNA amplification efficiencies of the *pfmdr1* and β-tubulin genes. *Pfmdr1* copy number was calculated according to the following formula: $\Delta C_t = C_tR - C_tG$, where C_tR is the reference β-tubulin C_t, and C_tG is that of *pfmdr1*. Each TaqMan run included three reference DNA samples from clones 3D7, K1, and Dd2 having *pfmdr1* copy numbers of 1, 1, and 3 respectively.

Statistical analyses

All statistical analyses were performed with SPSS 21. Chemoreversal assays were assessed with Student's t test, 2-tailed. Multiple linear regression was performed with the stepwise method, using the log of IC_{50}s and with dummy coded values for the respective amino acid residues.

Ethics statement

The blood collection protocol for *in vitro* malaria culture was approved by the Institutional Review Board (NUS-IRB Reference Code: 11–383, Approval Number: NUS-1475) of the National University of Singapore (NUS). All participants provided written informed consent. The clinical isolates used were obtained under ethical guidelines in the approved protocol: OXTREC Reference Number 29–09 (Center for Clinical Vaccinology and Tropical

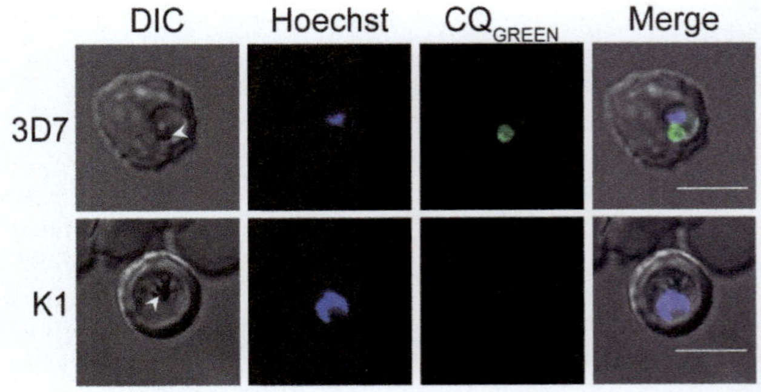

Figure 1. CQ_{GREEN} localization in *P. falciparum* 3D7. Parasites were stained with CQ_{GREEN} and Hoechst and visualized via confocal microscopy under a 100× objective. CQ_{GREEN} accumulates in the DV but also slightly stains parasite cytosol; erythrocyte cytosol is not stained. Arrowheads denote the DV. Scale bars represent 5 μm.

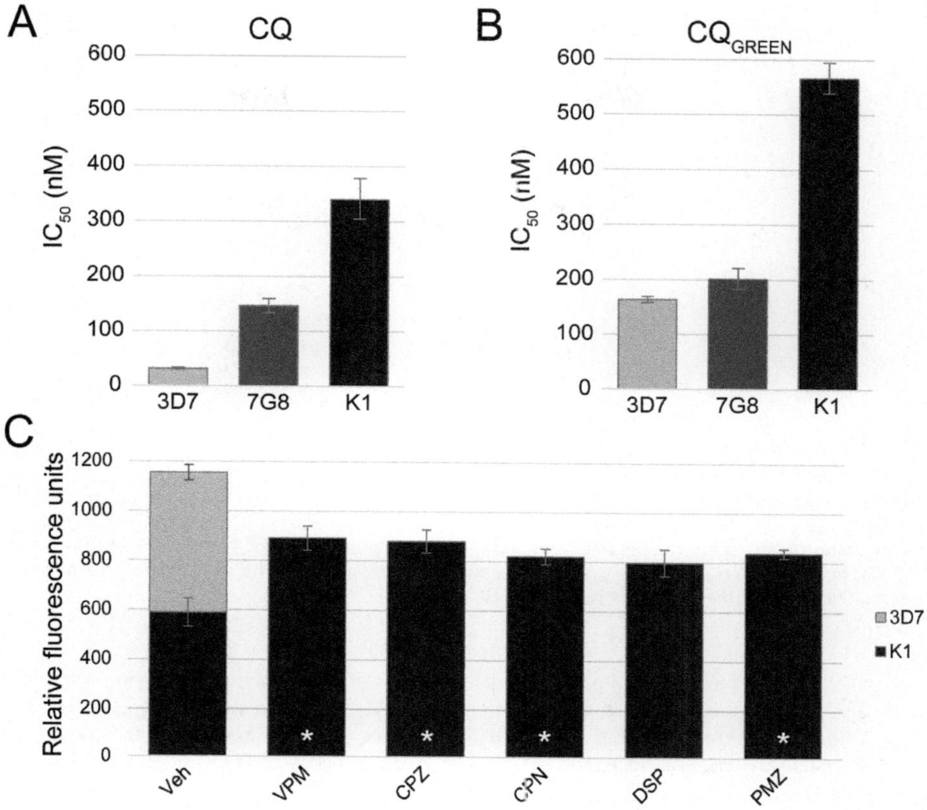

Figure 2. Validation of CQ_{GREEN} activity and uptake in laboratory strains. (A, B) CQ_{GREEN} IC_{50}s recapitulates CQ IC_{50}s in three *P. falciparum* laboratory strains, 3D7, 7G8 and K1. 3D7 is a CQ-susceptible strain, 7G8 is moderately resistant, while K1 is highly CQ-resistant. Data shown are geometric means from at least 3 experiments. As the error from IC_{50}s are not symmetrical, error bars indicate 95% confidence intervals instead of the standard error of the mean (S.E.M.). (C) CQ_{GREEN} uptake as measured by fluorescence is increased in CQ-resistant K1 when pretreated with chemoreversal agents. 3D7 exhibits highest uptake of CQ_{GREEN}. Data are means from at least 3 experiments; error bars are S.E.M. Veh: vehicle control; VPM: verapamil; CPZ: chlorpromazine; CPN: chlorpheniramine; DSP: desipramine; PMZ: promethazine. *: $p<0.05$.

Medicine, University of Oxford, Oxford, United Kingdom). Use of field isolates in NUS was in accordance with NUS IRB (Reference Code: 12–369E).

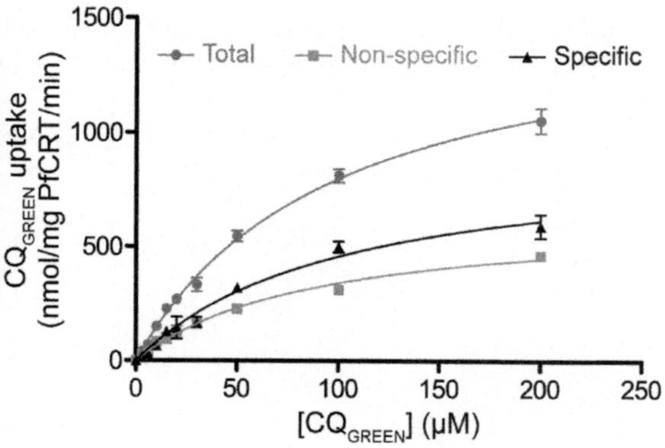

Figure 3. CQ_{GREEN} uptake by Dd2 PfCRT microsomes. Total uptake of CQ_{GREEN} was measured at various CQ_{GREEN} concentrations in microsomes carrying Dd2 PfCRT. Non-specific uptake of CQ_{GREEN} was measured with pre-treatment of excess unlabelled CQ. Specific uptake was estimated as the difference between total and non-specific uptake. V_{max} and K_m of the specific uptake was 938.5 nmol/mg PfCRT/min and 105.1 µM respectively. Data are means ± S.E.M.; n≥3.

Results and Discussion

Validation of CQ_{GREEN} localization and antimalarial activity

CQ is generally believed to accumulate in the DV as a result of ion-trapping [35,36]. Confocal microscopy was therefore performed to ascertain the localization of CQ_{GREEN} in the parasite. Cultures of *P. falciparum* 3D7 were co-stained with Hoechst dye and CQ_{GREEN}, revealing a preferential accumulation of CQ_{GREEN} in the parasite DV (Fig. 1). Interestingly, CQ_{GREEN} fluorescence was also observed in the parasite cytosol but not in the erythrocyte cytosol. As CQ_{GREEN} is a CQ analog, the antimalarial potency of CQ_{GREEN} should be similar to that of CQ. To assess this, reinvasion IC_{50}s of CQ and CQ_{GREEN} on the laboratory strains 3D7, 7G8 and K1 were determined. CQ_{GREEN} showed the same general trend of antimalarial activity as CQ, in that it is most potent against 3D7, followed by 7G8, then K1 (Fig. 2A, 2B).

CQ_{GREEN} fluorescence as a proxy for CQ uptake in parasites

Next, we determined if CQ_{GREEN} uptake by the highly CQ-resistant strain K1 can be increased by pre-treatment with chemosensitizers. Verapamil, chlorpromazine, chlorpheniramine, desipramine, and promethazine have previously been reported to reverse CQ resistance and increase CQ uptake in CQ-resistant

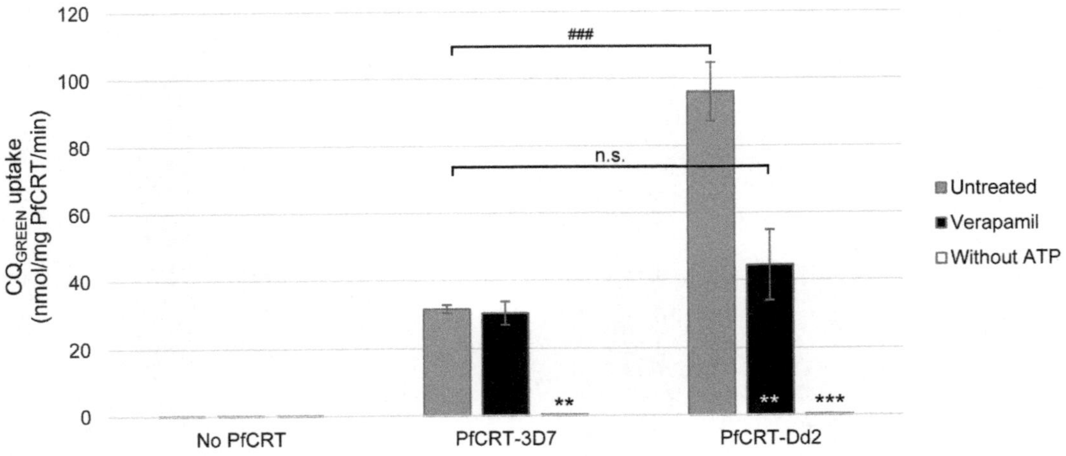

Figure 4. ATP-dependent, verapamil-sensitive uptake of CQ_{GREEN} in microsomes. Yeast microsomes expressing CQ-sensitive or -resistant PfCRT ("PfCRT-3D7" and "PfCRT-Dd2" respectively), or microsomes from plasmid vector control ("No PfCRT"), were incubated with CQ_{GREEN} under different conditions. Preincubation with 150 μM verapamil abrogated CQ_{GREEN} uptake from PfCRT-Dd2 but did not affect uptake in PfCRT-3D7 microsomes. Removal of ATP from buffer abolished CQ_{GREEN} uptake entirely. **, ***: $p<0.005$ and $p<0.001$ respectively, against untreated control. ###: $p<0.001$. N.s.: not significant. Data presented are means ± S.E.M.; n≥3.

strains [37–41]. CQ-sensitive 3D7 was included as a reference for complete reversal. All reversal agents except desipramine induced a significant increase in CQ_{GREEN} fluorescence (Fig. 2C). Desipramine is in fact a less potent reversal agent compared to verapamil when tested in the resistant strain Dd2, and two CQ-resistant field isolates [42]; this may explain why desipramine's effect on CQ_{GREEN} uptake did not achieve statistical significance. Taken together with the CQ_{GREEN} IC_{50} data, we believe that the reversibility of CQ_{GREEN} uptake by known chemoreversal agents suggests that CQ_{GREEN} shares similar structural properties with CQ.

Uptake of CQ_{GREEN} in microsomes bearing PfCRT

To test if CQ_{GREEN} can be transported by PfCRT, we have expressed PfCRT in *Pichia pastoris* and used the microsomes derived to study CQ_{GREEN} uptake. Figure 3 shows that CQ_{GREEN} uptake in Dd2 PfCRT-expressing microsomes is specific. At the highest concentration of CQ_{GREEN} used (200 μM), uptake was close to saturated. Michaelis-Menten approximation of CQ_{GREEN} uptake kinetics in Dd2 microsomes (Figure 3) yields a V_{max} and K_m of 938.5 nmol/mg PfCRT/min and 105.1 μM respectively, which are approximately 2000 and 500 times higher compared to when ^{3}H-CQ was used [33]. Conjugation of CQ with the BODIPY fluorophore may have altered the affinity of PfCRT for the molecule. However, both the high (micromolar) K_m and non-saturation of CQ_{GREEN} transport are consistent with a previous report in a *Xenopus* oocyte system using ^{3}H-CQ [43]. We have also compared CQ_{GREEN} uptake in microsomes with PfCRT originating from either CQ-sensitive 3D7 or CQ-resistant Dd2. CQ_{GREEN} uptake in Dd2 PfCRT-expressing microsome was 96.07 nmol/mg PfCRT/min, which was about three times that of 3D7 PfCRT (31.64 nmol/mg PfCRT/min) (Figure 4). PfCRT-

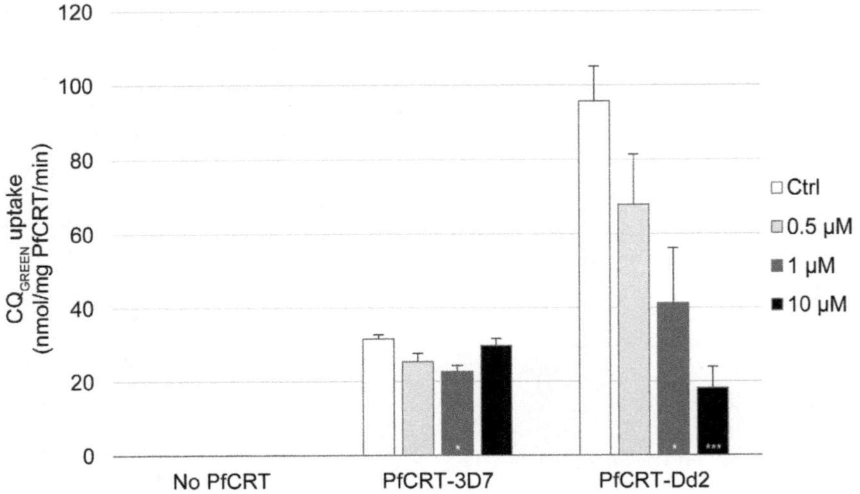

Figure 5. CQ_{GREEN} uptake by resistant-type PfCRT is inhibited by mibefradil in a dose-dependent manner. Microsomes were preincubated with varying concentrations of the PfCRT inhibitor mibefradil prior to addition of CQ_{GREEN}. At the highest concentration of 10 μM, mibefradil drastically suppressed CQ_{GREEN} uptake in PfCRT-Dd2 microsomes but had no significant effect on uptake in PfCRT-3D7 microsomes. *, ***: $p<0.05$ and $p<0.001$ respectively, against no mibefradil control (Ctrl). Data presented are means ± S.E.M.; n≥3.

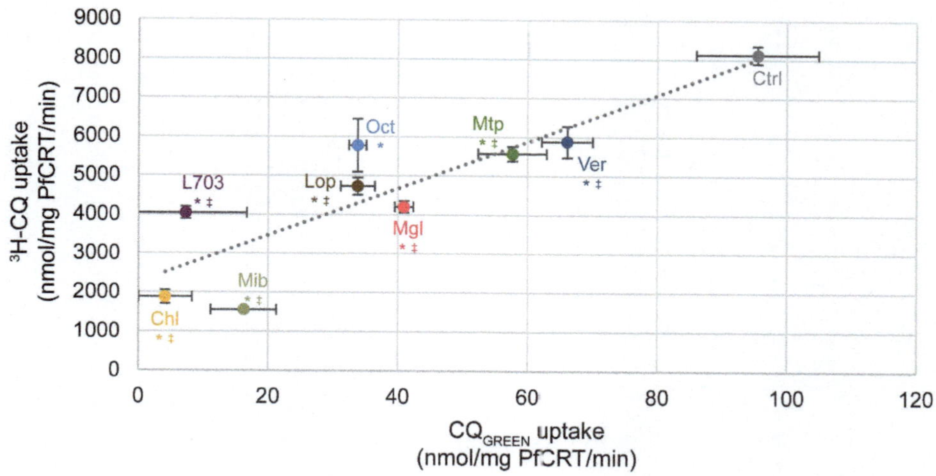

Figure 6. Accumulation of CQ$_{GREEN}$ and ^{3}H-CQ in PfCRT-Dd2 microsomes. Microsomes were incubated with 10 μM of chemosensitizers before addition of CQ$_{GREEN}$ or ^{3}H-CQ. Ctrl: negative control; Ver: verapamil; Mtp: methiothepin; Mgl: metergoline; Lop: loperamide; Oct: octoclothepin; Mib: mibefradil; L703: L703,606; Chl: chlorprothixene. *: $p<0.05$, comparing CQ$_{GREEN}$ uptake against control. ‡: $p<0.05$, comparing ^{3}H-CQ uptake against control. Data presented are means ± S.E.M.; n≥3.

mediated transport of CQ is thought to be ATP-dependent [25,44]. Here we demonstrated that removal of ATP completely abolished CQ$_{GREEN}$ uptake in both Dd2 and 3D7 PfCRT microsomes (Figure 4). Verapamil, a known CQ resistance chemosensitizer which has no effect on CQ-sensitive strains, can reverse CQ$_{GREEN}$ uptake in Dd2 PfCRT to almost that of 3D7 level but has almost no effect on 3D7 PfCRT (Figure 4). These results suggest that CQ$_{GREEN}$ is similar to CQ in that (1) it is differentially recognized by CQ-resistant versus CQ-sensitive PfCRT, (2) its uptake by PfCRT is ATP-dependent, and (3) its uptake by CQ-resistant PfCRT is verapamil-reversible. Others have shown that extensive CQ side-chain modifications can render the CQ analogues not transportable by PfCRT and abolish their verapamil sensitivity [45,46]. Our findings show that the additional BODIPY moiety in CQ$_{GREEN}$ still allows CQ$_{GREEN}$ to be differentially recognized by resistant versus sensitive PfCRT and these transport activities are still sensitive to verapamil. To demonstrate the usefulness of CQ$_{GREEN}$ in screening for PfCRT inhibitors, we found that CQ$_{GREEN}$ uptake by Dd2 PfCRT microsomes can be inhibited by mibefradil, a novel potent chemosensitizer [21], in a dose-dependent manner (Figure 5). Inhibition of CQ$_{GREEN}$ uptake by a panel of known chemoreversal agents was compared to that using ^{3}H-CQ (Figure 6). Mean uptake of CQ$_{GREEN}$ was positively correlated with that of ^{3}H-CQ (β of 60.24, $p=0.002$), showing moderately good agreement between CQ$_{GREEN}$ and ^{3}H-CQ uptake ($R^2=0.766$). However, ^{3}H-CQ uptake was roughly 100 times greater than that of CQ$_{GREEN}$.

Polymorphisms and copy number variation in *pfcrt* and *pfmdr1*

Eight laboratory strains and twelve clinical isolates were sequenced for polymorphisms in PfCRT and Pgh1, the proteins encoded by the genes *pfcrt* and *pfmdr1* respectively. Residues examined were 72, 74, 75, 76, 220, 271, 326, 356, and 371 for PfCRT, and 86, 184, 1034, 1042, 1226, and 1246 for Pgh1. These residues were chosen for analysis as they were previously implicated in modulating multidrug resistance as well as resistance against CQ [47,48]. Copy number of *pfmdr1* was also determined for each strain and isolate. All clinical isolates showed Dd2-type PfCRT mutations, whereas Pgh1 mutations and *pfmdr1* copy numbers were more varied (Table 1).

Genetic correlates of CQ$_{GREEN}$ uptake

All strains and isolates were assayed for CQ$_{GREEN}$ uptake in the trophozoite stage, and their CQ IC_{50}s determined with the standard reinvasion assay. For the entire data set, CQ$_{GREEN}$ uptake was inversely correlated to the log of CQ IC_{50}, with an R^2 of 0.53 (Fig. 7A). However, multiple linear regression with sequencing and copy number data revealed that CQ$_{GREEN}$ uptake was significantly correlated with not only CQ logIC_{50} but also a F1226Y substitution in Pgh1 (β of -587.32 and 178.70, $p<0.001$ and $p=0.024$ respectively; adjusted R^2 of 0.615). None of the other mutations was significantly correlated with CQ$_{GREEN}$ uptake. Separating the data set to two subpopulations improved the R^2 to 0.72 in the Pgh1 1226F group and 0.676 in the Pgh1 1226Y group (Fig. 7B). It is tempting to conclude that the Pgh1 F1226Y substitution plays a significant causal role in modulating CQ$_{GREEN}$ uptake, given that it is also correlated with resistance to artemisinin, mefloquine and lumefantrine [48] and Pgh1's putative sequestration of cytosol-active drugs in the DV [49]. However, it must be kept in mind that the F1226Y mutation was only detected in the clinical isolates, and given the localized collection of these isolates within a small geographical region, F1226Y is likely to be strongly correlated with other undiscovered mutations. In fact, the PfCRT and Pgh1 haplotypes of the F1226Y mutants examined were identical, apart from SMRU0501 which had an additional Pgh1 Y184F substitution (Table 1). It is notable that for the F1226Y mutants, CQ$_{GREEN}$ uptake could range as high as that of the CQ-susceptible strains (Fig. 7B) while still maintaining CQ resistance. One possible gene modulating CQ$_{GREEN}$ uptake could be *pfmrp*, which has been proposed to efflux drugs across the parasite plasma membrane into the parasitophorous vacuolar lumen [49], which would still contribute to CQ$_{GREEN}$ fluorescence but sequester the drug from its site of action. Knock-out mutants of this gene exhibit increased susceptibility to CQ, quinine, artemisinin, piperaquine, and primaquine [50]. Alternative mechanisms besides efflux of CQ could perhaps also contribute to CQ resistance, such as decreased susceptibility to CQ-induced apoptosis-like cell death [51].

Table 1. CQ IC_{50}s, PfCRT and Pgh1 polymorphisms, and *pfmdr1* copy number.

		PfCRT residue no.									Pgh1 residue no.						*pfmdr1* copy number
Laboratory strains	**CQ IC_{50} (nM)**	**72**	**74**	**75**	**76**	**220**	**271**	**326**	**356**	**371**	**86**	**184**	**1034**	**1042**	**1226**	**1246**	
T9/96	24	C	M	N	K	A	Q	N	I	R	N	Y	S	N	F	D	1
3D7	31	C	M	N	K	A	Q	N	I	R	N	Y	S	N	F	D	1
HB3	42	C	M	N	K	A	Q	N	I	R	N	**F**	S	**D**	F	D	1
CS2	115	C	**I**	**E**	**T**	**S**	**E**	**S**	I	**I**	**Y**	Y	S	N	F	D	3
T9-94	146	C	**I**	**E**	**T**	**S**	**E**	**S**	I	**I**	**Y**	Y	S	N	F	D	3
7G8	146	**S**	M	N	**T**	**S**	Q	**D**	**L**	R	N	**F**	**C**	**D**	F	**Y**	1
Dd2	276	C	**I**	**E**	**T**	**S**	**E**	**S**	**T**	**I**	**Y**	Y	S	N	F	D	3
K1	340	C	**I**	**E**	**T**	**S**	**E**	**S**	I	**I**	**Y**	Y	S	N	F	D	1
Clinical isolates	**CQ IC_{50} (nM)**	**72**	**74**	**75**	**76**	**220**	**271**	**326**	**356**	**371**	**86**	**184**	**1034**	**1042**	**1226**	**1246**	***pfmdr1* copy number**
SMRU0233	142	C	**I**	**E**	**T**	**S**	**E**	**S**	**T**	**I**	N	Y	S	N	**Y**	D	2
SMRU0116	147	C	**I**	**E**	**T**	**S**	**E**	**S**	**T**	**I**	N	Y	S	N	**Y**	D	2
SMRU0101	156	C	**I**	**E**	**T**	**S**	**E**	**S**	**T**	**I**	N	Y	S	N	F	D	2
SMRU1116	172	C	**I**	**E**	**T**	**S**	**E**	**S**	**T**	**I**	N	Y	S	N	F	D	1
SMRU0270	178	C	**I**	**E**	**T**	**S**	**E**	**S**	**T**	**I**	N	Y	S	N	F	D	2
SMRU0402	196	C	**I**	**E**	**T**	**S**	**E**	**S**	**T**	**I**	N	Y	S	N	**Y**	D	2
SMRU1093	231	C	**I**	**E**	**T**	**S**	**E**	**S**	**T**	**I**	N	Y	S	N	**Y**	D	2
SMRU0201	249	C	**I**	**E**	**T**	**S**	**E**	**S**	**T**	**I**	N	Y	S	N	F	D	1
SMRU0501	287	C	**I**	**E**	**T**	**S**	**E**	**S**	**T**	**I**	N	**F**	S	N	**Y**	D	1
SMRU0272	332	C	**I**	**E**	**T**	**S**	**E**	**S**	**T**	**I**	N	Y	S	N	**Y**	D	3
SMRU0002	377	C	**I**	**E**	**T**	**S**	**E**	**S**	**T**	**I**	N	Y	S	N	**Y**	D	1
SMRU0279	473	C	**I**	**E**	**T**	**S**	**E**	**S**	**T**	**I**	N	Y	S	N	**Y**	D	1

Bolded residues indicate deviation from 3D7 haplotype. CQ IC_{50}s are geometric means of at least 3 measurements.

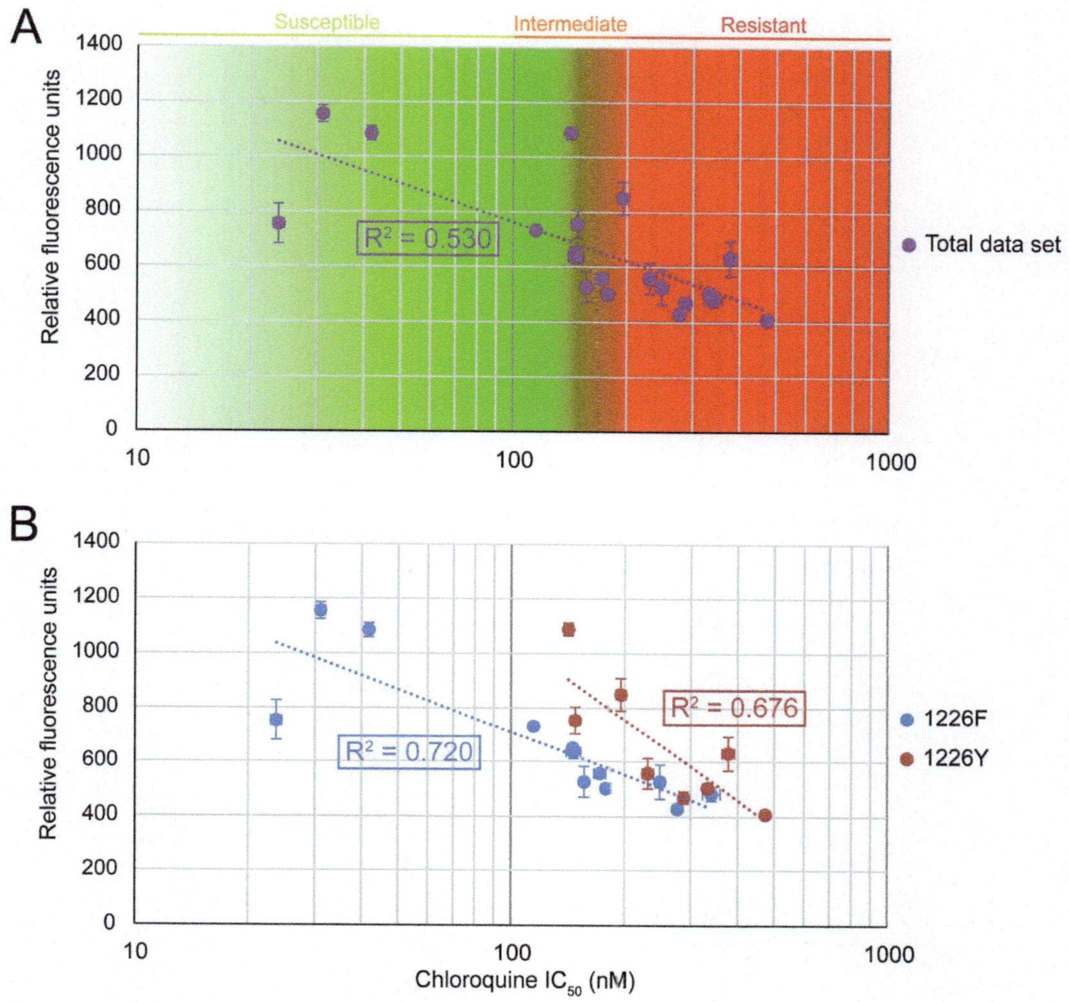

Figure 7. CQ_{GREEN} uptake correlates with CQ resistance. (A) CQ_{GREEN} fluorescence is inversely correlated with CQ $\log IC_{50}$. Total data set shows moderate R^2 of 0.53. (B) When split into subpopulations on the basis of Pgh1 residue 1226, R^2 is improved. Data shown are means of at least 3 experiments. Error bars represent S.E.M.

Conclusions

CQ_{GREEN} is a commercially available fluorescent CQ analog that interacts with the parasite in a similar fashion to CQ. CQ_{GREEN} presents some advantages over traditional radiolabelled CQ in uptake studies: it is safer to handle as it is not radioactive, and its fluorescence properties allows it to be monitored by common fluorescence equipment. Using a defined microsomal platform, we showed that CQ_{GREEN} interacts with PfCRT in a manner similar to CQ. Its use as a predictor of CQ susceptibility is enhanced if residue 1226 of Pgh1 is known. Unlike a typical reinvasion assay which may require 48 h or more, the use of CQ_{GREEN} allows for measurement of CQ susceptibility within several hours. We believe that CQ_{GREEN} could be a valuable tool in future drug discovery projects or used in the identification of factors involved in drug resistance.

Acknowledgments

We thank the patients and staff of Shoklo Malaria Research Unit (SMRU) for their contributions to this project. We would also like to extend our gratitude to the following for *P. falciparum* strains obtained through MR4's BEI Resources Repository, NIAID, NIH: DJ Carucci (3D7, MRA-102), TE Wellems (HB3, MRA-155; Dd2, MRA-156), SJ Rogerson (CS2, MRA-96), D Walliker (T9-94, MRA-153), and DE Kyle (7G8, MRA-154; K1, MRA-159). *P. falciparum* T9/96 was obtained from The European Malaria Reagent Repository (http://www.malariaresearch.eu).

Author Contributions

Conceived and designed the experiments: KSWT BR FHN LR MJL LMCC. Performed the experiments: YQL CCYL KWKC KYC RS. Analyzed the data: YQL CCYL RS. Contributed reagents/materials/analysis tools: KSWT LR FHN. Wrote the paper: YQL KSWT BR FHN LR MJL LMCC CCYL KWKC KYC RS.

References

1. Murray CJ, Rosenfeld LC, Lim SS, Andrews KG, Foreman KJ, et al. (2012) Global malaria mortality between 1980 and 2010: a systematic analysis. The Lancet 379: 413–431. doi:10.1016/S0140-6736(12)60034-8.

2. World Health Organization (2012) World Malaria Report 2012. Available: http://www.who.int/malaria/publications/world_malaria_report_2012/en/index.html. Accessed 2013 Mar 11.

3. Krafts K, Hempelmann E, Skórska-Stania A (2012) From methylene blue to chloroquine: a brief review of the development of an antimalarial therapy. Parasitol Res 111: 1–6. doi:10.1007/s00436-012-2886-x.
4. Coatney GR (1963) Pitfalls in a Discovery: The Chronicle of Chloroquine. Am J Trop Med Hyg 12: 121–128.
5. Butler A, Khan S, Ferguson E (2010) A brief history of malaria chemotherapy. J R Coll Physicians Edinb 40: 172–177. doi:10.4997/JRCPE.2010.216.
6. Wellems TE, Plowe CV (2001) Chloroquine-Resistant Malaria. J Infect Dis 184: 770–776. doi:10.1086/322858.
7. Sullivan DJ Jr, Gluzman IY, Goldberg DE (1996) Plasmodium hemozoin formation mediated by histidine-rich proteins. Science 271: 219–222.
8. Hempelmann E (2007) Hemozoin Biocrystallization in *Plasmodium falciparum* and the antimalarial activity of crystallization inhibitors. Parasitol Res 100: 671–676. doi:10.1007/s00436-006-0313-x.
9. Gildenhuys J, Roex T le, Egan TJ, de Villiers KA (2013) The Single Crystal X-ray Structure of β-Hematin DMSO Solvate Grown in the Presence of Chloroquine, a β-Hematin Growth-Rate Inhibitor. J Am Chem Soc 135: 1037–1047. doi:10.1021/ja308741e.
10. Rathore D, Jani D, Nagarkatti R, Kumar S (2006) Heme detoxification and antimalarial drugs – Known mechanisms and future prospects. Drug Discov Today Ther Strateg 3: 153–158. doi:10.1016/j.ddstr.2006.06.003.
11. Summers RL, Martin RE (2010) Functional characteristics of the malaria parasite's "chloroquine resistance transporter": implications for chemotherapy. Virulence 1: 304–308. doi:10.4161/viru.1.4.12012.
12. Pradines B, Bertaux L, Pomares C, Delaunay P, Marty P (2011) Reduced *in vitro* susceptibility to artemisinin derivatives associated with multi-resistance in a traveller returning from South-East Asia. Malar J 10: 268. doi:10.1186/1475-2875-10-268.
13. Miotto O, Almagro-Garcia J, Manske M, MacInnis B, Campino S, et al. (2013) Multiple populations of artemisinin-resistant *Plasmodium falciparum* in Cambodia. Nat Genet 45: 648–655. doi:10.1038/ng.2624.
14. Takala-Harrison S, Clark TG, Jacob CG, Cummings MP, Miotto O, et al. (2013) Genetic loci associated with delayed clearance of *Plasmodium falciparum* following artemisinin treatment in Southeast Asia. Proc Natl Acad Sci 110: 240–245. doi:10.1073/pnas.1211205110.
15. Ariey F, Witkowski B, Amaratunga C, Beghain J, Langlois A-C, et al. (2014) A molecular marker of artemisinin-resistant *Plasmodium falciparum* malaria. Nature 505: 50–55. doi:10.1038/nature12876.
16. Mok S, Imwong M, Mackinnon MJ, Sim J, Ramadoss R, et al. (2011) Artemisinin resistance in *Plasmodium falciparum* is associated with an altered temporal pattern of transcription. BMC Genomics 12: 391. doi:10.1186/1471-2164-12-391.
17. Witkowski B, Amaratunga C, Khim N, Sreng S, Chim P, et al. (2013) Novel phenotypic assays for the detection of artemisinin-resistant *Plasmodium falciparum* malaria in Cambodia: *in-vitro* and *ex-vivo* drug-response studies. Lancet Infect Dis. doi:10.1016/S1473-3099(13)70252-4.
18. Peyton DH (2012) Reversed chloroquine molecules as a strategy to overcome resistance in malaria. Curr Top Med Chem 12: 400–407.
19. Martin SK, Oduola AMJ, Milhous WK (1987) Reversal of Chloroquine Resistance in *Plasmodium falciparum* by Verapamil. Science 235: 899–901.
20. Burgess SJ, Selzer A, Kelly JX, Smilkstein MJ, Riscoe MK, et al. (2006) A Chloroquine-like Molecule Designed to Reverse Resistance in *Plasmodium falciparum*. J Med Chem 49: 5623–5625. doi:10.1021/jm060399n.
21. Ch'ng J-H, Mok S, Bozdech Z, Lear MJ, Boudhar A, et al. (2013) A Whole Cell Pathway Screen Reveals Seven Novel Chemosensitizers to Combat Chloroquine Resistant Malaria. Sci Rep 3. doi:10.1038/srep01734.
22. Martin RE, Butterworth AS, Gardiner DL, Kirk K, McCarthy JS, et al. (2012) Saquinavir Inhibits the Malaria Parasite's Chloroquine Resistance Transporter. Antimicrob Agents Chemother 56: 2283–2289. doi:10.1128/AAC.00166-12.
23. Adovelande J, Delèze J, Schrével J (1998) Synergy between two calcium channel blockers, verapamil and fantofarone (SR33557), in reversing chloroquine resistance in *Plasmodium falciparum*. Biochem Pharmacol 55: 433–440. doi:10.1016/S0006-2952(97)00482-6.
24. Kelly JX, Smilkstein MJ, Cooper RA, Lane KD, Johnson RA, et al. (2007) Design, Synthesis, and Evaluation of 10-N-Substituted Acridones as Novel Chemosensitizers in *Plasmodium falciparum*. Antimicrob Agents Chemother 51: 4133–4140. doi:10.1128/AAC.00669-07.
25. Sanchez CP, Stein W, Lanzer M (2003) Trans Stimulation Provides Evidence for a Drug Efflux Carrier as the Mechanism of Chloroquine Resistance in *Plasmodium falciparum*. Biochemistry (Mosc) 42: 9383–9394. doi:10.1021/bi034269h.
26. Sanchez CP, McLean JE, Stein W, Lanzer M (2004) Evidence for a Substrate Specific and Inhibitable Drug Efflux System in Chloroquine Resistant *Plasmodium falciparum* Strains. Biochemistry (Mosc) 43: 16365–16373. doi:10.1021/bi048241x.
27. Martin RE, Marchetti RV, Cowan AI, Howitt SM, Broer S, et al. (2009) Chloroquine Transport via the Malaria Parasite's Chloroquine Resistance Transporter. Science 325: 1680–1682. doi:10.1126/science.1175667.
28. Zishiri VK, Hunter R, Smith PJ, Taylor D, Summers R, et al. (2011) A series of structurally simple chloroquine chemosensitizing dibemethin derivatives that inhibit chloroquine transport by PfCRT. Eur J Med Chem 46: 1729–1742. doi:10.1016/j.ejmech.2011.02.026.
29. Cabrera M, Natarajan J, Paguio MF, Wolf C, Urbach JS, et al. (2009) Chloroquine Transport in *Plasmodium falciparum*. 1. Influx and Efflux Kinetics for Live Trophozoite Parasites Using a Novel Fluorescent Chloroquine Probe. Biochemistry (Mosc) 48: 9471–9481. doi:10.1021/bi901034r.
30. Ch'ng J-H, Kotturi SR, Chong AG-L, Lear MJ, Tan KS-W (2010) A programmed cell death pathway in the malaria parasite *Plasmodium falciparum* has general features of mammalian apoptosis but is mediated by clan CA cysteine proteases. Cell Death Dis 1: e26. doi:10.1038/cddis.2010.2.
31. Alcantara LM, Kim J, Moraes CB, Franco CH, Franzoi KD, et al. (2013) Chemosensitization potential of P-glycoprotein inhibitors in malaria parasites. Exp Parasitol 134: 235–243. doi:10.1016/j.exppara.2013.03.022.
32. Tram K, Yan H, Jenkins HA, Vassiliev S, Bruce D (2009) The synthesis and crystal structure of unsubstituted 4, 4-difluoro-4-bora-3a, 4a-diaza-s-indacene (BODIPY). Dyes Pigments 82: 392–395. doi:10.1016/j.dyepig.2009.03.001.
33. Tan W, Gou DM, Tai E, Zhao YZ, Chow LMC (2006) Functional reconstitution of purified chloroquine resistance membrane transporter expressed in yeast. Arch Biochem Biophys 452: 119–128. doi:10.1016/j.abb.2006.06.017.
34. Price RN, Uhlemann A-C, Brockman A, McGready R, Ashley E, et al. (2004) Mefloquine resistance in *Plasmodium falciparum* and increased *pfmdr1* gene copy number. The Lancet 364: 438–447. doi:10.1016/S0140-6736(04)16767-6.
35. Homewood CA, Warhurst DC, Peters W, Baggaley VC (1972) Lysosomes, pH and the anti-malarial action of chloroquine. Nature 235: 50–52.
36. Yayon A, Cabantchik ZI, Ginsburg H (1984) Identification of the acidic compartment of *Plasmodium falciparum*-infected human erythrocytes as the target of the antimalarial drug chloroquine. EMBO J 3: 2695–2700.
37. Van Schalkwyk DA, Walden JC, Smith PJ (2001) Reversal of Chloroquine Resistance in *Plasmodium falciparum* Using Combinations of Chemosensitizers. Antimicrob Agents Chemother 45: 3171–3174. doi:10.1128/AAC.45.11.3171-3174.2001.
38. Kyle DE, Oduola AMJ, Martin SK, Milhous WK (1990) *Plasmodium falciparum*: modulation by calcium antagonists of resistance to chloroquine, desethylchloroquine, quinine, and quinidine *in vitro*. Trans R Soc Trop Med Hyg 84: 474–478. doi:10.1016/0035-9203(90)90004-X.
39. Martiney JA, Cerami A, Slater AFG (1995) Verapamil Reversal of Chloroquine Resistance in the Malaria Parasite *Plasmodium falciparum* Is Specific for Resistant Parasites and Independent of the Weak Base Effect. J Biol Chem 270: 22393–22398. doi:10.1074/jbc.270.38.22393.
40. Oduola AM, Sowunmi A, Milhous WK, Brewer TG, Kyle DE, et al. (1998) *In vitro* and *in vivo* reversal of chloroquine resistance in *Plasmodium falciparum* with promethazine. Am J Trop Med Hyg 58: 625–629.
41. Egan T, Kaschula C (2007) Strategies to reverse drug resistance in malaria. Curr Opin Infect Dis 20: 598–604. doi:10.1097/QCO.0b013e3282f1673a.
42. Bayoumi RAL, Babiker HA, Arnot DE (1994) Uptake and efflux of chloroquine by chloroquine-resistant *Plasmodium falciparum* clones recently isolated in Africa. Acta Trop 58: 141–149. doi:10.1016/0001-706X(94)90053-1.
43. Summers RL, Dave A, Dolstra TJ, Bellanca S, Marchetti RV, et al. (2014) Diverse mutational pathways converge on saturable chloroquine transport via the malaria parasite's chloroquine resistance transporter. Proc Natl Acad Sci 111: E1759–E1767. doi:10.1073/pnas.1322965111.
44. Krogstad DJ, Gluzman IY, Herwaldt BL, Schlesinger PH, Wellems TE (1992) Energy dependence of chloroquine accumulation and chloroquine efflux in Plasmodium falciparum. Biochem Pharmacol 43: 57–62. doi:10.1016/0006-2952(92)90661-2.
45. De D, Krogstad FM, Cogswell FB, Krogstad DJ (1996) Aminoquinolines That Circumvent Resistance in *Plasmodium falciparum in Vitro*. Am J Trop Med Hyg 55: 579–583.
46. Lakshmanan V, Bray PG, Verdier-Pinard D, Johnson DJ, Horrocks P, et al. (2005) A critical role for PfCRT K76T in Plasmodium falciparum verapamil-reversible chloroquine resistance. EMBO J 24: 2294–2305. doi:10.1038/sj.emboj.7600681.
47. Valderramos SG, Valderramos J-C, Musset L, Purcell LA, Mercereau-Puijalon O, et al. (2010) Identification of a Mutant PfCRT-Mediated Chloroquine Tolerance Phenotype in *Plasmodium falciparum*. PLoS Pathog 6: e1000887. doi:10.1371/journal.ppat.1000887.
48. Veiga MI, Ferreira PE, Jörnhagen L, Malmberg M, Kone A, et al. (2011) Novel Polymorphisms in *Plasmodium falciparum* ABC Transporter Genes Are Associated with Major ACT Antimalarial Drug Resistance. PLoS ONE 6: e20212. doi:10.1371/journal.pone.0020212.
49. Sanchez CP, Dave A, Stein WD, Lanzer M (2010) Transporters as mediators of drug resistance in *Plasmodium falciparum*. Int J Parasitol 40: 1109–1118. doi:10.1016/j.ijpara.2010.04.001.
50. Raj DK, Mu J, Jiang H, Kabat J, Singh S, et al. (2009) Disruption of a *Plasmodium falciparum* Multidrug Resistance-associated Protein (PfMRP) Alters Its Fitness and Transport of Antimalarial Drugs and Glutathione. J Biol Chem 284: 7687–7696. doi:10.1074/jbc.M806944200.
51. Ch'ng J-H, Liew K, Goh AS-P, Sidhartha E, Tan KS-W (2011) Drug-induced permeabilization of parasite's digestive vacuole is a key trigger of programmed cell death in *Plasmodium falciparum*. Cell Death Dis 2: e216. doi:10.1038/cddis.2011.97.

The Prevalence of HIV-1 Drug Resistance among Antiretroviral Treatment Naïve Individuals in Mainland China: A Meta-Analysis

Yingying Su[1], Fujie Zhang[1], Huixin Liu[1], M. Kumi Smith[2], Lin Zhu[1], Jing Wu[1], Ning Wang[1]*

1 National Center for AIDS/STD Control and Prevention, Chinese Center for Disease Contro and Prevention, Beijing, China, **2** Department of Epidemiology, University of North Carolina, Chapel Hill, North Carolina, United States of America

Abstract

Background: Surveillance of drug resistance in antiretroviral treatment-naïve patients in China is needed to ensure optimal treatment outcomes and control of the human immunodeficiency virus (HIV) epidemic.

Methods: A systematic literature search was conducted in English and Chinese through PubMed (English), China National Knowledge Infrastructure (Chinese), Chinese Biomedical Literature Database (Chinese), and Wanfang (Chinese). Random effects models were used to calculate the pooled prevalence of transmitted drug resistance and subgroup analyses examined prevalence estimates across time periods, study locations, and study populations.

Results: Analysis of data from 71 studies (47 in Chinese and 24 in English) yielded a pooled prevalence of transmitted HIV drug resistance to any antiretroviral drug class of 3.64% (95% confidence interval [CI]: 3.00%–4.32%). Rates were significantly high at initial stage of free ART program from 2003 to 2005 (5.18%, 95%CI: 3.13%–7.63%), and were much lower among studies conducted in 2006–2008 (3.02%, 95%CI: 2.03%–4.16%). A slight increase was observed again in the most recent study period from 2009 to 2012 (3.68%, 95%CI: 2.78%–4.69%). Subgroup analysis revealed highest prevalence levels of transmitted drug resistance in Beijing city, and Henan and Hubei provinces (above 5%), and although differences in prevalence rates among risk groups were negligible, men who have sex with men were unique in their relatively large portion of protease inhibitor resistance, a second-line drug of limited availability in China.

Conclusions: Overall prevalence of transmitted HIV drug resistance in China is classified as "low" by the World Health Organization. However regional and temporal variability suggest a more complex epidemic for which closer HIV drug resistance surveillance is needed. A nationwide HIV drug resistance surveillance system to monitor both treatment-experienced and treatment-naïve patients will be a cornerstone to ensure the effectiveness of treatment scale-up, particularly as China seeks to expand a national policy of antiretroviral treatment as prevention.

Editor: Xia Jin, University of Rochester, United States of America

Funding: This work was supported by funding from the Mega-projects of national science research for the 12th Five-Year Plan (2012ZX10001-001) of China. The funders had no role in study design, data collection and analysis, decision to publish, or preparation of the manuscript.

Competing Interests: The authors have declared that no competing interests exist.

* Email: wangnbj@163.com

Introduction

China has one of the largest populations of persons infected with human immunodeficiency virus type-1(HIV-1). As of 2011 an estimated 780,000 people living with HIV/AIDS and among them 154,000 living with AIDS in China [1]. The HIV epidemic is largely concentrated in injecting drug users (IDU), as well as in groups with higher risk sexual exposures including female sex workers and men who have sex with men (MSM) [2]. As of 2007, however, data from the national case report system has suggested that heterosexual transmission (HST) has become the primary mode of transmission in China, signaling a potential transmission to a more generalized epidemic [1].Of the about 70,000 new HIV diagnoses from January to September in 2013, those believe to be acquired through unprotected sex accounted for 89.9% (heterosexual transmission for 69.1%, homosexual transmission for 20.8%) [3].

To combat the spread of HIV through the general population, the Chinese Center for Disease Control has launched a policy initiative to harness suppressive antiretroviral therapy (ART) to slow the spread of HIV in China. The existing national free ART program established in 2003, is expected to provide a strong foundation for such an effort. Initially started as a pilot program for patients who were former plasma donors (FPD), the program has undergone rapid scale up [4,5]. And by the end of September 2013, the national free ART program has treated about 260,000 cumulative patients [3], and has reduced AIDS associated mortality from 39.3 to 14.2 deaths per 100 persons from 2002 to 2009 [6].

Rapid scale up of ART access has been accompanied by a concomitant rise in drug resistance among treated HIV patients. This has been largely attributed to the limited selection of antiviral agents through the national ART program on which most patients rely, and the lack of critical disease monitoring tools like CD4 or viral load testing in most treatment settings [5,7,8]. Though the relationship between poor treatment outcomes, insufficient viral suppression, and emerge of drug resistance is well understood [9,10], little is known about onward transmission of resistant strains in Chinese settings. Primary infection with drug resistant strains severely limits new patients' treatment options and shortens time to treatment failure. The prevalence of transmitted drug resistance (TDR) in wider circulation could also undermine program efficacy of the national treatment efforts [11]. A better understanding of the prevalence of drug resistance in treatment naïve HIV/AIDS patients in China will provide insight critical for both clinical management and for broader disease control efforts [12].

The goal of this systematic review is to provide an overview and pooled prevalence estimate of HIV drug resistance among treatment-naïve patients in China. We also provide a simple analysis of several trends and distributions available from the data in order to provide more insight into better treatment and control of HIV in China.

Methods

Search strategy

We conducted a systematic literature in English and Chinese using the following four databases: PubMed, China National Knowledge Infrastructure (CNKI), Chinese Biomedical Literature Database (CBM), and Wanfang, from January 1, 1990 through February 28, 2014. Search terms included ("HIV" OR "AIDS" OR "human immunodeficiency virus" OR "acquired immunodeficiency syndrome") AND ("drug resistance" or "drug resistant") AND ("China"). Details on search strategy can be found in the Appendix S1. We further reviewed the reference lists for other relevant articles not captured in our initial search.

Selection Criteria

Investigators YYS and HXL independently assessed eligibility of articles and performed the data extraction. Discrepancies between the two investigators were resolved by consulting the other authors (NW and MKS). Eligible articles all met the following inclusion criteria: (1) inclusion of data on HIV-1-infected and treatment-naïve individuals; (2) availability of basic demographic information for study subjects such as age, sex, or most likely HIV transmission route; (3) availability of information on study design, location, and period; and (4) HIV drug resistance patterns were interpreted using the WHO recommended Stanford HIV Drug Resistance Database (http://hivdb.stanford.edu). Articles that restricted study subjects to HIV patients of a single subtype were excluded, as were those without information on patient treatment status or that pooled results of treated and untreated patients. Studies based outside of mainland China such as Taiwan, Hong Kong and Macao were excluded due to the fact that these regions maintain autonomous and separate HIV/AIDS treatment programs. Articles based on the same or overlapping study populations were included only once to avoid duplication, with preference given to the article with more detailed information about the population, or, all else equal, with the larger sample size.

Data extraction

A standardized form was used to assess article eligibility and to extract data including author names, year of publication, study period, language of article, study location, age range and sex of participants, most likely routes of primary HIV infection, time of infection(chronic or new diagnose), study design (cohort vs. cross section), recruitment method, system of HIV drug resistance typing, portion of tested samples whose HIV RNA was successfully amplified and sequenced, and characteristics of the sequences genotype including number of mutations with at least low-level resistance 1) to any antiretroviral agent, 2) to nucleoside reverse transcriptase inhibitor(NRTI), 3) to non-nucleoside reverse transcriptase inhibitor(NNRTI), 4) to protease inhibitor(PI).

Statistical analysis

The proportion of TDR was defined as number of samples exhibiting at least a lower-level resistance to any antiretroviral agent, divided by the total number of samples successfully amplified and sequenced [13]. This definition of TDR prevalence is consistent with that used for the WHO TDR surveillance mutation list [14]. The Freeman-Tukey Double arcsine transformation of proportions was used to calculate an overall proportion. Model choice (fixed versus random effects models) was based on heterogeneity assessment by examining the Cochran Q statistic (p-value of <0.1 was considered a statistically significant) [15] and the I^2 statistic (values of 25%, 50% and 75% are considered to represent low, medium and high heterogeneity respectively) [16]. The DerSimonian-Laird estimate was used in case of random effects. Subgroup analyses were conducted according to study period, study location and study population to assess the heterogeneity between studies. Publication bias was assessed with the funnel plot and Egger's test, which found no evidence of bias (Figure S1). Analyses were carried out using R EpiTools package [17] and SAS(version 9.3, SAS Institute).

Results

Study selection and characteristics

Of the 4964 articles (4656 in Chinese, 308 in English) initially identified, of which 1934 were excluded as duplicates, another 2858 were removed for relevance based on a title and abstract screening. Following a closer full text review of the remaining 172 articles we identified 71 studies for inclusion in the final analysis, of which 47 were in Chinese, 24 in English (Figure 1). Seventeen of these employed the WHO HIV drug resistance survey method which recommends use of truncated sequential sampling (TSS) to classify levels of transmitted drug resistance for each drug class in resource poor settings [18]. Among the remainder, 54 were cross-sectional studies. Details on resistance to specific classes of antiretroviral drug were available in 68 studies (Table S1).

Transmitted drug resistance prevalence rates

The 71 studies included in this analysis represented results from a collective total of 11,633 HIV-1 treatment naïve individuals, among whom 9,167 HIV RNA sequences were successfully amplified and analyzed. The random effects estimate for overall pooled prevalence of HIV-1 TDR to any antiretroviral drug was 3.64% (95% confidence interval [CI]: 3.00%–4.32%). The pooled prevalence of mutations resistant to NRTI's, NNRTI's, and PI's was 0.91% (95%CI: 0.60%–1.27%), 1.14% (95%CI: 0.75%–1.59%), 0.68% (95%CI: 0.41%–1.01%), respectively.

TDR prevalence was further explored within identifiable subgroups in studies where relevant information was provided (Table 1). Figure 2 shows temporal trends in TDR prevalence

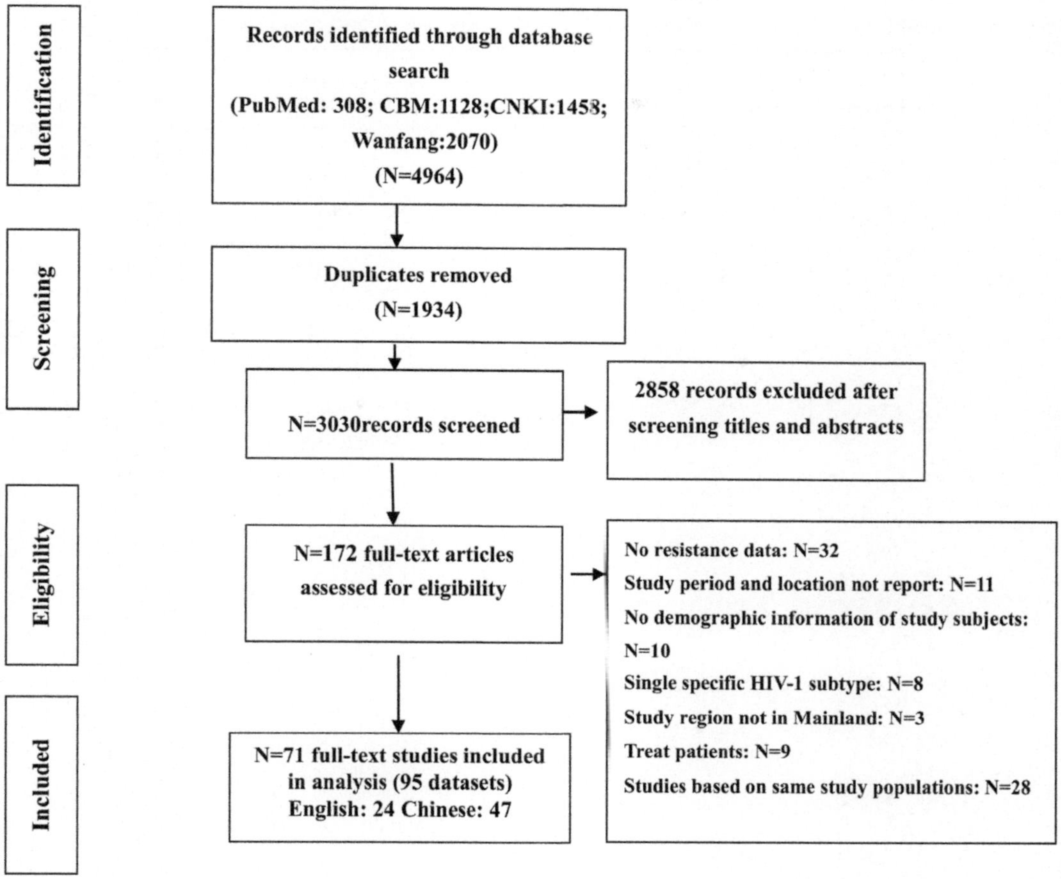

Figure 1. Flow chart of the process of studies selection.

according to the midpoint of the year in which study participants were recruited. A slightly upward trend can be detected in the composite time trend, though the trend was not statistically significant. We further categorized study times into four time periods roughly corresponding to the following phases of the national free ART program: pilot (before 2003), initiation (2003 to 2005), scale-up (2006 to 2008), and introduction of second line therapy (2009–2012) [5,8]. Results of the analyses showed that the overall TDR prevalence for any antiretroviral drug class was significantly high at initial stage of free ART program (5.18%, 95%CI: 3.13%–7.63%), but much lower during the scale-up phase (3.02%, 95%CI: 2.03%–4.16%). TDR prevalence rose again slightly after 2008 (3.68%, 95%CI: 2.78%–4.69%). Geographic subgroup analyses excluded studies in which the study represented the sole report on TDR from a given province [19–23] or in which results were reported as pooled prevalence of patients from multiple provinces [24–32]. Stratified analyses showed that TDR prevalence rates in the city of Beijing and the provinces of Henan and Hubei were all above 5%, but the rest were below this threshold. Finally, where information on most likely routes of HIV infection was provided, we estimated TDR prevalence for the following groups: FPD (5 articles), MSM (17 articles), IDU (5 articles), and those otherwise described as heterosexual individuals (3 articles). Pooled prevalence of TDR in these four risk groups were highest among MSM (4.03%, 95%CI:2.81%–5.43%), and slightly lower among FPD (3.78%, 95% CI: 2.48%–5.31%) and heterosexuals (3.73%, 95% CI: 1.85%–6.20%). Rates were lowest among IDU (2.78%, 95% CI: 1.04%–5.12%), but substantial overlap of the 95% confidence intervals suggests no significant difference in TDR rates among these groups. Figure 3 shows the prevalence of TDR to specific classes of antiretroviral drug within each subgroup, in which a disproportionately large prevalence of PIs resistance can be seen in MSM.

Discussion

Our meta-analysis showed that estimated prevalence of TDR in ART-naïve patients in China was 3.64% (95%CI: 3.00%–4.32%). Such a pooled prevalence rate of primary drug resistance is similar to those from other developing country settings such as Thailand (0.5%, 95% CI: 0.1%–1.4%) and India (2.7%, 95% CI: 1.1%–4.7%) [33], but lower than those estimated in Europe (8.4%, 95%CI: 7.4%–9.5%) [34] and the United States (14.6%, 95% CI: 12.9%–16.0%) [35]. The far lower rates of TDR in Chinese HIV patients as compared to those in western countries may be due to differences in sampling strategies, but may also reflect real differences in the prevalence of circulating strains of drug resistance HIV in the two settings. China's HIV epidemic is at a relatively earlier phase compared to those in Africa and the west, as has a shorter history of population level exposure to ART, leading to less drug resistance in the population.

Changes in estimated TDR prevalence across populations sampled over distinct time periods likely reflect population level response to the antiretroviral drug used most widely by beneficiaries of China's free ART program at each phase. Higher

Table 1. Result of stratified meta-analysis based on time period, study region and risk group.

Category	Number of estimates	Pooled prevalence (%)	95%CI (%)	Heterogeneity p-value[a]
Time period[b]				
before 2003	3	2.65	0.89–5.17	0.15
2003–2005	5	5.18	3.13–7.63	0.01
2006–2008	20	3.02	2.03–4.16	0.18
2009–2012	53	3.68	2.78–4.69	<0.0001
Region[c]				
Hubei	3	7.37	4.61–10.65	0.42
Henan	7	7.11	4.10–10.80	0.00
Beijing	5	6.52	4.35–9.05	0.27
Shanghai	3	4.91	2.25–8.38	0.87
Tianjin	2	4.90	1.25–10.31	0.65
Zhejiang	2	4.76	2.35–7.86	0.51
Liaoning	2	4.40	2.26–7.15	0.95
Guangdong	7	3.43	2.17–4.91	0.70
Guangxi	6	2.72	1.70–3.94	0.14
Hunan	2	2.69	0.91–5.22	0.83
Yunnan	11	2.44	1.14–4.11	0.08
Guizhou	2	2.25	0.00–14.08	0.04
Anhui	3	1.89	0.02–5.51	0.60
Shandong	2	1.56	0.07–4.24	0.67
Sichuan	3	0.49	0.00–2.16	0.24
Risk group[d]				
MSM	17	4.03	2.81–5.43	0.05
FPD	5	3.78	2.48–5.31	0.19
HST	3	3.73	1.85–6.20	0.05
IDU	5	2.78	1.04–5.12	0.22

Note: [a]Heterogeneity p value was calculated by examining the Cochran *Q* statistic (p-value of <0.1 was considered a statistically significant); [b]For time period subgroup analysis, 81 HIV TDR prevalence estimates from 71 articles were used; [c]For study region subgroup analysis, 60 HIV TDR prevalence estimates from 58 articles were used; [d]For risk group subgroup analysis, 30 HIV TDR prevalence estimates from 28 articles were used. MSM: men who have sex with men; FPD: former plasma donors; HST: Heterosexual transmission; IDU: Injecting drug users.

TDR prevalence estimates in the earliest phase of the national free ART program may be due to the widespread use of generic agents including zidovudine, didanosine, and nevirapine [5]. Subsequent survey work as shown a strong association between didanosine use and virological failure in the earliest participants of the free ART program, suggesting high drug resistance rates due to antiretroviral drug choice [36,37]. In addition, ART dispensation in the early phase of the epidemic relied on village clinics where clinicians who often lacked formal medical training and who worked without the benefit of routine laboratory monitoring to aid their patient management [5,7,38]. After 2005, routinization of ART practices [39] and availability of branded drugs including lamivudine and efavirenz may have led to lower population level prevalence of drug resistant strains [4]. Slightly higher rates of TDR following 2008 may be due to broad expansion of ART access during this phase [8,40] or due to a cumulative increase in the prevalence of drug strains in HIV infected person on therapy [41]. However the lack of a statistically significant difference in pooled prevalence of TDR in these two periods may reflect a spurious difference.

Geographic sub-analyses showed that pooled estimates of TDR were highest in Beijing (a provincial-level city) and Henan and Hubei provinces. Higher rates of TDR in Beijing may be due to the fact that three out of the five studies based in Beijing were of MSM [42–44]. Past reports of TDR in MSM in Beijing have been high [42]—as compared to rates among MSM recruited in Guangzhou city [45], Liaoning province [46] and Yunnan province [47]—but for reasons that are still poorly understood. Higher rates in Henan province likely reflect the fact that it has the largest concentration of treated HIV patients in China with the earliest access to ART due to the piloting of the national treatment program in this population [4]. Finally, the three articles conducted in Hubei province were carried out in the early phase of the free ART program, a possible explanation for the higher prevalence of TDR.

Though drug resistance patterns did not vary significantly across risk populations, one noteworthy finding was the elevated prevalence of PI-specific TDR among MSM. The fact that PI's were not available through China's free ART program until 2008 [48] and its limited use even today [25] may suggest acquisition of PI-specific TDR from individuals treated outside of

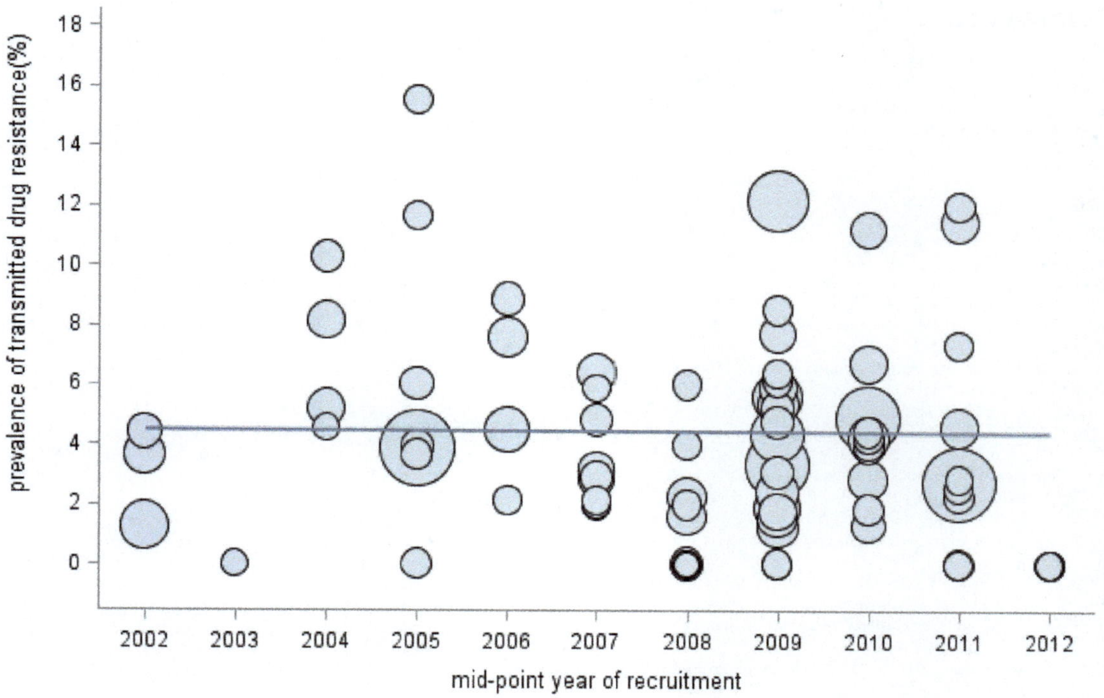

Figure 2. Temporal trends of prevalence of drug resistance in HIV ART-naïve patients in China. The size of circle correspond to sample size of every study, 81 estimates from 71 articles were used in this subgroup analysis.

the free ART program or foreign nationals treated abroad. Mixing between foreign-born and Chinese men in MSM sexual networks, and the relative affluence of MSM (or the subset captured in epidemiological studies) which allows for increased international travel [25] may be partial explanations for this phenomenon.

Our analysis faces several important limitations. Convenience sampling was the most common recruitment method for most studies included in this analysis which limits representativeness of findings. Selection bias may also have led to oversampling of studies in regions with high levels of ART penetration such as Guangdong, Beijing, Yunnan, Henan and Guangxi.

Despite these limitations, findings from this study represent the first comprehensive investigation of HIV TDR in treatment naïve individuals in China. Though estimated rates are lower than many western settings, the relative infancy of the Chinese ART program underscores the importance of consistent monitoring of onward transmission of drug resistant HIV at the population level. Beyond the clinical considerations of limited therapeutic options for TDR in treatment naïve patients, China's focus on treatment as prevention as part of its newly articulated HIV control strategy in the 12th Five-Year Action Plan [12] underscores the HIV prevention tradeoffs of expanded ART access.

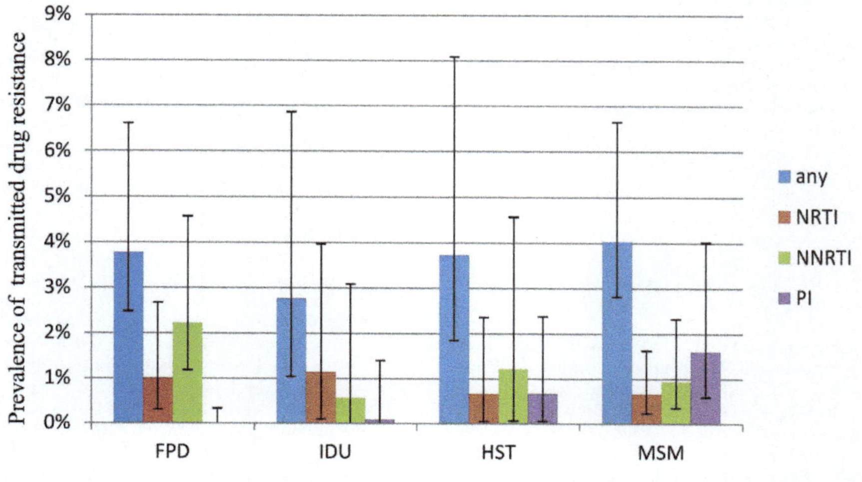

Figure 3. Prevalence of transmitted drug resistance in difference risk groups. Note: MSM: men who have sex with men; HST: Heterosexual transmission; FPD: former plasma donors; IDU: Injecting drug users; NRTI: nucleoside reverse transcriptase inhibitor; NNRTI: non-nucleoside reverse transcriptase inhibitor; PI: protease inhibitor.

Author Contributions

Conceived and designed the experiments: YYS NW. Performed the experiments: YYS HXL LZ MKS. Analyzed the data: YYS. Contributed reagents/materials/analysis tools: FJZ MKS LZ JW. Wrote the paper: YYS FJZ HXL MKS LZ JW NW.

References

1. Ministry of Health, People's Republic of China, Joint United Nations Programme on HIV/AIDS, World Health Organization (2011) 2011 Estimates for the HIV/AIDS Epidemic in China. Beijing, China.
2. Zhang L, Chow EP, Jing J, Zhuang X, Li X, et al. (2013) HIV prevalence in China: integration of surveillance data and a systematic review. Lancet Infect Dis 13: 955–963.
3. National Health and Family Planning Commission of the People's Republic of China (2013) Background of HIV/AIDS prevention and control in China. Available: http://www.nhfpc.gov.cn/jkj/s3586/201312/2b871ccd2ef446eb954 2875d3d68bbca.shtml. Accessed March 31, 2014.
4. Dou Z, Chen RY, Xu J, Ma Y, Jiao JH, et al. (2010) Changing baseline characteristics among patients in the China National Free Antiretroviral Treatment Program, 2002-09. Int J Epidemiol 39 Suppl 2: ii56–64.
5. Zhang F, Haberer JE, Wang Y, Zhao Y, Ma Y, et al. (2007) The Chinese free antiretroviral treatment program: challenges and responses. AIDS 21 Suppl 8: S143–148.
6. Zhang F, Dou Z, Ma Y, Zhang Y, Zhao Y, et al. (2011) Effect of earlier initiation of antiretroviral treatment and increased treatment coverage on HIV-related mortality in China: a national observational cohort study. Lancet Infect Dis 11: 516–524.
7. Zhang FJ, Pan J, Yu L, Wen Y, Zhao Y (2005) Current progress of China's free ART program. Cell Res 15: 877–882.
8. Zhao DC, Wen Y, Ma Y, Zhao Y, Zhang Y, et al. (2012) Expansion of China's free antiretroviral treatment program. Chin Med J (Engl) 125: 3514–3521.
9. Little SJ (2000) Transmission and prevalence of HIV resistance among treatment-naive subjects. Antivir Ther 5: 33–40.
10. Bennett DE, Bertagnolio S, Sutherland D, Gilks CF (2008) The World Health Organization's global strategy for prevention and assessment of HIV drug resistance. Antivir Ther 13 Suppl 2: 1–13.
11. Wittkop L, Gunthard HF, de Wolf F, Dunn D, Cozzi-Lepri A, et al. (2011) Effect of transmitted drug resistance on virological and immunological response to initial combination antiretroviral therapy for HIV (EuroCoord-CHAIN joint project): a European multicohort study. Lancet Infect Dis 11: 363–371.
12. Zhao Y, Poundstone KE, Montaner J, Wu ZY (2012) New policies and strategies to tackle HIV/AIDS in China. Chin Med J (Engl) 125: 1331–1337.
13. Avila-Rios S, Garcia-Morales C, Garrido-Rodriguez D, Ormsby CE, Hernandez-Juan R, et al. (2011) National prevalence and trends of HIV transmitted drug resistance in Mexico. PLoS One 6: e27812.
14. Bennett DE, Camacho RJ, Otelea D, Kuritzkes DR, Fleury H, et al. (2009) Drug resistance mutations for surveillance of transmitted HIV-1 drug-resistance: 2009 update. PLoS One 4: e4724.
15. Higgins J, Thompson SG (2002) Quantifying heterogeneity in a meta-analysis. Statistics in medicine 21: 1539–1558.
16. Huedo-Medina TB, Sanchez-Meca J, Marin-Martinez F, Botella J (2006) Assessing heterogeneity in meta-analysis: Q statistic or I2 index? Psychol Methods 11: 193–206.
17. R Core Team (2013) R: A language and environment for statistical computing. R Foundation for Statistical Computing, Vienna, Austria. Available: http://www.R-project.org/. Accessed April 7, 2014.
18. Bennett DE, Myatt M, Bertagnolio S, Sutherland D, Gilks CF (2008) Recommendations for surveillance of transmitted HIV drug resistance in countries scaling up antiretroviral treatment. Antivir Ther 13 Suppl 2: 25–36.
19. Qiu LJ, Wu SL, Liu XH, Xie MR, Yan PP, et al. (2013) Study on drug resistant mutations for HIV-1 strains in Fujian Province. Chin J AIDS STD 19: 6–9.
20. Li L, Lu X, Li H, Chen L, Wang Z, et al. (2011) High genetic diversity of HIV-1 was found in men who have sex with men in Shijiazhuang, China. Infect Genet Evol 11: 1487–1492.
21. Li WJ, Li H, Wang FX, Li YG (2012) Drug resistance mutations in HIV-1 strains of treatment-naive patients in Harbin, China. Chin J Viral Dis 2: 40–45.
22. Yang HT, Xiao ZP, Huang XP, Hu HY, Xu XQ, et al. (2012) Threshold survey of transmitted HIV-1 drug resistance in Jiangsu province. Acta Universitatis Medicinalis Nanjing (Natural Science) 2012: 1–5.
23. Han XX, Zhao B, Sun F, An MM, Yin LL, et al. (2012) Primary HIV-1drug resistance among injecting drug users in Xinjiang. Chin J Public Health 28: 810–811.
24. Liao L, Xing H, Li X, Ruan Y, Zhang Y, et al. (2007) Genotypic analysis of the protease and reverse transcriptase of HIV type 1 isolates from recently infected injecting drug users in western China. AIDS Res Hum Retroviruses 23: 1062–1065.
25. Yang J, Xing H, Niu J, Liao L, Ruan Y, et al. (2013) The emergence of HIV-1 primary drug resistance genotypes among treatment-naive men who have sex with men in high-prevalence areas in China. Arch Virol 158: 839–844.
26. Kuang JQ (2007) Genotypic HIV-1 Drug Resistance and Factors associated with the effect of Antiretroviral therapy Among Chinese HIV/AIDS Patients: Peking Union Medical College.
27. Liao L, Xing H, Shang H, Li J, Zhong P, et al. (2010) The prevalence of transmitted antiretroviral drug resistance in treatment-naive HIV-infected individuals in China. J Acquir Immune Defic Syndr 53 Suppl 1: S10–14.
28. Wang X, He C, Xing H, Liao L, Xu X, et al. (2012) Short communication: emerging transmitted HIV type 1 drug resistance mutations among patients prior to start of first-line antiretroviral therapy in middle and low prevalence sites in China. AIDS Res Hum Retroviruses 28: 1637–1639.
29. Li L, Sun G, Liang S, Li J, Li T, et al. (2013) Different Distribution of HIV-1 Subtype and Drug Resistance Were Found among Treatment Naive Individuals in Henan, Guangxi, and Yunnan Province of China. PLoS One 8: e75777.
30. Si XF, Huang HL, Wei M, Guan Q, Song YH, et al. (2004) Prevalence of drug resistance mutations among antiretroviral drug-naive HIV-1-infected patients in China. Chinese J Exp Chin Virol 18: 308–311.
31. Zeng P, Wang J, Huang Y, Guo X, Li J, et al. (2012) The human immunodeficiency virus-1 genotype diversity and drug resistance mutations profile of volunteer blood donors from Chinese blood centers. Transfusion 52: 1041–1049.
32. Yin CY, Lu HZ, Huang XX, Li XO, Lou GQ, et al. (2011) HIV-1 drug-resistance mutations in treatment-naive patients in China. Chin J Clin Infect Dis 4: 201–205.
33. Gupta RK, Jordan MR, Sultan BJ, Hill A, Davis DH, et al. (2012) Global trends in antiretroviral resistance in treatment-naive individuals with HIV after rollout of antiretroviral treatment in resource-limited settings: a global collaborative study and meta-regression analysis. Lancet 380: 1250–1258.
34. Vercauteren J, Wensing AM, van de Vijver DA, Albert J, Balotta C, et al. (2009) Transmission of drug-resistant HIV-1 is stabilizing in Europe. J Infect Dis 200: 1503–1508.
35. Wheeler WH, Ziebell RA, Zabina H, Pieniazek D, Prejean J, et al. (2010) Prevalence of transmitted drug resistance associated mutations and HIV-1 subtypes in new HIV-1 diagnoses, U.S.-2006. AIDS 24: 1203–1212.
36. Li T, Dai Y, Kuang J, Jiang J, Han Y, et al. (2008) Three generic nevirapine-based antiretroviral treatments in Chinese HIV/AIDS patients: multicentric observation cohort. PLoS One 3: e3918.
37. Xing H, Wang X, Liao L, Ma Y, Su B, et al. (2013) Incidence and associated factors of HIV drug resistance in Chinese HIV-infected patients receiving antiretroviral treatment. PLoS One 8: e62408.
38. Zhang FJ, Maria A, Haberer J, Zhao Y (2006) Overview of HIV drug resistance and its implications for China. Chin Med J (Engl) 119: 1999–2004.
39. Manual for National free HIV antiretroviral drug treatment: first edition (2005) Beijing: People's Medical Publishing House.
40. World Health Organization (WHO) (2012) WHO HIV drug resistance report 2012. Available: http://www.who.int/hiv/pub/drugresistance/report2012/en/ . Accessed January 31, 2014.
41. Liao L, Xing H, Su B, Wang Z, Ruan Y, et al. (2013) Impact of HIV drug resistance on virologic and immunologic failure and mortality in a cohort of patients on antiretroviral therapy in China. AIDS 27: 1815–1824.
42. Zhang X, Li S, Li X, Li X, Xu J, et al. (2007) Characterization of HIV-1 subtypes and viral antiretroviral drug resistance in men who have sex with men in Beijing, China. AIDS 21 Suppl 8: S59–65.
43. Wu YS, Zhang T, Wei FL, Li DM, Zhang WW, et al. (2011) Study on transmitted drug resistance in new infected or recently infected patients. Chinese Medical Association-Fifth national academic conferences of HIV/AIDS, viral hepatitis C and tropical diseases. Wu Han, Hubei.
44. Li L, Han N, Lu J, Li T, Zhong X, et al. (2013) Genetic characterization and transmitted drug resistance of the HIV type 1 epidemic in men who have sex with men in Beijing, China. AIDS Res Hum Retroviruses 29: 633–637.
45. Zhao GL, Feng TJ, Hong FC, Wang F, Cai YM, et al. (2009) Study on drug-resistant gene mutation in HIV-1 infected MSM population in Shenzhen. Chin J AIDS STD 15: 589–591.

46. Zhao B, Han X, Dai D, Liu J, Ding H, et al. (2011) New trends of primary drug resistance among HIV type 1-infected men who have sex with men in Liaoning Province, China. AIDS Res Hum Retroviruses 27: 1047–1053.
47. Chen M, Ma Y, Su Y, Yang L, Zhang R, et al. (2014) HIV-1 Genetic Characteristics and Transmitted Drug Resistance among Men Who Have Sex with Men in Kunming, China. PLoS One 9: e87033.
48. Manual for National free HIV antiretroviral drug treatment: second edition (2008) Beijing: People's Medical Publishing House.

14

Tuberculosis Treatment Managed by Providers outside the Public Health Department: Lessons for the Affordable Care Act

Melissa Ehman[1,2]*, Jennifer Flood[1], Pennan M. Barry[1]

1 Tuberculosis Control Branch, Division of Communicable Disease Control, Center for Infectious Diseases, California Department of Public Health, Richmond, California, United States of America, **2** Institute of Global Health, Global Health Sciences, University of California San Francisco, San Francisco, California, United States of America

Abstract

Introduction: Tuberculosis (TB) requires at least six months of multidrug treatment and necessitates monitoring for response to treatment. Historically, public health departments (HDs) have cared for most TB patients in the United States. The Affordable Care Act (ACA) provides coverage for uninsured persons and may increase the proportion of TB patients cared for by private medical providers and other providers outside HDs (PMPs). We sought to determine whether there were differences in care provided by HDs and PMPs to inform public health planning under the ACA.

Methods: We conducted a retrospective, cross-sectional analysis of California TB registry data. We included adult TB patients with culture-positive, pulmonary TB reported in California during 2007–2011. We examined trends, described case characteristics, and created multivariate models measuring two standards of TB care in PMP- and HD-managed patients: documented culture conversion within 60 days, and use of directly observed therapy (DOT).

Results: The proportion of PMP-managed TB patients increased during 2007–2011 ($p = 0.002$). On univariable analysis ($N = 4{,}606$), older age, white, black or Asian/Pacific Islander race, and birth in the United States were significantly associated with PMP care ($p < 0.05$). Younger age, Hispanic ethnicity, homelessness, drug or alcohol use, and cavitary and/or smear-positive TB disease, were associated with HD care. Multivariable analysis showed PMP care was associated with lack of documented culture conversion (adjusted relative risk [aRR] = 1.37, confidence interval [CI] 1.25–1.51) and lack of DOT (aRR = 8.56, CI 6.59–11.1).

Conclusion: While HDs cared for TB cases with more social and clinical complexities, patients under PMP care were less likely to receive DOT and have documented culture conversion. This indicates a need for close collaboration between PMPs and HDs to ensure that optimal care is provided to all TB patients and TB transmission is halted. Strategies to enhance collaboration between HDs and PMPs should be included in ACA implementation.

Editor: John Z. Metcalfe, University of California, San Francisco, United States of America

Funding: The authors have no support or funding to report.

Competing Interests: The authors have declared that no competing interests exist.

* Email: Melissa.Ehman@cdph.ca.gov

Introduction

Despite a decline in tuberculosis (TB) in the United States (U.S.) in the past two decades, TB remains a significant public health problem and is a challenging, resource-intensive disease to diagnose and treat. Treatment of active disease requires at least six months of a multidrug regimen and necessitates systematic monitoring for side effects and response to treatment. Because most TB patients have historically been managed by publicly funded local and state TB programs, [1] these programs have substantial expertise to successfully detect and treat TB disease in the U.S. However, the private sector plays an increasingly important role in diagnosing and treating TB. [2] As TB cases continue to decline in the U.S., [3] community health care providers may not see enough cases to build or maintain expertise in managing cases of TB. Regardless of the source of direct patient care, public health programs are responsible for oversight of TB patient treatment, to ensure that transmission is prevented. This need to protect the public from TB makes public-private collaboration crucial for effective management of TB. [2,4,5]

Effective management of TB should ensure timely conversion of sputum cultures to negative and prevent acquired drug resistance (ensure adherence to treatment). [6–8] Documenting prompt culture conversion also allows for the use of short-course TB therapy. [9] The practice of directly observed therapy (DOT) does not simply ensure treatment adherence, but also facilitates overall monitoring of treatment efficacy and provides patient support through structured contact with the health care system. [4,9]

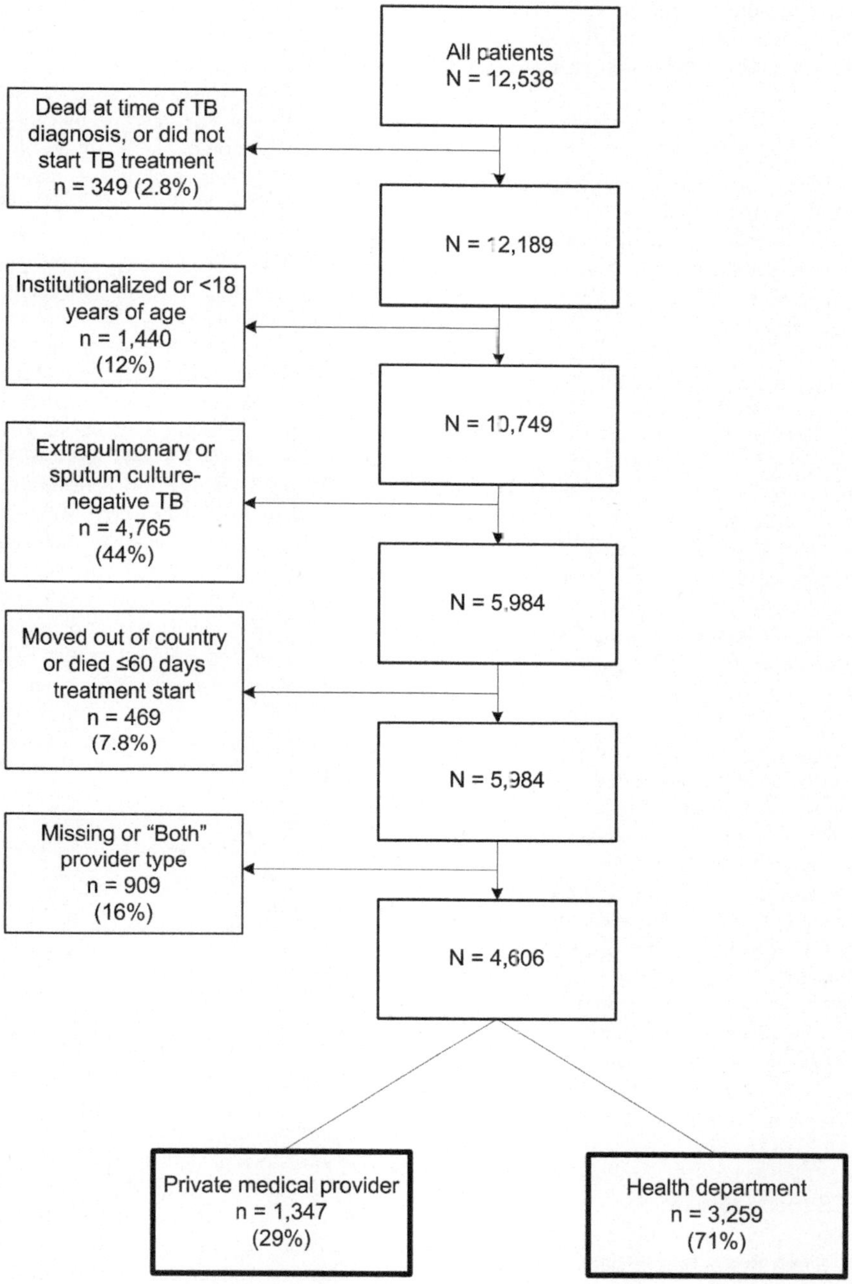

Figure 1. Analytic cohort, 2007–2011. TB = tuberculosis.

The Patient Protection and Affordable Care Act (ACA) [10] expands opportunities for patients to obtain health insurance and may increase health care provision in the private sector. In order to understand the potential impact of a shift in TB care from public TB programs to the private sector, we examined trends in providers caring for California TB patients over time, and examined differences in demographic and clinical characteristics of these two patient populations. We also sought to determine whether differences exist between care practices, including documenting that a patient has converted sputum cultures to negative and providing DOT to prevent acquisition of drug resistance.

Materials and Methods

Ethics statement

The California Department of Public Health (CDPH) routinely collects surveillance data, performs analyses and monitors trends for public health purposes. This analysis was determined to be a non-research public health analysis, and not subject to human subjects review. [11] All patient data were anonymized and de-identified prior to analysis.

Analytic design

We used TB surveillance data in a retrospective, cross-sectional analysis to model the relationships between the provider type for TB care – within the public health department or outside the health department (e.g. private and other providers) – and two measures of optimal TB management: documenting culture conversion to negative, and ensuring treatment adherence through DOT.

Data sources

TB surveillance data were captured through mandatory reporting by public health departments (HDs) of all TB cases to CDPH, using a standard report form containing demographic, clinical, and management information, including the type of clinical provider that managed the TB care. [12,13] On the TB reporting form, a case was classified as "Health Department," "Private/Other," or "Both." "Health Department" refers to patient care in a clinic directly managed by the public health department; for the vast majority of TB patients under HD care, this was a clinic devoted solely to TB diagnosis and treatment. "Private/Other" (hereafter private medical provider, or PMP) designates any other type of provider outside the public health department, including health maintenance organizations (HMOs), and county hospitals and clinics not directly managed by the HD. "Both" means that the patient received both HD and PMP care. Human immunodeficiency virus (HIV) infection status was determined by matching TB case records with the CDPH HIV/acquired immune deficiency syndrome registry.

Inclusion criteria

For analysis of TB management, TB cases reported to CDPH during 2007–2011 among persons 18 years and older with a sputum culture positive for *M. tuberculosis* and no extrapulmonary disease were selected (Figure 1). The study population inclusion criteria were designed to create a cohort of cases for which TB case management standards are best defined: culture-positive pulmonary TB in adults. [9] Patients who did not start TB treatment, died within 60 days of starting treatment, moved out of the U.S. during treatment, or were diagnosed in an institutional setting, i.e., a correctional or long-term care facility, were excluded. Because exclusive management of patients by PMPs or HDs are the most clearly defined and consistently reported provider types across California, [14] we limited the analysis to PMP and HD provider types, and excluded those designated as Both.

Measures of optimal TB management

We examined two measures of TB management because of their importance for protecting the public's health: evidence of converting sputum culture from positive to negative during the initial treatment phase, and use of DOT. Calculations for these measures were based on indicators routinely used by CDPH [15] and the Centers for Disease Control and Prevention (CDC) [16] to monitor performance of TB control programs.

Culture conversion was defined as a report of a documented sputum culture-negative specimen which was collected within 60 days of the patient's treatment start date. We selected 60 days as the standard window used in clinical guidelines to assess treatment response and need for extension of therapy. [9] A patient was determined to have received DOT if therapy was reported to have been either totally directly observed, or both directly observed and self-administered.

Data analysis

We examined trends, stratified by type of provider for TB cases, using Joinpoint version 4.0.4. [17] We fit the regression lines to the counts of PMP and HD patients during 1993–2011, identified join points that best fit the data using a Monte Carlo permutation test, and compared the trends for TB patients managed by PMPs compared to HDs, using a permutation test for parallelism. [18]

For all other calculations, we used SAS version 9.3. [19] We tested trends for statistical significance with the Cochrane-Armitage test for trend, for the study cohort and also for a cohort that included patients reported as managed by both PMPs and HDs. We compared PMP-managed patients to HD-managed patients for a range of clinical and demographic factors, and tested associations using the Mantel-Haenszel chi square test, or Fisher's exact test where any expected cell count was less than five. We assessed correlation with phi coefficient less than 0.30 for multivariable models. We combined race and ethnicity into mutually exclusive categories, where Hispanic ethnicity was coded as Hispanic, regardless of race; and white, black, and Asian race categories each excluded persons of Hispanic ethnicity. In the multivariable models, we used birth in the U.S. instead of race/ethnicity categories due to correlation of these two variables. Other correlated variables were combined into categories: smear-positive and cavitary TB disease, and homelessness and substance use (alcohol, injecting and non-injecting). We modeled the association of provider type with the two measures of TB management in univariable and multivariable analyses. We calculated crude and adjusted relative risk using modified Poisson regression with robust error variance. [20,21] Covariables associated with both provider type and measures of TB case management, and factors cited in the literature as possible confounders, were included *a priori* in each multivariable model. We compared models with additional covariables in different combinations using the QIC statistic (quasi-likelihood under the independence model criterion). [22] We selected the model with the lowest QIC as the final model. To assess the impact of excluding patients reported with a provider type of Both (PMP and HD), we analyzed a cohort that included patients managed by Both, and with all other study exclusions. We constructed multivariate models using this cohort in two ways: 1) PMP+Both vs. HD, and 2) PMP vs. HD+Both. We conducted a sensitivity analysis of the relationship between PMP care, DOT, and death during therapy, by excluding patients who died at varying times, to assess the robustness of the 60-day cut point for excluding deaths. We created two separate models to investigate documented culture conversion: documented culture conversion in greater than 60 days after starting treatment, and lack of documented culture conversion ever, to assess whether PMP care was associated with never having a culture conversion documented, or with culture conversion documented after 60 days of treatment. Because California guidelines prioritize key patient groups for DOT when resources do not allow for universal DOT, [23] such as homeless or those with drug-resistant disease, we conducted a subanalysis of predictors of DOT with a cohort restricted to patients with at least one indication for DOT.

Results

Among the 12,538 TB patients reported in California during 2007–2011, 4,606 were included in the analysis (Figure 1). While most patients were managed by the HD (3,259 or 71%), 29% (1,347) were privately managed.

The number of TB patients reported in California steadily declined during 1993–2011, [24] and TB cases meeting our study population criteria also declined during this time. Figure 2 shows the differences in TB incidence trends between public and private sectors. During 1993–2006, the incidence of HD TB patients declined 3.1% per year (95% confidence interval [CI] −3.6, −

2.7); whereas PMP patients declined 5.8% per year (CI −7.0, −4.7). During 2006–2011, however, the decline in PMP-managed TB cases leveled off (annual percent change [APC] = 1.9, CI −4.2, 8.3), whereas HD-managed patients continued to decline at the previous rate. This difference in the rate of incidence decline was statistically significant ($p = 0.001$).

In our analytic cohort, the proportion of cases with PMP care increased, rising from 25% in 2007 to 32% in 2011 ($p = 0.002$). In the modified study cohort that included patients with a provider type of Both, the proportion of patients categorized as Both declined from 18% in 2007 to 8.8% in 2011 ($p<0.001$).

Univariable analysis

Table 1 shows the characteristics of cases managed by PMPs compared with cases managed by HDs. Older patients, those of white, black or Asian/Pacific Islander race, or born in the U.S., were all significantly more likely to have PMP care. Patients cared for by HDs were more likely to be young, Hispanic, homeless, excess alcohol users, non-intravenous drug users, or have cavitary and/or smear-positive TB disease. No patients reported stopping treatment or having treatment extension beyond 12 months due to adverse reactions to TB medication. Documented culture conversion and DOT were observed less frequently among patients managed by PMPs.

Multivariable analysis

A total of 4,604 (99.9%) patients in the analytic cohort had complete information on culture conversion. The absence of documented culture conversion within 60 days of treatment start was significantly associated with privately managed care in multivariable analysis (adjusted relative risk [aRR] 1.37, CI 1.25–1.51). This model adjusted for ten covariables including age, cavitary and/or smear-positive TB disease, and drug resistance. In addition to private medical care, factors predicting lack of documented culture conversion in the adjusted model included: cavitary and/or smear-positive TB disease, moving during treatment, no DOT, birth in the U.S., male sex, and any homelessness or drug use (Table 2). Privately managed care was a significant predictor of both never having culture conversion documented (aRR 3.34, CI 2.54–4.36) and of delayed culture conversion, documented later than 60 days (aRR 1.24, CI 1.12–1.39).

Nearly all patients (99.5%) had complete information on therapy administration method (DOT or self-administered therapy [SAT]). In the adjusted model, patients under private care were significantly more likely to have received SAT during the course of TB treatment, and PMP care was the strongest predictor of SAT among all covariables (aRR 8.56, CI 6.59–11.1). Lack of documented culture conversion within 60 days was also associated with SAT. Patient characteristics associated with receipt of DOT included homelessness or drug use, cavitary and/or smear-positive TB, a history of TB disease, resistance to any first-line drug used to treat TB, and age 65 years old or greater (Table 3). Similar results were found when excluding patients who died during therapy at 30 or 90 days. Among patients who received DOT, 85% received at least 21 weeks of DOT, and 94% received at least 10 weeks of DOT.

In a subanalysis, limited to patient groups prioritized for DOT under California guidelines, PMP care also had the greatest association with lack of DOT (aRR 6.46, CI 4.61–9.06). Patient groups for whom DOT is a priority in California include the homeless, drug or alcohol users, HIV infected, and patients with drug-resistant TB, smear-positive TB, delayed culture conversion, or prior history of TB.

Multivariable models using the modified study cohort including Both HD and PMP patients showed similar results for each study outcome, with no change in the direction nor significance of association for any covariable.

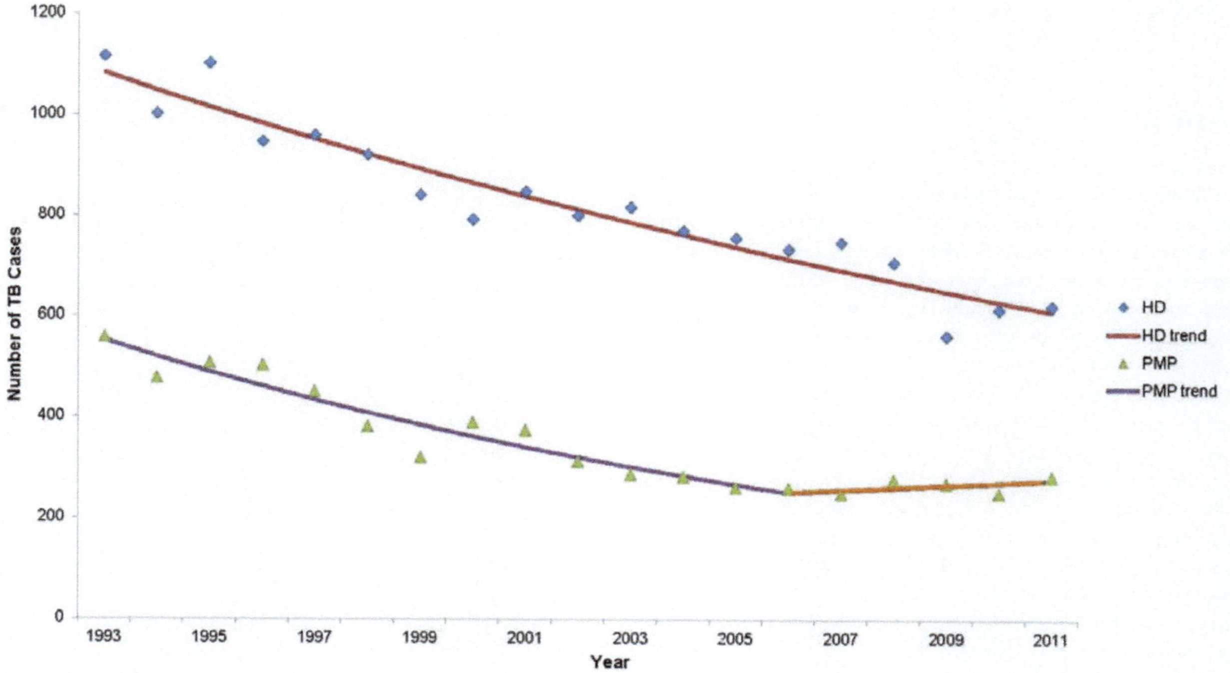

Figure 2. Tuberculosis case count trend, by provider type, California 1993–2011. HD = health department, PMP = private medical provider. Annual percent change (APC): HD 1993–2011 = −3.13, PMP 1993–2006 = −5.83, PMP 2006–2011 = +1.85.

Table 1. Characteristics of TB Patients by Provider Type, Reported in California, 2007–2011 (n = 4606).

	All patients[a]	PMP (n = 1347)	HD (n = 3259)	PMP vs. HD
Characteristic	n (col %)	n (col %)	n (col %)	P value[b]
Male	2893 (63)	819 (61)	2074 (64)	0.070
18–24 years old	456 (9.9)	100 (7.4)	356 (11)	<0.001
25–44 years old	1421 (31)	366 (27)	1055 (32)	<0.001
45–64 years old	1611 (35)	423 (31)	1188 (36)	0.001
Age ≥65 years old	1118 (24)	458 (34)	660 (20)	<0.001
White	369 (8.0)	147 (11)	222 (6.8)	<0.001
Black	283 (6.2)	110 (8.2)	173 (5.3)	<0.001
Hispanic	1586 (34)	293 (22)	1293 (37)	<0.001
Asian/Pacific Islander	2359 (51)	797 (59)	1562 (48)	<0.001
American Indian/Alaskan Native	8 (0.2)	0 (0)	8 (100)	0.114
Born in United States	792 (17)	297 (22)	495 (15)	<0.001
Moved ever during treatment	234 (5.1)	72 (5.4)	162 (5.0)	0.599
Homeless	273 (5.9)	53 (3.9)	220 (6.8)	<0.001
Excessive alcohol use	431 (9.4)	98 (7.3)	333 (10)	0.002
Intravenous drug use	40 (0.9)	11 (0.8)	29 (0.9)	0.804
Nonintravenous drug use	250 (5.4)	51 (3.8)	199 (6.2)	0.002
Homelessness, alcohol or drug use	682 (15)	143 (11)	539 (17)	<0.001
History of TB disease	286 (6.3)	89 (6.7)	197 (6.1)	0.448
HIV-positive[c]	143 (3.1)	47 (3.4)	96 (3.0)	0.333
Cavitary and/or smear-positive TB disease	3183 (70)	881 (66)	2302 (71)	<0.001
Resistance to any first-line TB drug	651 (14)	183 (14)	468 (14)	0.506
No documented culture conversion ≤60 days	1469 (32)	533 (40)	936 (29)	<0.001
No directly observed therapy	342 (7.4)	270 (20)	72 (2.2)	<0.001

[a]alive at diagnosis and starting treatment, pulmonary only, culture-positive TB, age ≥18 years old; excluding missing or Both provider type, moved out of United States, deaths ≤60 days after treatment start.
[b]for Mantel-Haenszel chi-square, or Fisher's Exact Test where any expected cell count is <5.
[c]vs. HIV-negative or unknown status.
TB = tuberculosis, PMP = private medical provider, HD = health department, HIV = human immunodeficiency virus.

Discussion

We found that during 2007–2011, before implementation of the ACA, an increasing proportion of TB cases in California were exclusively managed by PMPs. We also found that patients managed by PMPs were less likely to have received optimal management of their pulmonary TB even after adjusting for possible confounding clinical and social factors. These findings may have implications for TB care under the ACA, when a more dramatic shift of TB care to the private sector is expected. If PMP patients are less likely to receive optimal management, then shifting toward PMP care without a concomitant increase in HD partnership with PMPs could be a threat to TB control.

Our findings are consistent with evidence showing that patients in public health TB clinics are more likely to receive care recommended in TB control guidelines. [25–31] Similar findings have been published in which clinicians and clinics with the most experience and volume provide care resulting in better outcomes to patients for other diagnoses such as HIV [32] and *Staphylococcus aureus* bacteremia. [33–35] The results also highlight the concept that public health TB clinics and programs warrant consideration as expert, specialty referral clinics, and not simply government sector safety net providers. In fact, specialty TB care has been established across the U.S. since the early 1900s, and has been intentionally centered in public health TB clinics since the 1960s. [1] These TB specialty clinics provide patient-centered clinical care, case management, and monitoring for persons with complex health issues over months of treatment. However, TB control programs are charged with "ensuring the quality and completeness" [4] of care for all patients and these findings also point out that there should be close collaboration between PMPs and TB control programs to ensure that patients managed exclusively by PMPs receive the same optimal care that patients managed exclusively by the HD do.

Further highlighting the important role of HD TB care, we found that HDs care for a disproportionate burden of complex patients, such as homeless or drug users, who may be more difficult to treat. HDs also will remain critical "safety net" providers for populations uncovered by ACA, such as undocumented immigrants, who comprise an estimated 23% of persons with active TB disease in the United States. [36] Regardless of where TB patients access care, any lack of optimal care that results in increased transmission would increase the risk of TB in the community, and require heightened response from HDs in California, because they will retain responsibility for outbreak management and contact investigations. Shifting patients from HD to PMP care may therefore increase the need for HD resources, rather than lessen it. Studies have shown that increased

Table 2. No Documented Culture Conversion within 60 Days of TB Treatment Start, TB Cases Reported in California, 2007–2011 (n = 4604).

	All patients[a]	No CC (n = 1469)	CC (n = 3135)	Univariable		Multivariable	
Characteristic	(col %)	(col %)	(col %)	RR	CI	aRR	CI
Private provider/other	1346 (29)	533 (36)	813 (26)	1.38	1.27–1.50	1.37	1.25–1.51
Male	2893 (63)	994 (68)	1899 (61)	1.24	1.13–1.36	1.19	1.08–1.31
Age ≥65 years old	1118 (24)	341 (23)	777 (25)	0.94	0.85–1.04	0.95	0.86–1.06
Born in United States	792 (17)	313 (21)	479 (15)	1.31	1.19–1.44	1.18	1.06–1.31
Moved ever during treatment	234 (5.1)	94 (6.4)	140 (4.4)	1.28	1.09–1.50	1.23	1.04–1.46
Homelessness, alcohol or drug use	692 (15)	269 (18)	413 (13)	1.29	1.16–1.43	1.16	1.04–1.30
History of TB disease	286 (6.3)	97 (6.7)	189 (6.1)	1.07	0.91–1.27	1.09	0.92–1.29
HIV-positive[b]	143 (3.1)	41 (2.8)	102 (3.3)	0.90	0.69–1.16	0.82	0.63–1.08
Cavitary and/or smear-positive TB	3181 (70)	1168 (80)	2013 (65)	1.73	1.54–1.93	1.80	1.61–2.02
Resistance to any first-line TB drug	651 (14)	182 (13)	469 (15)	0.86	0.76–0.98	0.91	0.80–1.04
No directly observed therapy	342 (7.4)	138 (9.4)	204 (6.5)	1.30	1.13–1.48	1.24	1.07–1.45

[a]alive at diagnosis and starting treatment, pulmonary only, culture-positive TB, age ≥18 years old; excluding missing or Both provider type, moved out of United States, deaths ≤60 days after treatment start.
[b]vs. HIV-negative or unknown status.
TB = tuberculosis, CC = culture conversion, RR = relative risk, CI = 95% confidence interval, aRR = adjusted relative risk, HIV = human immunodeficiency virus.

Table 3. No Directly Observed Therapy throughout TB Treatment, TB Cases Reported in California, 2007–2011 (n = 4585).

	All patients[a]	No DOT (n = 342)	DOT (n = 4226)	Univariable		Multivariable	
Characteristic	**(col %)**	**(col %)**	**(col %)**	**RR**	**CI**	**aRR**	**CI**
Private provider/other	1335 (29)	270 (79)	1065 (25)	9.13	7.10–11.7	8.56	6.59–11.1
Male	2880 (63)	196 (57)	2684 (63)	0.79	0.64–0.98	NA	NA
Age ≥65 years old	1108 (24)	97 (28)	1011 (24)	1.24	0.99–1.56	0.79	0.64–0.99
Born in United States	791 (17)	68 (20)	723 (17)	1.20	0.93–1.54	NA	NA
Moved ever during treatment	228 (5.0)	23 (6.7)	205 (4.8)	1.38	0.92–2.06	NA	NA
Homelessness, alcohol or drug use	682 (15)	16 (4.7)	666 (16)	0.28	0.17–0.46	0.39	0.24–0.66
History of TB disease	284 (6.3)	14 (4.1)	270 (6.4)	0.64	0.39–1.09	0.53	0.30–0.94
HIV-positive[b]	143 (3.1)	5 (1.4)	138 (3.3)	0.46	0.19–1.10	0.51	0.21–1.22
Cavitary and/or smear-positive TB disease	3168 (70)	150 (45)	3018 (72)	0.35	0.29–0.43	0.41	0.33–0.49
Resistance to any first-line TB drug	648 (14)	31 (9.2)	617 (15)	0.61	0.43–0.88	0.64	0.46–0.91
No documented culture conversion ≤60 days	1459 (32)	138 (40)	1321 (31)	1.44	1.18–1.78	1.23	1.01–1.49

[a]alive at diagnosis and starting treatment, pulmonary only, culture-positive TB, age ≥18 years old; excluding missing or Both provider type, moved out of United States, deaths ≤60 days after treatment start.
[b]vs. HIV-negative or unknown status.
TB = tuberculosis, DOT = directly observed therapy, RR = relative risk, CI = 95% confidence interval, aRR = adjusted relative risk, NA = not in final multivariate model, HIV = human immunodeficiency virus.

funding is associated with better public health performance, [37] specifically for infectious disease morbidity. [38]

Our analysis assessed two aspects of clinical TB care that are modifiable and important. Documenting sputum culture conversion to negative during the initiation phase of treatment is a key clinical measure of treatment efficacy, and allows determination of appropriate length of therapy. [9] Directly observed therapy is a strategy widely used throughout the world to ensure treatment adherence. [9,39] Providing DOT has been shown to reduce rates of acquired drug resistance. [6]

There are many reasons PMPs may not provide DOT or collect sputum cultures for TB patients, including 1) financial resources, 2) a lack of knowledge of how to provide DOT and sputum induction, and 3) lack of understanding of the value of DOT and need to document sputum culture results during treatment. There is also a tradition of HDs providing DOT for patients. [1] However, HDs with limited resources may not be able to provide DOT or sputum induction to PMP patients, or may have policies preventing this. In California, sputum induction and HD-provided DOT are reimbursable procedures under Medicaid, [40] but capacity and agreements to bill private insurance for services provided by the health department are not in place in many HDs. Encouragement under the ACA for accountable care organizations to include HD TB specialists in their group of providers might promote HD-PMP collaboration in a way that is reimbursable for the HD. Ensuring that HDs have policies and procedures in place to facilitate access to sputum induction and DOT for PMP patients is an important first step.

With an anticipated increase in private care, and a history of uneven assurance of private TB care by HDs, how will TB care be strengthened for privately-managed patients? Despite barriers, improvement is possible and has been demonstrated by California TB control programs that systematically and routinely review PMP-managed patients during case conferences. [41] Health departments have reported success in working with PMPs on TB prevention when HDs have administrative support, have a toolkit of educational information, protocols and forms, and work to develop relationships with PMPs in the community. Alerts generated by electronic health records that could trigger PMP action such as collecting sputum or notifying the HD may also contribute to success. These lessons from TB prevention could be applied to HD-PMP partnership for active TB treatment as well. [42] These interventions can be replicated in HDs undergoing fiscal cuts. The HD staff time needed to put oversight, tracking and collaborative measures in place can save time downstream as follow-up and patient care issues are averted.

Two additional points deserve comment. First, the measures of patient care analyzed in this paper are interrelated. Lack of documented culture conversion and not receiving DOT were each far more likely among PMP patients, and each also predicted the other in multivariable models. This was expected because DOT and monitoring culture conversion are case management activities overseen by the HD for both privately and publicly managed patients. These activities are likely to occur together as part of routine monitoring. We adjusted for each practice in the multivariable models to account for confounding. Second, several groups of patients were found to be more likely to receive DOT, including those with previous TB, cavitary or smear-positive TB, drug resistance, or social factors complicating treatment. This was not surprising because in California, patients with these factors are prioritized for DOT. In both the main analysis, and the subanalysis which included only patients with at least one indication for DOT in California, PMP care remained the strongest independent predictor of not receiving DOT.

Our analysis is subject to some limitations. First, our exclusion criteria were constructed to focus on adults with culture-positive, pulmonary TB, and as a result may not be generalizable to extrapulmonary, pediatric, or culture-negative TB cases. However, the relationship between PMPs and the measures of interest held even among excluded groups of cases where sample sizes were sufficient and analysis was possible, e.g., we could not assess culture conversion among culture-negative cases. Second, information on provider type in TB surveillance data may be subject to misclassification because HDs may not apply a uniform definition. [14] Patients reported as managed by Both HD and PMP may be co-managed, or sequentially managed, depending on local policies and procedures or individual situation; these distinctions would affect the interpretation of results for this category of patients. However, we restricted our analytic cohort to cases managed solely by either private or public health providers, types for which the definitions are clearest. We also showed that the proportion of patients reported as Both HD and PMP was small and declining, and that grouping these patients with either HD or PMP patients did not affect the overall results. Third, we were unable to distinguish between types of private medical provider settings (e.g., HMOs, private medical practices, non-profit clinics, and private or non-HD public hospitals). Aggregating disparate provider groups in our analysis may have masked important findings specific to one type of private provider. Stratified analysis by these private provider types, or even by specific large provider groups, would be an important next step in determining tailored interventions. Last, the TB report form was unable to distinguish whether a recorded delay in culture conversion was due to a delay in sputum collection, or persistent culture-positive sputum. Therefore, it may be that PMPs are equally able to render patients culture negative but PMP patients are less likely to have this documented in the public health record in a consistent and timely way. However, PMP care predicted both delayed culture conversion and never documented conversion in our analysis. Regardless of the reason for delayed culture conversion, our findings show the need for increasing PMP education and collaboration between HDs and PMPs.

In conclusion, our findings point to an opportunity for improvement in clinical care provided to TB patients managed exclusively in the private sector. Implementation of the ACA should take into account the TB expertise found in public sector TB programs, reinforce the role of public health departments to ensure optimal treatment of persons with TB disease, and make resources for key TB management practices available and accessible to all patients, regardless of where care is received.

Acknowledgments

The authors wish to acknowledge Alex Golden, for assistance with manuscript preparation; Lisa Pascopella, for statistical methodology consultation; Cathy Miller, Peter Oh, Jan Young, and Stephanie Spencer for key reference materials; and the California Tuberculosis Control Branch Registry, for data quality assurance and database preparation.

Author Contributions

Conceived and designed the experiments: ME JF PMB. Analyzed the data: ME. Wrote the paper: ME JF PMB.

References

1. Sbarbaro JA (1970) The public health tuberculosis clinic: its place in comprehensive health care. Am Rev Respir Dis 101: 463–465.
2. Binkin NJ, Vernon AA, Simone PM, McCray E, Miller BI, et al. (1999) Tuberculosis prevention and control activities in the United States: an overview of the organization of tuberculosis services. Int J Tuberc Lung Dis 3: 663–674.
3. Centers for Disease Control and Prevention (2012) Reported Tuberculosis in the United States, 2011. Available at: http://www.cdc.gov/tb/statistics/reports/2011. Accessed on October 31, 2013.
4. Taylor Z, Nolan CM, Blumberg HM (2005) Controlling tuberculosis in the United States. Recommendations from the American Thoracic Society, CDC, and the Infectious Diseases Society of America. MMWR Recomm Rep 54: 1–81. rr5412a1 [pii].
5. Institute of Medicine (2000) Ending Neglect: The Elimination of Tuberculosis in the United States. Washington, D.C.: National Academy Press.
6. Porco TC, Oh P, Flood JM (2013) Antituberculosis drug resistance acquired during treatment: an analysis of cases reported in California, 1994–2006. Clin Infect Dis 56: 761–769. cis989 [pii]; 10.1093/cid/cis989 [doi].
7. Moonan PK, Quitugua TN, Pogoda JM, Woo G, Drewyer G, et al. (2011) Does directly observed therapy (DOT) reduce drug resistant tuberculosis? BMC Public Health 11: 19, . 1471-2458-11-19 [pii]; 10.1186/1471-2458-11-19 [doi].
8. Weis SE, Slocum PC, Blais FX, King B, Nunn M, et al. (1994) The effect of directly observed therapy on the rates of drug resistance and relapse in tuberculosis. N Engl J Med 330: 1179–1184. 10.1056/NEJM199404283301702 [doi].
9. American Thoracic Society, Centers for Disease Control and Prevention, Infectious Diseases Society of America (2003) Treatment of tuberculosis. MMWR Recomm Rep 52: 1–77.
10. US Government Printing Office (2012) Public Law 111–148: Patient Protection and Affordable Care Act.
11. U.S. Department of Health and Human Services (2009) U.S. Code of Federal Regulations, 45 CFR 46.101.
12. Centers for Disease Control and Prevention (2003) Report of Verified Case of Tuberculosis (RVCT). CDC 72.9A.
13. Centers for Disease Control and Prevention (2009) Report of Verified Case of Tuberculosis (RVCT). CDC 72.9A.
14. Sprinson JE, Lawton ES, Porco TC, Flood JM, Westenhouse JL (2006) Assessing the validity of tuberculosis surveillance data in California. BMC Public Health 6: 217.
15. Ehman M, Shaw T, Cass A, Lawton E, Westenhouse J, et al. (2013) Developing and Using Performance Measures Based on Surveillance Data for Program Improvement in Tuberculosis Control. J Public Health Manag Pract 19: E29–E37. 10.1097/PHH.0b013e3182751d6f [doi].
16. Centers for Disease Control and Prevention (2010) Monitoring tuberculosis programs - National Tuberculosis Indicator Project, United States, 2002–2008. MMWR Morb Mortal Wkly Rep 59: 295–298. mm5910a3 [pii].
17. National Cancer Institute (2013) Joinpoint Regression Program, Version 4.0.4. Available at: http://srab.cancer.gov/joinpoint. Accessed on October 31, 2013.
18. Kim HJ, Fay MP, Feuer EJ, Midthune DN (2000) Permutation tests for joinpoint regression with applications to cancer rates. Stat Med 19: 335–351. 10.1002/(SICI)1097-0258(20000215)19:3<335:AID-SIM336>3.0.CO; 2-Z [pii].
19. SAS Institute Inc. (2012) SAS, version 9.3 [Software].
20. McNutt LA, Wu C, Xue X, Hafner JP (2003) Estimating the relative risk in cohort studies and clinical trials of common outcomes. Am J Epidemiol 157: 940–943.
21. Zou G (2004) A modified poisson regression approach to prospective studies with binary data. Am J Epidemiol 159: 702–706.
22. Pan W (2001) Akaike's information criterion in generalized estimating equations. Biometrics 57: 120–125.
23. California Department of Public Health, California Tuberculosis Controller's Association (2011) Guidelines for Directly Observed Therapy Program Protocols in California. http://www.ctca.org/fileLibrary/file_347_pdf.
24. Tuberculosis Control Branch (2013) Report on Tuberculosis in California, 2012. http://www.cdph.ca.gov/data/statistics/Pages/TuberculosisDiseaseData_aspx.
25. Golub JE, Bur S, Cronin WA, Gange S, Baruch N, et al. (2005) Patient and health care system delays in pulmonary tuberculosis diagnosis in a low-incidence state. Int J Tuberc Lung Dis 9: 992–998.
26. Liu Z, Shilkret KL, Finelli L (1998) Initial drug regimens for the treatment of tuberculosis: evaluation of physician prescribing practices in New Jersey, 1994 to 1995. Chest 113: 1446–1451.
27. Liu Z, Shilkret KL, Ellis HM (1999) Predictors of sputum culture conversion among patients with tuberculosis in the era of tuberculosis resurgence. Arch Intern Med 159: 1110–1116.
28. Rao SN, Mookerjee AL, Obasanjo OO, Chaisson RE (2000) Errors in the treatment of tuberculosis in Baltimore. Chest 117: 734–737.
29. Richardson NL (2000) Evaluating provider prescribing practices for the treatment of tuberculosis in Virginia, 1995 to 1998: an assessment of educational need. J Contin Educ Health Prof 20: 146–155. 10.1002/chp.1340200303 [doi].
30. Silin M, Laraque F, Munsiff SS, Crossa A, Harris TG (2010) The impact of monitoring tuberculosis reporting delays in New York City. J Public Health Manag Pract 16: E09–E17. 10.1097/PHH.0b013e3181c87ae5 [doi]; 00124784-201009000-00018 [pii].
31. Kong DG, Watt JP, Marks S, Flood J (2013) HIV status determination among tuberculosis patients from California during 2008. J Public Health Manag Pract 19: 169–177. 10.1097/PHH.0b013e3182550a83 [doi]; 00124784-201303000-00010 [pii].
32. Handford CD, Rackal JM, Tynan AM, Rzeznikiewiz D, Glazier RH (2012) The association of hospital, clinic and provider volume with HIV/AIDS care and mortality: systematic review and meta-analysis. AIDS Care 24: 267–282. 10.1080/09540121.2011.608419 [doi].
33. Honda H, Krauss MJ, Jones JC, Olsen MA, Warren DK (2010) The value of infectious diseases consultation in Staphylococcus aureus bacteremia. Am J Med 123: 631–637. S0002-9343(10)00128-2 [pii]; 10.1016/j.amjmed.2010.01.015 [doi].
34. Jenkins TC, Price CS, Sabel AL, Mehler PS, Burman WJ (2008) Impact of routine infectious diseases service consultation on the evaluation, management, and outcomes of Staphylococcus aureus bacteremia. Clin Infect Dis 46: 1000–1008. 10.1086/529190 [doi].
35. Schmitt S, McQuillen DP, Nahass R, Martinelli L, Rubin M, et al. (2013) Infectious Diseases Specialty Intervention is Associated with Decreased Mortality and Lower Healthcare Costs. Clin Infect Dis. cit610 [pii]; 10.1093/cid/cit610 [doi].
36. Davidow AL, Katz D, Reves R, Bethel J, Ngong L (2009) The challenge of multisite epidemiologic studies in diverse populations: design and implementation of a 22-site study of tuberculosis in foreign-born people. Public Health Rep 124: 391–399.
37. Mays GP, McHugh MC, Shim K, Lenaway D, Halverson PK, et al. (2004) Getting what you pay for: public health spending and the performance of essential public health services. J Public Health Manag Pract 10: 435–443.
38. Erwin PC, Greene SB, Mays GP, Ricketts TC, Davis MV (2011) The association of changes in local health department resources with changes in state-level health outcomes. Am J Public Health 101: 609–615. AJPH.2009.177451 [pii]; 10.2105/AJPH.2009.177451 [doi].
39. World Health Organization (2010) Treatment of tuberculosis: guidelines. Geneva: WHO Press.
40. California Department of Health Services (CDHS) (2013) Medi-Cal Rates. Available at: http://files.medi-cal.ca.gov/pubsdoco/rates/rateshome.asp. Accessed on October 31, 2013.
41. Cass A, Shaw T, Ehman M, Young J, Flood JM, et al. (2013) Improved Outcomes Found After Implementing a Systematic Evaluation and Program Improvement Process for Tuberculosis. Public Health Rep 128: 367–376.
42. Curry International Tuberculosis Center (2012) Engaging the Private Sector in TB Prevention. Available at: http://www.nationaltbcenter.ucsf.edu/training/webarchive/private/arch_private.cfm. Accessed on October 31, 2013.

Epidemiological and Virological Characteristics of Influenza Viruses Circulating in Cambodia from 2009 to 2011

Srey Viseth Horm[1]⁹, Sek Mardy[1,2]⁹, Sareth Rith[1]⁹, Sovann Ly[3], Seng Heng[3], Sirenda Vong[1], Paul Kitsutani[4], Vannra Ieng[2], Arnaud Tarantola[1], Sowath Ly[1], Borann Sar[5], Nora Chea[5], Buth Sokhal[6], Ian Barr[7], Anne Kelso[7], Paul F. Horwood[1], Ans Timmermans[8], Aeron Hurt[7], Chanthap Lon[8], David Saunders[8], Sam An Ung[6], Nima Asgari[2], Maria Concepcion Roces[2], Sok Touch[3], Naomi Komadina[7], Philippe Buchy[1]*

1 Institut Pasteur du Cambodge, Réseau International des Instituts Pasteur, Phnom Penh, Cambodia, **2** World Health Organization, Phnom Penh, Cambodia, **3** Communicable Disease Control Department, Ministry of Health, Phnom Penh, Cambodia, **4** Influenza Division, National Center for Immunization and Respiratory Disease, Center for Disease Control and Prevention, Atlanta, Georgia, United States of America, **5** Centers for Disease Control and Prevention, Cambodia Office, Phnom Penh, Cambodia, **6** National Institute of Public Health, Phnom Penh, Cambodia, **7** WHO Collaborating Centre for Reference and Research on Influenza, Melbourne, Australia, **8** Armed Forces Research Institute of Medical Sciences, Bangkok, Thailand

Abstract

Background: The Cambodian National Influenza Center (NIC) monitored and characterized circulating influenza strains from 2009 to 2011.

Methodology/Principal Findings: Sentinel and study sites collected nasopharyngeal specimens for diagnostic detection, virus isolation, antigenic characterization, sequencing and antiviral susceptibility analysis from patients who fulfilled case definitions for influenza-like illness, acute lower respiratory infections and event-based surveillance. Each year in Cambodia, influenza viruses were detected mainly from June to November, during the rainy season. Antigenic analysis show that A/H1N1pdm09 isolates belonged to the A/California/7/2009-like group. Circulating A/H3N2 strains were A/Brisbane/10/2007-like in 2009 before drifting to A/Perth/16/2009-like in 2010 and 2011. The Cambodian influenza B isolates from 2009 to 2011 all belonged to the B/Victoria lineage represented by the vaccine strains B/Brisbane/60/2008 and B/Malaysia/2506/2004. Sequences of the M2 gene obtained from representative 2009–2011 A/H3N2 and A/H1N1pdm09 strains all contained the S31N mutation associated with adamantanes resistance except for one A/H1N1pdm09 strain isolated in 2011 that lacked this mutation. No reduction in the susceptibility to neuraminidase inhibitors was observed among the influenza viruses circulating from 2009 to 2011. Phylogenetic analysis revealed that A/H3N2 strains clustered each year to a distinct group while most A/H1N1pdm09 isolates belonged to the S203T clade.

Conclusions/Significance: In Cambodia, from 2009 to 2011, influenza activity occurred throughout the year with peak seasonality during the rainy season from June to November. Seasonal influenza epidemics were due to multiple genetically distinct viruses, even though all of the isolates were antigenically similar to the reference vaccine strains. The drug susceptibility profile of Cambodian influenza strains revealed that neuraminidase inhibitors would be the drug of choice for influenza treatment and chemoprophylaxis in Cambodia, as adamantanes are no longer expected to be effective.

Editor: Florian Krammer, Icahn School of Medicine at Mount Sinai, United States of America

Funding: The study was funded by the World Health Organization office in Cambodia, the French Agency for Development (SISEA project) and by the Office of the Assistant Secretary for Preparedness and Response within the U.S. Department of Health and Human Services. The funders had no role in study design, data collection and analysis, decision to publish, or preparation of the manuscript.

Competing Interests: The authors have declared that no competing interests exist.

* Email: buchyphilippe@hotmail.com

⁹ These authors contributed equally to this work.

Background

Influenza is a major human pathogen associated with high morbidity and mortality, both in the temperate and subtropical/tropical regions. It is characterized by epidemics that occur seasonally throughout the world every year, with occasional pandemics arising from novel subtypes of the virus causing a considerable economic burden and significant cumulative morbidity and mortality [1–3]. Despite a plethora of information on influenza epidemiology and seasonality, which remains important in planning prevention and treatment strategies, overall patterns of infection have not been fully described on broad geographic scales and for specific types and subtypes of the influenza virus, thus highlighting the need for more countries to conduct year-round viral surveillance and report reliable incidence data at the type and subtype level, especially in the tropics [4].

In temperate regions influenza viruses typically circulate during the winter period [5]. In tropical areas, influenza activity usually occurs all year round with annual/biannual peaks in relation with rainy seasons and/or winter months, but infections can also occur without a clear seasonality [6–11]. Cambodia is a South-East Asian tropical country, which lies geographically in the Northern hemisphere but its influenza season occurs during June-December, each year [12]. Northern hemisphere countries usually experience influenza season from November to March/April, whereas the influenza season of southern hemisphere countries usually occur from May to September [13,14]. Hence, Cambodia's influenza seasonality appears unusual and knowledge of the epidemiological and virological characteristics of such influenza circulation is important for public health preparedness.

We have previously reported preliminary data and described the circulation and seasonality of influenza viruses in Cambodia during three consecutive years following the establishment of the Cambodian National Influenza Centre (NIC) in 2006 [12]. In the present study, we documented the dynamics of influenza activity, performed antigenic and drug susceptibility analyses of influenza virus strains and conducted phylogenetic analysis of influenza A strains isolated between 2009 and 2011, which included the 2009 pandemic.

Materials and Methods

Geographic background

Cambodia is a tropical country of almost 15 million people, with a land area of 181,035 square kilometers in the southwestern part of the Indochina peninsula [15]. International borders are shared with Thailand and Laos on the West and the North, and Vietnam on the East and the Southeast. As the country is affected by monsoon, it is hot and humid with a mean temperature of 27°C and mean relative humidity of 77.5%. There are two distinct seasons: the dry season runs generally from November to April and the rainy season starts in May-June and ends in October-November.

Patients

The Cambodian NIC was established in August 2006 as a joint collaboration between the Virology Unit at the Institut Pasteur in Cambodia (IPC), the Communicable Disease Control Department of the Ministry of Health (CDC/MoH) and the World Health Organization (WHO) office in Cambodia for the purpose of documenting the dynamics of influenza disease and to virologically characterize the circulating strains. To continuously monitor influenza activity, an outpatient sentinel surveillance system for influenza-like illness (ILI) with a weekly reporting and sampling scheme was initially established in five hospital sites in 2006. In addition, hospital-based surveillance of acute lower respiratory infection (ALRI) cases was established in two sites in Takeo and Kampong Cham provincial hospitals (SISEA project, French Agency for Development). Three additional ILI sentinel surveillance sites were then opened in 2009 and operated by the National Institute of Public Health (NIPH) in collaboration with the US CDC local office in the referral hospitals of Mondulkiri (Eastern Cambodia), Svayreang (South-East Cambodia) and Kampot (South-West Cambodia) Provinces. The Armed Forces Research Institute of Medical Sciences (AFRIMS) operated one sentinel site starting in 2009 in Western Cambodia in Thmor Kol (Battambang Province), and opened two additional sentinel sites in 2011 in Anlong Veng (Oddar Meancheay Province) and Pailin. During the pandemic of influenza A/H1N1 2009 virus, an event-based surveillance system was developed by CDC/MoH and WHO. The system was designed for rapid detection, reporting and confirmation of clusters of the disease or suspected cases admitted in both private clinics and hospitals [16].

An ILI case was defined by the sudden onset of fever (≥38°C axillary temperature) and cough or sore throat in the absence of other diagnosis. For the ALRI study, in children under 5, a suspect case was defined as an illness of <10 days duration with cough or breathing difficulties plus tachypnea. For the 5–14 years age group, case definition included the above symptoms plus fever (≥38°C axillary temperature) on admission. For patients over 15 years old, a case was defined as a person with fever (≥38°C axillary temperature) on admission plus tachypnea or chest pain or auscultatory crackles. For event-based surveillance of pandemic influenza A/H1N1 2009 virus (A/H1N1pdm09 virus) infection conducted from 2009 to 2010, inclusion criteria included any person with acute febrile respiratory illness (fever ≥38°C and respiratory symptoms, e.g. cough, sore throat, difficulty breathing) with no other apparent diagnosis and one or more of the following exposures to the risk of A/H1N1pdm09 virus infection within 7 days prior to symptoms onset: a) Close contact with a probable or confirmed case of A/H1N1pdm09 virus infection; b) Residing in or travelled to a province or foreign country with confirmed community transmission of A/H1N1pdm09 virus; c) Is part of a cluster of ILI cases; d) Handled specimens suspected of containing A/H1N1pdm09 virus.

An overview of the all of the surveillance and research programs involved with the collection and testing of strains for this study is provided in Table S1.

Ethical statement

The ALRI study was approved by the National Ethics Committee of the Kingdom of Cambodia. All patients or parents of sick children who participated in the studies provided written informed consent. The ILI and event-based surveillance systems are public health activities organized by the Ministry of Health in Cambodia and as such have a standing authorization from the National Ethics Committee. Samples were all anonymized for the purpose of this study.

Specimen collection

The sentinel sites obtained weekly epidemiological data from patients who fulfilled the ILI case definition, and collected naso-pharyngeal specimens from 5 to 10 cases per week. Respiratory samples together with clinical records were obtained from patients hospitalized with ALRI during 2009 and 2010 in Takeo and Kampong Cham hospitals. Naso-pharyngeal and throat specimens were placed in a vial containing 2 ml of virus transport medium (VTM). The specimens collected from the initial 5 sites of ILI sentinel surveillance and the 2 ALRI study sites were then immediately frozen in liquid nitrogen and the containers were sent on a weekly basis to the IPC's Virology Unit where specimens were stored at −80°C prior to testing. For the specimens collected from the 6 ILI sentinel surveillance sites newly established in 2009 and onward, the VTM tubes containing the naso-pharyngeal swab samples were stored at 4°C and shipped to the NIPH and/or AFRIMS laboratory for testing within 48 to 72 hours. For the A/H1N1pdm09 event-based surveillance, the specimens were collected mainly from hospitals and private clinics in and around the capital, placed into VTM, and immediately sent in a cool box to IPC's laboratory for testing.

Laboratory methods

Viral RNA was extracted at the participating national laboratories (NIC/IPC, NIPH and AFRIMS) using commercial extraction kits (e.g. MagNa Pure LC, Qiagen Viral RNA mini kits) according to the manufacturer's instructions. RNA was amplified using real-time RT-PCR to detect influenza A and B viruses using standard protocols. Influenza A virus subtypes H1, H1pdm, H3 and H5 were detected using subtype-specific real-time RT-PCR assays. All samples that tested positive for influenza were sent to the NIC for confirmation and further characterization.

At the NIC's lab, viral RNA was extracted from 140 μL of virus transport medium containing the nasopharyngeal swab by using the MagNa Pure LC system (Roche), according to the manufacturer's instructions. Seasonal influenza viruses (A/H3N2, A/H1N1 and influenza B virus) were detected and subtyped by a multiplexed conventional reverse transcriptase-polymerase chain reaction (RT-PCR) as described previously [12]. The detection of A/H1N1pdm09 viruses was performed by using a one-step real-time RT-PCR according to the CDC protocol for detection and characterization of influenza A 2009 H1N1, with the use of a 96-well format IQ5 instrument (BioRad). Primer and probe reagents, reaction master-mix, and cycling parameters were used as described in the protocol [17]. The H5N1 virus was detected by a one-step real-time RT-PCR developed by the laboratory at the National Institute of Infectious Diseases (NIID), Tokyo, Japan, and recommended by WHO as the protocol for laboratory procedures to detect avian influenza A H5N1 virus in specimens from suspected human cases [18].

All of the influenza strains described in this study were isolated at the IPC laboratory by inoculation of the specimens that tested positive by molecular methods onto Madin-Darby canine kidney (MDCK) cells in a biosafety level 2+ laboratory for seasonal influenza viruses and in a biosafety level 3 laboratory for pandemic (H1N1) 2009 virus (until 2010, only) and for highly pathogenic avian influenza A/H5N1 virus. In brief, two hundred microlitres of each specimen was inoculated into MDCK cells, with 2 ml per well of maintenance medium containing TPCK trypsin (except for H5N1 virus) at a concentration of 2.0 μg/ml, in a 6-well plate. The plates were incubated at 35°C in a 5% CO_2 atmosphere for 1 week to assess cytopathic effects. The influenza isolates were characterized by a hemagglutination inhibition assay using reference antigens and anti-sera kindly provided by the WHO Collaborating Center (WHOCC) for Reference and Research on Influenza in Melbourne, Australia. A representative number of influenza isolates collected by the various surveillance and event-based systems were sent each year to the WHOCC in Melbourne for confirmation and further analysis (Table 1).

Genome sequencing and phylogenetic analysis

At the NIC's laboratory, viral RNA was extracted from 200 μl of MDCK supernatant using the QIAamp Viral RNA Mini Kit, according to the manufacturer's recommendations (Qiagen, Hilden, Germany). RNA was used as template for RT-PCR with subtype and segment-specific primers (reaction conditions and primer sequences are available from the authors upon request). Nucleotide sequencing reactions were performed with BigDye Terminator Cycle Sequencing kits (Applied Biosystems) following the protocols supplied by the manufacturer and the products were sequenced using an ABI 3730XL automatic DNA Analyser (Life Technologies, Carlsbad, CA, USA) in a commercial facility

Table 1. Influenza isolates collected in Cambodia by type and subtype in 2009-2011.

Year	N# of specimens tested	N# of positives (%)	Influenza Type		Subtype of Influenza A				WHOCC, Melbourne[a]		
			A (%)	B (%)	H1N1 (%)	H3N2 (%)	H5N1 (%)	H1N1pdm[b] (%)	H1N1pdm[b]	H3N2	IB[c]
2009	4601	747 (16.2)	653 (87.4)	94 (12.6)	2 (0.3)	266 (40.7)	1 (0.2)	384 (58.8)	10	15	15
2010	3507	446 (12.7)	343 (76.9)	103 (23.1)	0 (0)	155 (45.2)	1 (0.3)	187 (54.5)	31	17	41
2011	2705	491 (18.2)	173 (35.2)	318 (64.8)	0 (0)	44 (25.4)	8 (4.6)	121 (70)	36	14	72

[a]Specimens analyzed by the WHO Collaborating Center in Melbourne from 2009 to 2011.
[b]H1N1pdm09.
[c]Influenza B.

(Macrogen, Seoul, Korea). Multiple sequence alignment was conducted using ClustalW, version 2 [19]. Phylogenetic trees based on the HA gene sequences were obtained by using the Neighbor-joining method, HKY model, generated using PAUP software [20]. Bootstrap analysis ($n = 1000$) was carried out to determined the best-fitting tree. Vaccine strain sequences, as well as sequences from viruses collected in other countries were obtained from EpiFlu Database available via the GISAID website (www.gisaid.org), and were included in the analysis.

Nucleotide sequence accession numbers

All Cambodian influenza A/H3N2, A/H1N1pdm09 and Influenza B virus sequences included in the analysis were submitted to Genbank and/or to EpiFlu Database and all of these sequences are available via GISAID website (www.gisaid.org). Table S2 provides detailed information about all of the Cambodian strains and sequences analysed in this study.

Neuraminidase inhibitors susceptibility assays

Representative isolates, with respect to date of detection and province of origin of the patient, were randomly selected for neuraminidase drug testing. The NA-Star kit (Life Technologies, Carlsbad, CA, USA), a chemiluminescent NA inhibition assay which utilizes a 1,2-dioxetane derivative of sialic acid as substrate, was used according to the manufacturer's instructions. Oseltamivir carboxylate, zanamivir and 4-amino-4-deoxy-Neu5Ac2en were provided by the Institute for Glycomics, Griffith University, Gold Coast, Australia. Compounds were prepared in distilled water and stored at −20°C until time of use. The concentration of drug required to inhibit 50% of the NA activity (IC50) was calculated using the non-linear curve-fitting function in the Graphpad Prism 4 package (GraphPad Software, Inc., La Jolla, CA, USA). The average IC50 (nM) (± standard deviation) of two independent determinations was calculated for each virus. Outliers of more than 2 standard deviations from the overall mean were retested twice [21].

Susceptibility to adamantanes

The full M gene segments from representative influenza strains of the influenza A/H3N2, A/H1N1 and A/H1N1pdm09 subtypes virus strains isolated from humans were sequenced to detect mutations associated with resistance to adamantanes. Amino acids at positions 26, 27, 30 and 31 of the transmembrane region of the M2 protein are the residues most frequently associated with amantadine- and rimantadine-resistant strains.

Statistical analysis

The comparisons between percentages and two means were tested by chi-squared test and Student's t test respectively. A p value <0.05 was considered statistically significant. Proportions, means and all statistical analyses were performed using STATA 9.0 (StataCorp., College Station, TX, Texas).

Results

Influenza activity in Cambodia

The Cambodian ILI surveillance program collected a total of 7376 samples during 2009–2011. The average age of the ILI patients was 8.7 years (range, 2 weeks to 84 years) and 51.8% were male. Of these, 1246 (16.9%) tested positive for influenza virus: 17.6% in 2009, 14.5% in 2010 and 18.7% in 2011. The average age of influenza virus-infected patients was 7.3 years (range, 1 month to 65 years) and 53.9% were male. There was a statistical difference between the age of confirmed influenza-positive patients and those who tested negative (7.3 versus 8.9 years, p<0.001).

During 2009–2010, a total of 2248 patients presenting with ALRI symptoms were tested. Their mean age was 34.6 years (range, 3 weeks to 95 years) and 52% were male. Of these, 59 (2.6%) tested positive for influenza virus (3% in 2009 and 1.5% in 2010). The median age of influenza-positive cases among ALRI was 16.9 years (range, 3 months to 70 years), and 40.7% were male. The patients with influenza virus infection were significantly younger than the patients who tested negative in the ALRI study (16.9 versus 35 years, p<0.001).

During 2009 and 2010, of the 1005 patients identified through event-based surveillance for suspected A/H1N1pdm09 virus infection 347 (34.5%) were positive for A/H1N1 2009 virus by real-time RT-PCR.

From 2009 to 2011, influenza activity was observed mainly from June to December. The proportion of influenza virus-positive samples detected per month among ILI specimens between 2009 and 2011 varied from 0 to 21% during the dry seasons and from 3 to 61% during the rainy months, from June to November (Figure S1). Only 6 to 12% of the ALRI specimens were positive for influenza virus during the transmission season (Figure S1). The total number of human influenza strains detected during the 2009–2011 period by ILI, ALRI and A/H1N1pdm09 virus event-based surveillance are detailed in Table 1 and Figure 1.

The A/H1N1pdm09 virus was first detected on June 23rd 2009 and progressively became, along with H3N2, the most frequent influenza A subtype circulating in Cambodia. A/H1N1pdm09 activity, detected via event-based surveillance, demonstrated a peak during the period from September to November 2009, dropping to a very low level between January and May 2010 before increasing again to a small peak in June-July 2010. Community transmission of this virus began in August 2009, after a 2 month-period during which only imported cases were identified. The peak of incidence corresponded to the spread of the virus across the country when it started to be detected by both the ILI and the event-based surveillance systems (Figure 2). A/H3N2 virus was frequently isolated in 2009 and 2010 but in 2011 the incidence of this virus was lower.

Influenza B virus co-circulated with A/H3N2 and A/H1N1pdm09 viruses during the 2009–2011 period, with the peak detection in the rainy season of 2011.

Antigenic analysis

Antigenic analyses indicated that A/H3N2 strains belonged to the A/Brisbane/10/2007-like group in 2009 before drifting to the A/Perth/16/2009-like group in 2010 and 2011. All A/H1N1pdm09 strains detected from 2009 to 2011 belonged to the A/California/7/2009-like group. The influenza B strains isolated in 2009–2011 demonstrated the antigenic characteristics of the B/Brisbane/60/2008-like and B/Malaysia/2506/2004 viruses (B/Victoria/2/87 lineage) (Table 2).

Antiviral drug resistance

Among the influenza strains collected during 2009–2011, all 34 A/H3N2 and 39 of 40 (97.5%) A/H1N1pdm09 isolates showed genetic markers of resistance to adamantanes (Table S3). All of the isolated amantadine-resistant viruses contained the amino acid change from serine to asparagine at position 31 (Ser31Asn) in the M2 protein. All influenza A viruses that were sequenced indicated resistance to adamantanes, except for one A/H1N1pdm09 strain isolated in 2011. Analysis of susceptibility to the neuraminidase inhibitors demonstrated that all of the tested strains were sensitive to oseltamivir and zanamivir (Table 3).

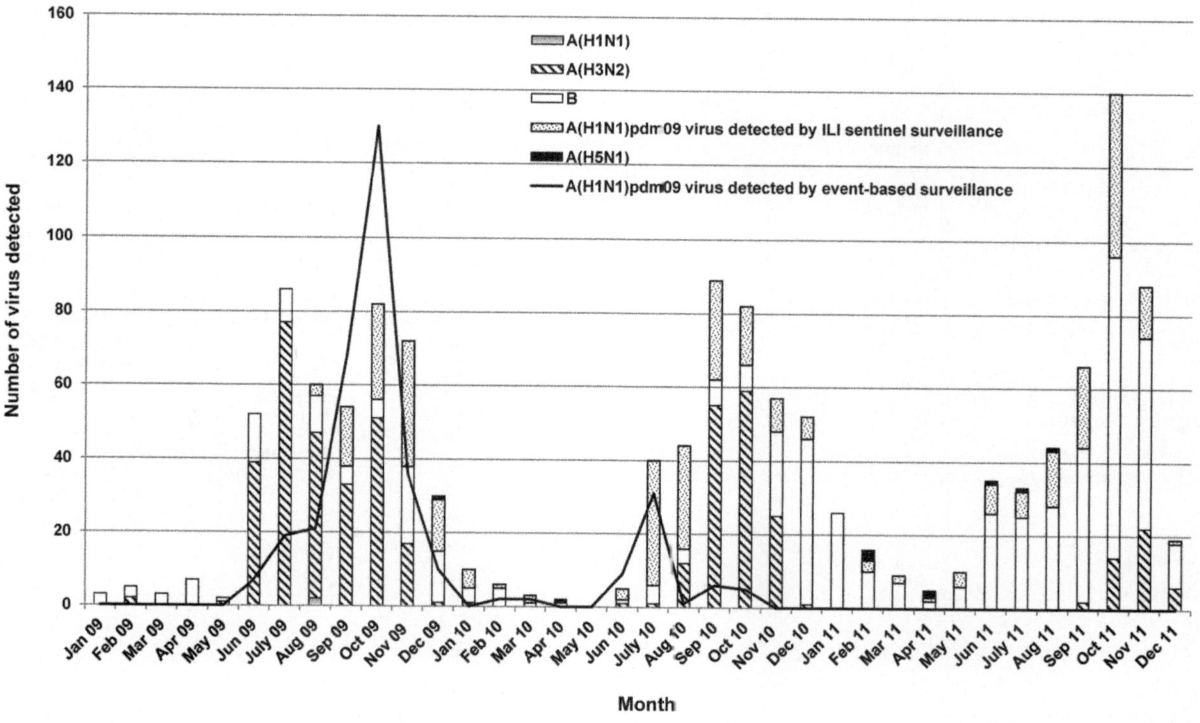

Figure 1. Monthly distribution of influenza virus detected from 2009 to 2011.

Molecular characterization of A/H3N2 isolates

Phylogenetic analysis was carried out for the sequences of the HA1 domain of 28 representative A/H3N2 strains isolated from 2009 to 2011 in Cambodia. Additional sequences retrieved from GenBank and/or from EpiFlu Database (via GISAID) and corresponding to vaccine seed strains, to viruses isolated in Australia and New Zealand (southern hemisphere), and various other South-East Asian countries (Vietnam, Laos, Thailand, Myanmar, and Philippines) were included in the analysis. As

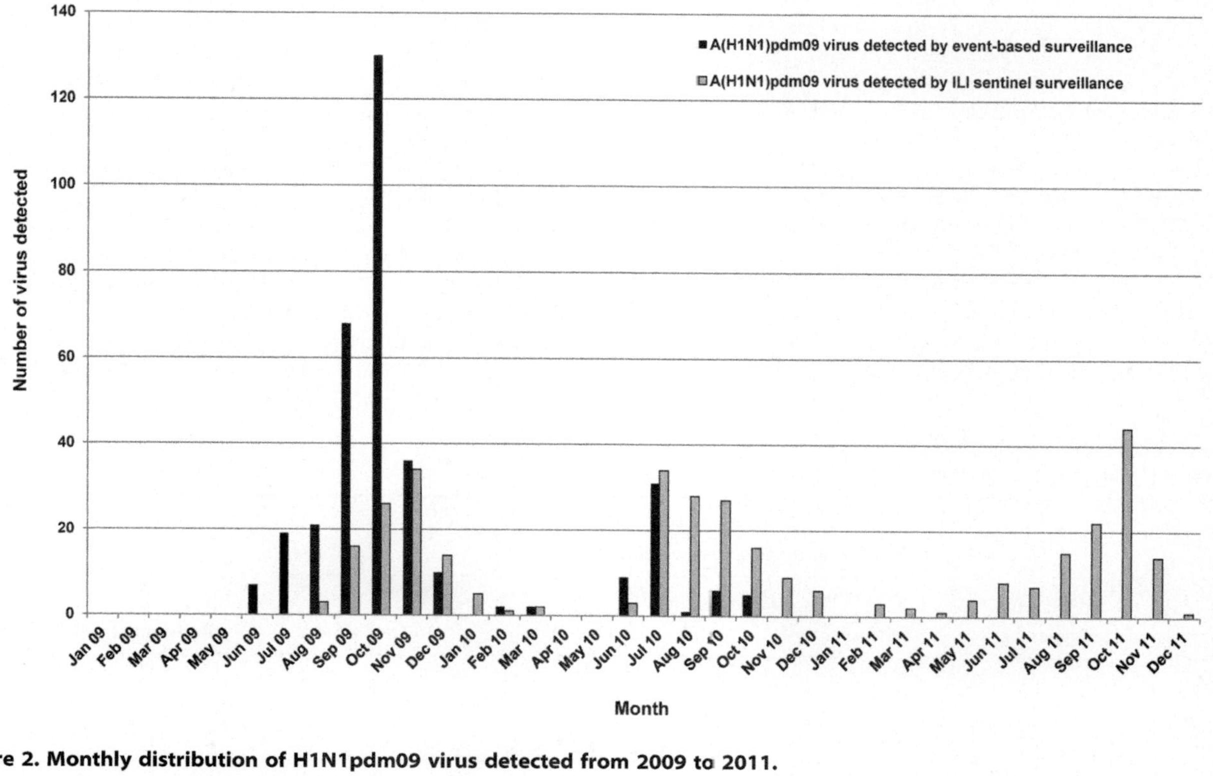

Figure 2. Monthly distribution of H1N1pdm09 virus detected from 2009 to 2011.

Table 2. Comparison between vaccine strains and circulating influenza strains in Cambodia from 2009 to 2011.

	A/H1N1pdm09		A/H3N2		Influenza B	
Season	Vaccine strains	Cambodian strains	Vaccine strains	Cambodian strains	Vaccine strains	Cambodian strains
2009	**A/California /7/2009-like**	**A/California /7/2009-like**	**A/Brisbane/ 10/2007-like**	**A/Brisbane/ 10/2007-like**	**B/Florida/ 4/2006-like**	**B/Brisbane/ 60/2008-like**
2009–2010	*A/California/ 7/2009-like*		*A/Brisbane/ 10/2007-like*		*B/Brisbane/ 60/2008-like*	
2010	**A/California /7/2009-like**	**A/California /7/2009-like**	**A/Perth/ 16/2009-like**	**A/Perth/ 16/2009-like**	**B/Brisbane/ 60/2008-like**	**B/Brisbane/ 60/2008-like; B/ Malaysia/2506/ 2004-like**
2010–2011	*A/California/ 7/2009-like*		*A/Perth/ 16/2009-like*		*B/Brisbane/ 60/2008-like*	
2011	**A/California /7/2009-like**	**A/California /7/2009-like**	**A/Perth/ 16/2009-like**	**A/Perth/ 16/2009-like**	**B/Brisbane/ 60/2008-like**	**B/Brisbane/ 60/2008-like; B/ Malaysia/2506/ 2004-like**
2011–2012	*A/California/ 7/2009-like*		*A/Perth/16/2009-like*		*B/Brisbane/ 60/2008-like*	

shown in Figure 3 (Table S4 and Figure S2), the HA sequences of the A/H3N2 viruses isolated during the three consecutive seasons fell into three distinct clusters also corresponding to strains detected during distinct influenza seasons: Cluster 1 (clade 1) with virus isolated in 2009 and with one strain detected in August 2010 (A/Cambodia/U0825342/2010); cluster 2 (clade 5) with isolates obtained in 2010; cluster 3 (clade 3C) with isolates obtained in 2011. The viruses from the cluster 1 belonged to the A/Perth/16/2009 clade 1 viruses and contained five characteristic amino acid changes, E62K, N144K, K158N, K173Q and N189K compared to that of A/Brisbane/10/2007 virus. The viruses from clusters 2 and 3 had multiple amino acid changes in common compared to that of the A/Brisbane/10/2007 virus, namely K158N, K173Q, N189K and T212A. The Cambodian A/H3N2 viruses isolated in 2010 and 2011 from these clusters diverged into two major genetic clades represented by A/Perth/10/2010 and A/Victoria/361/2011, respectively. The A/Perth/10/2010 clade 5 viruses had characteristic mutations, Y94H, I230V and E280A. The 2011 A/H3N2 Cambodian isolates belonged to clade 3C represented by A/Victoria/361/2011 defined by the mutations S45N, T48I, A198S, V223I and N312S.

Isolates from each group were interspersed with strains isolated from other tropical countries throughout the phylogenetic groups in which the strains from Australia (southern hemisphere) were clearly positioned basal to each viral clade.

Molecular characterization of A/H1N1pdm09 isolates

In order to gain insight into the degree of genetic variability of A/H1N1pdm09 isolates in Cambodia, a phylogenetic tree analysis was carried out for 33 HA sequences from A/H1N1pdm09 strains circulating from 2009 to 2011, comprising the region encoding the HA1 and HA2 domain of the HA protein. The trees were created with the corresponding regions of 21 HA sequences from strains isolated from various other South-East Asian countries (Vietnam, Laos, Thailand, Myanmar, and Singapore). As seen in Figure 4 almost all of the Cambodian isolates fell into the S203T clade. Although one of the Cambodian strains isolated in mid-July 2009 (A/Cambodia/T021/2009) did not cluster with this clade. The majority of strains collected in 2010 and all strains collected in 2011 shared an amino acid change at position E374K. Within this group, one subgroup of strains detected in 2010 shared an amino acid change at position N125D and another subgroup of strains collected in 2011 had the characteristic mutations S143G, S185T, A197T, S451N and E499K. The amino acid substitutions are outlined in detail in Figure 4 (Table S5 and Figure S3).

Compared to the vaccine strain A/California/07/2009 all of the Cambodian isolates possessed residues D187 and D222 in the receptor-binding site. The most common variations compared to the vaccine strain were P83S, S203T, R223Q and I321V, which were actually the substitutions defining clades. These mutations were observed in all Cambodian isolates except for A/Cambodia/T021/2009, which did not possess the amino acid change at

Table 3. Neuraminidase inhibitors resistance in influenza viruses, 2009–2011.

	Resistance to neuraminidase inhibitors			
	Oseltamivir		Zanamivir	
Virus type and subtype	Isolates tested	Resistant N# (%)	Isolates tested	Resistant N# (%)
A/H3N2	12	0	12	0
A/H1N1pdm09	61	0	61	0
B	81	0	81	0

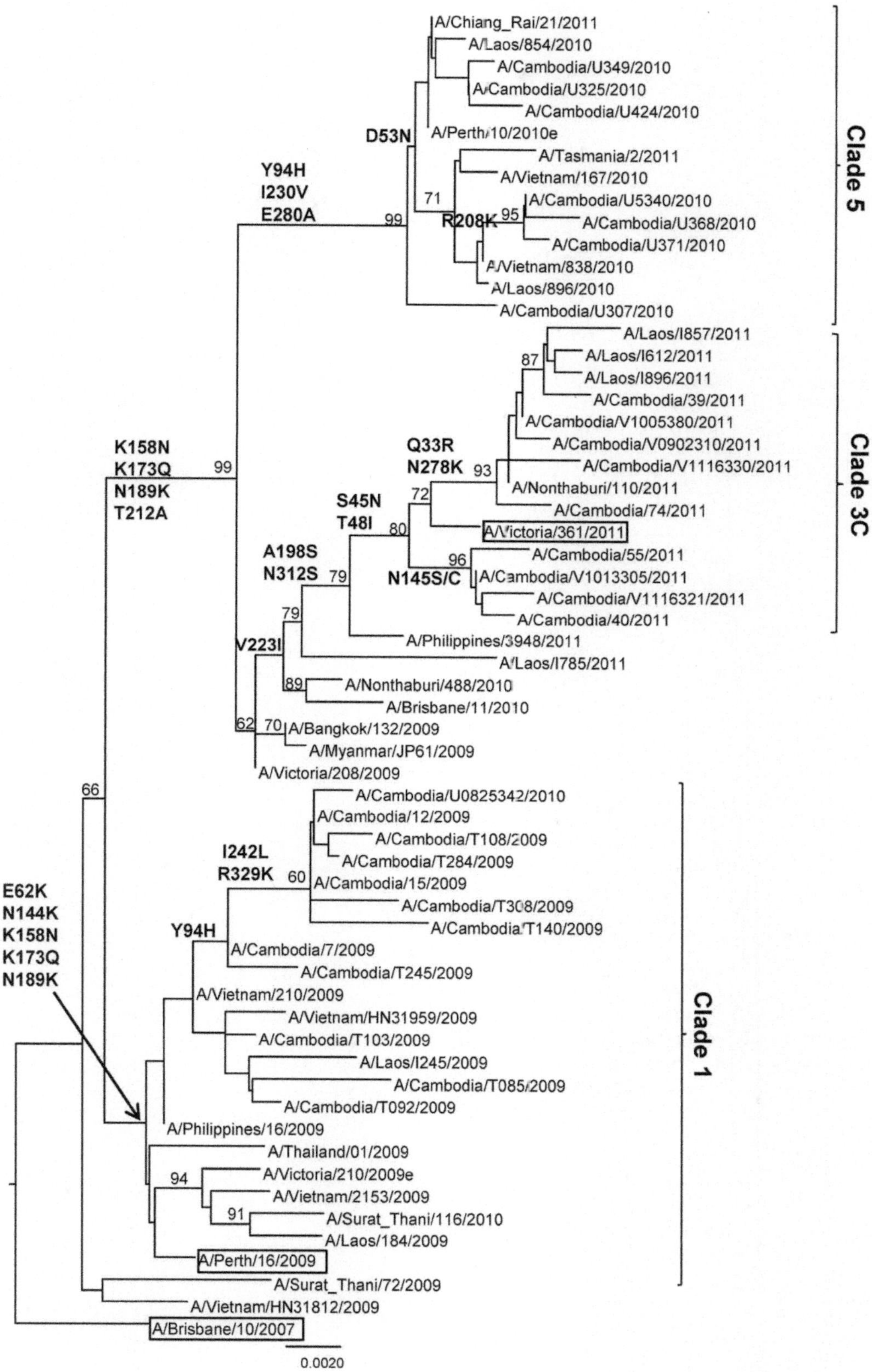

Figure 3. Phylogenetic analysis of the HA1 domains of the HA genes (sequences of 987 nucleotides (nt49–1035)) of influenza A/H3N2 virus isolates collected in Cambodia from 2009 to 2011. The phylogenetic analysis was conducted as a distance-based neighbour-joining phylogenetic tree of influenza using the HKY model and generated using PAUP software with 1,000 bootstrap replicates (values ≥60 shown on branch) and rooted to A/Brisbane/10/2007. Major amino acid changes are shown in block letters at the appropriate nodes. The vaccine reference strains are boxed. Scale bar indicates number of nucleotide substitution per site.

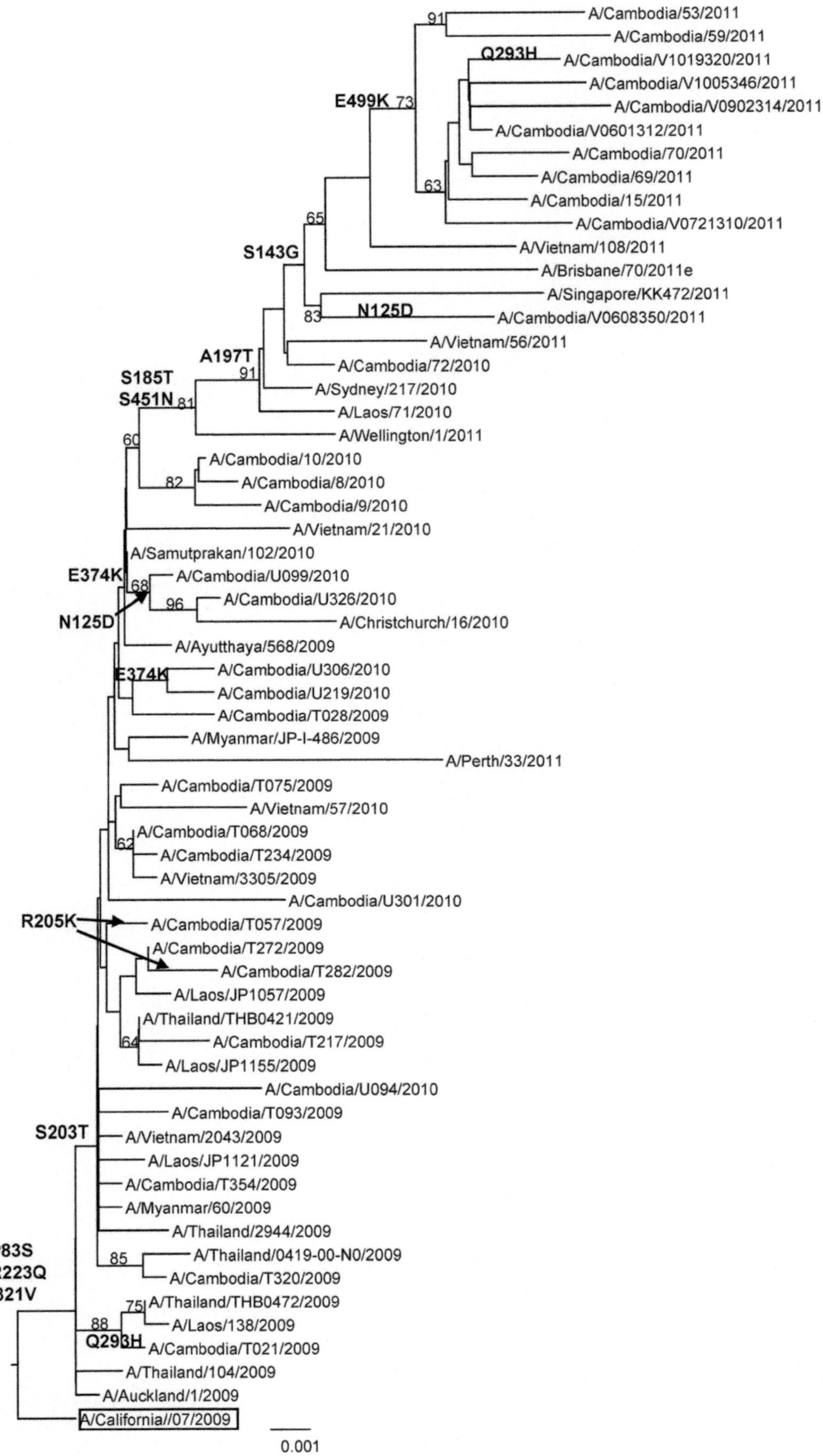
A/Cambodia/53/2011
A/Cambodia/59/2011
Q293H
A/Cambodia/V1019320/2011
A/Cambodia/V1005346/2011
A/Cambodia/V0902314/2011
A/Cambodia/V0601312/2011
A/Cambodia/70/2011
A/Cambodia/69/2011
A/Cambodia/15/2011
A/Cambodia/V0721310/2011
A/Vietnam/108/2011
A/Brisbane/70/2011e
A/Singapore/KK472/2011
N125D
A/Cambodia/V0608350/2011
A/Vietnam/56/2011
A/Cambodia/72/2010
A/Sydney/217/2010
A/Laos/71/2010
A/Wellington/1/2011
A/Cambodia/10/2010
A/Cambodia/8/2010
A/Cambodia/9/2010
A/Vietnam/21/2010
A/Samutprakan/102/2010
A/Cambodia/U099/2010
A/Cambodia/U326/2010
A/Christchurch/16/2010
A/Ayutthaya/568/2009
A/Cambodia/U306/2010
A/Cambodia/U219/2010
A/Cambodia/T028/2009
A/Myanmar/JP-I-486/2009
A/Perth/33/2011
A/Cambodia/T075/2009
A/Vietnam/57/2010
A/Cambodia/T068/2009
A/Cambodia/T234/2009
A/Vietnam/3305/2009
A/Cambodia/U301/2010
A/Cambodia/T057/2009
A/Cambodia/T272/2009
A/Cambodia/T282/2009
A/Laos/JP1057/2009
A/Thailand/THB0421/2009
A/Cambodia/T217/2009
A/Laos/JP1155/2009
A/Cambodia/U094/2010
A/Cambodia/T093/2009
A/Vietnam/2043/2009
A/Laos/JP1121/2009
A/Cambodia/T354/2009
A/Myanmar/60/2009
A/Thailand/2944/2009
A/Thailand/0419-00-N0/2009
A/Cambodia/T320/2009
A/Thailand/THB0472/2009
A/Laos/138/2009
A/Cambodia/T021/2009
A/Thailand/104/2009
A/Auckland/1/2009
A/California//07/2009
E499K
S143G
A197T
S185T
S451N
E374K
N125D
E374K
R205K
S203T
P83S
R223Q
I321V
Q293H
91
73
63
65
83
91
81
60
82
68
96
62
64
85
88
75
0.001

Figure 4. Phylogenetic analysis of the HA genes (1685nt (nt1–1685)) of influenza A/H1N1pdm09 virus isolates collected in Cambodia from 2009 to 2011. The phylogenetic analysis was conducted as a distance-based neighbour-joining phylogenetic tree of influenza using the HKY model and generated using PAUP software with 1,000 bootstrap replicates (values ≥60 shown on branch) and rooted to A/California/07/2009. Major amino acid changes are shown in block letter at the appropriate nodes. The vaccine strain is boxed. Scale bar indicates number of nucleotide substitution per site.

position S203T. However, this strain possessed the substitution Q293H. This mutation was also detected in one strain in 2011 (A/Cambodia/V1019320/2011). Variations were also observed on viral HA antigenic sites. Two of the Cambodian isolates, A/Cambodia/T282/2009 and A/Cambodia/T057/2009 possessed R205K substitutions at the Ca1 antigenic site of the haemagglutinin. Moreover, some of the Cambodian strains isolated in 2010 and one strain in 2011 contained the substitutions N125D involving the Sa haemagglutinin antigenic site. Substitutions in the HA2 domain of the HA protein were also observed, particularly the E374K, in the majority of Cambodian strains isolated in 2010 and in strains isolated in 2011.

Molecular characterization of influenza B isolates

The results of the phylogenetic analyses of the HA1 sequence data generated from 20 representative Cambodian influenza B viruses obtained from 2009 to 2011 are shown in Figure 5. It can be clearly seen that Cambodian influenza B isolates in 2009 to 2011 belonged to the B/Victoria-lineage represented by the vaccine strain B/Brisbane/60/2008 and B/Malaysia/2506/2004. All of the 2009 isolates, most of the 2010 isolates and more than two thirds of the 2011 isolates belonged to genetic group 1 (including influenza B vaccine strain B/Brisbane/60/2008), sharing amino acid changes at positions N75K, N165K and S172P. There were three subgroups within genetic group 1: one subgroup with isolates collected in 2009 and 2010, shared amino acid changes at K56R and V146I in the HA; another subgroup of strains isolated in 2011 (which had two different subgroups with mutation N129D and A154E respectively); and the other subgroup with some of the isolates collected in 2010 and one strain in 2011, shared an amino acid change at L58P in the HA. Some other viruses collected in 2011 and one 2010 isolate belonged to genetic group 2 sharing an amino acid change T37I which was a specific mutation of genetic group 2. No influenza B viruses isolated between 2009 and 2011 belonged to the B/Yamagata-lineage represented by vaccine strain B/Wisconsin/01/2010.

Regions of the Cambodian influenza B virus isolates containing 347 amino acids of HA were compared with those of reference vaccine strains in order to understand the variations within the Victoria lineage of influenza B viruses (Table S4). The Victoria lineage strains were divided into two distinct groups, B/Malaysia/2506/2004 and B/Brisbane/60/2008 (Table S4). All Cambodian influenza B viruses isolated between 2009 and 2011 presented in this study and the vaccine strain B/Brisbane/60/2008 had two amino acid changes (S134P and A199T) compared to B/Malaysia/2506/2004. The B viruses isolated from 2009, 2010 and 2011, which belong to genetic group 1 of Victoria lineage, had three common amino acid changes from B/Malaysia/2506/2004 at N75K, N165K and S172P in their HA genes. In this genetic group 1 of Cambodian influenza B viruses, the subgroup of isolates in 2009 and 2010, had amino acid changes at K56R and V146I. The subgroup of strains isolated in 2011 had two different subgroups with the mutations N129D and A154E respectively. The other subgroup, containing some of the isolates collected in 2010 and 1 strain of 2011, shared an amino acid change at L58P in the HA. The other amino acid changes of this genetic group 1, compared to the B/Malaysia/2506/2004 strain, occurred at I7V for the B/Cambodia/T1231151/2009 strain, K162N for the B/Cambodia/U0707330/2010 strain, S255P for the B/Cambodia/V1103320/2011 strain and K345R for the B/Cambodia/U1102389/2010 strain. Some other isolates from 2011 and one isolate from 2010 were clustered together in genetic group 2 of Victoria lineage. These strains had one common amino acid change from B/Malaysia/2506/2004 at T37I; and variation occurring at P31S and N197S for the B/Cambodia/30/2011 strain; and at V90I for the B/Cambodia/V0316324/2011 strain. No Cambodian B viruses isolated from 2009–2011 belonged to the Yamagata lineage (Figure 5, Table S6 and Figure S4).

Discussion

Seasonal influenza virus epidemics occur in temperate regions of the world each winter, from November to March in the northern hemisphere and from May to September in the southern hemisphere [27]. Although influenza virus has long been regarded as a "cold-weather" pathogen due to its marked winter epidemics in the temperate zones, recent studies show that tropical regions can have significant year-round influenza virus activity [5,8,11,13,28,29]. In contrast, this study combined with our previous analysis of influenza activity in Cambodia from 2006 to 2008 has demonstrated that the seasonality of influenza in Cambodia, which is located in the northern hemisphere, has a consistent pattern characteristic of influenza circulation in the southern hemisphere, peaking during the rainy season from June to November as we previously reported [12]. Cambodian influenza activity reveals a discrete peak for influenza A during June-December, whereas year round circulation of influenza B and rare cases of H5N1 and other influenza A viruses occur during the dry season. This pattern is intermediate between southern hemisphere and tropical transmission.

We found no statistically significant sex ratio difference between non-influenza and influenza infected patients. As previously reported in our study from 2006 to 2008 [12], the proportion of severe influenza infections detected among hospitalized patients with ALRI from 2009 to 2010 during the influenza season was lower compared to the number of patients positive for influenza virus observed in the ILI surveillance system (Figure S1). During 2009–2010 periods, only 2.6% of the patients recruited in ALRI surveillance were identified as positive for influenza virus infection, whereas 16.9% were observed among patients within the NIC's ILI surveillance. This low frequency of severe influenza cases, which was also found in another Cambodian study [22] could be related to yearly variations in influenza severity [23]; however, the true burden of severe influenza is likely to be underestimated. Although the cases detected were from hospital-based surveillance, there might be an unknown proportion of cases that were missed due to poor access to hospital care. This is particularly an issue in developing countries such as Cambodia [23]. Furthermore, some studies have shown that influenza viruses predispose to secondary bacterial infections [24,25] and we cannot exclude that bacterial lung infections diagnosed among patients recruited in our ALRI study were not associated with initial influenza disease. In developing countries such as Cambodia poor access to healthcare may result in late presentation to hospitals and clinics when influenza is no longer detectable in clinical specimens. To rule out

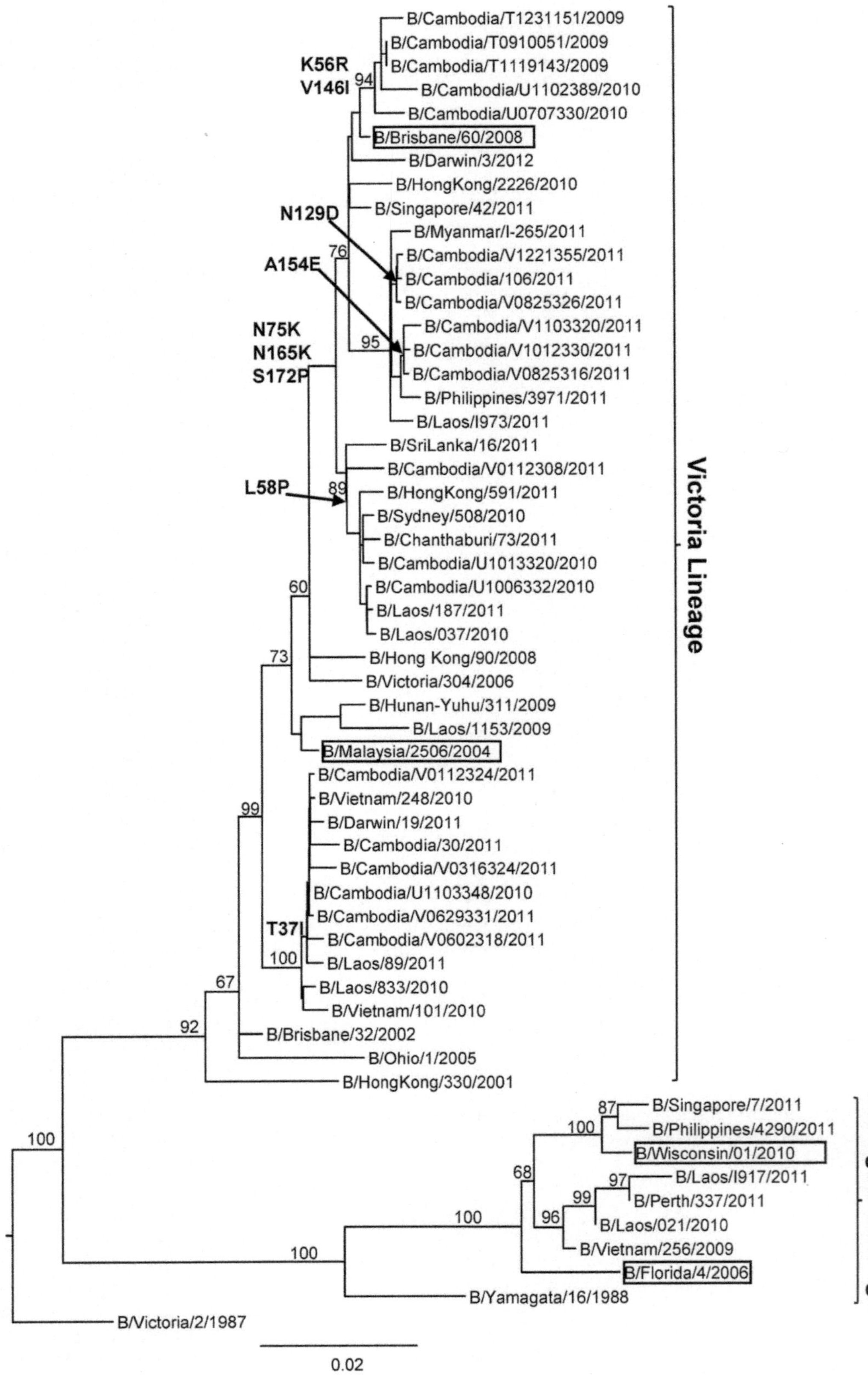

Figure 5. Phylogenetic analysis of the HA1 domains of the HA genes (sequences of 1041 nucleotides (nt 46–1086)) of influenza B virus isolates collected in Cambodia from 2009 to 2011. The phylogenetic analysis was conducted as a distance-based neighbour-joining phylogenetic tree of influenza using the HKY model and generated using PAUP software with 1,000 bootstrap replicates (values ≥60 shown on branch) and rooted to B/Victoria/2/1987. Major amino acid changes are shown in block letters at the appropriate nodes. The vaccine reference strains are boxed. Scale bar indicates number of nucleotide substitution per site.

this hypothesis, serological testing for influenza exposure is required among patients with bacterial respiratory infections during the seasonal influenza epidemics. The A/H1N1pdm09 virus infection detected both in event-based surveillance and sentinel-based surveillance systems occurred in all age groups. However, the median ages of event-based patients tended to be older than that of influenza positive patients detected in ILI surveillance. This finding is consistent with the characteristics of hospitalized cases observed in the USA [26].

The first imported A/H1N1pdm09 cases from Cambodia were detected from June to July 2009 through event-based surveillance in two areas where major international border crossings take place: the international air travel through the Phnom Penh International Airport and the international overland border crossings from Poipet, Thailand to Siem Reap Province. Positive cases of this virus were thereafter detected from communities around the capital and from the ILI sentinel sites in Siem Reap and Battambang Province by both event-based and sentinel surveillance from August 2009. In October 2009, the virus spread throughout the country as indicated by ILI and ALRI surveillance data. The number of A/H1N1pdm09 virus isolates detected by the event-based surveillance compared to the number detected by the sentinel (ILI and ALRI) surveillance systems during the highest point of the outbreak in October 2009 was 5:1 (Figure 2), suggesting that when it comes to the timely detection of outbreaks and important public health events, event-based surveillance is more sensitive than sentinel surveillance systems, which are not suited to the detection of rare but high-impact outbreaks of emerging and unknown diseases [16]. However, at the start of the influenza season in June 2010, the A/H1N1pdm09 isolates were first detected by both the event-based and sentinel surveillance systems before being mostly detected by the sentinel surveillance system for the rest of the season. This finding is consistent with the WHO's declaration on 10 August 2010 that the World was moving into the post-pandemic period and confirmed that this virus was taking on the behavior of a seasonal influenza virus [30].

Worldwide, there continued to be co-circulation of influenza A/H1N1pdm09, A/H3N2 and B viruses in 2010, with the latter two being predominant [31]. In Cambodia, since September 2009, there was a sustained circulation of predominantly A/H1N1pdm09 viruses, but also to a lesser extent, seasonal influenza H3N2 and B virus, with the latter being the most frequently detected strain from November and December 2010 through 2011. During this period, limited data from Vietnam indicated community transmission of influenza there as well, predominantly influenza type B, but the remainder of southern Asia including India, Bangladesh, Thailand, Singapore, China Hong Kong Special Administrative Region, Southern China and Chinese Taipei reported small numbers of a mixture of all three circulating types [32]. The circulation of type B virus tends to rise as overall influenza activity declines. There was a surge of influenza B cases near the end of each influenza season, usually from November to December, as shown in our previous report from 2006 to 2008 [12]. There are frequent bottleneck years during which prevalence of B strains is low, which usually was correlated with high prevalence of influenza A strains. Indeed, low level of B viruses in the year 2009 was observed during the pandemic of A/H1N1pdm09 virus. Once the bottleneck is relieved, there are usually changes in the prevalence of B strains which could explain the rise of influenza B virus at the end of 2010 and in 2011 after the A/H1N1pdm09 subsided (Figure 1). From 2007 to 2011, except in 2009, the influenza B strains identified in Cambodia were well matched by the recommended influenza B vaccine virus (Table 2) [12]. Currently, seasonal influenza vaccines are only available through private market purchase and Cambodians are not accustomed to influenza vaccination [33]. However, it could be problematic if the B strain that was circulating did not match the one contained in the seasonal influenza vaccine. Mismatches between influenza B strains circulating in Cambodia and strains included in the trivalent vaccine should lead to the use of a quadrivalent vaccine that contains both influenza B lineages.

The emergence of resistance following treatment with amantadine has occurred only transiently and the levels of circulating resistant viruses were low until 2004 [34,35]. In Cambodia, no amantadine-resistant strains were found in seasonal influenza viruses in 2006, but the prevalence increased to 100% in 2007 and was maintained at this level through 2011 [12]. This finding supports that of a study conducted by another group reporting that adamantane resistance in influenza A viruses increased in 2007 in South East Asia [36]. Surveillance for antiviral susceptibility of influenza viruses in the Asia-Pacific region during 2011, showed that >99% of influenza A strains continued to be resistant to the adamantane drugs [37]. The high levels of amantadine resistance in influenza A virus strains in the community were not related to domestic amantadine use, as antiviral drugs for upper respiratory infection is uncommon among Cambodians. Following the outbreaks of severe acute respiratory syndrome and avian influenza, an unusually high prevalence of amantadine resistance has been observed most prominently in China, Hong Kong, and Taiwan since 2003, probably due to widespread use of amantadine [34,38]. However, there was also evidence that the spread of resistant A/H3N2 virus was related to linked fitness mutations and not related to drug pressure [39,40]. Cambodia's rise of resistance in 2007 may therefore reflect a strong influence from China or Taiwan in terms of influenza transmission.

A single Cambodian A/H1N1pdm09 strain, isolated from a 10 year old girl patient in Takeo Province in mid-May 2011, did not contain the mutation Ser31Asn nor any of the transmembrane region of the M2 protein mutations known to be associated with adamantane resistance [34].

Alongside neuraminidase inhibitors, oseltamivir and zanamivir have never been used to any large extent in the country and there is no evidence that any of the Cambodian patients were exposed to the drugs before or during influenza infection. The general lack of primary resistance to this class of antiviral has been previously reported before introduction of the neuraminidase inhibitors anywhere in the world [41]. In 2008, a low prevalence of neuraminidase inhibitors resistance (<1%) was detected in isolates from untreated patients [42]. During 2011, neuraminidase inhibitors resistance in A/H1N1pdm09, A/H3N2 and B viruses in the Asia-Pacific region remained low (<3% and <1% for influenza A and B strains respectively) [37]. Therefore, it is not surprising that all Cambodian isolates tested retained sensitivity to oseltamivir and zanamivir. However, in our previous study, we detected two influenza A/H1N1 strains with the H275Y mutation, associated with a high level of resistance to oseltamivir [43]. One of these strains detected in October 2007 showed high resistance to neuraminidase inhibitors by antiviral drug susceptibility assay, even though the other strain could not be grown in Madin-Darby canine kidney cells. Following the emergence in Europe in late 2007 of a transmissible human A/H1N1 influenza strain that was highly resistant to the neuraminidase inhibitor oseltamivir, the emergence of similar viruses was observed in South Africa and several countries in Oceania and South East Asia, especially from May 2008 onwards [44]. However, the mutant H275Y virus found in Cambodia is not a variant of the emergent H275Y strain detected elsewhere, but instead has arisen independently, as occasionally reported previously [42,45].

The phylogenetic analysis of the HA1 genes of A/H3N2 strains from Cambodia showed three groups corresponding to three consecutive influenza seasons from 2009 to 2011. Each group, except the group from 2010–2011, belonged to the phylogenetic group represented by the corresponding vaccine strain. We previously demonstrated that there was co-circulation of different lineages of viruses isolated in 2005 and probable re-emergence of A/H3N2 viruses in the two consecutive seasons of 2005–2006 [42]. The HA sequences from the A/H3N2 isolates from 2009 to 2011 fell within three distinct phylogenetic clades. These findings are consistent with the dynamic evolution of A/H3N2 viruses that present a characteristic co-circulation of several variants for up to three years, which are subsequently replaced by new emerging variants with different antigenic features. This constant evolution of A/H3N2 viruses leads to the emergence of new viral variants that can elude the human immune response, causing outbreaks [46,47]. This viral reintroduction plays a key role in the genesis of new clades and the global spread of these novel influenza virus variants. The Cambodian isolates collected from each season seem to have evolved from common ancestor isolates that are also probably at the origin of Australian srains [43,48]. Additional research is required to more precisely define factors that contribute to this influenza seasonality in Cambodia such as the role of climate in triggering seasonal epidemics and the role for natural selection and host susceptibility.

The earliest A/H1N1pdm09 isolates from Cambodia (A/Cambodia/T028/2009, A/Cambodia/T075/2009, A/Cambodia/T272/2009, A/Cambodia/T282/2009, A/Cambodia/T093/2009 and A/Cambodia/T021/2009) were detected in the period from the end of June to mid-July 2009 through event-based surveillance focused on American students visiting Phnom Penh, the capital city of Cambodia. Positive cases of this virus were thereafter detected from communities all over the country from August 2009 by both the event-based and sentinel surveillance systems. Except for A/Cambodia/T021/2009, all of the other strains detected in that period clustered into the S203T clade. This finding suggested that the A/H1N1pdm09 viruses circulating in Cambodia in that period were introduced into the country, possibly from the USA. Since August 2009, all strains isolated in 2009, 2010 and 2011 were assigned to this clade. This is in agreement with recent results obtained in other regions of the world where it was the most commonly detected clade [49,50].

All of the Cambodian isolates possessed residues D187 and D222 in the receptor-binding site, which is known to confer binding of H1 viruses to the human receptor, supporting efficient transmission of these viruses in humans [51]. In other studies the D222G substitution was detected more frequently in viruses isolated from patients with fatal outcomes and in isolates from lungs [50,52]. This residue was well conserved in all Cambodian strains isolated from patients, but no fatal cases were examined in these studies. The substitution Q293H had been observed recently in a substantial proportion of virus strains isolated from postmortem samples from patients who died from laboratory-confirmed cases of A/H1N1pdm09 infection [52]. Two Cambodian isolates (A/Cambodia/T021/2009 and A/Cambodia/V1019320/2011) possessed this mutation and caused pneumonia requiring hospital admission, although there was no fatal complication. The known antigenic sites in the HA of the 1918 pandemic virus have been conserved in the pandemic A/H1N1pdm09 virus [53]. Molecular analysis showed that these sites were relatively conserved in Cambodian isolates. Two isolates, A/Cambodia/T282/2009 and A/Cambodia/T057/2009, which caused mild symptoms, possessed the R205K variation at the Ca1 antigenic site of the hemagglutinin without altering the properties of the polar and basic amino acid. The other two isolates from 2010 (A/Cambodia/U099/2010 and A/Cambodia/U326/2010) and one isolate from 2011 (A/Cambodia/V0608350/2011), which caused severe illness, had a substitution N125D at the hemagglutinin Sa antigenic site, which slightly modified the properties of the neutral side chain to a basic one. Immunological pressure may have driven virus evolution as shown by the displacement of E374E by E374K, a site important for membrane fusion, and the emergence of the two R205K and N125D amino acid substitutions involving the antigenic sites Ca1 and Sa respectively [54]. These antigenic sites contain many amino acids involved in neutralising epitopes near the receptor-binding pockets [53]. Although a single amino acid substitution involving one antigenic site may be sufficient to cause antigenic change, more commonly antigenic drift variants of epidemiological importance have resulted from changes of at least four amino acids across two or more antigenic sites [55]. Thus the variations observed in the hemagglutinin region of Cambodian A/H1N1pdm09 virus isolates most likely did not affect the antigenicity of the viruses. Although the results from antigenic analysis of Cambodian strains support this hypothesis, *in vitro* results cannot always predict *in vivo* effects.

There have been two distinct evolutionary lineages of influenza B viruses in recent years, represented by the two epidemic strains, B/Victoria/2/87 and B/Yamagata/16/1988. Phylogenetic analyses of influenza B isolates from Cambodia indicated that the circulating influenza B viruses in 2009 to 2011 belonged to the Victoria lineage; and the Yamagata lineage viruses did not circulate in Cambodia during those years (Figure 5). Some strains from 2010 and 2011 were closely related to the vaccine reference strain B/Malaysia/2506/2004, whereas the remainder of strains from 2009–2011 were closely related to the vaccine strain B/Brisbane/60/2008 (Figure 5 and Table S4). Our results indicated that a mismatch between the vaccine and epidemic strains in Cambodia occurred in the Southern Hemisphere World Health Organization 2009 influenza B virus vaccine recommendations; B/Brisbane/60/2008-like should have been the candidate strain for Cambodia (Table 2). Interestingly, the influenza B viruses that circulated in 2010 were found to have evolved from different genetic groups, indicating that the outbreak was caused by multiple sources. All of the influenza B viruses collected in 2009 and the majority of the 2010 and 2011 viruses possessed amino acid substitutions N75K, N165K and S172P that defined genetic group 1 in which the vaccine B/Brisbane/60/2008 strain was grouped. All of the 2009 and some of the 2010 viruses carried the amino acid substitution K56R compared with the vaccine B/Brisbane/60/2008 strain; some strains from 2010–2011 viruses carried the substitution T37I and L58P; some isolates from 2010 and all strains from 2011 contained amino acid I146V compared with the vaccine B/Brisbane/60/2008. None of these substitutions had a marked effect on the antigenicity of the viruses. This is not unusual, as influenza B virus activity is highly variable world-wide, with no established pattern [56]. However, this could lead to the emergence of influenza B reassortants in the next influenza season [57].

This manuscript presents the characterization of influenza strains collected from numerous surveillance programs and research studies, and as such has several limitations. As the collection and initial analysis of samples was conducted in numerous clinics and laboratories respectively, the influenza strains included in this study were subject to any sampling or testing biases that were inherent in the different studies. The analyses presented in the study were also limited to the data available to the NIC and did not include epidemiological information such as detailed admission data and rates of infection. In addition, samples were not collected throughout 2009–2011 for some of the studies (such as event-based surveillance for

H1N1pdm09) and some of the study sites did not contribute samples throughout the entire study period. As a large number of samples were tested throughout the study (n = 10,813) only a representative number of isolates could be further analysed for genetic, antigenic and antiviral drug resistance characteristics. However, a large proportion of isolates were also sent to WHOCCs for further characterization (data not shown).

In summary, this study coupled with our previous study of influenza activity in Cambodia over a six year period from 2006 to 2011 provides strong evidence that influenza was prevalent in Cambodia and influenza activity occurred throughout the year with annual peaks during the rainy season from June to November. Since September 2009, there has been sustained circulation of predominantly A/H1N1pdm09 viruses. Seasonal influenza A/H3N2 and B virus have also circulated, although to a lesser extent. In 2010 there continued to be co-circulation of influenza A/H1N1pdm09, A/H3N2 and B viruses, with the latter two being predominant from November and December 2010 through 2011. The influenza virus epidemics in Cambodia occurred through genetically distinct multiple viruses, even if all of the isolates had similar antigenicity to the reference vaccine strain. The drug susceptibility profile of Cambodian influenza strains showed that neuraminidase inhibitors would be the drug of choice for influenza treatment and chemoprophylaxis in Cambodia, as adamantanes are no longer expected to be effective and not recommended for use [58]. Despite the reduced number of influenza vaccines used in Cambodia, policy makers should consider the use of a quadrivalent vaccine (when available) to counter the co-circulation of the two influenza A lineages.

Supporting Information

Figure S1 Monthly proportion of positive influenza samples among ILI specimens tested from 2009 to 2011 and among ALRI specimens tested from 2009 to July 2010.

Figure S2 Amino acid alignment of HA1 domain of 28 representative A/H3N2 strains isolated from 2009 to 2011 in Cambodia with the vaccine strains A/Brisbane/10/2007 and A/Perth/16/2009.

Figure S3 Amino acid alignment of 33 HA sequences from A/H1N1pdm09 strains circulating from 2009 to 2011 in Cambodia with the vaccine strain A/California/07/2009.

Figure S4 Amino acid alignment of 20 HA1 sequences from influenza B strains circulating from 2009 to 2011 in Cambodia with the vaccine strain B/Malaysia/2506/2004 and B/Brisbane/60/2008.

Table S1 Outline of all of the surveillance programs and research studies that contributed to the present study.

Table S2 All Cambodian influenza A/H3N2, A/H1N1pdm09 and Influenza B virus sequences included in the analysis are available via GISAID website (www.gisaid.org).

Table S3 Resistance to adamantanes predicted following M gene sequencing of influenza A viruses, 2009–2011.

Table S4 Amino acid substitutions in A/H3N2 viruses isolated in Cambodia from 2009 to 2011.

Table S5 Amino acid substitutions in A/H1N1pdm09 viruses isolated in Cambodia from 2009 to 2011.

Table S6 Variations in HA1 amino acid sequences in influenza B-Victoria lineage viruses isolated in Cambodia in 2009, 2010 and 2011 by comparison with vaccine strain B/Malaysia/2506/2004.

Acknowledgments

We would like to thank the following Institutions/Organizations/Project for their contribution: The World Health Organization (WHO) office in Cambodia, assisting the implementation and supporting the monitoring of influenza activity and the establishment the sentinel surveillance system for influenza-like illness (ILI); WHO Collaborating Centre for Reference and Research on Influenza, Melbourne, Australia; The Office of the Assistant Secretary for Preparedness and Response within the U.S. Department of Health and Human Services (DHHS, US); The SISEA project of French Agency for Development (AFD); Influenza Division, National Center for immunization and Respiratory Disease. Centers for Disease Control and Prevention, USA; Centers for Disease Control and Prevention, Cambodia office; National Institute of Public Health (NIPH), Phnom Penh, Cambodia; Communicable Disease Control Department, Ministry of Health, Phnom Penh, Cambodia; The Armed Forces Research Institute of Medical Sciences (AFRIMS), Bangkok, Thailand. The Melbourne WHO Collaborating Centre for Reference and Research on Influenza is supported by the Australian Government Department of Health.

Author Contributions

Conceived and designed the experiments: PB. Performed the experiments: SM SVH SR. Analyzed the data: SM SVH SR PB NK. Contributed reagents/materials/analysis tools: SL SH SV PK VI AT SL BS NC BS IB AK PFH AT AH CL DS SAU NA MCR ST NK. Wrote the paper: SVH SM SR FFH NK PB.

References

1. Simonsen L, Clarke MJ, Williamson GD, Stroup DF, Arden HA, et al. (1997) The impact of influenza epidemics on mortality: introducing a severity index. Am J Public Health 87: 1944–1950.
2. Thompson WW, Shay DK, Weintraub E, Brammer L, Cox N, et al. (2003) Mortality associated with influenza and respiratory syncytial virus in the United States. JAMA 289: 179–186.
3. Dawood FS, Luliano AD, Reed C, Meltzer MI, Shay DK, et al. (2012) Estimated global mortality associated with the first 12 months of 2009 pandemic influenza A H1N1 virus circulation: a modeling study. Lancet Infect Dis 12: 687–695.
4. Finkelman BS, Viboud C, Koelle K, Ferrari MJ, Bharti N, et al. (2007) Global Patterns in Seasonal Activity of Influenza A/H3N2, A/H1N1, and B from 1997 to 2005: Viral Coexistence and Latitudinal Gradients. PLos One 12: e1296.
5. Shek LP, Lee BW (2003) Epidemiology and seasonality of respiratory tract infections in the tropics. Paediatr Respir Rev 4: 104–111.
6. Beckett CG, Hosasih H, Ma'roef C, Listiyonengsih E, Elyazar IRF, et al. (2004) Influenza surveillance in Indonesia: 1999–2003. Clin Infect Dis 39: 443–449.
7. Doraisingham S, Goh KT, Ling AE, Yu M (1988) Influenza surveillance in Singapore: 1972–86. Bull World Health Organ 66: 57–63.

8. Nguyen HLK, Saito R, Ngiem HK, Nishikawa M, Shobugawa Y, et al. (2007) Epidemiology of influenza in Hanoi, Vietnam, from 2001 to 2003. J Infect 55: 58–63.
9. Rao BL, Banerjee K (1993) Influenza surveillance in Pune, India. Bull World Health Organ, 71: 177–181.
10. Tsai HP, Kuo PH, Liu CC, Wang JR et al. (2001) Respiratory viral infections among inpatients and outpatients in Taiwan from 1997 to 1999. J Clin Microbiol 39: 111–118.
11. Members of the Western Pacific Region Global Influenza Surveillance and Response System (2012) Epidemiological and Virological Characteristics of Influenza in the Western Pacific Region of the World Health Organization, 2006–2010. PLoS ONE 7: e37568.
12. Mardy S, Ly S, Heng S, Vong S, Huch C, et al. (2009) Influenza activity in Cambodia during 2006–2008. BMC Infect Dis 9: 168.
13. Viboud C, Alonso WJ, Simonsen L (2006) Influenza in tropical regions. PLoS Med 3: 0468–0471.
14. Webster RG, Bean WJ, Gorman OT, Chambers TM, Kawaoka Y (1992) Evolution and ecology of influenza A viruses. Microbiol Rev 56: 152–179.
15. National Institute of Statistics, Phnom Penh, Cambodia. Final census results (2nd ed.) (2002) General population census of Cambodia 1998.
16. World Health Organization (2008) A guide to establishing event-based surveillance. Available: http://www.wpro.who.int/internet/resources.ashx/CSR/Publications/eventbasedsurv.pdf. Accessed 21 March 2011.
17. World Health Organization (2009) CDC protocol of realtime RT-PCR for influenza A (H1N1). Available: http://www.who.int/csr/resources/publications/swineflu/realtimertpcr/en/index/html. Accessed 21 March 2011.
18. World Health Organization (2007) Recommendations for laboratory procedures to detect avian influenza A H5N1 virus in specimens from suspected human cases. Available: http://www.who.int/csr/disease/avian_influenza/guidelines/RecAIlabtestsAug07.pdf Accessed 21 March 2011.
19. Larkin MA, Blackshields G, Brown NP, Chenna R, McGettigan PA, et al. (2007) Clustal W and Clustal X version 2.0. Bioinformatics, 23: 2947–2948.
20. Swofford DL (2002) PAUP*: Phylogenetic Analysis Using Parsimony (and other methods) 4.0 Beta. Available: paup.csit.fsu.edu/about.html. Accessed 21 March 2011.
21. Naughtin M, Dyason JC, Mardy S, Sorn S, von Itzstein M, et al. (2011) Neuraminidase inhibitor sensitivity and receptor-binding specificity of Cambodian clade 1 highly pathogenic H5N1 virus. Antimicrob Agents Chemother 55: 2004–2010.
22. Blair PJ, Wierzba TF, Touch S, Vonthanak S, Xu X, et al. (2010) Influenza epidemiology and characterization of influenza viruses in patients seeking treatment for acute fever in Cambodia. Epidemiol Infect 138: 199–209.
23. Nair H, Brooks WA, Katz M, Roca A, Berkley JA, et al. (2011) Global burden of respiratory infections due to seasonal influenza in young children: a systematic review and meta-analysis. Lancet 378: 1917–1930.
24. O'Brien KL, Walters MI, Sellman J (2000) Severe pneumococcal pneumonia in previously healthy children: the role of preceding influenza infection. Clin Infect Dis 30: 784–789.
25. McCullers JA, McAuley JL, Browall S, Iverson AR, Boyd KL, et al. (2010) Influenza enhances susceptibility to natural acquisition of and disease due to Streptococcus pneumoniae in ferrets. J Infect Dis 202: 1287–1295.
26. World Health Organization (2009) Human infection with pandemic A(H1N1) 2009 influenza virus: clinical observations in hospitalized patients, Americas, July 2009 – update. Wkly Epidemiol Rec 84: 301–308.
27. Hope-Simpson RE (1981) The role of season in the epidemiology of influenza. J Hyg 86: 35–47.
28. Chittaganpitch M, Supawat K, Olsen SJ, Waicharoen S, Patthamadilok S, et al. (2012) Influenza viruses in Thailand: 7 years of sentinel surveillance data, 2004–2010. Influenza Other Respir Viruses 6: 276–283.
29. Jian JW, Chen JW, Lai CT, Hsu LC, Chen PJ, et al. (2008) Genetic and epidemiological analysis of influenza virus epidemics in Taiwan during 2003 to 2006. J Clinic Microbiol 46: 1426–1434.
30. World Health Organization (2011) Influenza A(H1N1) 2009 virus: current situation and post-pandemic recommendations. Wkly Epidemiol Rec 86: 61–72.
31. World Health Organization (2010) Recommended viruses of influenza vaccines for use in the 2011 influenza season (Southern hemisphere). Wkly Epidemiol Rec 85: 61–72.
32. World Health Organization (2010) Global Influenza Program, Influenza update. Available: http://www.who.int/csr/disease/influenza/2010_12_30_GIP_surveillance/en/index.html Accessed 22 March 2011.
33. Members of the Western Pacific Region Global Influenza Surveillance and Response System (2013) Seasonal influenza vaccine policies, recommendations and use in the World Health Organization's Western Pacific Region. WPSAR 4: doi: 10.5365.
34. Bright RA, Medina MJ, Xu X, Perez-Oronoz G, Wallis TR, et al. (2005) Incidence of amantadine resistance among influenza A(H3N2) viruses isolated worldwide from 1994 to 2005: a cause for concerne. Lancet 366: 1175–1181.
35. Bright R, Shay D, Shu B, Cox N, Klimov A (2006) Adamantane resistance among influenza A viruses isolated early during the 2005–2006 influenza season in the United States. JAMA 295: 891–894.
36. Barr I, Deng Y, Iannello P, Hurt A, Komadina N (2008) Adamantane resistance in influenza A(H1) viruses increased in 2007 in South East Asia but decreased in Australia and some other countries. Antivir. Res 80: 200–205.
37. Leang SK, Deng YM, Shaw R, Caldwell N, Iannello P, et al. (2013) Influenza antiviral resistance in the Asia-Pacific region during 2011. Antiviral Res 97: 206–210.
38. He G, Qiao J, Dong C, He C, Zhao L, et al. (2008) Amantadine resistance among H5N1 avian influenza viruses isolated in Northern China. Antivir Res 77: 72–76.
39. Simonsen L, Viboud C, Grenfell BT, Dushoff J, Jennings L, et al. (2007) The genesis and spread of reassortment human influenza A/H3N2 viruses conferring adamantane resistance. Mol Biol Evol 24: 1811–1820.
40. Ilyushina N, Govorkova E, Webster R (2005) Detection of amantadine-resistant variants among avian influenza viruses isolated in North America and Asia. Virology 341: 102–106.
41. McKimm-Breschkin JL, Trivedi T, Hampson A, Hay A, Klimov A, et al. (2003) Neuraminidase sequence analysis and susceptibilities of influenza virus clinical isolates to zanamivir and oseltamivir. Antimicrob. Agents Chemother 47: 2264–2272.
42. Hurt AC, Barr IG (2008) Influenza viruses with reduced sensitivity to the NA inhibitor drugs in untreated young children. Commun Dis Intell 32: 57–62.
43. Fourment M, Mardy S, Channa M, Buchy P (2010) Evidence for persistence of and antiviral resistance and reassortment events in seasonal influenza virus strains circulating in Cambodia. J Clin Microbiol 48: 295–297.
44. Hurt AC, Ernest J, Deng YM, Iannello P, Besselaar TG, et al. (2009) Emergence and spread of oseltamivir-resistant A(H1N1) influenza viruses in Oceania, South East Asia and South Africa. Antivir Res 83: 90–93.
45. Monto AS, McKimm-Breschkin JL, Macken C, Hampson AW, Hay A, et al. (2006) Detection of influenza viruses resistant to neuraminidase inhibitors in global surveillance during the first 3 years of their use. Antimicrob Agents Chemother 50: 2395–2402.
46. Holmes EC, Ghedin E, Miller N, Taylor J, Bao Y, et al. (2005) Whole-genome analysis of human influenza A virus reveals multiple persistent lineages and reassortment among recent H3N2 viruses. PLoS Biol 3: 1579–1588.
47. Pariani E, Fratt ER, Amendola A, Zappa A, Bianchi S, et al. (2009) Molecular characterization and phlylogenetic analysis of human influenza A viruses in three consecutive seasons with different epidemiological profiles. J Prev Med Hyg 50: 113–116.
48. Nelson MI, Simonsen L, Viboud C, Miller MA, Holmes EC (2007) Phylogenetic analysis reveals the global migration of seasonal influenza A viruses. PLos Pathogens 3: 1220–1228.
49. Nelson M, Spiro D, Wentworth D, Fan J, Beck E, et al. (2009) The early diversification of influenza A/H1N1pdm. PLoS Curr 1: RRN1126.
50. Potdar VA, Chadha MS, Jadhav SM, Mullick J, Cherian SS, et al. (2010) Genetic characterization of the influenza A pandemic (H1N1) 2009 virus isolates from India. PLoS ONE 3: e9693.
51. Stevens J, Corper AL, Basler CF, Taubenberger JK, Palese P, et al. (2004) Structure of the uncleaved human H1 hemagglutinin from the extinct 1918 influenza virus. Science 303: 1866–1870.
52. Glinsky GV (2010) Genomic analysis of pandemic (H1N1) 2009 reveals association of increasing disease severity with emergence of novel hemagglutinin mutations. Cell Cycle 9: 958–970.
53. Igarashi M, Ito K, Yoshida R, Tomabechi D, Kida H, et al. (2010) Predicting the antigenic structure of the pandemic (H1N1) 2009 influenza virus hemagglutinin. PLoS ONE 1: e8553.
54. Maurer-Stroh S, Lee RT, Eisenhaber F, Cui L, Phuah SP, et al. (2010) A new common mutation in the hemagglutinin of the 2009 (H1N1) influenza A virus. PLoS Curr 2: RRN1162
55. Ferguson NM, Galvani AP, Bush RM (2003) Ecological and immunological determinants of influenza evolution. Nature 422: 428–433.
56. McCullers JA, Wang GC, He S, Webster RG (1999) Reassortment and insertion-deletion are strategies for the evolution of influenza B viruses in nature. J Virol 73: 7343–7348.
57. Tsai HP, Wang HC, Kiang D, Huang SW, Kuo PH, et al. (2006) Increasing appearance of reassortant influenza B virus in Taiwan from 2002 to 2005. J Clin Microbiol 44: 2705–2713.
58. CDC (2011) Influenza Antiviral Medications: A Summary for Clinicians. Available: http://www.cdc.gov/flu/professionals/antivirals/summary-clinicians.htm>.

16

Evidence of Carbon Fixation Pathway in a Bacterium from Candidate Phylum SBR1093 Revealed with Genomic Analysis

Zhiping Wang[1,2], Feng Guo[1], Lili Liu[1,3], Tong Zhang[1]*

1 Environmental Biotechnology Laboratory, The University of Hong Kong, Hong Kong, **2** School of Environmental Science and Engineering, Shanghai Jiao Tong University, Shanghai, China, **3** State Environmental Protection Key Laboratory of Environmental Risk Assessment and Control on Chemical Process, East China University of Science and Technology, Shanghai, China

Abstract

Autotrophic CO_2 fixation is the most important biotransformation process in the biosphere. Research focusing on the diversity and distribution of relevant autotrophs is significant to our comprehension of the biosphere. In this study, a draft genome of a bacterium from candidate phylum SBR1093 was reconstructed with the metagenome of an industrial activated sludge. Based on comparative genomics, this autotrophy may occur via a newly discovered carbon fixation path, the hydroxypropionate-hydroxybutyrate (HPHB) cycle, which was demonstrated in a previous work to be uniquely possessed by some genera from *Archaea*. This bacterium possesses all of the thirteen enzymes required for the HPHB cycle; these enzymes share 30~50% identity with those in the autotrophic species of *Archaea* that undergo the HPHB cycle and 30~80% identity with the corresponding enzymes of the mixotrophic species within *Bradyrhizobiaceae*. Thus, this bacterium might have an autotrophic growth mode in certain conditions. A phylogenetic analysis based on the 16S rRNA gene reveals that the phylotypes within candidate phylum SBR1093 are primarily clustered into 5 clades with a shallow branching pattern. This bacterium is clustered with phylotypes from organically contaminated environments, implying a demand for organics in heterotrophic metabolism. Considering the types of regulators, such as FnR, Fur, and ArsR, this bacterium might be a facultative aerobic mixotroph with potential multi-antibiotic and heavy metal resistances. This is the first report on *Bacteria* that may perform potential carbon fixation via the HPHB cycle, thus may expand our knowledge of the distribution and importance of the HPHB cycle in the biosphere.

Editor: Gabriel Moreno-Hagelsieb, Wilfrid Laurier University, Canada

Funding: This research work is financed by the Research Grants Council of Hong Kong (GRF7190/12E) and National Natural Science Foundation of China (51108262). Dr. Wang would thank Hong Kong Scholar Program for financial support (XJ2012030), and Dr. Guo would thank the University of Hong Kong (HKU) for the postdoctoral fellowship. The funders had no role in study design, data collection and analysis, decision to publish, or preparation of the manuscript.

Competing Interests: The authors have declared that no competing interests exist.

* Email: zhangt@hku.hk

Introduction

As the most diverse and abundant cellular life forms in the biosphere, microorganisms play key roles in nearly all biogeochemical processes. However, most microorganisms are not available in pure cultures and can only be detected with culture-independent molecular surveys, which greatly inhibits our comprehension of their roles in ecological and biogeochemical processes. The genomic sequencing of these microorganisms is significant in the construction of blueprints for evolutionary and metabolic diversity [1]. With advances in next generation sequencing (NGS) and bioinformatics, draft genomes of uncultured bacteria can be reconstructed from various complex environmental samples via single-cell genome sequencing [2] or genome binning [3]. Therefore, metabolic deductions and evolutionary analyses can be performed based on the reconstructed genomes and comparative genomics [4], which may greatly expand our understanding of microbial metabolism and its potential role in ecology and biogeochemistry.

SBR 1093 was established as a candidate phylum using several 16S rRNA gene clones in phosphate-removing activated sludge from a sequencing batch reactor [5] that was supplied with sodium acetate for phosphate removal. Thereafter, they were continuously detected in an industrial wastewater treatment system receiving low-molecular-weight organic acids and short-chain alcohols [6], activated sludge from coking wastewater treatment, chlorinated hydrocarbon-contaminated soil and hydrocarbon-contaminated soil [7]. All of these environments were associated with short-chain fatty acids, which implied that the bacteria within this candidate phylum may proliferate effectively with short-chain fatty acids. In addition to the contaminated environment, 16S rRNA clones within candidate phylum SBR1093 were also detected in samples from ocean environments, such as ocean crust from the East Pacific Rise [8], polymetallic nodules and the surrounding sediments, oceanic surface sediment [9], sponges [10], etc. Considering these specific niches, deficiency of light, O_2 and organics, the most probable metabolism for these bacteria may be chemoautotrophy rather than heterotrophy. This is consistent with a report on a stalactite microbial community found in a desert cave [11] in which SBR1093-like 16S rRNA gene sequences comprised up to 10% of the total bacterial 16S rRNA gene sequences. Thus far, the metabolism of bacteria within candidate phylum SBR1093 remains elusive because there are no available pure cultures or enrichments from experiments or genomes. Because their abundance in the known microbial community is very low (less than 1% [11]), the metabolism of SBR1093 in these artificial and biogeochemical processes is difficult to deduce. Therefore, genome binning using the metagenome of a microbial community enriched with a member from this phylum could shed light on its metabolic properties and ecological functions.

As opposed to microbial communities in municipal wastewater treatment plants, which are fed with a mixture of natural organics and dominated by bacteria within *Proteobacteria*, *Bacteroidetes*, *Actinobacteria*, etc. [12], those in industrial wastewater treatment plants show unique populations in each plant [13]. Shaped by the specific substrates and physical-chemical conditions, microbial communities in industrial wastewater treatment plants are often enriched with uncultured microorganisms with specific metabolisms [14], and their metabolisms are associated with the biotransformation and biodegradation of specific substrates. Considering their relatively high abundance in these systems, draft genomes of the dominant populations could be reconstructed via the genome binning of the metagenome [15,16] in an attempt to elucidate their physiological and ecological functions in the microbial community (as well as their taxonomy) [17]. Based on a survey of 454 pyrosequencing for the microbial community pyrosequencings in industrial-activated sludge (data not shown here), a bacterium of candidate phylum SBR1093 was enriched in a full industrial wastewater treatment plant (WWTP), which fed with morpholine distilling-wastewater and performed an alternating anoxic/aerobic process. The objective of this study is to reconstruct the draft genome of a bacterium from candidate phylum SBR1093 with the metagenome of activated sludge from this WWTP. This may shed light on its taxonomic identity, metabolic properties and ecological role, thus be helpful in determining potential conditions for its cultivation and isolation.

Materials and Methods

Sample collection and DNA extraction

Activated sludge samples were collected from a local industrial wastewater treatment plant: two samples from anoxic and aerobic tanks, respectively, which were fed with morpholine distilling-wastewater and that operated in alternate anoxic/aerobic processes (There is no specific permission required for the collection of sludge samples. This sampling site is located at the Shanghai Industrial Park, N31.01, E121.41, and the field studies did not involve endangered or protected species.). The collected samples were fixed onsite with absolute ethanol at a volume ratio of 1:1, and then transported in an icebox and stored at −20°C prior to DNA extraction. For the DNA extraction, the microbial cells in the samples were collected after centrifugation and washed twice with phosphate-buffered solution (PBS). The DNA extraction was performed according to the protocol of the FastDNA SPIN Kit for soil (Qbiogene Inc., CA, USA), which was verified as the most suitable method to extract DNA from the activated sludge samples [18].

Metagenomic sequencing

With the extracted DNA, libraries with insert sizes of 200 bp and 800 bp were constructed for each sample according to the manufacturer's instructions (Illumina, San Diego, USA). Then, the metagenomic sequencing was performed with an Illumina HiSeq 2000 Platform (Illumina, San Diego, USA) using the 101 bp paired-end (PE101) strategy (BGI, Shenzhen, China). With a 2/5 Illumina sequencing run for the 200-bp library and a 1/5 sequencing run for the 800-bp library, approximately 101 and 35 million sequencing reads (100 bp) were obtained, respectively. Raw reads containing any ambiguous bases or those with an average quality score lower than 20 were removed prior to the following analysis.

De novo assembly

The filtered reads were imported into the CLC genomic workbench (version 4.9), and the paired-end sequences were used for the following *de novo* assembly in the CLC genomic workbench. The K parameter (k-mer size) was set to 51 (half of the PE sequencing length) during the assembly. Only contigs longer than 500 bp were output as well as the corresponding mapping reads for further analysis. More than 50% of the reads were assembled into contigs >500 bp (98,505 contigs), with a maximum length of 349,894 bp. As a test to examine the potential errors in the assembly, the coverage consistence of the assembled contigs was checked according to the previous report [15].

Genome binning

Genome binning was performed according to the previous work [19], based on a plot of coverage and GC ratio of contigs, including PE-tracking and reassembly, which was further refined with Metacluster 4.0 [20]. Then, the integrity and redundancy of the binning draft genome were assessed via the comparison of essential single copy genes (ESCGs) of most organisms in the domain *Bacteria* [15,19].

Gene annotation and comparison

Open reading frames (ORFs) were predicted online with MetaGeneMark [21], and the deduced amino acid (AA) sequences were obtained for BLASTp against the NCBI nr database (released on July 18, 2013) with an E-value of 10^{-5} and minimum alignment of 50 AA, respectively [22]. The results were taxonomically assigned with the lowest common ancestor (LCA) algorithm and functionally annotated based on KEGG using MEGAN 4.0 [23]. A Pfam search with an E-value of 10^{-5} was performed based on the Hidden Markol Model and against PfamA database version 26.0 [24], which could also be used for the comparison of gene clusters and verification of MEGAN annotation.

Results and Discussion

Genome binning and completeness assessment

A draft genome containing 94 scaffolds with a total size of 3,099,643 bp with GC contents of 56.4% was reconstructed (Figure 1). According to the prediction of MetaGeneMark, 3,228 ORFs were presented and 3,037 were in full length with the sole initiator and terminator, implying that the *de novo* assembly was accurately performed and that only the ORFs at the ends of the contigs were incomplete. Considering the functional assignment, 2,532 ORFs shared a mean similarity of 51.2% with the known enzymes in the nr-database (released on July 18, 2013), which is nearly at full align length (with a mean cover ratio of 0.87). All of the 40 universally occurring clusters of orthologous groups (COGs) [25] and tRNAs for all 20 amino acids are presented in this draft genome (Table S1 in File S1), which implies that it is near completeness. Additionally, based on the Hidden Markov Model (HMM) search, 102 unique ESCGs were found in this draft genome (Table S2 in File S1), indicating a completeness

>96% [26]. Among the four suspected repetitive ESCGs, TIGR00436 and PF00750 are not always a single copy gene [19], and TIGR02350 is hit with the duplicate genes located within contig_838, implying no contamination from other bacteria. The only suspected duplicate ESCGs (PF00162) are distributed in different contigs and in best hit on the NCBI web to sequences in heterotrophic *Anoxybacillus* sp. (YP002316858) and autotrophic *Nitrososphaera* sp. (YP006862459), respectively. These two sequences are assigned as glyceraldehyde 3-phosphate dehydrogenase and triosephosphate isomerase, which are responsible for glycolysis and gluconeogenesis, respectively, and thus may coexist in an autotrophic or mixotrophic bacterium (consistent with the following metabolism analysis). In summary, this draft genome has no verifiable contamination from other bacteria.

Phylogenetic and biogeographic characterization

This draft genome contains a complete rRNA operon (16S, 5S and 23S, 3,379–8,941 bp) on contig_439, and the 16S rRNA gene (1,567 bp, 6,992–8,558 bp) is used for the taxonomic identification with BLASTn against NCBI and the Greengenes 16S rRNA gene database (released at May, 2013). The genome shares only 85.9% similarity with the 16S rRNA gene of pure culture *Vibrio* sp. Gp-3–5.1 (HF912444) but approximately 94.9% similarity with the first nominated SBR1093 sequence (AF269002) and 99.9% with the uncultured bacterium (HE646343). Therefore, this draft genome should represent a bacterium from candidate phylum SBR1093, named as SBR1093 HKSP. The neighbor-joining and maximum-likelihood phylogenetic tree of the 16S rRNA gene in this bacterium and strains from relative phyla revealed that the candidate phylum SBR1093 represented by this bacterium is close to *Proteobacteria* (Figure S1 in File S1). Additionally, as shown in the phylogenetic tree of this 16S rRNA gene and the reference sequences (Figure 2), phylotypes from candidate phylum SBR1093 are primarily divided into two subdivisions, terrestrial and marine. They are further clustered into 5 clades within which SBR1093 HKSP belongs to clade I. The shallow branching pattern within this phylum (the largest distance between these phylotypes is less than 0.12) implies that bacteria within SBR1093 are recently radiated [27]. It should be noted that some sequences named SBR1093 (GQ348518, 350258, 350948, AY907765, EF573230) are clustered with *Vibrio* sp. (HF912444) and should be excluded from this candidate phylum. Clades within this candidate phylum are distinguished clearly according to their biogeographic distributions, which are differed from dissolve oxygen, salinity, pH, as well as organic nutrients. Bacteria in clades I and II are primarily found in terrestrial environments associated with fertile organics, whereas those in clade IV and V are primarily found in barren marine environments. The 16S rRNA gene of this draft genome SBR1093 HKSP is clustered with phylotypes found in the activated sludge or contaminated environment (with similarity > 99.9%), conditions with plenty of organics, thus implying the potential of organic metabolism (see details below).

In addition to the taxonomy of the 16S rRNA gene, proteins encoding in the genome may be another important resource in determining the phylogenetic position of an unknown bacterium. With the results of BLASTp checked against those in the nr-database, MEGAN may be used to classify these proteins using an LCA algorithm. Therefore, the ancestry of this bacterium can be speculated according to the phylogenetic relationship of proteins encoding in this draft genome. Of the 2,532 proteins in SBR1093 HKSP that have homologs with proteins in the nr-database, the largest section is clustered within the phylum level of *Proteobacteria* (n = 549), which is followed by *Firmicutes* (n = 70), *Cyanobacteria* (n = 53) and *Bacteroidetes* (n = 42) (Figure S2 in File S1). On the genus level, the most hits (n = 104) belong to *Geobacter* in *Deltaproteobacteria*, and the rests are evenly distributed among more than 50 genera. Therefore, candidate phylum SBR1093 represented by this bacterium may be close to *Proteobacteria* and *Firmicutes* but nonetheless separate from them, which is consistent with the taxonomy analysis of the 16S rRNA gene.

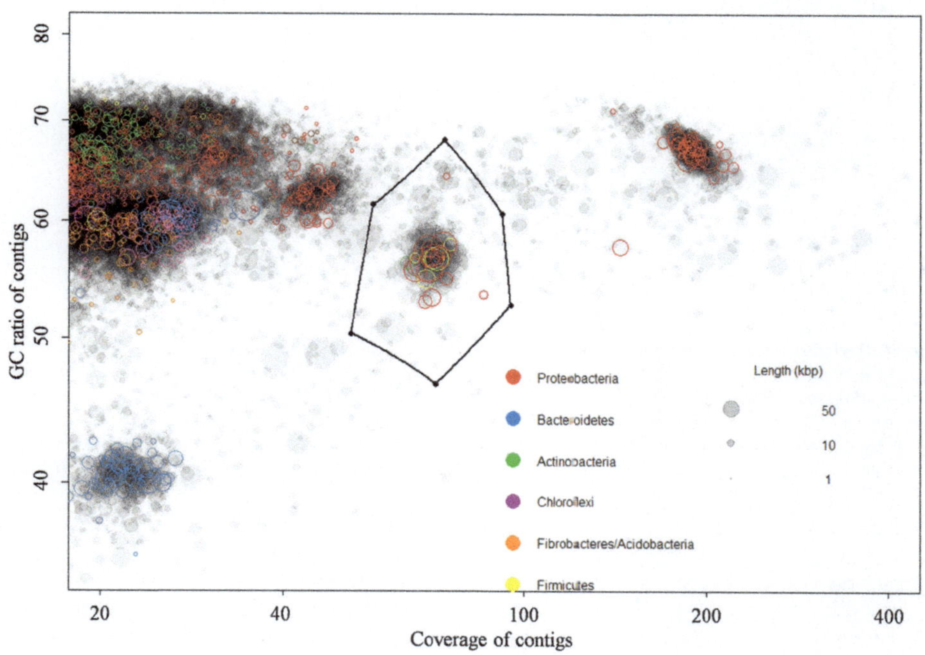

Figure 1. Genome binning of the dominant population with a plot of assembly contigs (based on coverage versus GC ratio). The circles represent the contigs with the size of the square root of their length. Clusters of contigs with similar color present potential genome bins, and contigs cluster with a coverage of approximately 80 (enclosed with black line) were collected for genome binning in this study.

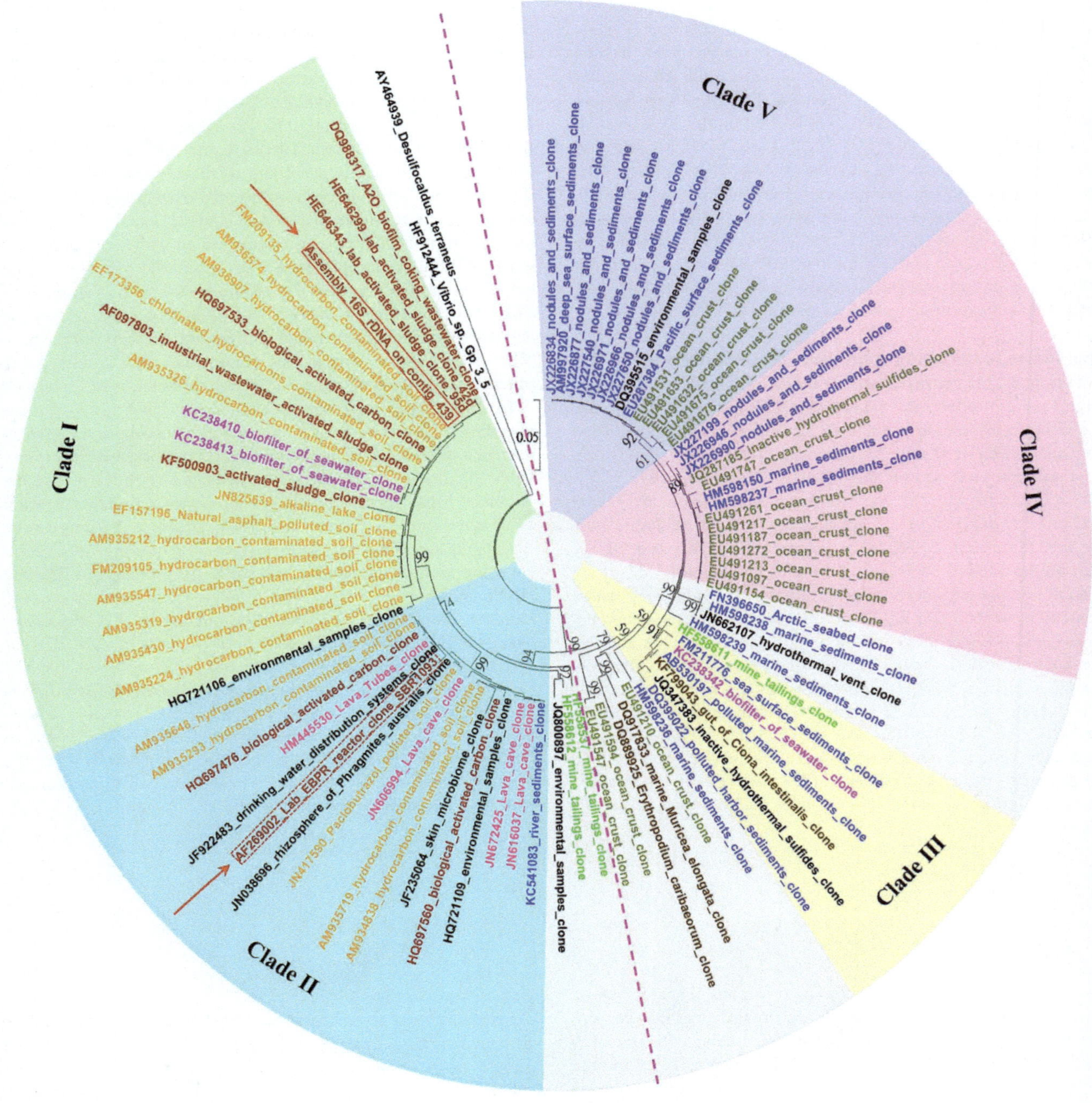

Figure 2. Phylogeny of phylotypes affiliated with the candidate phylum SBR1093. This phylogenetic tree is constructed with 16S rRNA gene sequences based on the neighbor-joining method with Jukes and Cantor distances. The main clades with nodes supported by a bootstrap value of >50% are labeled and marked with different background colors (Clade I green, Clade II blue, Clade III yellow, Clade IV pink and Clade V purple). The phylotypes derived from different sources are labeled with the following: dark red, activated sludge; orange, soil; blue, sediments; dark green, ocean crust; pink, lava; purple, seawater; green, mine tailings; brown, marine organisms; black, others. The phylotype SRB1093 HKSP obtained in this study is enclosed with a solid red line, whereas the first reported phylotype is enclosed with a red dashed line. The scale bar represents 0.05 nucleotide substitutions per site.

Bacterial morphology and cell wall type

This SBR1093 bacterium might be rod-shape, for there is a complete set of genes for rod shape proteins identified in this draft genome (gene_1559–1563). Considering with the cell wall type, a complete set of genes responsible for peptidoglycan biosynthesis (gene_1066–1072) and outer membrane lipoprotein encoding (gene_41, 44, 411 and 2121) are identified, which are only present in Gram-negative bacteria, suggesting this bacterium might be Gram negative. During the life cycle, the bacterium may form spores in adverse conditions and germinate in more suitable conditions, because there is a complete set of genes for spore formation, maturation and germination (gene_1062, 1522 and 265).

Primary metabolism

According to the KEGG annotation in MEGAN, approximately half of the predicted proteins (1,153 of the total 2,329) belong to metabolism, including 268 for carbohydrate metabolism, 223 for amino acid metabolism and 76 for lipid metabolism. The genes responsible for glycolysis/gluconeogenesis and the citrate cycle are complete in this draft genome, as well as a set for oxidative phosphorylation (Table S3 in File S1), indicating that this bacterium should be an aerobic heterotroph. However, two genes encoding the CRP/FNR family regulator and five genes for the Fur family regulators were also identified in this draft genome, indicating the potential anoxic metabolism of this bacterium. Therefore, it appears to be a facultative aerobic bacterium. This is consistent with the conditions of the anoxic/aerobic process from which these samples were collected, as well as the biogeographic distribution of the phylotypes from candidate phylum SBR1093. Additionally, full genes responsible for the upstream metabolism of glycolysis, such as the metabolism of glycogen/starch and cellulose, are also identified in this draft genome, but there were no genes related to the uptake of glucose or metabolism of sucrose, maltose or xylose. Therefore, this bacterium may use glycolysis only for the catabolism of self-producing sugar rather than external sugar.

Based on the genetic analysis, this bacterium may reduce nitrate via an assimilation path and nitrite via an assimilation or dissimilation path (Figure S3 in File S1), which indirectly verifies the facultative aerobic metabolism. Although most of the amino acid can be synthetized with the assimilated ammonia, no genes for asparagine synthesis were detected, implying the necessity for asparagine supplementation in the enrichment of this bacterium. This is consistent with the branched-chain amino acid ABC transporters. Additionally, only genes encoding sulfite reductase exist in this draft genome (none for sulfate reductase), indicating that this bacterium may survive in niches with relatively high redox potential rather than strictly anaerobic conditions. This is also consistent with the conditions in which the microbial community enriched this bacterium (i.e., cycling between the anoxic and aerobic tanks).

Regarding the utilization of the substrates, microbes enriched in a specific condition may adapt to this environment and interact closely with their habitat through metabolic reactions [28]. Because this bacterium is enriched in an anoxic/aerobic process and is fed with morpholine distilling wastewater, it may be involved in the metabolism of the main organics in this wastewater. However, the organics in this influent are primarily in the form of morpholine (or its derivatives), which is reported to decompose into glyoxylate and glycolate only by some species of *Mycobacterium* containing a complex of cytochrome p450, ferredoxin and ferredoxin reductase [29]. Based on the genetic analysis, there is no p450 gene in this draft genome. Therefore, this bacterium may not be active in the ring-opening of morpholine and is more likely involved in the metabolism of the intermediates of morpholine decomposition such as glycolate and glyoxylate. This is consistent with multiple genes responsible for the conversion of glycolate (gene_1890, 1891, 2884, 2885, 2887) and glyoxylate (gene_55), as well as the fusion protein responsible for the glyoxylate shunt (Table S3 in File S1, gene_720), which combines the glyoxylate with acetyl-CoA for the synthesis of malate.

Identification of the carbon fixation pathway

Interestingly, this draft genome possesses genes encoding phosphoenolpyruvate (PEP), acetyl-CoA and propionyl-CoA carboxylase, which catalyze carbon fixation, indicating that this bacterium might have autotrophic metabolism potential. Further comparative genetics reveal that this draft genome contains all of the 13 enzymes required for the hydroxypropionate–hydroxybutyrate (HPHB) cycle (Figure 3), a carbon fixation pathway possessed only by *Archaea* [30]. The enzymes in this draft genome share 30~50% amino acid identity with those involved in the HPHB cycle in *Metallosphaera sedula*, a typical strain of *Archaea* that undergoes the HPHB cycle (Table 1). In this carbon fixation pathway, two molecules CO_2 are assimilated in form of bicarbonate with the carboxylation of a bifunctional biotin-dependent acetyl-CoA/propionyl-CoA carboxylase, and the produced acetyl-CoA is transferred to the citrate cycle for the synthesis of the amino acids and fatty acids or alternatively glycogen via gluconeogenesis. For other carbon fixation pathways, some key enzymes are absent from this draft genome (Figure S4 in File S1), such as pyruvate synthase (EC:1.2.7.1) for the 3-dicarboxylate-hydroxybutyrate (DCHP) pathway, ATP-citric lyase (EC:2.3.3.8) and 2-oxoglutarate synthase (EC:1.2.7.3) for the reductive citric acid cycle pathway, NADP+ dependent formate dehydrogenase (EC:1.2.1.43) and formyltetrahydrofolate synthetase (EC:6.3.4.3) for the reductive acetyl-CoA pathway, malyl-CoA lyase (EC:4.1.3.24) for the 3-hydroxypropionate cycle pathway, and ribulose-bisphosphate carboxylase (EC:4.1.1.39) for the reductive pentose phosphate cycle pathway [31]. Therefore, SBR1093 HKSP may have autotrophic metabolism via the HPHB carbon fixation pathway, whereas the phosphoenolpyruvate carboxylase may function to maintain the balance of intermediates within the citrate cycle [32]. Additionally, enzymes employed for this carbon fixation pathway in *Archaea* are oxygen-tolerant, which is also consistent with the living conditions of this bacterium.

However, microbes possessing the HPHB pathway may evolve as facultative autotrophs rather than obligate autotrophs because this carbon fixation pathway is the most energy-consuming pathway, requiring nine ATP equivalents for one pyruvate [33]. Organisms harboring this carbon fixation pathway may proliferate via a heterotrophic rather than autotrophic pathway as long as there are available organic substrates, which is the nutritional strategy of microbes [34]. For example, although many species within *Sulfolobales* are described as autotrophs or mixotrophs via this carbon fixation pathway, some strains deposited in the public culture collection could lose their autotrophic ability after continuous laboratory cultivation in nutrient-rich media [33]. According to the above discussion, SBR1093 HKSP may possess potential of both autotrophic and heterotrophic metabolisms, thus should be a mixotroph. This is consistent with the enriching conditions, as well as the habitats of strains in the same taxonomic clade. Mixotrophy has been demonstrated to be a widespread and important phenomenon in the biosphere [35], which combines the traits of autotrophs and heterotrophs to utilize both inorganic and organic resources to enable the host survive in oligotrophic or transient environments. In this study, SBR1093 HKSP is enriched in an anoxic/aerobic process, which cycled microbes continuously from feast to fast according to anoxic and aerobic conditions. The microorganisms thus adapt to the transient environment and strive for the dominant position in this microbial community.

Additionally, in this draft genome, a key enzyme for this HPHB pathway, 4-hydroxybutyryl-CoA dehydratase (gene_213), is diverged from those in *Archaea* possessed the HPHB pathway, but clustered with facultative autotrophic bacteria such as *Bradyrhizobium* and *Afipia* within *Bradyrhizobiaceae* (Figure 4). This suggests that this gene is unlikely to be horizontally obtained across a recent domain for SBR1093. A further comparison analysis reveals that in addition to the identified autotrophic

Table 1. Genes responsible for the HPHB cycle in the identified genomes and their identity with those in SBR1093.

SBR1093 HKSP	Functional enzymes	*Ac. hospitalis*		*M. sedula*		*S. tokodaii*		*N. gargensis*		*Af. felis*		*B. sp.* STM3843		*O.carboxidovorans*	
		Acc. No.	Identity	Acc. No.	Identity	Acc. No.	Identity	Acc. No.	Identity	Acc. No.	Identity	Acc. No.	Identity	Acc. No.	Identity
gene_764	ACR	58889	32%	91508	31%	78610	32%	63583	37%	19219	37%	39404	35%	31729	36%
gene_2584	**ACC**	**59288**	**44%**	**90248**	**45%**	**76481**	**45%**	**60794**	**51%**	**19087**	**53%**	**33415**	**51%**	**32993**	**52%**
gene_2098	HPCD	58277	45%	92065	47%	77478	44%	62785	46%	19777	41%	36654	42%	32328	35%
gene_2045	MCE	59386	40%	90738	38%	76441	44%	63084	43%	18664	42%	33379	38%	33030	42%
gene_2581	MSR	59445	41%	91505	41%	78088	43%	61994	39%	21469	44%	35006	54%	34115	43%
gene_5	MCM	59385	47%	90737	47%	76440	48%	63082	48%	//	//	36638	63%	//	//
gene_1024	HBCS	58189	29%	91107	31%	77036	30%	62874	25%	20234	36%	37419	37%	31206	37%
gene_451	HPCS	58189	48%	91435	53%	78009	52%	62874	54%	20930	57%	37438	59%	31167	55%
gene_2450	ACK	59345	41%	90755	43%	76400	41%	62339	28%	21237	59%	33768	59%	31369	58%
gene_1582	**MCR**	**58545**	**28%**	**91779**	**26%**	**77174**	**29%**	**60718**	**31%**	**20991**	**55%**	**38304**	**55%**	**31211**	**55%**
gene_595	**CCH**	**59445**	**24%**	**90500**	**29%**	**75917**	**30%**	**61994**	**23%**	**21942**	**27%**	**35636**	**29%**	**33680**	**27%**
gene_213	HBCD	59205	30%	91403	31%	77631	30%	64012	34%	19793	82%	33076	80%	32741	82%
gene_2933	SSR	59446	31%	91506	30%	78341	29%	63668	37%	19498	32%	38465	35%	32182	30%

The identified genomes are downloaded from NCBI FTP with the accession names *Acidianus hospitalis* W1 uid66875 (*Ac. hospitalis*), *Metallosphaera sedula* DSM 5348 uid58717 (*M. sedula*), *Sulfolobus tokodaii* 7 uid57807 (*S. tokodaii*), Candidatus *Nitrososphaera gargensis* Ga9 2 uid176707 (*N. gargensis*), *Afipia felis* ATCC 53690 uid179396 (*Af. felis*), Bradyrhizobium STM 3843 uid80711 (*B. sp.* STM3843), *Oligotropha carboxidovorans* OM5 uid72795 (*O. carboxidovorans*). The enzymes in bold are bifunctional enzymes in the HPHB pathway, and columns filled with//indicate that there were no hit enzymes in the genome. *M. sedula*, *S. tokodaii* and *Ac. hospitalis* are members of *Sulfolobales* and have the potential for autotrophic metabolism via the reported HPHB pathway, whereas *B. sp.* STM3843, *Af. felis* and *O. carboxidovorans* are members from *Bradyrhizobiaceae* with facultative autotrophic metabolism via an unknown carbon fixation pathway. The accession number for *Acidianus hospitalis* should have 'YP0044' before the presented numbers, and *Metallosphaera* should have 'YP0011', *Sulfolobus* 'NP3', *Nitrososphaera* 'YP0068', *Afipia* 'ZP114', *Bradyrhizobium* 'ZP0943', and *Oligotropha* 'YP0046'.

Abbreviations for functional enzymes: Acryloyl-CoA reductase (ACR); Acetyl/propionyl-CoA carboxylase (ACC); 3-Hydroxypropionyl-CoA dehydratase (HPCD); Methylmalonyl-CoA epimerase (MCE); Malonic semialdehyde reductase (MSR); Methylmalonyl-CoA mutase (MCM); 4-Hydroxybutyrate-CoA ligase (HBCS); 3-Hydroxypropionate-CoA ligase (HPCS); Acetoacetyl-CoA β-ketothiolase (ACK); Succinyl/Malonyl-CoA reductase (MCR); Crotonyl-CoA hydratase (CCH); 4-Hydroxybutyryl-CoA dehydratase (HBCD); Succinic semialdehyde reductase (SSR).

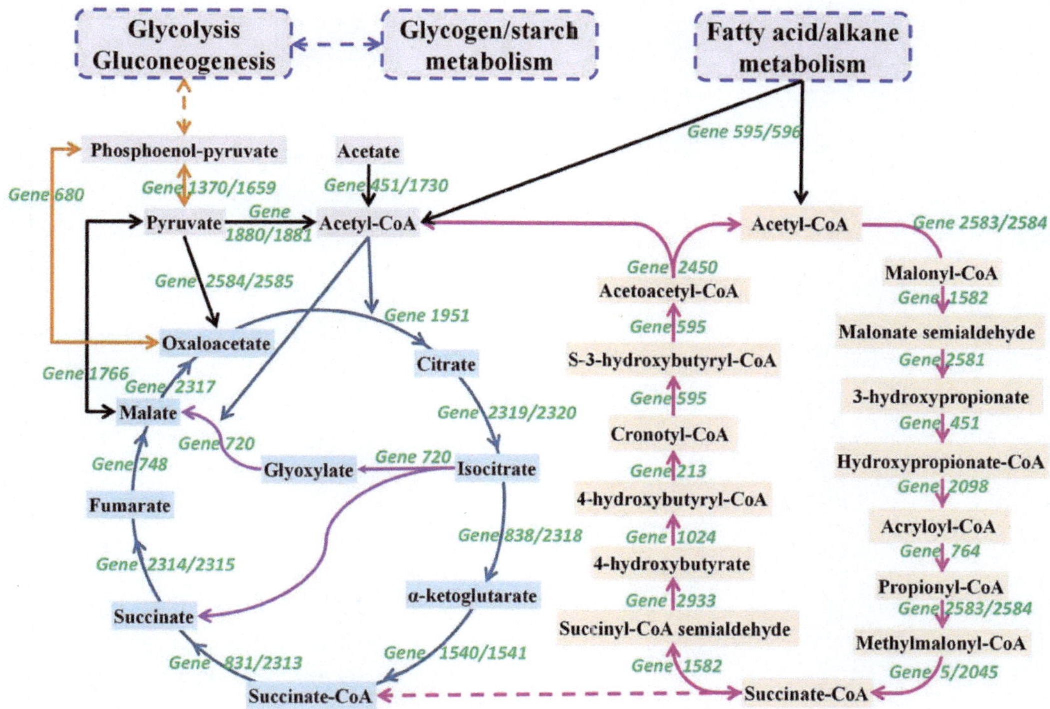

Figure 3. Putative metabolic pathway of SBR1093 HKSP (based on the genetic analysis). Carbon fixation with the HPHB cycle is used for the biomass synthesis via the transfer of acetyl-CoA to the citrate cycle or gluconeogenesis, and the genes responsible for each step are marked with green words. Intermediates connected with colored lines represent different metabolic pathways.

Archaea, all of the 13 enzymes involved in the HPHB pathway are also identified in some species within *Bradyrhizobiaceae* (Table 1), implying the wide distribution of the HPHB pathway in the biosphere and the potential importance of carbon fixation in soils and oceans. A phylogenetic analysis based on concatenated amino acid sequences, that are responsible for the transfer of succinyl-CoA to acetoacetyl-CoA and shared by DCHP & HPHB pathway, also reveals that this bacterium is closer to bacteria within *Bradyrhizobiaceae* than *Archaea* (those responsible for the transfer of crotonyl-CoA to acetoacetyl-CoA are not included because they are fusion proteins of enoyl-CoA hydratase and hydroxybutyryl-CoA dehydrogenase in bacteria but in reverse order in *Archaea*). Therefore, this bacterium may have a nutrition pattern similar to that of bacteria within *Bradyrhizobiaceae*.

Compared with the oligotrophic niches in which *Archaea* reside, *Bradyrhizobiaceae* bacteria are often found in fertile habitats rich in nutrients. Bacteria within *Bradyrhizobiaceae* often appear as heterotrophs rather than autotrophs, although most can grow chemolithoautotrophically using hydrogen, thiosulfate or sulfide as an electron donor [36]. Similarly, the bacterium SBR1093 HKSP is enriched in an industrial WWTP with an influent chemical oxygen demand (COD) concentration of up to 7,000 mg L^{-1} and ammonia concentration up to 100 mg L^{-1}; thus, it may also be a facultative heterotroph or obligate mixotroph.

Antibiotics and heavy metal resistance

It is also interesting to note that this bacterium may have multidrug resistance, such as penicillin, tetracycline, methylenomycin A, chloramphenicol and some macrolide-specific drugs, because abundance of genes associated with drug resistance and efflux transporters are identified in the draft genome. For example, 13 genes encoding a putative drug exporter in the RND superfamily may catalyze antibiotics efflux via an H+ antiport mechanism (Table S4 in File S1), and 19 genes encoding MFS transporter may have antibiotic resistance potential.

Additionally, a complete set gene associated with arsenate reduction and transport has also been identified in this draft genome, as well as 12 genes encoding the ArsR family transcriptional regulator, which may repress the expression of operons linked to stress of di- and multivalent heavy metal ions [37]. Therefore, a variety of heavy metal resistances and transporter genes can be expected in the genome, including the resistance genes for copper and mercury and transport genes for Cu^{2+} and chromate (Table S4 in File S1). The variety resistance to a variety of heavy metals is consistent with the relevant geographical distribution, ocean crust, marine sediment, contaminated soil and activated sludge.

Conclusions

In summary, a draft genome of uncultivated bacteria from candidate phylum SBR1093 was reconstructed with the metagenome of a microbial community from a full-scale industrial wastewater treatment plant. According to the phylogenetic analysis, this bacterium belongs to clade I of candidate phylum SBR1093, which is associated with a contaminated environment and indicates the demand of organics for metabolism. Genome analysis indicates that the bacterium SBR1093 in this phylum may grow autotrophically via the HPHB cycle, a carbon fixation pathway recently found only in some genera from *Archaea*. Enzymes in this draft genome involved in carbon fixation are diverged from those in *Archaea* but share obvious homology to those found in *Bradyrhizobiaceae*. Therefore, this indicates that this bacterium may be a mixotroph. So far, all of the metabolic properties of this SBR1093 HKSP are deduced only based on the

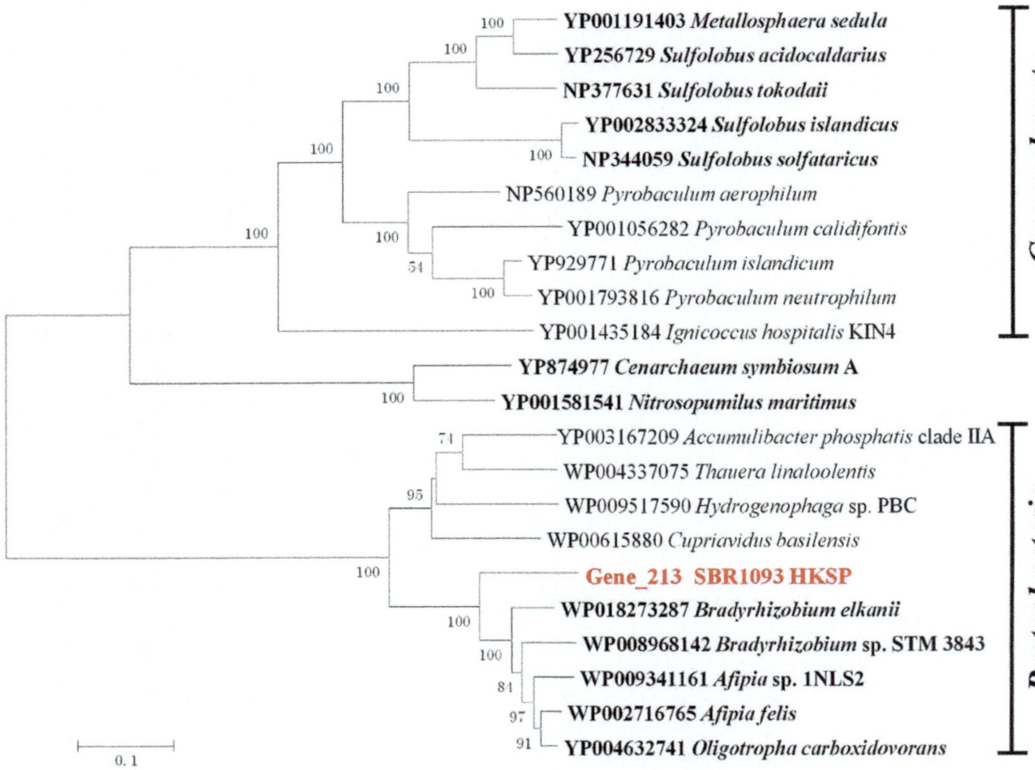

Figure 4. Phylogenetic tree of 4-hydroxybutyryl-CoA dehydratase proteins. The 4-hydroxybutyryl-CoA dehydratases in this draft genome and others in identified autotrophic *Achaea* were retrieved from NCBI to build the phylogenetic tree. The hosts of these marked (in bold) are suspected autotrophic microbes that undergo the HPHB cycle. The number in front of the taxonomy presents the accession number in NCBI. The tree topography and evolutionary distances are given via the neighbor-joining method with a Poisson correction. The numbers at the nodes indicate the percentage bootstrap values for the clade of this group in 1,000 replications. The scale bar represents a difference of 0.1 substitutions per site.

genomic analysis and comparative genomics. Further understanding of the ecological role of this candidate phylum will be obtained through its effective enrichment in the laboratory and the investigation on pure culture.

Supporting Information

File S1 Supporting Figures and Tables. Figure S1. Phylogenetic tree of sequences of SBR1093 and the reference sequences from representative phyla. This phylogenetic tree is constructed with 16S rRNA gene sequences based on the maximum-likelihood method with Jukes and Cantor distances. Only bootstrap value of >50% is labeled and the phylotypes of SRB1093 are marked with red in bold. The scale bar represents 0.1 nucleotide substitutions per site. **Figure S2. Taxonomy of proteins in this SBR1093 HKSP.** Proteins are converted with the predicted gene, and performed BLASTp against NCBI nr database (released at July 18, 2013). Therefore, it is imported into Megan for taxonomic classification. **Figure S3. Putative nitrogen metabolic pathway of this SBR1093 HKSP (Adapted from KEGG). Figure S4. Suspected carbon fixation pathway possessed by this SBR1093 HKSP (Adapted from KEGG).** Columns with solid fill indicate these enzymes are identified in this draft genome, which are connected with colorful lines and arrows and each color represents a type of carbon fixation pathway. Only the red one are full filled, which represents the HPHB cycle. **Table S1. Universally occurring clusters of orthologous groups and tRNAs for amino acids identified in this draft genome. Table S2. List of ESCGs identified in this draft genome and their taxonomy. Table S3. Suspected genes in this draft genome that involving in the primary metabolism. Table S4. Suspected genes in this draft genome that involving in the antibiotics and heavy metal resistance or export.**

Author Contributions

Conceived and designed the experiments: ZW TZ. Performed the experiments: ZW LL. Analyzed the data: ZW FG. Contributed reagents/materials/analysis tools: ZW LL TZ. Contributed to the writing of the manuscript: ZW FG LL TZ.

References

1. Rinke C, Schwientek P, Sczyrba A, Ivanova NN, Anderson IJ, et al. (2013) Insights into the phylogeny and coding potential of microbial dark matter. Nature 499: 431–437.
2. Lasken RS (2012) Genomic sequencing of uncultured microorganisms from single cells. Nat Rev Microbiol 10(9): 631–640.
3. Tyson GW, Chapman J, Hugenholtz P, Allen EE, Ram RJ, et al. (2004) Community structure and metabolism through reconstruction of microbial genomes from the environment. Nature 428: 37–43.
4. Flowers JJ, He SM, Malfatti S, del Rio TG, Tringe SG, et al. (2013) Comparative genomics of two *Candidatus Accumulibacter*' clades performing biological phosphorus removal. ISME J 7: 2301–2314.

5. Bond PL, Hugenholtz P, Keller J, Blackall LL (1995) Bacterial Community Structures of phosphate-removing and non-phosphate-removing activated sludges from sequencing batch reactors. Appl Environ Microbiol 61(5): 1910–1916.
6. Layton AC, Karanth PN, Lajoie CA, Meyers AJ, Gregory IR, et al. (2000) Quantification of *Hyphomicrobium* populations in activated sludge from an Industrial wastewater treatment system as determined by 16S rRNA analysis. Appl Environ Microbiol 66(3): 1167–1174.
7. Militon C, Boucher D, Vachelard C, Perchet G, Barra V, et al. (2010) Bacterial community changes during bioremediation of aliphatic hydrocarbon- contaminated soil. FEMS Microbiol Ecol 74: 669–681.
8. Santelli CM, Orcutt BN, Banning E, Bach W, Moyer CL, et al. (2008) Abundance and diversity of microbial life in ocean crust. Nature 453: 653–656.
9. Li H, Yu Y, Luo W, Zeng Y, Chen B (2009) Bacterial diversity in surface sediments from the Pacific Arctic Ocean. Extremophiles 13(2): 233–246.
10. Schmitt S, Tsai P, Bell J, Fromont J, Ilan M, et al. (2012) Assessing the complex sponge microbiota: core, variable and species-specific bacterial communities in marine sponges. ISME J 6: 564–576.
11. Ortiz-Ortiz M (2012) Kartchner caverns: habitat scale community diversity and function in a carbonate cave. Thesis. The University of Arizona. 65: 99.
12. Zhang T, Shao MF, Ye L (2012) 454 Pyrosequencing reveals bacterial diversity of activated sludge from 14 sewage treatment plants. ISME J 6: 1137–1147.
13. Ju F, Xia Y, Guo F, Wang ZP, Zhang T (2014) Taxonomic relatedness shapes bacterial assembly in activated sludge of globally distributed wastewater treatment plants. Environmental Microbiology DOI: 10.1111/1462-2920.12355.
14. Werner JJ, Knights D, Garcia ML, Scalfone NB, Smith S, et al. (2010) Bacterial community structures are unique and resilient in full-scale bioenergy systems. PNAS 108(10): 4158–4163.
15. Hess M, Sczyrba A, Egan R, Kim TW, Chokhawala H, et al. (2011) Metagenomic discovery of biomass-degrading genes and genomes from cow rumen. Science 331: 463–467.
16. Sharon I, Banfield JF (2013) Genomes from metagenomics. Science 342: 1057–1058.
17. Strous M, Pelletier E, Mangenot S, Rattei T, Lehner A, et al. (2006) Deciphering the evolution and metabolism of an anammox bacterium from a community genome. Nature 440: 790–794.
18. Guo F, Zhang T (2013) Biases during DNA extraction of activated sludge samples revealed by high throughput sequencing. Appl Microbiol Biotechnol 97: 4607–4616.
19. Albertsen M, Hugenholtz P, Skarshewski A, Nielsen KL, Tyson GW, et al. (2013) Genome sequences of rare, uncultured bacteria obtained by differential coverage binning of multiple metagenomes. Nat Biotechnol 31: 533–538.
20. Wang Y, Leung HCM, Yiu S, Chin FYL (2012) MetaCluster 4.0: a novel binning algorithm for NGS reads and huge number of species. J Comput Biol 19: 241–249.
21. Zhu W, Lomsadze A, Borodovsky M (2010) Ab initio gene identification in metagenomic sequences. Nucleic Acids Res 38(12): e132.
22. Wong MT, Zhang D, Li J, Hui RK, Tun HM, et al. (2013) Towards a metagenomic understanding on enhanced biomethane production from waste activated sludge after pH 10 pretreatment. Biotechnol Biofuels 6(1): 38.
23. Yu K, Zhang T (2013) Construction of customized sub-databases from NCBI-nr database for rapid annotation of huge metagenomic datasets using a combined BLAST and MEGAN Approach. PLoS ONE 8(4): e59831.
24. Sonnhammer EL, Eddy SR, Birney E, Bateman A, Durbin R (1998) Pfam: multiple sequence alignments and HMM-profiles of protein domains. Nucleic Acids Res 26(1): 320–322.
25. Stark M, Berger SA, Stamatakis A, von Mering C (2010) MLTreeMap-accurate Maximum Likelihood placement of environmental DNA sequences into taxonomic and functional reference phylogenies. BMC Genomics 11: 461.
26. Dupont CL, Rusch DB, Yooseph S, Lombardo MJ, Richter RA, et al. (2012) Genomic insights to SAR86, an abundant and uncultivated marine bacterial lineage. ISME J 6: 1186–1199.
27. Hunt DE, David LA, Gevers D, Preheim SP, Alm EJ, et al. (2008) Resource partitioning and sympatric differentiation among closely related *Bacterioplankton*. 320: 1080–1085.
28. Finlay B, Clark K (1999) Ubiquitous dispersal of microbial species. Nature 400: 828–828.
29. Trigui MPS, Truffaut N, Thomas D, Poupin P (2004) Molecular cloning, nucleotide sequencing and expression of genes encoding a cytochrome P450 system involved in secondary amine utilization in *Mycobacterium* sp. strain RP1. Res Microbiol 155: 1–9.
30. Ramos-Vera WH, Berg IA, Fuchs G (2009) Autotrophic carbon dioxide assimilation in *Thermoproteales* revisited. J Bacteriol 191: 4286–4297.
31. Berg IA (2011) Ecological aspects of the distribution of different autotrophic CO2 fixation pathways. Appl Environ Microbiol 77(6): 1925–1936.
32. Dunn MF (2011) Anaplerotic function of phosphoenolpyruvate carboxylase in *Bradyrhizobium japonicum* USDA110. Curr Microbiol 62: 1782–1788.
33. Berg IA, Ramos-Vera WH, Petri A, Huber H, Fuchs G (2010) Study of the distribution of autotrophic CO_2 fixation cycles in *Crenarchaeota*. Microbiology 156: 256–269.
34. Tittel J, Bissinger V, Zippel B, Gaedke U, Bell E, et al. (2003) Mixotrophs combine resource use to outcompete specialists: Implications for aquatic food webs. PNAS 100(22): 12776–12781.
35. Zubkov MV, Tarran GA (2008) High bacterivory by the smallest phytoplankton in the North Atlantic Ocean. Nature 455: 224–227.
36. Fuhrmann S, Ferner M, Jeffke T, Henne A, Gottschalk G, et al. (2003) Complete nucleotide sequence of the circular megaplasmid pHCG3 of *Oligotropha carboxidovorans*: function in the chemolithoautotrophic utilization of CO, H_2 and CO_2. Gene 322: 67–75.
37. Busenlehner LS, Pennella MA, Giedroc DP (2003) The SmtB/ArsR family of metalloregulatory transcriptional repressors: structural insights into prokaryotic metal resistance. FEMS Microbiol Rev 17: 131–143.

Effect of the Influenza Virus Rapid Antigen Test on a Physician's Decision to Prescribe Antibiotics and on Patient Length of Stay in the Emergency Department

Hye Won Jeong[1]*, Jung Yeon Heo[1], Jung Soo Park[2], Woo Joo Kim[3,4]

1 Department of Internal Medicine, Chungbuk National University College of Medicine, Heungduk-gu, Cheongju, Republic of Korea, 2 Department of Emergency Medicine, Chungbuk National University College of Medicine, Heungduk-gu, Cheongju, Republic of Korea, 3 Division of Infections Disease, Department of Internal Medicine, Korea University College of Medicine, Guro-dong, Guro-gu, Seoul, Republic of Korea, 4 Transgovernmental Enterprise for Pandemic Influenza in Korea, Seoul, Republic of Korea

Abstract

Background: Influenza virus infection is a common reason for visits to the emergency department (ED) during the influenza season. A rapid and accurate diagnosis of influenza virus infection is important to reduce unnecessary antibiotic prescription and to improve patient care. The aim of this study was to examine whether using the Influenza Virus Rapid Antigen Test (IVRAT) in the ED affects the decision to prescribe antibiotics or the length of hospital stay (LOS).

Methods: Data from patients suffering from an influenza-like illness (ILI) and who were discharged after visiting the ED at Chungbuk National University Hospital were reviewed over two influenza seasons: 2010–2011, when IVRAT was not used in the ED, and 2011–2012, when it was. The numbers of antibiotic prescriptions issued and the ED LOS during these two seasons were then compared.

Results: The number of antibiotic prescriptions was significantly lower in 2011–2012 (54/216, 25.0%) than in 2010–2011 (97/221, 43.9%; $P<0.01$). However, the median ED LOS for patients in 2011–2012 was much longer than that of patients in 2010–2011 (213 minutes *vs.* 257 minutes; $P<0.01$). During the 2011–2012 influenza season, 73 ILI patients showed a positive IVRAT result whereas 123 showed a negative result. Upon discharge, antibiotics were given to 42/123 (34.1%) ILI patients with a negative IVRAT result, but to only 7/73 (9.6%) patients with a positive IVRAT result ($P<0.01$).

Conclusions: Performing IVRAT in the ED reduced the prescription of antibiotics to ILI patients discharged after ED care. However, the ED LOS for patients who underwent IVRAT was longer than that for patients who did not. Thus, performing IVRAT in the ED reduces the unnecessary prescription of antibiotics to ILI patients during the influenza season.

Editor: Benjamin J. Cowling, University of Hong Kong, Hong Kong

Funding: This study was supported by the Health and Medical Technology Research and Development Project of Ministry of Health and Welfare (grant number: A103001). The funder had no role in study design, data collection and analysis, decision to publish, or preparation of the manuscript.

Competing Interests: The authors have declared that no competing interests exist.

* Email: hwjeong@chungbuk.ac.kr

Introduction

Influenza virus infection is a common reason for visits to the emergency department (ED) during annual influenza epidemics. The incidence of medically-attended illness (influenza-like illness (ILI) severe enough to cause the patient to seek medical care) in unvaccinated populations is estimated to be 10–20%, although the rates can be as high as 40–50% [1–3]. Non-specific symptoms such as high fever and a rapid deterioration in the patients' general condition mean that the clinical manifestations of influenza virus infection may mimic bacterial sepsis, particularly in young infants and the elderly [4,5]. A diagnosis of influenza can be made by viral culture or by RT-PCR analysis of nasal or throat swab specimens [6,7]. Various techniques are also available for the rapid diagnosis of influenza at the point-of-patient care [8]. Most are based on specific antibodies that detect viral antigens in respiratory secretions. The test results are visualized on a filter paper, an optical device, or a dipstick.

The rapid and specific diagnosis of influenza virus infection in the ED is important for optimal patient care and infection control, since patients can then receive appropriate antiviral therapy. The influenza virus rapid antigen test (IVRAT) is valuable in the ED setting because it is simple to perform and has a fast turnaround time. However, it has low sensitivity (40–80%) though its specificity is very high [9,10]. Despite this, some small studies of pediatric ILI patients show that a positive IVRAT test results in a fall in the number of antibiotic prescriptions issued in the ED, reduces the need for additional diagnostic tests, and reduces the ED length of stay (LOS) [11–13]. However, a large systematic review of pediatric patients in the ED showed that the routine use of IVRAT had no significant effect on the rate of antibiotic prescription; it also showed that even though patients that received IVRAT were less likely to undergo chest radiography, the ED LOS was no different [14].

It is not clear whether the use of IVRAT in the ED affects a physician's decision making. No studies have examined whether testing adult ILI patients with IVRAT reduces antibiotic prescription and ED LOS. Therefore, the aim of this study was to examine the effect of IVRAT on the prescription of antibiotics and ED LOS in adult ILI patients.

Materials and Methods

Study design and patient population

Chungbuk national university hospital, a 620-bed teaching hospital in Cheongju, Republic of Korea, began performing IVRAT in the ED during the 2011–2012 influenza season, when it became a participant in a sentinel hospital-based surveillance system for influenza (Hospital-based Influenza Morbidity and Mortality; HIMM) [15]. During the 2011–2012 season, all ILI patients who visited the ED were encouraged to provide consent for IVRAT. Nasal or throat swab specimens from ILI patients managed in the ED were tested at the bedside using influenza detections kits (SD Bioline Influenza Antigen Test, Standard Diagnostics, Kyonggi, Korea). At the same time, an additional swab specimen was taken and transported to the central HIMM laboratory. ILI was defined as the presence of fever (temperature ≥38°C, as measured in the ED) accompanied by at least one of the following respiratory symptoms (recorded by the attending physician): cough, sore throat, or rhinorrhea [15].

The medical records of ILI patients who visited the ED at Chungbuk National University Hospital during the 2010–2011 (pre-IVRAT) and 2011–2012 (IVRAT) influenza seasons were reviewed by infectious disease specialists. The following data were collected: the patients' reason for visiting the ED; the medical diagnosis made by the ED primary physicians; the time of arrival at the ED; the time at which the patient was seen by an ED physician; the decision regarding whether to hospitalize the patient; and the time at which the patient was discharged from the ED. If the patient was discharged home after ED care, the patient's discharge medications were also reviewed. The influenza season was roughly defined as the period from November to February. All ILI patients discharged from hospital after ED care were included in the study. The rates of antibiotic prescription and ED LOS (the time interval between the patient being seen by the physician and their departure from the ED) during each of the two influenza seasons were then compared. The characteristics of ILI patients with a positive IVRAT result and of those with a negative result were also compared.

Patients hospitalized after ED care were excluded from the study. This is because the most common reason for hospitalization was pneumonia, which requires treatment with antibiotics (in most cases). Another reason is because other factors, such as waiting for an available room on the ward, would affect the ED LOS for this group.

Ethics statement

The study protocol was approved by the Institutional Review Board of Chungbuk National University Hospital (IRB No. 2012-02-012) and all participants (or next of kin/caregiver in the case of children) who attended the ED during the 2011–2012 (IVRAT) season provided written informed consent to their participation in the study. Consent was waived in the case of ILI patients attending the ED during the 2010–2011 (pre-IVRAT) season. Patient records were anonymized and de-identified prior to analysis.

Statistical analysis

Categorical variables were compared using the Chi-squared or Fisher's exact tests. Continuous variables that did not show a normal distribution were compared using the Mann-Whitney test. SPSS version 13.0 (SPSS, Inc., Chicago, IL) was used for all statistical analyses. All tests were two-tailed. A P value <0.05 was considered statistically significant.

Results

During the 2010–2011 influenza season, 493 ILI patients visited the ED at Chungbuk National University Hospital. Of these, 221 were discharged after ED care (group A) and 264 were hospitalized. During the 2011–2012 influenza season (when the hospital began using IVRAT in the ED), 448 ILI patients visited the ED and 216 patients were discharged after ED care (group B). Data from the patients who were discharged after ED care in each year (groups A and B) were analyzed and the results were compared (Figure 1).

The age-range of the patients in group A was from newborn to 88 years and that of group B was from newborn to 100 years. The median age of the patients group B was greater than that of patients in group A (9.0 years *vs.* 34.5 years, respectively; $P<0.01$), and 47.5% (105/221) of the patients in group A and 54% (117/216) in group B were female. Antibiotics were prescribed to 97 patients (43.9%) in group A and to 54 patients (25.0%) in group B (Table 1). Thus, the proportion of antibiotic prescription was significantly lower in 2011–2012 than in 2010–2011 ($P<0.01$). ILI patients were most often prescribed cephalosporins and amoxicillin/clavulanate (Table 2). The median ED LOS for patients in group B was much longer than that of patients in group A (213 minutes *vs.* 257 minutes; $P<0.01$; Table 1). The overall ED LOS of all patients who visited ED was not significantly different between the two seasons (median 152 minutes vs. 156 minutes; $P=0.35$).

Next, we examined the number of antibiotic prescriptions and ED LOS in group B patients according to the IVRAT results. Of the 216 patients eligible for testing, 73 showed a positive result and seven (9.6%) were prescribed antibiotics; however, antibiotics were prescribed to 42/123 (34.1%) of the ILI patients with a negative IVRAT result. Of the 20 patients who refused the IVRAT test, five received antibiotics (25.0%). Thus, a positive IVRAT result led to a significant reduction in the number of antibiotics prescriptions issued ($P<0.01$; Table 1). The median ED LOS for the 73 patients with a positive IVRAT result was 265 minutes, which was longer than (but not significantly different from) that of the 123 patients with a negative IVRAT result (253 minutes; $P=0.15$).

Discussion

Here, we examined whether performing IVRAT in the ED affected the number of antibiotic prescriptions issued to ILI patients discharged home after ED treatment. We also examined whether IVRAT had an effect on the ED LOS for these patients. We found that performing IVRAT in the ED reduced the number of antibiotics prescriptions issued to ILI patients who were not hospitalized. However, the ED LOS for ILI patients discharged home after ED care was actually longer during the influenza season in which IVRAT was performed than in the previous season when IVRAT was not performed. Taken together, these findings suggest that IVRAT affects a physician's decision making and reduces the prescription of antibiotics to ILI patients.

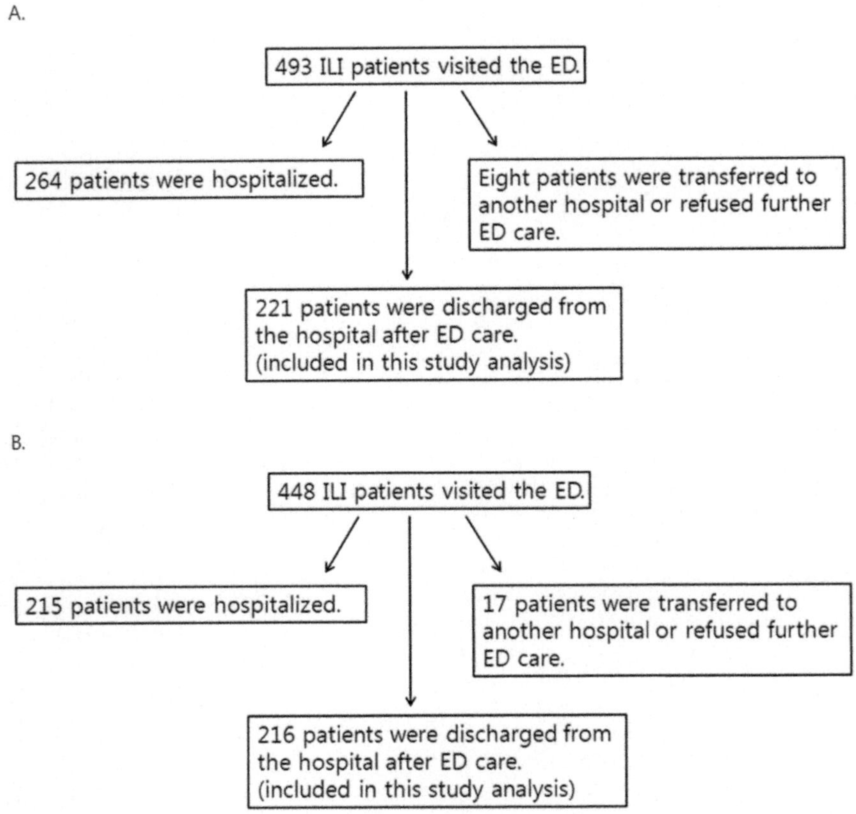

Figure 1. ILI patients examined in the present study. A. 2010–2011 influenza season (pre-IVRAT year). B. 2011–2012 influenza season (IVRAT year).

A diagnosis of influenza can be made on epidemiologic grounds. When the influenza virus is known to be circulating within a community, predictive clinical symptoms and signs are diagnostic in around 80% of cases [16–18]. The most reliable predictive symptoms are acute fever and cough [17]. Due to the high accuracy of a clinical diagnosis and the relatively low sensitivity of IVRAT, some believe that IVRAT is not necessary in clinical settings that utilize local influenza surveillance systems. However, the present study clearly shows that performing IVRAT in the ED led to a reduction in the number of antibiotics prescriptions issued to non-hospitalized ILI patients by ED physicians. This suggests that IVRAT can play an important role in distinguishing true influenza patients from those with other acute febrile illnesses. Making a clinical diagnosis of influenza during the annual influenza season is not difficult for specialists in infectious or respiratory disease, even in the absence of laboratory tests. However, it can be difficult in the ED setting, which is usually overcrowded and in which ILI patients without complications are usually triaged and are regarded as "less-urgent" than other patients [19]. The present study shows that performing IVRAT in the ED can improve influenza diagnosis and reduce the unnecessary prescription of antibiotics. Although we did not review patient data regarding the prescription of antiviral medicines, a previous study shows that a positive IVRAT test results in the early treatment of influenza patients with appropriate antiviral medicines [11].

Some studies show that prescribing antibiotics to patients with upper respiratory infections prevents secondary infectious compli-

Table 1. The average length of stay and antibiotic prescription rates for patients with influenza-like illness discharged after emergency department care during the 2010–2011 and 2011–2012 influenza seasons.

	Group A (n = 221)	Group B (n = 216)			*P* value
		IVRAT-positive (n = 73)	IVRAT- negative (n = 123)	IVRAT not performed (n = 20)	
Proportions of antibiotic-prescribed patients (%)	97/221 (43.9)	54/216 (25.0)			$P<0.01$
		7/73 (9.6)	42/123 (34.1)	5/20 (25.0)	$P<0.01$
Median ED LOS (min) (range)	213 (11–1464)	257 (16–1755)			$P<0.01$

Abbreviations: ED, emergency department; IVRAT, influenza virus rapid antigen test; LOS, length of stay; n, number.

Table 2. Antibiotics prescribed to patients discharged after ED care during the two influenza seasons.

Antibiotic	2010–2011 influenza season	2011–2012 influenza season
Cephalosporins	53	21
Amoxicillin/clavulanate	38	18
Quinolones	3	13
Macrolides	3	2
Total	97	54

cations and reduces the severity of symptoms [20,21]. However, considering the possible adverse effect of antibiotics and their role in the development of antibiotic resistance, the benefit derived from prescribing antibiotics for upper respiratory infections is probably very small. Indeed, a previous study shows that immediate prescription of antibiotics to treat uncomplicated upper respiratory infections is not associated with better clinical outcomes or improved patient satisfaction when compared with delayed or no treatment [22]. However, we found that antibiotics were prescribed to 43.9% of ILI patients in group A and to 25.0% of ILI patients in group B. This high percentage of antibiotics prescription may be related to patients' expectations and/or to concerns about secondary infectious complications. We believe that further study of antibiotics prescription in the ED is necessary to reduce usage in this setting.

LOS is an important variable to consider when evaluating the quality and efficacy of ED care. Most hospitals make a huge effort to reduce ED LOS. The results of the present study show that the ED LOS was actually longer when IVRAT was performed in the ED. The reasons of this are not clear. ED LOS is influenced by many factors, including a shortage of hospital beds, overcrowding, patient admission and discharge procedures, and the speed at which the ED physicians work [23,24]. Thus, it may be that factors other than IVRAT led to the increase in ED LOS observed for group B. Another consideration is that the mean age of the patients in group B was higher than that in group A; therefore, these patients may have had comorbidities that increased the work-up time in the ED. The overall ED LOS of all patients who had visited ED was not significantly different between the two seasons. Although the actual time taken to get informed consents and perform IVRAT to ILI patients was not so long, that may be one reason of longer ED LOS in 2011–2012 season. Further studies are needed to clarify this.

The present study is the first to examine the effects of IVRAT on antibiotic prescription and ED LOS in adult ILI patients. In contrast to a large meta-analysis of pediatric ILI patients, which showed that IVRAT had no effect on antibiotic prescription to ILI patients [14], we found that IVRAT reduced the number of antibiotic prescriptions issued to influenza patients who were discharged home from the ED. The design of the present study is different from that of other studies that compared data from IVRAT-positive and IVRAT-negative patients [13], or compared data from patients in whom IVRAT was performed with those from patients in whom IVRAT was not performed during the same influenza season [11]. The design of these studies may introduce bias by only performing IVRAT on patients who showed clear clinical symptoms and signs of influenza; therefore, the results might not represent the true effects of IVRAT on a physician's decision making. The results of the present study provide a more accurate picture of how IVRAT affects the decision to prescribe antibiotics.

In summary, performing IVRAT in the ED reduces the number of antibiotic prescriptions issued to ILI patients discharged home after ED care. Also, IVRAT is very useful for differentiating influenza from other acute febrile illnesses. These findings suggest that widespread use of IVRAT in the ED will reduce the unnecessary prescription of antibiotics to ILI patients during influenza seasons, which is important in a climate of increasing antibiotic resistance.

Acknowledgments

We thank the Transgovernmental Enterprise for Pandemic Influenza in Korea (TEPIK) for help with setting up IVRAT in the ED at Chungbuk National University Hospital. This study was supported by the Health and Medical Technology Research and Development Project of Ministry of Health and Welfare (grant number: A103001).

Author Contributions

Conceived and designed the experiments: HWJ JYH JSP WJK. Performed the experiments: HWJ JSP. Analyzed the data: HWJ WJK. Contributed reagents/materials/analysis tools: WJK. Wrote the paper: HWJ JYH.

References

1. Clezen WP, Keitel WA, Taber LH, Piedra PA, Clover RD, et al. (1991) Age distribution of patients with medically-attended illness caused by sequential variants of influenza A/H1N1: comparison to age-specific infection rates, 1978-1989. Am J Epidemiol 133: 296–304.
2. Shrestha SS, Swerdlow DL, Borse RH, Prabhu VS, Finelli L, et al. (2011) Estimating the burdem of 2009 pandemic influenza A (H1N1) in the United States (April 2009 – April 2010). CID (Suppl 1): S75–S82.
3. Glezen WP (1996) Emerginc infections: pandemic influenza. Epidemiol Rev 18: 64–76.
4. Dagan R, Hall CB (1984) Influenza A virus infection imitating bacterial sepsis in early infancy. Pediatr Infect Dis 3: 218–221.
5. Lenzi L, Wiens A, Pontarolo R (2013) The characteristics, clinical manifestations and outcomes of pandemic influenza A(H1N1) 2009 in the elderly. Rev Soc Bras Med Trop 46: 135–140.
6. Reina J, Munar M, Blanco I (1996) Evaluation of a direct immunofluorescence assay, dot-blot enzyme immunoassay, and shell vial culture in the diagnosis of lower respiratory tract infections caused by influenza A virus. Diagn Microbiol Infect Dis 25: 143–145.
7. He J, Bose ME, Beck ET, Fan J, Tiwari S, et al. (2009) Rapid multiplex reverse transcription-PCR typing of influenza A and B virus, and subtyping of influenza A virus into H1, 2, 3, 5, 7, 9, N1 (human), N1 (animal), N2, and N7, including typing of novel swine origin influenza A (H1N1) virus, during the 2009 outbreak in Milwaukee, Wisconsin. J Clin Microbiol 47: 2772–2778.
8. Rapid Diagnostic Testing for Influenza: Information for clinical laboratory directors. Available: http://www.cdc.gov/flu/professionals/diagnosis/rapidlab.htm. Accessed 2014 Sep 6.
9. Agoritsas K, Mack K, Bonus BK, Goodman D, Salamon D, et al. (2006) Evaluation of the Quidel QuickVue test for detection of influenza A and B viruses in the pediatric emergency medicine setting by use of three specimen collection methods. J Clin Microbiol 44: 2638–2641.
10. Cho CH, Woo MK, Kim JY, Cheong S, Lee CK, et al. (2013) Evaluation of five rapid diagnostic kits for influenza A/B virus. J Virol Methods 187: 51–56.

11. Noyola DE, Demmler GJ (2000) Effect of rapid diagnosis on management of influenza A infections. Pediatr Infect Dis J 19: 303–307.
12. Sharma V, Dowd MD, Slaughter AJ, Simon SD (2002) Effect of rapid diagnosis of influenza virus type A on the emergency department management of febrile infants and toddlers. Arch Pediatr Adolesc Med 156: 41–43.
13. Hojat K, Duppenthaler A, Aebi C (2013) Impact of the availability of an influenza virus rapid antigen test on diagnostic decision making in a pediatric emergency department. Pediatr Emerg Care 29: 696–698.
14. Doan Q, Enarson P, Kissoon N, Klassen TP, Johnson DW (2012) Rapid viral diagnosis for acute febrile respiratory illness in children in the emergency department. Cochrane Database Syst Rev 5: CD006452.
15. Song JY, Cheong HJ, Choi SH, Baek JH, Han SB, et al. (2013) Hospital-based influenza surveillance in Korea: hospital-based influenza morbidity and mortality study group. J Med Virol 85: 910–917.
16. Boivin G, Hardy I, Tellier G, Maziade J (2000) Predicting influenza infections during epidemics with use of a clinical case definition. CID 31: 1166–1169.
17. Monto AS, Gravenstein S, Elliott M, Colopy M, Schweinle J (2000) Clinical signs and symptoms predicting influenza infection. Arch Intern Med 160: 3243–3247.
18. Zambon M, Hays J, Webster A, Newman R, Keene O (2001) Diagnosis of influenza in the community. Arch Intern Med 161: 2116–2122.
19. Murray M, Bullard M, Grafstein E (2004) Revisions to the Canadian Emergency Department Triage and Acuity Scale Implementation Guidelines. CJEM 6: 421–427.
20. Petersen I, JohnSon AM, Islam A, Duckworth G, Livermore DM, et al. (2007) Protective effect of antibiotics against serious complications of common respiratory tract infections: retrospective cohort study with the UK general practice research database. BMJ 335: 982.
21. Smucny J, Fahey T, Becker L, Glazier R (2004) Antibiotics for acute bronchitis. Cochrane Database Syst Rev 18: CD000245.
22. Spurling GK, Del Mar CB, Dooley L, Foxlee R, Farley R (2013) Delayed antibiotics for respiratory infections. Cochrane Database Syst Rev 4: CD004417.
23. Kyriacou DN, Ricketts V, Dyne PL, McCollough MD, Talan DA (1999) A 5-year time study analysis of emergency department patient care efficiency. Ann Emerg Med 34: 326–335.
24. Richardson DB (2002) The access-block effect: relationship between delay to reaching an inpatient bed and inpatient length of stay. Med J Aust 177: 492–495.

18

The Effect of Probiotics Supplementation on *Helicobacter pylori* Eradication Rates and Side Effects during Eradication Therapy

Yini Dang[1], Jan D. Reinhardt[2,3,4], Xiaoying Zhou[1], Guoxin Zhang[1]*

1 Department of Gastroenterology, The First Affiliated Hospital of Nanjing Medical University, Nanjing, China, and First Clinical Medical College of Nanjing Medical University, Nanjing, China, **2** Institute for Disaster Management and Reconstruction, Sichuan University, Chengdu, China, and Hong Kong Polytechnical University, Hung Hom, Hong Kong, China, **3** Department of Health Sciences, University of Lucerne, Lucerne, Switzerland, **4** Swiss Paraplegic Research, Nottwil, Switzerland

Abstract

Background: Previous meta-analyses reported that probiotics improve the effectiveness of *Helicobacter pylori* (*H. pylori*) eradication during antibiotic therapy, while results regarding a possible reduction of side effects remained inconclusive. Moreover, the effectiveness of different strains of probiotics has not been studied so far. It is further conceivable that probiotics will produce additional effects only if antibiotics are relatively ineffective.

Methods: This meta-analysis includes eligible randomized controlled trials examining effects of probiotics supplementation on eradication rates (ER) and side effects, published up to May 2014. Sub-group analysis was performed to compare different probiotic strains and antibiotic therapies with different effectiveness in controls (ER <80% vs.>80%). Publication bias was assessed with funnel plots and Harbord's test. The quality of the trials was assessed with the Cochrane risk of bias tool.

Results: Thirty-three RCTs involving a total of 4459 patients met the inclusion criteria in case of eradication rates of which 20 assessed total side effects in addition. Overall, the pooled eradication rate in probiotics supplementation groups was significantly higher than in controls (ITT analysis: RR 1.122, 95% CI 1.086–1.159, PP analysis: RR 1.114, 95% CI 1.070–1.159). Sub group-analysis could, however, confirm this finding only for four individual strains (*Lactobacillus acidophilus, Lactobacillus casei* DN-114001, *Lactobacillus gasseri*, and *Bifidobacterium infantis* 2036) and for relatively ineffective antibiotic therapies. There was a significant difference between groups in the overall incidence of side effects (RR 0.735, 95% CI 0.598–0.902). This result was, however, only confirmed for non-blinded trials.

Conclusions: The pooled data suggest that supplementation with specific strains of probiotics compared with eradication therapy may be considered an option for increasing eradication rates, particularly when antibiotic therapies are relatively ineffective. The impact on side effects remains unclear and more high quality trials on specific probiotic strains and side effects are thus needed.

Editor: Pere-Joan Cardona, Fundació Institut d'Investigació en Ciències de la Salut Germans Trias i Pujol. Universitat Autònoma de Barcelona. CIBERES, Spain

Funding: This work was supported by grants from the National Natural Science Foundation of China (No. 81072032 and No. 81270476). The funders had no role in study design, data collection and analysis, decision to publish, or preparation of the manuscript.

Competing Interests: The authors have declared that no competing interests exist.

* Email: guoxinz@njmu.edu.cn

Introduction

'*Helicobacter pylori* (*H. pylori*) is a common bacterium infecting about half of the world's population. It is causally associated with a diverse spectrum of gastrointestinal disorders' [1]. Eradication of *H. pylori* is necessary for the management of *H. pylori*-related complications. The recommended first approach for *H. pylori* eradication is standard triple antibiotic therapy. Other choices include sequential therapy and quadruple therapy [2]. However, due to antibiotic resistance and patient non-compliance, several studies showed widespread failure of antibiotic therapy [3,4]. Driven by the growing necessity for alternative solutions to eradication regimens, some studies have started to focus on probiotics [5], i.e. 'live microorganisms which when administered in adequate amounts confer a health benefit on the host' [6]. The most commonly used probiotic bacteria belong to the genera *Lactobacillus* and *Bifidobacterium*, and these also include several yeasts such as *Saccharomyces boulardii* [7]. It has been hypothesized that probiotics could improve *H. pylori* eradication and reduce side effects during therapy.

Although seven meta-analyses [8–14] on this topic have been conducted over the past seven years, we perform an additional meta-analysis based on the following considerations: 1) The number of articles included in six of the seven previously

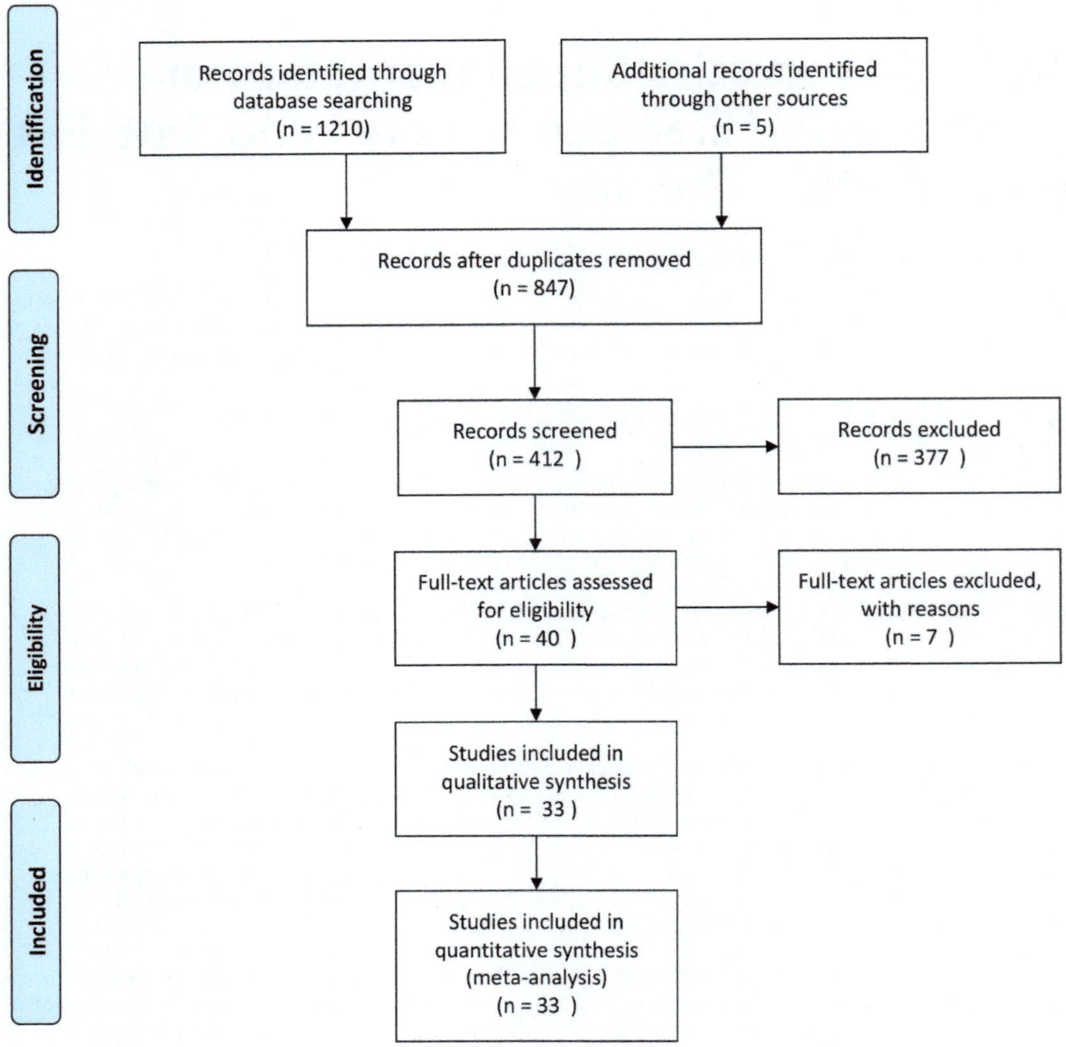

Figure 1. Flow of article selection.

conducted meta-analysis was ten or less [9–14]. In 2010–2014, many new studies were performed to evaluate the effectiveness of probiotics supplementation increasing the evidence base. 2) There are various genera of probiotics used in clinical practice and most meta-analyses [9,11,12,13] solely concentrated on one specific strain. 3) Opposite conclusions were drawn regarding side effects: Sachdeva [10], Zou [9], and Zheng [13] report no significant reduction of overall side effects in probiotics groups. However, Tong [8], Szajewska [11], Wang [12], and Li [14] arrive at opposite conclusions. Moreover, no sensitivity analysis regarding side effects has been conducted so far. 4) While all meta-analysis conducted so far reported increased eradication rates due to probiotics supplementation, no sensitivity analysis with regard to the effectiveness of the antibiotic therapy has been conducted, i.e. it is conceivable that probiotics supplementation will solely increase effectiveness in relatively ineffective antibiotic regimens.

This meta-analysis is thus designed to evaluate the current evidence regarding effects of probiotics supplementation compared with eradication triple therapy only on H. pylori eradication rates and side effects. Moreover, we aim to compare different strains of probiotics as well as differentially effective antibiotic therapies and to evaluate the quality of the trials conducted so far.

Methods

Search Strategy

We performed a literature search in Pubmed and Web of Science, covering papers published up to May 2014. A combination of the following keywords was used: (probiotics OR prebiotics OR *Bifidobacterium* OR *Lactobacillus* OR *Saccharomyces*) AND (*Helicobacter pylori* OR *H. pylori*). The reference lists of the selected papers and the seven previous meta-analyses were also screened for other eligible articles that may have been missed in the initial search. Next we scanned the titles and abstracts of the trials identified in the computerized search to exclude studies that were obviously irrelevant and scrutinized the full-texts of the remaining studies.

Inclusion and Exclusion Criteria

The following criteria were used for the selection of relevant articles: 1) studies should be randomized controlled trials (RCTs); 2) study populations should never have been treated for *H. pylori* infection before; 3) studies should include at least two branches of treatment consisting of (a) studies should have patients in a control group who received antibiotic therapy, (b) there should be patients

Table 1. Characteristica of studies included in the meta-analysis.

Authors (references)	Year	Area	Case Number (treatment/ control)	Patients	Diagnositic Methods	Eradication Therapy	Probiotics Supplementation	% Eradication of Treatment (no. of patients)	% Rradication of Control (no. of patients)	% Side Effects (no. of patients) (treatment/ control)
Canducci [15]	2000	Italy	120 (60/60)	symptomatic adults	UBT +Histology/ UBT-Histology (6 weeks later)	rabeprazole 20 mg bid, clarithromycin 250 mg tid, amoxicillin 500 mg tid, 7 days	Lacteol Fort (Bruschettini s.r.l., Genoa, Italy), containing 5×10^9 heat-killed *L. acidophilus* per capsule, tid, 10 days	ITT 86.7% (52/60) PP 89.7% (52/58)	ITT 70.0% (42/60) PP 71.2% (42/59)	10.0% (6/60) 10.0% (6/60)
Armuzzi a[16]	2001	Italy	60 (30/30)	asymptomatic adults	UBT +ELISA/UBT (6 weeks later)	rabeprazole 20 mg bid, clarithromycin 500 mg bid, tinidazole 500 mg bid, 7 days	a probiotic preparation containing 6×10^9 *Lactobacillus GG* (Giflorex, Errekappa Euroterapici S.p.A, Milan, Italy), bid, 14 days	ITT 83.3% (25/30)	ITT 80.0% (24/30)	40.0% (12/30) 33.3% (10/30)
Armuzzi b 17]	2001	Italy	120 (60/60)	symptomatic adults	UBT +ELISA/UBT (6 weeks later)	pantoprazole 40 mg bid, clarithromycin 500 mg bid, tinidazole 500 mg bid, 7 days	a probiotic preparation containing 6×10^9 *Lactobacillus GG* (Giflorex, Errekappa Euroterapici S.p.A, Milan, Italy), bid, 14 days	ITT 80.0% (48/60) PP 80.0% (48/60)	ITT 70.0% (42/60) PP 80.7% (46/57)	43.3% (26/60) 61.7% (37/60)
Cremonini LB[18]	2002	Italy	42 (21/21)	asymptomatic adults	UBT/UBT (5–7 weeks later)	rabeprazole 20 mg bid, clarithromycin 500 mg bid, tinidazole 500 mg bid, 7 days	a probiotic preparation containing 6×10^9 *Lactobacillus GG* (Giflorex, Errekappa Euroterapici, Milan, Italy), bid, 14 days	ITT 76.2% (16/21) PP 76.2% (16/21)	ITT 76.2% (16/21) PP 80.0% (16/20)	14.3% (3/21) 57.1% (12/20)
Cremonini SM[18]	2002	Italy	42 (22/21)	asymptomatic adults	UBT/UBT (5–7 weeks later)	rabeprazole 20 mg bid, clarithromycin 500 mg bid, tinidazole 500 mg bid, 7 days	a probiotic preparation containing 5×10^9 *Saccharomyces boulardii* (Codex, SmithKline Beecham, Milan, Italy), bid, 14 days	ITT 77.3% (17/22) PP 85.0% (17/20)	ITT 76.2% (16/21) PP 80.0% (16/20)	13.6% (3/22) (57.1% (12/20)
Cremonini MS[18]	2002	Italy	42 (21/21)	asymptomatic adults	UBT/UBT (5–7 weeks later)	rabeprazole 20 mg bid, clarithromycin 500 mg bid, tinidazole 500 mg bid, 7 days	a combination of 5×10^9 live *Lactobacillus acidophilus* and *Biphidobacterium lactis* (Ferzym, Specchiasol, Milan, Italy), bid, 14 days	ITT 85.7% (18/21) PP 85.7% (18/21)	ITT 76.2% (16/21) PP 80.0% (16/20)	23.8% (5/21) 57.1% (12/20)
Sheu [19]	2002	China (Taiwan)	160 (80/80)	symptomatic adults	Histology+RUT/UBT (8 weeks later)	lansoprazole 30 mg bid, clarithromycin 500 mg bid, amoxicillin 1 g tid, 7 days	200 mL AB-Yogurt containing 5×10^9 *Lactobacillus* and *Bifidobacterium* (President Corp., Tainan, Taiwan), bid, 5 weeks	ITT 91.3% (73/80) PP 94.8% (73/77)	ITT 78.8% (63/80) PP 87.5% (63/72)	/
Nista [20]	2004	Italy	106 (54/52)	asymptomatic adults	UBT/UBT (6 weeks later)	rabeprazole 20 mg bid, clarithromycin 500 mg bid, amoxicillin 1 g bid, 7 days	a probiotic preparation containing 2×10^9 *B. clausii* (Enterogermina, Sanofi–Synthelabo OTC, Milan, Italy), tid, 14 days	ITT 72.2% (39/54) PP 78.0% (39/50)	ITT 71.2% (37/52) PP 74.0% (37/50)	74.0% (40/54) 82.7% (43/50)
Myllyuom [21]	2005	Finland	47 (23/24)	asymptomatic adults	UBT +ELISA/UBT+ ELISA (6 weeks later)	lansoprazole 30 mg bid, clarithromycin 500 mg bid, amoxicillin 1 g tid, 7 days	65 mL milk-based fruit drink containing 1×10^9 cfu/mL of *Lactobacillus GG, L. rhamnosus LC705, Bifidobacterium breve Bb99* and *Propionibacterium freudenreichii ssp. shermanii JS* (Valio Ltd, Helsinki, Finland), bid during the eradication treatment and qd for the following 3 weeks	ITT 91.3% (21/23) PP 91.3% (21/23)	ITT 79.2% (19/24) PP 79.2% (19/24)	86.9% (20/23) 91.7% (22/24)
Sykora[22]	2005	UK	86 (39/47)	symptomatic children	Histology +RUT+ HpSA/UBT + HpSA (4 weeks later)	omeprazole 10 mg (15–30 kg) or 20 mg (30 kg) bid, clarithromycin 7.5 mg/kg bid, amoxicillin 25 mg/kg bid, 7 days	100 mL of fermented milk containing 10^10 cfu/mL *L. casei DN-114001*, qd, 14 days	ITT 84.6% (33/39) PP 91.7% (33/36)	ITT 57.4% (27/47) PP 79.2% (27/44)	23.1% (9/39) 21.3% (10/47)

Table 1. Cont.

Authors (references)	Year	Area	Case Number (treatment/ control)	Patients	Diagnositic Methods	Eradication Therapy	Probiotics Supplementation	% Eradication of Treatment (no. of patients)	% Rradication of Control (no. of patients)	% Side Effects (no. of patients) (treatment/ control)
Goldman[23]	2006	Argentina	65 (33/32)	symptomatic children	UBT/UBT (4 weeks later)	omeprazole 1 mg/kg qd, clarithromycin 15 mg/kg qd, amoxicillin 50 mg/kg qd, 7 days	250 mL commercial yogurt containing 10^7 cfu/mL *Bifidobacterium animalis* and *Lactobacillus casei*	ITT 45.5% (15/33) PP 45.5% (15/33)	ITT 37.5% (12/32) PP 37.5% (12/32)	/
Lionetti[24]	2006	Italy	40 (20/20)	symptomatic children	Histology +RUT+ UBT/UBT (20 days later)	omeprazole 1 mg/kg/d plus amoxycillin 50 mg/kg/d for 5 days; omeprazole 1 mg/kg/d, clarithromycin 15 mg/kg/d and tinidazole 20 mg/kg/d for the next 5 days	each pill containing 10^8 cfu of *L. reuteri ATCC 55730* (SD2112) (BioGaia, Sweden), qd, 20 days	ITT 85.0% (17/20) PP 89.5% (17/19)	ITT 80.0% (16/20) PP 84.2% (16/19)	/
Zieminak[25]	2006	Poland	245 (53/192)	symptomatic adults	UBT/UBT (6 weeks later)	pantoprazole 40 mg bid, clarithromycin 500 mg bid, amoxicillin 1 g bid, 10 days	Lacidofil containing *Lactobacillus acidophilus* and *Lactobacillus rhamnosus*, bid, 10 days	ITT 96.2% (51/53) PP 96.2% (51/53)	ITT 85.9% (165/192) PP 85.9% (165/192)	/
Cindoruk[26]	2007	Turkey	124 (62/62)	symptomatic adults	Histology/UBT (6 weeks later)	lansoprazole 30 mg bid, clarithromycin 500 mg bid, amoxicillin 1 g bid, 14 days	250 mg sachets of *Saccharomyces boulardii* (Reflor, Sanofi-Synthelabo Ilac A.S., Istanbul, Turkey), bid, 14 days	ITT 71.0% (44/62) PP 71.0% (44/62)	ITT 59.7% (37/62) PP 59.7% (37/62)	22.6% (14/62) 59.7% (37/62)
De Bortoli N [27]	2007	Italy	206 (105/101)	symptomatic adults	HpSA+ UBT/UBT (8 weeks later)	esomeprazole 20 mg bid, clarithromycin 500 mg bid, amoxicillin 1 g bid, 7 days	bLf 20 mg bid; Pb containing 5×10^9 *Lactobacillus plantarum*, 2×10^9 L. reuterii, 2×10^9 *L. caseisubsp.rhamnosus*, 2×10^9 *Bifidobacterium infantis* and *B.longum*, 1×10^9 L. *salivarius*, 1×10^9 *L. acidophilus*, 5×10^9 *Streptococcus termophilus*, and 1×10^9 L. *sporogenes*, bid, 7 days	ITT 88.6% (93/105) PP 92.1% (93/101)	ITT 72.3% (73/101) PP 76.0% (73/96)	9.5% (10/105) 40.6% (41/101)
Francavill [28]	2008	Italy	40 (20/20)	symptomatic adults	Histology+HpSA+ UBT/HpSA+ UBT (4 weeks later)	rabeprazole 20 mg bid plus amoxicillin 1 g, bid for 5 days; rabeprazole 20 mg bid, clarithromycin 500 mg bid and tinidazole 500 mg bid for the next 5 days	each tablet containing 10^8 *L. reuteri* ATCC 55730 (Reuterin, Nóos), qd, 4 weeks	ITT 85.0% (17/20) PP 85.0% (17/20)	ITT 80.0% (16/20) PP 80.0% (16/20)	/
Kim[29]	2008	Korea	347 (168/179)	symptomatic adults	Histology+UBT/UBT (4 weeks later)	standard PPI bid, clarithromycin 500 mg bid, amoxicillin 1 g bid, 7 days	150 mL Will yogurt containing *L. acidophilus* HY 2177 (> 10^5 cfu/mL), *L. casei* HY 2743 (> 10^5 cfu/mL), *B. longum* HY 8001 (> 10^6 cfu/mL), and *S. thermophilus* B-1 (> 10^8 cfu/mL) (Korea Yakult Company Limited; Chunan-Si, Chungnam, S. Korea), qd, 3 weeks	ITT 79.2% (133/168) PP 87.5% (133/152)	ITT 72.1% (129/179) PP 78.6% (129/164)	41.1% (69/168) 26.3% (47/179)
Scaccianoce LB[30]	2008	Italy	33 (17/16)	symptomatic adults	Histology/UBT (4–6 weeks later)	lansoprazole 30 mg bid, clarithromycin 500 mg bid, amoxicillin 1 g bid, 7 days	each tablet containing 10^8 *L. reuteri* ATCC 55730 (Reuterin, Nóos), bid, 7 days	ITT 52.9% (9/17) PP 52.9% (9/17)	ITT 62.5% (10/16) PP 66.7% (10/15)	5.9% (1/17) 25.0% (4/16)

Table 1. Cont.

Authors (references)	Year	Area	Case Number (treatment/ control)	Patients	Diagnositic Methods	Eradication Therapy	Probiotics Supplementation	% Eradication of Treatment (no. of patients)	% Rradication of Control (no. of patients)	% Side Effects (no. of patients) (treatment/ control)
Scaccianoce MS[30]	2008	Italy	31 (15/16)	symptomatic adults	Histology/UBT (4–6 weeks later)	lansoprazole 30 mg bid, clarithromycin 500 mg bid, amoxicillin 1 g bid, 7 days	a probiotic mixture with *Lactobacillus plantarum* (5×10^9), *L. reuteri* (2×10^9), *Lactobacillus casei subsp. Rhamnosus* (2×10^9), *Bifidobacterium infantis* (2×10^9), *Bifidobacterium longum* (2×10^9), *Lactobacillus salivarius* (1×10^9), *Lactobacillus acidophilus* (1×10^9), *Streptococcus termophilus* (5×10^9), and *Lactobacillus sporogenes* (1×10^9); 5 g/dose, bid, 7 days	ITT 53.3% (8/15) PP 53.3% (8/15)	ITT 62.5% (10/16) PP 66.7% (10/15)	20.0% (3/15) 25.0% (4/16)
Hurduc[31]	2009	Romania	90 (48/42)	symptomatic children	Histology+RUT/ Histology+RUT (4–6 weeks later)	omeprozole or esomeprazole 0.5 mg/kg bid, clarothromycin 7.5 mg/kg bid, amoxicillin 25 mg/kg bid, 7–10 days	*Saccharomyces boulardii* (Enterol; Biocodex, Gentilly Cedex, France), 250 mg/day, bid, 4 weeks	ITT 93.8% (45/48) PP 100% (45/45)	ITT 80.9% (34/42) PP 75.6% (34/42)	8.3% (4/48) 30.9% (13/42)
Szajewsk[32]	2009	Poland	66 (34/32)	asymptomatic children	Histology+ RUT+UBT/ UBT (4–6 weeks later)	omeprozole 0.5 mg/kg bid, clarithromycin 10 mg/kg bid, amoxicillin 25 mg/kg bid, 7 days	10^9 cfu/mL *Lactobacillus GG*, bid, 7 days	ITT 67.6% (23/34) PP 67.6% (23/34)	ITT 68.8% (22/32) PP 68.8% (22/32)	52.9% (18/34) 40.6% (13/32)
Song[33]	2010	Korea	661 (330/331)	symptomatic adults	UBT/UBT (4 weeks later)	omeprazole 20 mg bid, clarithromycin 500 mg bid, amoxicillin 1 g bid, 7 days	3×10^9 cfu/g *Saccharomyces boulardii* (Bioflor250, Kuhnil Pharmacy, Seoul, Korea), tid, 4 weeks	ITT 80.0% (264/330) PP 85.4% (264/309)	ITT 71.6% (237/331) PP 80.0% (237/296)	/
Yasar[34]	2010	Turkey	76 (38/38)	symptomatic children	Histology/UBT (4 weeks later)	pantoprazole 40 mg bid, clarithromycin 500 mg bid, amoxicillin 1 g bid, 14 days	125 ml yogurt containing 10^10 cfu/g *Bifidobacterium* DN-173 010, qd, 14 d	ITT 65.8% (25/38) PP 65.8% (25/38)	ITT 52.6% (20/38) PP 52.6% (20/38)	/
Bekar[35]	2011	Turkey	82 (46/36)	symptomatic adults	UBT/UBT (45 days later)	lansoprazole 30 mg bid, clarithromycin 500 mg bid, amoxicillin 1 g bid, 14 days	250 mL kefir, bid, 141d	ITT 78.3% (36/46) PP 78.3% (36/46)	ITT 50.0% (18/36) PP 50.0% (18/36)	/
Madeiros[36]	2011	France	62 (31/31)	asymptomatic adults	Histology+ culture/ UBT (≥6 weeks later)	esomeprazole 20 mg bid, clarithromycin 500 mg bid, amoxicillin 1 g bid, 8 days	Lacteol (BioSaúde laboratories, Portugal), capsule containing 5×10^9 *L. acidophilus*, bid, 8 days	ITT 83.9% (26/31) PP 83.9% (26/31)	ITT 77.4% (24/31) PP 77.4% (24/31)	/
Ozdil[37]	2011	Turkey	193 (98/95)	symptomatic adults	Histology/HpSA (5 weeks later)	lansoprazole 30 mg bid, clarithromycin 500 mg bid, amoxicillin 1 g bid, 14 days	250 mg capsules containing *saccharomyces boulardii*, qd, 14 days	ITT 72.4% (71/98) PP 77.2% (71/92)	ITT 86.3% (82/95) PP 89.1% (82/92)	/
Deguchi[38]	2012	Japan	229 (115/114)	symptomatic adults	Histology-RUT-culture/ UBT+HpSA (8 weeks later)	rabeprazole 10 mg bid, clarithromycin 200 mg bid, amoxicillin 750 mg bid, 7 days	yogurt containing 112 g *L. gasseri* OLL2716 (>10 cfu/g), bid, 4 weeks	ITT 82.6% (95/115) PP 89.6% (95/106)	ITT 69.3% (79/114) PP 71.2% (79/111)	5.2% (6/115) 3.5% (4/114)

Table 1. Cont.

Authors (references)	Year	Area	Case Number (treatment/ control)	Patients	Diagnositic Methods	Eradication Therapy	Probiotics Supplementation	% Eradication of Treatment (no. of patients)	% Rradication of Control (no. of patients)	% Side Effects (no. of patients) (treatment/ control)
Du YQ[39]	2012	China	157 (78/79)	symptomatic adults	Histology+RUT+UBT/ UBT (4 weeks later)	omeprazole 20 mg bid, clarithromycin 500 mg bid, amoxicillin 1 g bid, 7 days	probiotics, containing 3×10^7 *Lactobacillus acidophilus*, qd, 14 d	ITT 79.5% (62/78) PP 81.6% (62/76)	ITT 60.7% (48/79) PP 61.5% (48/78)	1.3% (1/78) 0.0% (0/79)
Manfredi[40]	2012	Italy	149 (73/76)	symptomatic adults	Histology+ HpSA/HpSA (8–10 weeks later)	esomeprazole 20 mg bid plus amoxicillin 1 g, bid for 5 days; esomeprazole 20 mg bid, clarithromycin 500 mg bid and tinidazole 500 mg bid for the next 5 days	Lactogermine plus (Humana Italia s.p.a., Milano, Italy), containing 10^9 *Lactobacillus acidophilus*, 5×10^8 *Bifidobacterium bifidum*, 10^9 *Streptococcus thermophilus*, 10^9 *Lactobacillus bulgaricus*	ITT 91.8% (67/73) PP 94.4% (67/71)	ITT 85.5% (65/76) PP 92.8% (65/70)	39.7% (29/73) 65.8% (50/76)
Mirzaee[41]	2012	Iran	68 (34/34)	symptomatic adults	UBT/UBT (4 weeks later)	pantoprazole 40 mg qd, clarithromycin 500 mg bid, amoxicillin 1 g bid, 7 days	150 mg probiotic yogurt, bid, 7 days	ITT 55.9% (19/34) PP 61.3% (19/31)	ITT 58.8% (20/34) PP 64.5% (20/31)	64.7% (22/34) 67.6% (23/34)
Tolone[42]	2012	Italy	68 (34/34)	symptomatic children	Histology+ UBT/UBT (4 weeks later)	omeprazole 1 mg/kg qd, amoxicillin 50 mg/kg bid, clarithromycin 15 mg/kg bid, 7 days	the PB supplement containing 5×10^9 *Lactobacillus plantarum*, 2×10^9 *L. reuterii*, 2×10^9 *L.casei subsp. rhamnosus*, 2×10^9 *Bifidobacterium infantis* and *B. longum*, 1×10^9 *L. salivarius*, 1×10^9 *L. acidophilus*, 5×10^9 *Streptococcus termophilus*, and 1×10^9 *L. sporogenes*, qd, 7 days	ITT 88.2% (30/34) PP 88.2% (30/34)	ITT 76.5% (26/34) PP 76.5% (26/34)	/
Ahmad[43]	2013	Iran	66 (33/33)	symptomatic children	Histology+RUT/HpSA (4–8 weeks later)	omeprazole 0.5 mg/kg bid, amoxicillin 25 mg/kg bid, furazolidone 3 mg/kg bid, 7 days	probiotic combination (restore, 10^9 cfu/ each sachet, Protexin Co, UK) containing *Lactobacillus acidophilus*, *Lactobacillus rhamnosus*, *Lactobacillus bulgaricus*, *Lactobacillus casei*, *Streptococcus thermophilus*, *Bifidobacterium infantis* and *Bifidobacterium breve*, qd, 7 days	ITT 90.9% (30/33) PP 90.9% (30/33)	ITT 69.7% (23/33) PP 69.7% (23/33)	/
Navarro[44]	2013	Brazil	107 (55/52)	symptomatic adults	Histology+RUT+UBT/ UBT (8 weeks later)	lansoprazole 30 mg bid, tetracycline 500 mg bid, furazolidone 200 mg bid, 7 days	the probiotic compound containing 1.25×10^9 CFUs *Lactobacillus acidophilus*, 1.25×10^9 CFUs *Lactobacillus rhamnosus*, 1.25×10^9 CFUs *Bifidobacterium bifidum* and 1.25×10^9 CFUs *Streptococcus faecium*, bid, 30 days	ITT 89.1% (49/55) PP 96.1% (49/51)	ITT 84.6% (44/52) PP 89.8% (44/49)	58.2% (32/55) 71.2% (37/52)
Shavakhi[45]	2013	Iran	180 (90/90)	symptomatic adults	Histology+ RUT/UBT (4 weeks later)	omeprazole 20 mg bid, bismuth subcitrate 240 mg bid, amoxicillin 1000 mg bid, clarithromycin 500 mg bid, 14 days	Balance capsue containing *Lactobacillus* strains (*L. casei, L. rhamnosus, L. acidophilus*, and *L. bulgaricus*), *Bifidobacterium* strains (*B. breve* and *B. longum*), and *Streptococcus thermophiles*, total viable count is 10^8 CFU/per capsule, bid, 14 days	ITT 76.7% (69/90) PP 82.1% (69/84)	ITT 81.1% (73/90) PP 84.9% (73/86)	18.9% (17/90) 16.7% (15/90)
Dajani[46]	2013	United Arab Emirates	206 (100/106)	symptomatic adults	Histology+RUT+UBT+ HpSA/UBT (6–8 weeks later)	omeprazole 20 mg bid, clarithromycin 500 mg bid, amoxicillin 1 g bid, 7 days	3×10^9 *B. infantis* 2036, bid, 7 days	ITT 83% (83/100) PP 83% (83/100)	ITT 68.9% (73/106) PP 68.9% (73/106)	/

Table 1. Cont.

Authors (references)	Year	Area	Case Number (treatment/ control)	Patients	Diagnositic Methods	Eradication Therapy	Probiotics Supplementation	% Eradication of Treatment (no. of patients)	% Rradication of Control (no. of patients)	% Side Effects (no. of patients) (treatment/ control)
Wang YH[47]	2013	China	100 (49/51)	symptomatic children	UBT/UBT (6 weeks later)	PPI 0.6–0.8 mg/kg bid, clarithromycin 10–15 mg/kg bid, amoxicillin 30–50 mg/kg bid, and metronidazole 15–20 mg/kg bid for penicillin allergic children, 14 days	each packs of probiotics containing 2 g *L. acidophilus-5* (4.7×10^9 cfu/100 g) and 2 g *B. bifidum-12* (4.3×10^8 cfu/100 g), <5 years old, qd; >5 years old, bid, 14 days	ITT 73.5% (36/49) PP 83.7% (26/43)	ITT 56.9% (29/51) PP 64.4% (29/45)	10.2% (5/49) 23.5% (12/51)

in an intervention group who received probiotics plus the identical antibiotic therapy; 4) there should be data on successful eradication rates available; 5) for the analysis of side effects data on occurrence of side effects during treatment were required as well. Exclusion criteria were as follows: 1) the design and the definition of the trials were obviously different from those specified above; 2) essential information was not provided; 3) papers were letters, commentaries, editorials, reviews and duplicate publications; 4) articles were written in a language other than English.

Data Extraction

Two authors extracted data independently from all eligible papers. For conflicting evaluations, another author was consulted to solve the dispute and a final decision was made by the majority of the votes. The data extracted included authors, year of publication, base characteristics of the patients, details of the *H. pylori* eradication therapy, details related to interventions, primary and secondary outcomes, and confirmation methods of *H. pylori* infection. Extracted information was entered into a database.

Assessment of Study Quality

We used the Cochrane Tool of Bias [15] to assess study quality. Two authors independently evaluated all studies. Results were then compared and discussed to form consensus. If consensus could not be reached, another author was consulted and a decision made by the majority of the votes.

Statistical Analysis

Statistical heterogeneity was analyzed with Chi-squared distributed Chochran's Q and the I-squared statistics ($I^2 = 100\% \times (Q-df)/Q$). I-square indicates the percentage of variation across studies due to hetereogeneity as opposed to chance. We assumed sufficient homogeneity to apply a fixed-effects model (Mantel-Haenszel) for the meta-analysis if I-square was under 40% and/or Q was not significant at $p<0.05$. Otherwise, we opted for a random-effects model (DerSimonian and Laird). The jacknife was used to assess the influence of individual studies, i.e. estimation of the overall effect was repeated while omitting one study at each time. Funnel plots and Harbord's modified test for small study effects were used to assess publication bias [16]. The influence of the probiotic strain applied (multi-strain interventions were excluded from this analysis), blinded vs. non-blinded trials, and in the case of eradication rates, the effectiveness of antibiotic therapy in the control group (eradication rate >80% vs. <80%) was assessed with sub-group analysis. Concerning eradication rates, meta-regression was used to determine the influence of studies in pediatric vs. adult as well as symptomatic and asymptomatic populations. All statistical analysis was performed with STATA 12.0 (Stata Corporation, Texas, USA).

Results

Study Characteristics

The bibliographic search yielded 1114 articles, 40 of which were reviewed in full text (Figure 1) [17–56]. Of these studies, 33 RCTs [17–49] met the inclusion criteria for the analysis of eradication rates and 20 were eligible for analyzing side effects. These trials randomized a total of 4459 patients, 4261 of which were followed up. Nine studies enrolled only children [24–26,33,34,36,44,45,49] and 24 were undertaken exclusively with adults [17–23,27–32,35,37–43,46–48]. From all the included trials, 29 used PPI-triple therapy [17–25,27–29,31–41,43–46,48,49], three used sequential therapy [26,30,42] and one used bismuth-quadruple therapy [47]. These studies were undertaken

A
Random sequence generation (selection bias)
Allocation concealment (selection bias)
Blinding of participants and personnel (performance bias)
Blinding of outcome assessment (detection bias)
Incomplete outcome data (attrition bias)
Selective reporting (reporting bias)
Other bias
0%
25%
50%
75%
100%
Low risk of bias
Unclear risk of bias
High risk of bias
B
Random sequence generation (selection bias)
Allocation concealment (selection bias)
Blinding of participants and personnel (performance bias)
Blinding of outcome assessment (detection bias)
Incomplete outcome data (attrition bias)
Selective reporting (reporting bias)
Other bias
2000 Canducci
2001Armuzzi a
2001Armuzzi b
2002 Cremonini
2002 Sheu
2004 Nista
2005 Myllyuom
2005 Sykora
2006 Goldman
2006 Lionetti
2006 Zieminak
2007 Cindoruk
2007 de Bortoli N
2008 Francavill
2008 Kim
2008 Scaccianoce
2009 Hurduc
2009 Szajewsk
2010 Song
2010Yasar
2011 Bekar
2011 Madeiros
2011 Ozdil
2012 Deguchi
2012 Du YQ
2012 Manfredi
2012 Mirzaee
2012 Tolone
2013 Dajani
2013 Khodada
2013 Navarro
2013 Shavakhi
2013 Wang

Figure 2. Risk of bias graph (A): review authors' judgements about each risk of bias item presented as percentages across all included studies. (B) Risk of bias summary: review authors' judgements about each risk of bias item for each included study. (+) = low risk of bias, (?) = unclear, (-) = high risk of bias.

in Argentina [25], Brazil [46], mainland China [41,49], China (Taiwan) [21], Finland [23], France [38], Iran [43,45,47], Italy [17–20,22,26,29,30,32,42,44], Japan [40], Korea [31,35], Poland [27,34], Romania [33], Turkey [28,36,37,39], United Kingdom [24] and the United Arabian Emirates [48].(Table 1)

Trial Quality Assessment

Figure 2A shows authors' judgements for each Cochrane risk of bias item and Figure 2B presents the Cochrane risk of bias score for each citation included.

Eradication Rates

Eradication risk ratios (RRs) were available for 4459 patients (2189 in the probiotics supplementation group and 2270 in the control group). Heterogeneity was found to be low for the overall incidence of eradication rates in ITT analysis ($\chi^2 = 42.97$, $p = 0.167$, $I^2 = 18.5\%$) but higher for PP analysis ($\chi^2 = 58.55$, $p = 0.006$, $I^2 = 41.9\%$). Therefore, a fixed effects model based on Mantel-Haenszel's estimation method was used in the case of ITT and a random effects model in the case of PP analysis. Nonetheless, the pooled RRs from intention-to-treat (ITT) and from pre-protocol (PP) analysis for the probiotics supplementation group over controls were very similar with 1.122 (95% CI 1.086–1.159) and 1.114 (95% CI 1.070–1.159) respectively (Figure 3A&B).

Omitting individual studies from the meta-analysis did not change occurrence of RRs being significantly higher than 1. Funnel-plot and Harbord's modified test showed no evidence for publication bias ($p = 0.64$). Neither the age of the population (pediatric vs. adult, $p = 0.127$) nor symptom status ($p = 0.314$) played a role according to the meta-regression.

Studies used different strains of probiotics with 13 applying *Lactobacillus* [17–20,24,26,27,30,32,34,38,40,41], two using *Bifidobacterium* [36,48], one using *Bacillus clausii* [22], five using *Saccharomyces* [20,28,33,35,39] and 15 using multistrain [20,21,23,26,29,31,32,37,42–47,49]. Among those studies applying individual probiotic supplementation, four specific species were effective: *Lactobacillus acidophilus* (pooled RR = 1.235, 95% CI 1.090–1.400), *Lactobacillus casei* DN-114001 (pooled RR = 1.473, 95% CI 1.13–1.949), *Lactobacillus gasseri* (pooled RR = 1.192, 95% CI 1.028–1.382), *Bifidobacterium infantis* 2036 (pooled RR = 1.205, 95% CI 1.031–1.408). While double-blinded (pooled RR = 1.118, 1.045–1.196) and non-blinded trials (pooled RR = 1.120, 95% CI 1.080–1.162) basically arrived at the same results, sub-group analysis by effectiveness of the antibiotic therapy in the control group revealed that supplementation with probiotics solely increased eradication rates in relatively ineffective therapies.

Another sub-group analysis was done according to the effectiveness of eradication regimens. The less effective antibiotic therapies were, the more useful probiotic supplementation was: when eradication rate <60%, pooled RR = 1.28, 1.12–1.45; when eradication rate within 60%–69%, pooled RR = 1.18, 1.10–1.27; when eradication rate within 70%–79%, pooled RR = 1.11, 1.06–1.17; while if eradication rate over 80%, the supplementation was useless (pooled RR = 1.01, 0.96–1.77). (Figure 4)

Side Effects

There were 20 trials (2487 patients, 1269 in the probiotics supplementation group and 1218 in the control group) which provided data on the overall incidence of side effects. Significant heterogeneity was found for the overall occurrence of side effects ($I^2 = 72.2\%$, $P<0.001$). Therefore, the random effects model was used. The pooled RR in the probiotics supplementation over control was 0.735 (95% CI 0.598–0.902) (Figure 5). While Harbord's modified test for publication bias was insignificant ($p = 0.17$), the Funnel Plot did reveal some asymmetry (Figure 6): Small studies showing a considerable reduction in side-effects occur more often than those only showing a small reduction. When stratified by probiotics, only the pooled RR for *Saccharomyces boulardii* indicates a significant reduction in side effects (0.335, 95% CI 0.220–0.510; LB: 0.892, 95% CI 0.632–1.259; MS: 0.760, 95% CI 0.568–1.017; BB: 0.895: 0.737–1.087).

Only eight studies reporting the overall incidence of side effects were blinded and solely results from non-blinded trials provided evidence for the reduction of side-effects (pooled RR = 0.589, 95% CI: 0.412–0.842), while double blinded studies did not (pooled RR = 0.889, 95% CI: 0.728–1.085) (Figure 7).

Discussion

This is the largest meta-analysis on the effects of probiotics on *H. pylori* eradication and side effects conducted to date. The quality of the trials was medium to low and no study had a low risk of bias across all Cochrane criteria. Overall, this analysis of 33 RCTs suggests that supplementation of antibiotic therapy with probiotics increases eradication rates compared to a placebo or no intervention. However, regarding individual probiotic strains, this could only be confirmed for several strains of *Lactobacilus* and one strain of *Bifidobacterium*. Moreover, probiotics only demonstrated additional effects if the eradication rates in the control groups were relatively low, i.e. cases where the antibiotic therapy was relatively ineffective. Although, we found an overall decrease of total side effects, this held true only for *Saccharomyces boulardii* and non-blinded RCTs.

In comparison with previously published meta-analysis [8,9,10,11,12,13,14], our updated study represents the most comprehensive analysis. Various potential influence factors were taken into consideration such as age, symptom status, eradication therapy regime, eradication rate, bacterial strain and blinding method. Moreover, from the previous studies, six [8,10,11,12,13,14] excluded trials using Quadruple therapy as the co-intervention and none performed subgroup analysis with regard to the efficacy of antibiotic therapy. Moreover, sensitivity analysis for side effects had not been conducted so far. Our results confirm previous meta-analyses only to some extent. At a first glance, probiotics seem effective in increasing eradication rates and decreasing side effects. At a closer look, however, evidence only supports these claims for specific probiotic strains, ineffective antibiotic therapies and low-quality trials (i.e. non-blinded studies) as far as side effects are concerned.

Before Marshall found *H. pylori* in 1984 [57], the stomach was considered to be a sterile organ due to its low pH level. In the aftermath, '*H. pylori* has been intensively studied and recent sequencing analysis of other gastric microbiota shows that *H. pylori* is not alone' [58]. Although there are only 10^2-4 cfu/g in the gastric mucosa [59], these commensal bacteria can play an important role in maintaining human health. Probiotic supple-

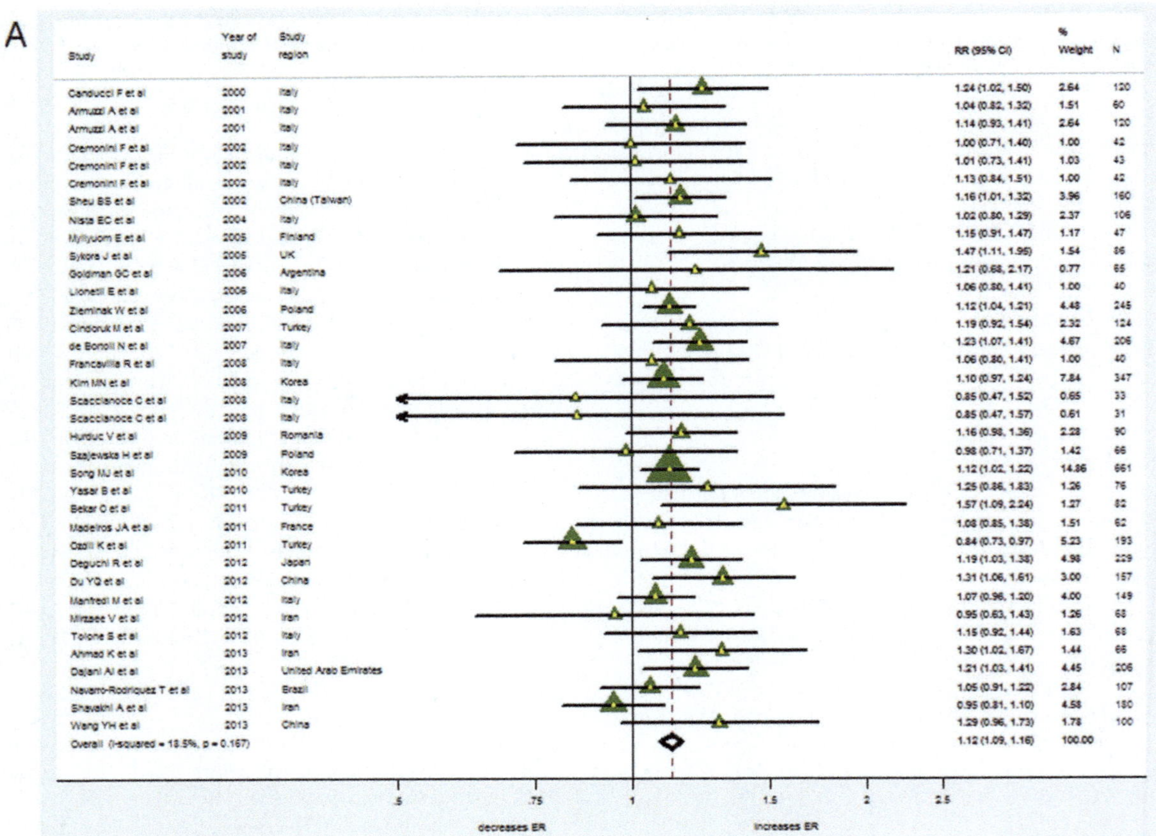
A
Study | Year of study | Study region | RR (95% CI) | % Weight | N
Canducci F et al 2000 Italy 1.24 (1.02, 1.50) 2.64 120
Armuzzi A et al 2001 Italy 1.04 (0.82, 1.32) 1.51 60
Armuzzi A et al 2001 Italy 1.14 (0.93, 1.41) 2.64 120
Cremonini F et al 2002 Italy 1.00 (0.71, 1.40) 1.00 42
Cremonini F et al 2002 Italy 1.01 (0.73, 1.41) 1.03 43
Cremonini F et al 2002 Italy 1.13 (0.84, 1.51) 1.00 42
Sheu BS et al 2002 China (Taiwan) 1.16 (1.01, 1.32) 3.96 160
Nista EC et al 2004 Italy 1.02 (0.80, 1.29) 2.37 106
Myllyluom E et al 2005 Finland 1.15 (0.91, 1.47) 1.17 47
Sykora J et al 2005 UK 1.47 (1.11, 1.95) 1.54 86
Goldman GC et al 2006 Argentina 1.21 (0.68, 2.17) 0.77 65
Lionetti E et al 2006 Italy 1.06 (0.80, 1.41) 1.00 40
Zieminak W et al 2006 Poland 1.12 (1.04, 1.21) 4.48 245
Cindoruk M et al 2007 Turkey 1.19 (0.92, 1.54) 2.32 124
de Bortoli N et al 2007 Italy 1.23 (1.07, 1.41) 4.67 206
Francavilla R et al 2008 Italy 1.06 (0.80, 1.41) 1.00 40
Kim MN et al 2008 Korea 1.10 (0.97, 1.24) 7.84 347
Scaccianoce G et al 2008 Italy 0.85 (0.47, 1.52) 0.65 33
Scaccianoce G et al 2008 Italy 0.85 (0.47, 1.57) 0.61 31
Hurduc V et al 2009 Romania 1.16 (0.98, 1.36) 2.28 90
Szajewska H et al 2009 Poland 0.98 (0.71, 1.37) 1.42 66
Song MJ et al 2010 Korea 1.12 (1.02, 1.22) 14.86 661
Yasar B et al 2010 Turkey 1.25 (0.86, 1.83) 1.26 76
Bekar O et al 2011 Turkey 1.57 (1.09, 2.24) 1.27 82
Medeiros JA et al 2011 France 1.08 (0.85, 1.38) 1.51 62
Ozdil K et al 2011 Turkey 0.84 (0.73, 0.97) 5.23 193
Deguchi R et al 2012 Japan 1.19 (1.03, 1.38) 4.98 229
Du YQ et al 2012 China 1.31 (1.06, 1.61) 3.00 157
Manfredi M et al 2012 Italy 1.07 (0.96, 1.20) 4.00 149
Mirzaee V et al 2012 Iran 0.95 (0.63, 1.43) 1.26 68
Tolone S et al 2012 Italy 1.15 (0.92, 1.44) 1.63 68
Ahmad K et al 2013 Iran 1.30 (1.02, 1.67) 1.44 66
Dajani AI et al 2013 United Arab Emirates 1.21 (1.03, 1.41) 4.45 206
Navarro-Rodriguez T et al 2013 Brazil 1.05 (0.91, 1.22) 2.84 107
Shavakhi A et al 2013 Iran 0.95 (0.81, 1.10) 4.58 180
Wang YH et al 2013 China 1.29 (0.96, 1.73) 1.78 100
Overall (I-squared = 18.5%, p = 0.167) 1.12 (1.09, 1.16) 100.00
.5 .75 1 1.5 2 2.5
decreases ER
increases ER

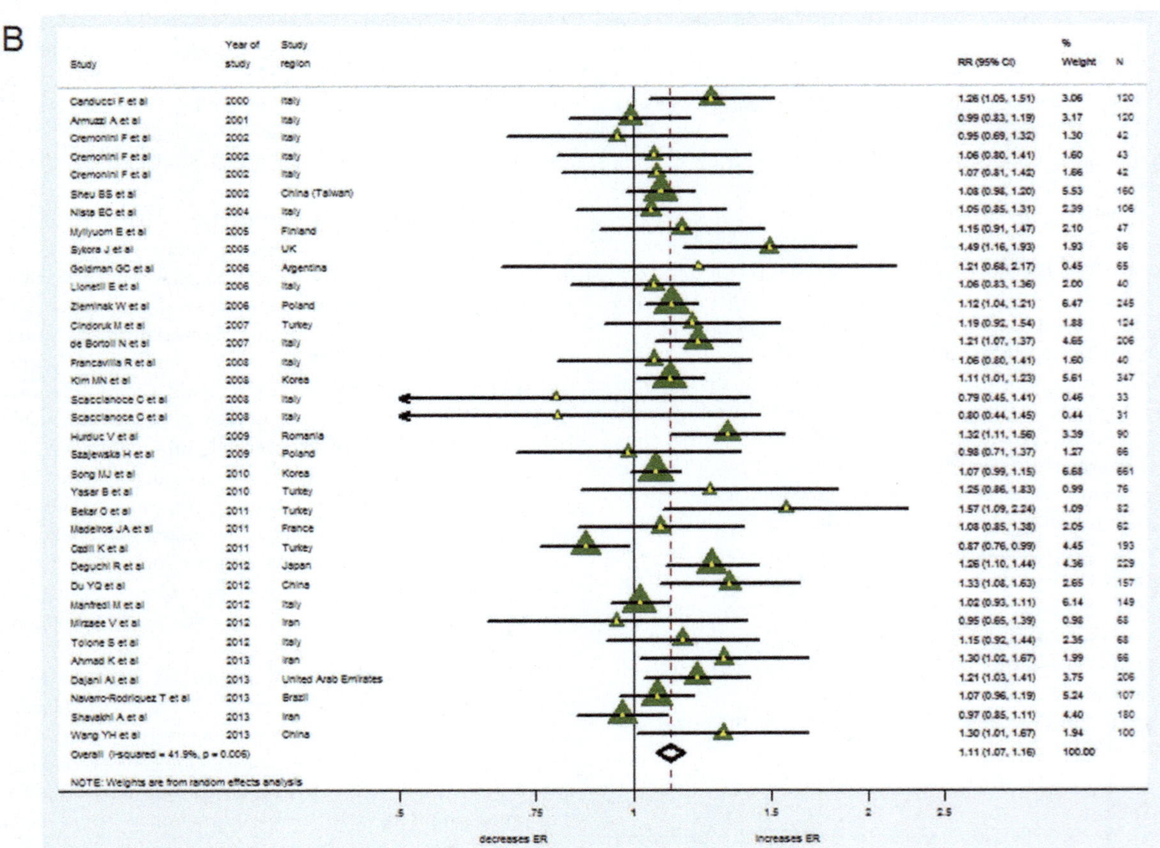
B
Study | Year of study | Study region | RR (95% CI) | % Weight | N
Canducci F et al 2000 Italy 1.26 (1.05, 1.51) 3.06 120
Armuzzi A et al 2001 Italy 0.99 (0.83, 1.19) 3.17 120
Cremonini F et al 2002 Italy 0.95 (0.69, 1.32) 1.30 42
Cremonini F et al 2002 Italy 1.06 (0.80, 1.41) 1.60 43
Cremonini F et al 2002 Italy 1.07 (0.81, 1.42) 1.66 42
Sheu BS et al 2002 China (Taiwan) 1.08 (0.98, 1.20) 5.53 160
Nista EC et al 2004 Italy 1.05 (0.85, 1.31) 2.39 106
Myllyluom E et al 2005 Finland 1.15 (0.91, 1.47) 2.10 47
Sykora J et al 2005 UK 1.49 (1.16, 1.93) 1.93 86
Goldman GC et al 2006 Argentina 1.21 (0.68, 2.17) 0.45 65
Lionetti E et al 2006 Italy 1.06 (0.83, 1.36) 2.00 40
Zieminak W et al 2006 Poland 1.12 (1.04, 1.21) 6.47 245
Cindoruk M et al 2007 Turkey 1.19 (0.92, 1.54) 1.88 124
de Bortoli N et al 2007 Italy 1.21 (1.07, 1.37) 4.65 206
Francavilla R et al 2008 Italy 1.06 (0.80, 1.41) 1.60 40
Kim MN et al 2008 Korea 1.11 (1.01, 1.23) 5.61 347
Scaccianoce G et al 2008 Italy 0.79 (0.45, 1.41) 0.46 33
Scaccianoce G et al 2008 Italy 0.80 (0.44, 1.45) 0.44 31
Hurduc V et al 2009 Romania 1.32 (1.11, 1.56) 3.39 90
Szajewska H et al 2009 Poland 0.98 (0.71, 1.37) 1.27 66
Song MJ et al 2010 Korea 1.07 (0.99, 1.15) 6.68 661
Yasar B et al 2010 Turkey 1.25 (0.86, 1.83) 0.99 76
Bekar O et al 2011 Turkey 1.57 (1.09, 2.24) 1.09 82
Medeiros JA et al 2011 France 1.08 (0.85, 1.38) 2.05 62
Ozdil K et al 2011 Turkey 0.87 (0.76, 0.99) 4.45 193
Deguchi R et al 2012 Japan 1.26 (1.10, 1.44) 4.36 229
Du YQ et al 2012 China 1.33 (1.08, 1.63) 2.65 157
Manfredi M et al 2012 Italy 1.02 (0.93, 1.11) 6.14 149
Mirzaee V et al 2012 Iran 0.95 (0.65, 1.39) 0.98 68
Tolone S et al 2012 Italy 1.15 (0.92, 1.44) 2.35 68
Ahmad K et al 2013 Iran 1.30 (1.02, 1.67) 1.99 66
Dajani AI et al 2013 United Arab Emirates 1.21 (1.03, 1.41) 3.75 206
Navarro-Rodriguez T et al 2013 Brazil 1.07 (0.96, 1.19) 5.24 107
Shavakhi A et al 2013 Iran 0.97 (0.85, 1.11) 4.40 180
Wang YH et al 2013 China 1.30 (1.01, 1.67) 1.94 100
Overall (I-squared = 41.9%, p = 0.006) 1.11 (1.07, 1.16) 100.00
NOTE: Weights are from random effects analysis
.5 .75 1 1.5 2 2.5
decreases ER
increases ER

Figure 3. The effect of probiotics supplementation vs. without probiotics on eradication rates by intention-to-treat analysis in (A). RR, risk ratio; CI, confidence interval. The triangles represent individual studies and the size of the triangle represents the weight given to each study in the meta-analysis. The diamond represents the combined results. (B) The effect of probiotics supplementation vs. without probiotics on eradication rates by pre-protocol analysis. RR, risk ratio; CI, confidence interval. The triangles represent individual studies and the size of the triangle represents the weight given to each study in the meta-analysis. The diamond represents the combined results.

mentation as an approach to manipulate gastrointestinal flora has been intensively studied.

While all previously conducted meta-analysis concluded that supplementation of antibiotic therapy with probiotics is effective in increasing eradication rates [8,9,10,11,12], our study identified only four effective individual strain: *Lactobacillus acidophilus* (Pubmed ID: 190198), *Lactobacillus casei* DN-114001 (Pubmed ID: 35628), *Lactobacillus gasseri* (Pubmed ID: 35528), *Bifidobacterium infantis* 2036 (Pubmed ID: 41468). However, more than one trial was conducted only for *Lactobacillus acidophilus* so far, limiting the generalizability of the findings regarding the other three strains. While *Lactobacillus GG* had previously been reported to be an effective strain [8], we do reach the same conclusion here.

Moreover, our results suggest that probiotic supplementation is only useful in less effective (eradication rate <80%) antibiotic

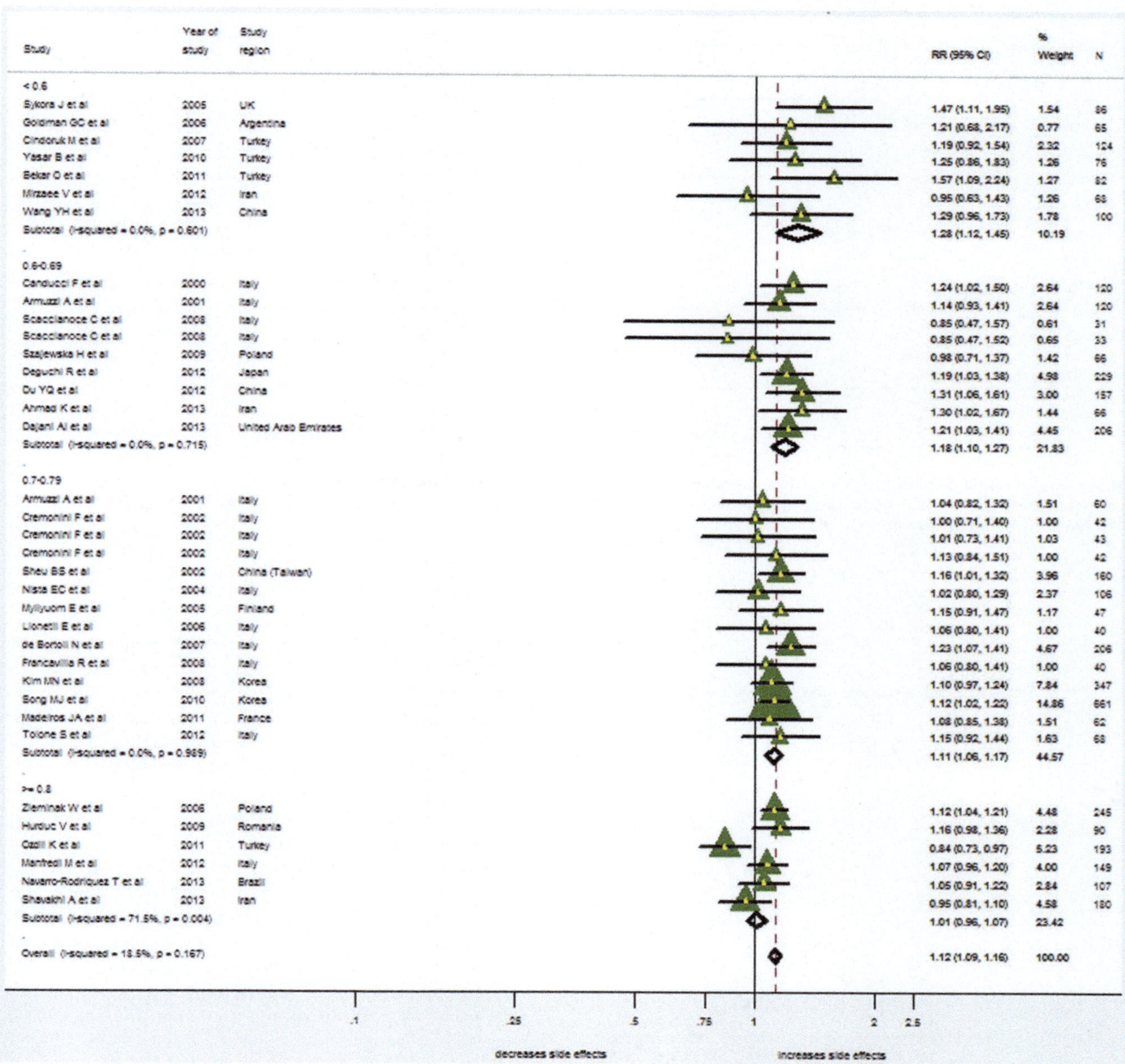

Figure 4. Meta-analysis of eradication rates by eradication rates due to antibiotic therapy only (control group). RR, risk ratio; CI, confidence interval. The triangles represent individual studies and the size of the triangle represents the weight given to each study in the meta-analysis. The diamond represents the combined results.

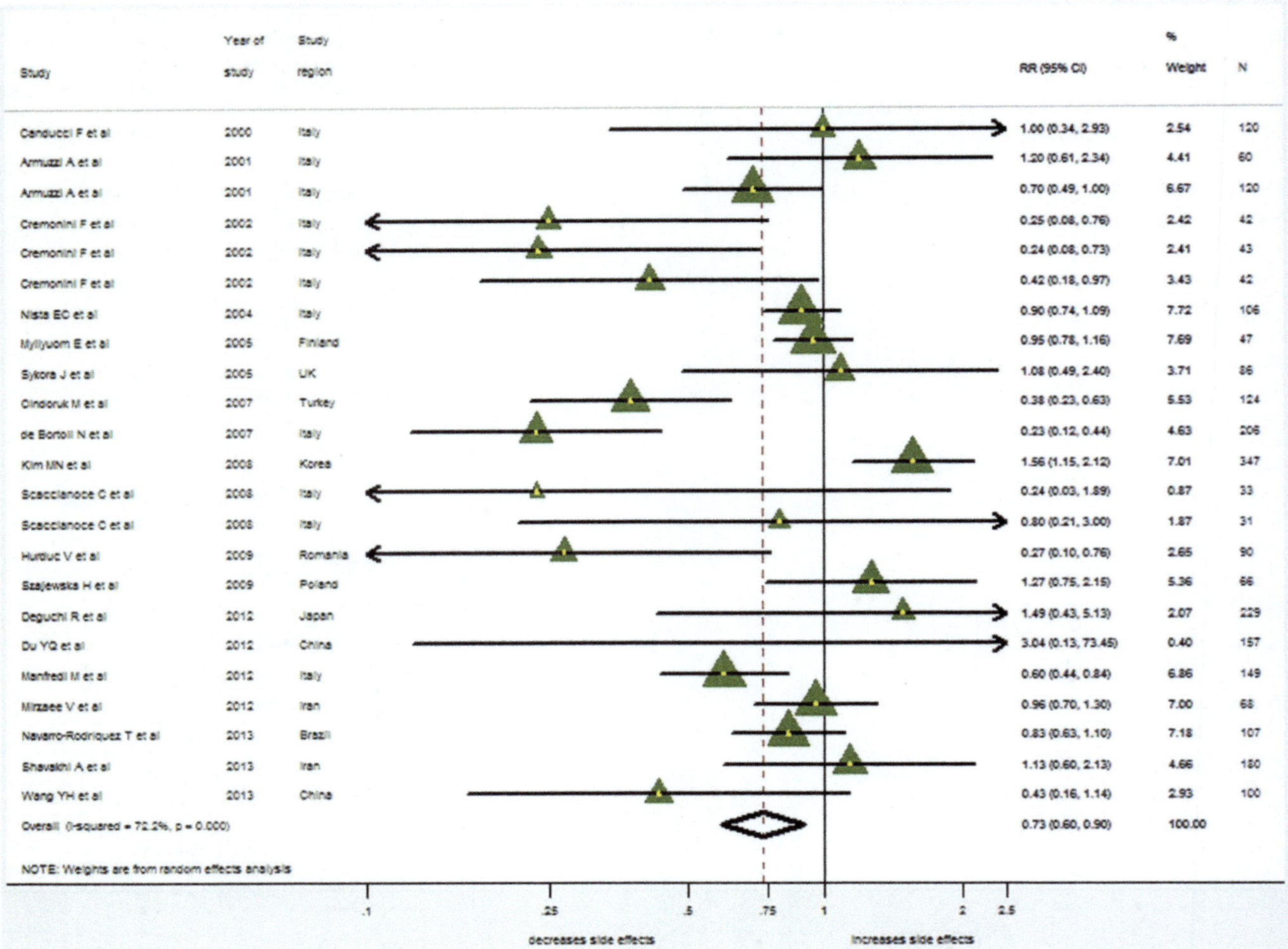

Figure 5. The effect of probiotics supplementation vs. without probiotics on total side effects. RR, risk ratio; CI, confidence interval. The triangles represent individual studies and the size of the triangle represents the weight given to each study in the meta-analysis. The diamond represents the combined results.

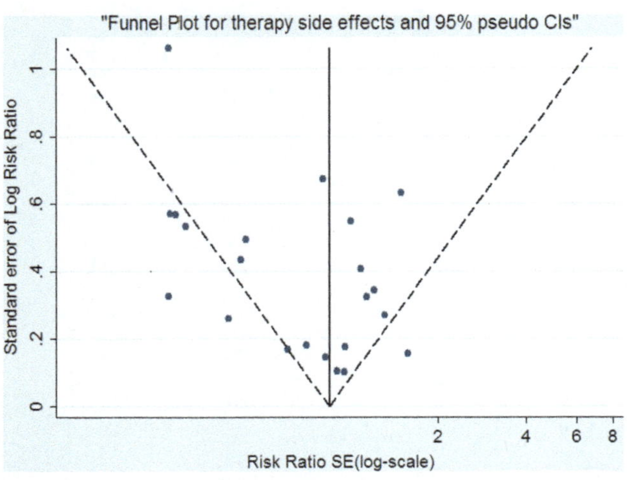

Figure 6. Funnel plot analysis of publication bias for side effects.

therapies. Acceptable success rates have often been defined as 80% or more on an ITT basis [2]. In those effective regimens, probiotic supplementation may not be needed. On the other hand, however, increasing resistance to antibiotics, for example clarithromycin, has been noted [2]. Probiotics supplementation may be a potential approach for eliminating resistant strains. However, no study included in this meta-analysis reported the detection of antibiotic resistance. Further research is warranted to clarify this issue.

The occurrence of adverse effects is one of the major drawbacks of antibiotic treatment. Although antibiotics may modify the composition of intestinal bacteria, broad spectrum antibiotics also often lead to gastrointestinal side effects [20,60]. Results regarding the effectiveness of probiotics in reducing side effects from previous meta-analyses have been inconclusive [8,9,10,11,12]. While we found that probiotics reduce the overall occurrence of side effects in the pooled data, this result must be taken with caution. Only one specific strain significantly reduced side effects and this overall result was only confirmed for non-blinded trials. While eradication rates can be determined with objective measures, assessment of side effects must rely on patients' subjective reporting. When patients are not blinded, their reports are likely to be biased due to awareness of a respective intervention.

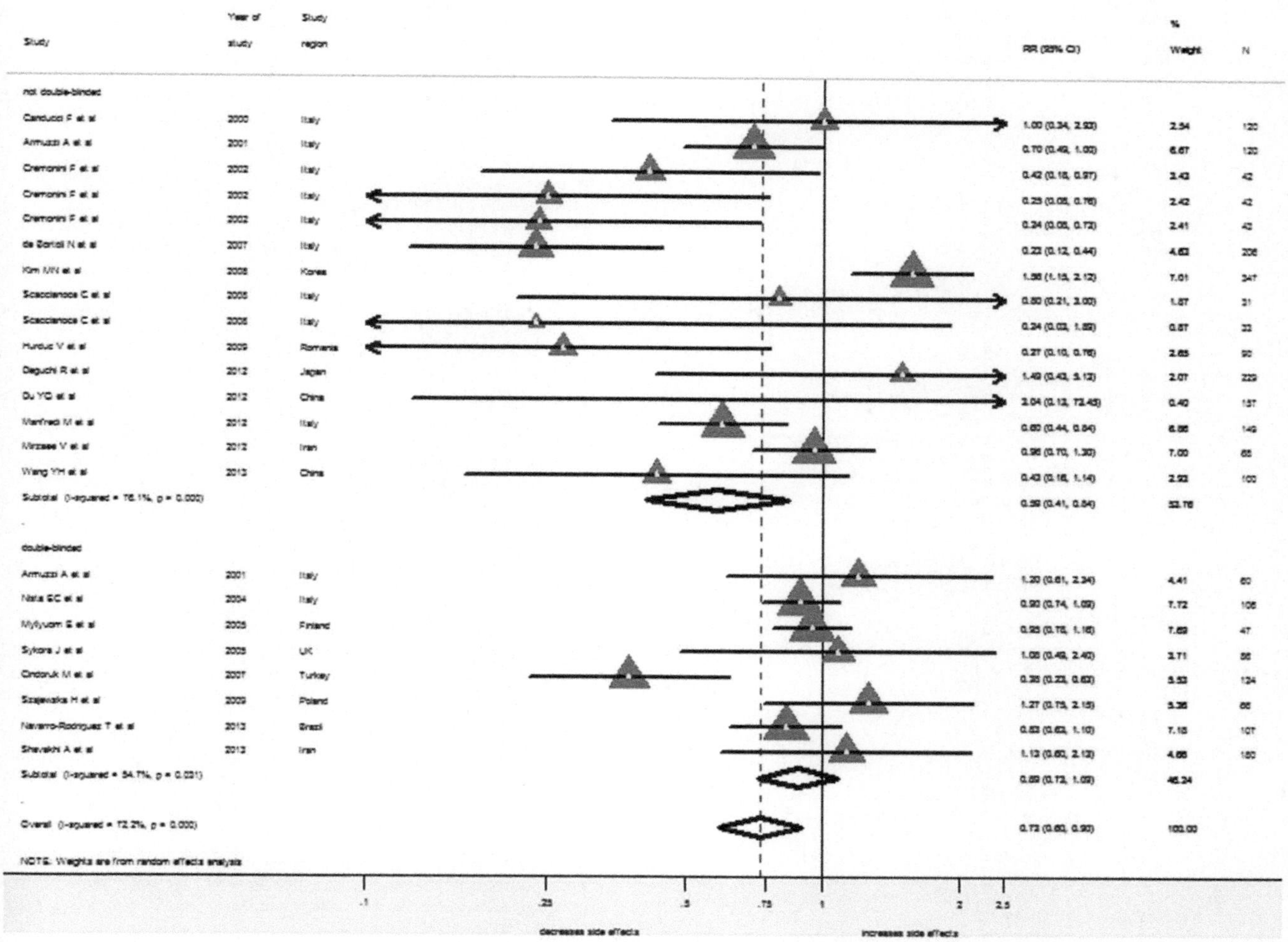

Figure 7. Meta-analysis of side effects by double-blinded and nondouble-blinded trials. RR, risk ratio; CI, confidence interval. The triangles represent individual studies and the size of the triangle represents the weight given to each study in the meta-analysis. The diamond represents the combined results.

Limitations of this meta-analysis include a limited number of high quality trials that could be analyzed, particularly regarding individual probiotic strains and side effects. Only one controlled trial was conducted for the following specific species *Bacillus clausii*, *Bifidobacterium* DN-173 010, *Lactobacillus casei* DN-114001, *Lactobacillus gasseri*, *Bifidobacterium infantis* 2036. Accordingly, results regarding those strains need to be interpreted with caution. Thirteen of the analyzed trials did not provide data on overall occurrence of side effects. Moreover, tools used for measuring side effects and reporting individual side effects largely differed across studies. From 20 studies reporting on the overall incidence of side effects, only eight were blinded.

Conclusions

The pooled data suggests that supplementation with specific strains of probiotics compared with eradication therapy may be considered as an option for increasing eradication rates of *H. pylori*, particularly when antibiotic therapies are relatively ineffective.

Acknowledgments

We acknowledge the help of Dr. Andrew Pennycott in correcting the manuscript for grammar and style.

Author Contributions

Conceived and designed the experiments: YD GZ. Performed the experiments: YD XZ. Analyzed the data: YD JDR. Contributed reagents/materials/analysis tools: XZ. Wrote the paper: YD JDR GZ.

References

1. Go MF (2002) Review article: natural history and epidemiology of *Helicobacter pylori* infection. Aliment Pharmacol Ther 16: 3–15.
2. Malfertheiner P, Meqraud F, O'Morain CA, Atherton J, Axon AT, et al. (2012) Management of *Helicobacter pylori* infection–the Masstricht IV/Florence Consensus Report. Gut 61: 646–64.
3. Meqraud F (2004) Basis of the management of drug-resistant *Helicobacter pylori* infection. Drugs 64: 1893–904.
4. Yoqueswaran K, Chen G, Cohen L, Cooper MA, Yong E, et al. (2011) How well is *Helicobacter pylori* treated in usual practice? Can J Gastroenterol 25: 543–6.

5. Lionetii E, Indrio F, Pavone L, Borrello G, Cavallo L, et al. (2010) Role of probiotics in pediatric patients with *Helicobacter pylori* infection: a comprehensive review of the literature. Helicobacter 15: 79–87.
6. FAO/WHO (2001) Report on Joint FAO/WHO Expert Consultation on Evaluation of Health and Nutritional Properties of Probiotics in Food Including Powder Milk with Live Lactic Acid Bacteria. Cordoba, FAO/WHO, 2001.
7. Saxelin M, Tynkkynen S, Mattila-Sandholm T, de Vos WM (2005) Probiotic and other functional microbes: from markets to mechanisms. Curr Opin Biotechnol 16: 204–11.
8. Tong JL, Ran ZH, Shen J, Zhang CX, Xiao SD (2007) Meta-analysis: the effect of supplementation with probiotics on eradication rates and adverse events during *Helicobacter pylori* eradication therapy. Aliment Pharmacol Ther 25: 155–68.
9. Zou J, Dong J, Yu X (2009) Meta-analysis: Lactobacillus containing quadruple therapy versus standard triple first-line therapy for *Helicobacter pylori* eradication. Helicobacter 14: 449–459.
10. Sachdeva A, Naqpal J (2009) Effect of fermented milk-based probiotic preparations on *Helicobacter pylori* eradication: a systematic review and meta-analysis of randomized-controlled trials. Eur J Gastroenterol Hepatol 21: 45–53.
11. Szajewska H, Horvath A, Piwowarczyk A (2010) Meta-analysis: the effects of Saccharomyces boulardii supplementation on *Helicobacter pylori* eradication rates and side effects during treatment. Aliment Pharmacol Ther 32: 1069–79.
12. Wang ZH, Gao QY, Fang JY (2013) Meta-analysis of the efficacy and safety of Lactobacillus-containing and Bifidobacterium-containning probiotic compound preparation in *Helicobacter pylori* eradication therapy. J Clin Gastroenterol 47: 25–32.
13. Zheng X, Lyu L, Mei Z (2013) Lactobacillus-containingprobiotic supplementation increases Helicobacter plyori eradication rate: evidence from a meta-analysis. Rev Esp Enferm Dig 105: 445–53.
14. Li S, Huang XL, Sui JZ, Chen SY, Xie YT, et al. (2014) Meta-analysis of randomized controlled trails on the efficacy of probiotics in Helicobacter pylori eradication therapy in children. Eur J Pediatr 173: 153–61.
15. Higgins JP, Altman DG, Gotzsche PC, Juni P, Moher D, et al. (2011) The Cochrane Collaboration's tool for assessing risk of bias in randomized trials. BMJ 343: d5928.
16. Roger MH, Ross JH, Jonathan AC (2009) Updated tests for small-study effects in meta-analyses. The Stata Journal 9: 197–210.
17. Canducci F, Armuzzi A, Cremonini F, Cammarota G, Bartolozzi F, et al. (2000) A lyophilized and inactivated culture of Lactobacillus acidophilus increases *Helicobacter pylori* eradication rates. Aliment Pharmacol Ther 14: 1625–9.
18. Armuzzi A, Cremonini F, Bartolozzi F, Canducci F, Candelli M, et al. (2001) The effect of oral administration of Lactobacillus GG on antibiotic-associated gastrointestinal side-effects during *Helicobacter pylori* eradication therapy. Aliment Pharmacol Ther 15: 163–9.
19. Armuzzi A, Cremonini F, Ojetti V, Bartolozzi F, Canducci F, et al. (2001) Effect of Lactobacillus GG supplementation on antibiotic-associated gastrointestinal side effects during *Helicobacter pylori* eradication therapy: a pilot study. Digestion 63: 1–7.
20. Cremonini F, Di Caro S, Covino M, Armuzzi A, Gabrielli M, et al. (2002) Effect of different probiotic preparations on anti-*Helicobacter pylori* therapy-related side effects: a parallel proup, triple blind, placebo-controlled study. Am J Gastroenterol 97: 2744–9.
21. Sheu BS, Wu JJ, Lo CY, Wu HW, Chen JH, et al. (2002) Impact of supplement with Lactobacillus- and Bifidobacterium-containing yogurt on triple therapy for *Helicobacter pylori* eradication. Aliment Pharmacol Ther 16: 1669–75.
22. Nista EC, Candelli M, Cremonini F, Cazzato IA, Zocco MA, et al. (2004) Bacillus clausii therapy to reduce side-effects of anti-*Helicobacter pylori* treatment: randomized, double-blind, placebo controlled trial. Aliment Pharmacol Ther 20: 1181–8.
23. Myllyluoma E, Veijola L, Ahlroos T, Tynkkynen S, Kankuri E, et al. (2005) Probiotic supplementation improves tolerance to *Helicobacter pylori* eradication therapy-a placebo-controlled, double-blind randomized pilot study. Aliment Pharmacol Ther 21: 1263–72.
24. Sykora J, Valeckova K, Amlerova J, Siala K, Dedek P, et al. (2005) Effects of a specially designed fermented milk product containing probiotic Lactobacillus casei DN-114 001 and the eradication of *H. pylori* in children: a prospective randomized double-blind study. J Clin Gastroenterol 39: 692–8.
25. Goldman CG, Barrado DA, Balcarce N, Rua EC, Oshiro M, et al. (2006) Effect of a probiotic food as an adjuvant to triple therapy for eradication of *Helicobacter pylori* infection in children. Nutrition 22: 984–8.
26. Lionetti E, Miniello VL, Castellaneta SP, Maqista AM, de Canio A, et al. (2006) Lactobacillus reuteri therapy to reduce side-effects during anti-*Helicobacter pylori* treatment in children: a randomized placebo controlled trial. Aliment Pharmacol Ther 24: 1461–8.
27. Ziemniak W (2006) Efficacy of *Helicobacter pylori* eradication taking into account its resistance to antibiotics. J Physiol Pharmacol 57: 123–41.
28. Cindoruk M, Erkan G, Karakan T, Dursun A, Unal S (2007) Efficacy and safety of Saccharomyces boulardii in the 14-day triple anti-*Helicobacter pylori* therapy: a prospective randomized placebo-contrilled double-blind study. Helicobacter 12: 309–16.
29. de Bortoli N, Leonardi G, Ciancia E, Merlo A, Bellini M, et al. (2007) *Helicobacter pylori* eradication: a randomized prospective study of triple therapy versus triple therapy plus lactoferrin and probiotics. Am J Gastroenterol 102: 951–6.
30. Francavilla R, Lionetii E, Casrellaneta SP, Maqusta AM, Mauroqivanni G, et al. (2008) Inhibition of *Helicobacter pylori* infection in humans by Lactobacillus reuteri ATCC 55730 and effect on eradication therapy: a pilot study. Helicobacter 13: 127–34.
31. Kim MN, Kim N, Lee SH, Park YS, Hwang JH, et al. (2008) The effects of probiotics on PPI-triple therapy for *Helicobacter pylori* eradication. Helicobacter 13: 261–8.
32. Scaccianoce G, Zullo A, Hassan C, Gentili F, Cristofari F, et al. (2008) Triple therapies plus different probiotics for *Helicobacter pylori* eradication. Eur Rev Med Pharmacol Sci. 12: 251–6.
33. Hurduc V, Plesca D, Dragomir D, Sajin M, Vandenplas Y (2009) A randomized, open trial evaluating the effect of Saccharomyces boulardii on the eradication rate of *Helicobacter pylori* infection children. Acta Paediatr 98: 127–31.
34. Szajewska H, Albrecht P, Topczewska-Cabanek A (2009) Randomized, double-blind, placebo-controlled trial: effect of lactobacillus GG supplementation on *Helicobacter pylori* eradication rates and side effects during treatment in children. J Pediatr Gastroenterol Nutr 48: 431–6.
35. Song MJ, Park DI, Park JH, Kim HJ, Cho YK, et al. (2010) The effect of probiotics and mucoprotective agents on PPI-based triple therapy for eradication of *Helicobacter pylori*. Helicobacter 15: 206–13.
36. Yasar B, Abut E, Kayadibi H, Toros B, Sezikli M, et al. (2010) Efficacy of probiotics in *Helicobacter pylori* eradication therapy. Turk J Gastroenterol 21: 212–7.
37. Bekar O, Yilmaz Y, Gulten M (2011) Kefir improves the efficacy and tolerability of triple therapy in eradicating *Helicobacter pylori* 14: 344–7.
38. Medeiros JA, Goncalves TM, Boyanova L, Pereira MI, de Carvalho JN, et al. (2011) Evaluation of *Helicobacter pylori* eradication by triple therapy plus Lactobacillus acidophilus compared to triple therapy alone. Eur J Clin Microbiol Infect Dis 30: 555–9.
39. Ozdil K, Calhan T, Sahin A, Senates E, Kahraman R, et al. (2011) Levofloxacin based sequential and triple therapy compared with standard plus probiotic combination for *Helicobacter pylori* eradication. Hepatogastroenterology 58: 1148–52.
40. Deguchi R, Nakaminami H, Rimbara E, Noquchi N, Sasatsu M, et al. (2012) Effect of pretreatment with Lactobacillus gasseri OLL2716 on first-line *Helicobacter pylori* eradication therapy. J Gastroenterol Hepatol 27: 888–92.
41. Du YQ, Su T, Fan JG, Lu YX, Zheng P, et al. (2012) Adjuvant probiotics improve the eradication effect of triple therapy for *Helicobacter pylori* infection. World J Gastroenterol 18: 6302–7.
42. Manfredi M, Bizzarri B, Sacchero RI, Maccari S, Calabrese L, et al. (2012) *Helicobacter pylori* infection in clinical practice: probiotics and a combination of probiotics + lactoferrin improve compliance, but not eradication, in sequential therapy. Helicobacter 17: 254–63.
43. Mizaee V, Rezahosseini O (2012) Randomized control trial: Comparison of Triple Therapy plus Probiotic Yogurt vs. Standard Triple Therapy on *Helicobacter pylori* Eradication. Iran Red Crescent Med J 14: 657–66.
44. Tolone S, Pellino V, Vitaliti G, Lanzafame A, Tolone C (2012) Evaluation of *Helicobacter pylori* eradication in pediatric patients by triple therapy plus lactoferrin and probiotics compared to triple therapy alone. Ital J Pediatr 38: 63.
45. Ahmad K, Fatemeh F, Mehri N, Maryam S (2013) Probiotics for the treatment of pediatric *Helicobacter pylori* infection: a randomized double blind clinical trial. Iran H Pediatr 23: 79–84.
46. Navarro-Rodriquez T, Silva FM, Barbuti RC, Mattar R, Moraes-Filho JP, et al. (2013) Association of a probiotic to a *Helicobacter pylori* eradication regimen does not increase efficacy or decreases the adverse effects of the treatment: a prospective, randomized, double-blind, placebo-controlled study. BMC Gastroenterol 13: 56.
47. Shavakhi A, Tabesh E, Yaqhoutkar A, Hashemi H, Tabesh F, et al. (2013) The effects of multistrain probiotics compound on bismuth-containing quadruple therapy for *Helicobacter pylori* infection: A randomized placebo-controlled triple-blind study. Helicobacter 18: 280–4.
48. Dajani AI, Abu Hammour AM, Yang DH, Chung PC, Nounou MA, et al. (2013) Do probiotics improve eradication response to *Helicobacter pylori* on standard triple or sequential therapy? Saudi J Gastroenterol 19: 113–120.
49. Wang YH, Huang Y (2013) Effect of Lactobacillus acidophilus and Bifidobacterium bifidum supplementation to standard triple therapy on Helicobacter pylori eradication and dynamic changes in intestinal flora. World J Microbiol Biotechnol 30: 847–53.
50. Sheu BS, Cheng HC, Kao AW, Wang ST, Yang YJ, et al. (2006) Pretreatment with Lactobacillus- and Bifidobacterium-containing yogurt can improve the efficacy of quadruple therapy in eradicating residual *Helicobacter pylori* infection after failed triple therapy. Am J Clin Nutr 83: 864–9.
51. Boonyaritichaikij S, Kuwabara K, Nagano J, Naqano J, Kobayashi K, et al. (2009) Long-term administration of probiotics to asymptomatic pre-school children for either the eradication or the prevention of *Helicobacter pylori* infection. Helicobacter 14: 202–7.
52. Yoon H, Kim N, Kim JY, Park SY, Park JH, et al. (2011) Effects of multistrain probiotic containing yogurt on second-line triple therapy for Hp infection. J Gastroenterol Hepatol 26: 44–8.
53. Selinger CP, Bell A, Cairns A, Lockett M, Sebastian S, et al. (2013) Probiotics VSL#3 prevents antibiotic-associated diarrhea in a double-blind, randomized, placebo-controlled clinical trial. J Hosp Infect 84: 159–65.

54. Mehling H, Busjahn A (2013) Non-viable Lactobacillus reuteri DSMZ 17648 (Pylopass™) as a new approach to Helicobacter pylori control in humans. Nutrients 5: 3062–73.
55. Dore MP, Cuccu M, Pes GM, Manca A, Graham DY (2014) Lactobacillus reuteri in the treatment of *Helicobacter pylori* infection. Intern Emerg Med 9: 649–54.
56. Francavilla R, Polimeno L, Demichina A, Mauroqiovanni G, Principi B, et al. (2014) Lactobacillus reuteri strain combination in *Helicobacter pylori* infection: a randomized, double-blind, placebo-controlled study. J Clin Gastroenterol 48: 407–13.
57. Marshall BJ, Warren JR (1984) Unidentified curved bacilli in the stomach of patients with gastritis and peptic ulceration. Lancet 1: 1311–5.
58. Walker MM, Talley NJ (2014) Review article: bacteria and pathogenesis of disease in the upper gastrointestinal tract—beyond the era of Helicobacter pylori. Aliment Pharmacol Ther 39: 767–79.
59. Delqado S, Cabrera-Rubio R, Mira A, Suarez A, Mayo B (2013) Microbiological survey of the human gastric ecosystem using culturing and pyrosequencing methods. Microb Ecol 65: 763–72.
60. Patel A, Shah N, Prajapati JB (2014) Clinical appliance of probiotics in the treatment of *Helicobacter pylori* infection-A brief review. J Microbiol Immunol Infect 47: 429–37.

19

Prevalence of Veterinary Antibiotics and Antibiotic-Resistant *Escherichia coli* in the Surface Water of a Livestock Production Region in Northern China

Xuelian Zhang, Yanxia Li*, Bei Liu, Jing Wang, Chenghong Feng, Min Gao, Lina Wang

State Key Laboratory of Water Environment Simulation, School of Environment, Beijing Normal University, Beijing, China

Abstract

This study investigated the occurrence of 12 veterinary antibiotics (VAs) and the susceptibility of *Escherichia coli* (*E. coli*) in a rural water system that was affected by livestock production in northern China. Each of the surveyed sites was determined with at least eight antibiotics with maximum concentration of up to 450 ng L^{-1}. The use of VAs in livestock farming probably was a primary source of antibiotics in the rivers. Increasing total antibiotics were measured from up- to mid- and downstream in the two tributaries. Eighty-eight percent of the 218 *E. coli* isolates that were derived from the study area exhibited, in total, 48 resistance profiles against the eight examined drugs. Significant correlations were found among the resistance rates of sulfamethoxazole-trimethoprim, chloromycetin and ampicillin as well as between tetracycline and chlortetracycline, suggesting a possible cross-selection for resistance among these drugs. The *E. coli* resistance frequency also increased from up- to midstream in the three rivers. *E. coli* isolates from different water systems showed varying drug numbers of resistance. No clear relationship was observed in the antibiotic resistance frequency with corresponding antibiotic concentration, indicating that the antibiotic resistance for *E. coli* in the aquatic environment might be affected by factors besides antibiotics. High numbers of resistant *E. coli* were also isolated from the conserved reservoir. These results suggest that rural surface water may become a large pool of VAs and resistant bacteria. This study contributes to current information on VAs and resistant bacteria contamination in aquatic environments particularly in areas under intensive agriculture. Moreover, this study indicates an urgent need to monitor the use of VAs in animal production, and to control the release of animal-originated antibiotics into the environment.

Editor: Tara C. Smith, Kent State University, United States of America

Funding: The research was funded by the Natural Science Foundation of China (No. 21277013, 20977010) and the Special Fund for Environmental Protection Research in the Public Interest (200909042). The funders had no role in study design, data collection and analysis, decision to publish, or preparation of the manuscript.

Competing Interests: The authors have declared that no competing interests exist.

* Email: liyxbnu@bnu.edu.cn

Introduction

Antibiotic and antibiotic-resistant microbial contamination is an issue of growing global concern, both in the public and in the research community. The rampant use of antibiotics in medicine and agriculture has resulted in the extensive detection of antibiotics, antibiotic-resistant bacteria and antibiotic resistance genes (ARGs) in the environment worldwide, including in 139 streams in the US, the Osaka area of Japan, the Haihe River and the Yangtze estuary in China [1–4].

Animals in concentrated feeding operations are the chief consumers of antimicrobial agents, which are administered for growth improvement and disease control. The amount of veterinary antibiotics (VAs) annually administered in the world reaches 10^5–10^6 tonnes [5–7]. Guidelines have existed that ban the use of certain antibiotics as growth promoters in both the European Union and the US since 1999, but this prohibition has led to a corresponding increased use of VAs for disease control and improving feed efficiency [8]. Considering Denmark as an example, the total veterinary use remained as high as 107 t in 2011 [9,10]. Antibiotics are poorly absorbed in the animal gut; as a result, approximately 40–90% of these antibiotics will be excreted as parent compounds or metabolites via urine or feces [11,12], and the residue of VAs in animal wastes has been widely reported [7,13–16].

Animal wastes are usually stored using a lagoon system and/or are composted instead of intensive treatment before being discharged from animal farms [10]. However, both of these methods are limited in their ability to completely remove antibiotics. For instance, compost can reduce antibiotics by 20%–99% [17,18], and a lagoon system can decrease tylosin by no more than 75% [19]. Thus, high amounts of antibiotics might remain in animal waste and are a potential pollution source of environmental antibiotics. Residual antibiotics in post-treated animal wastes can be disseminated into the surrounding aquatic environment via runoff when utilized as fertilizer on farmlands [20,21] or in some cases, when directly released to receiving watersheds through sewage discharge or occasional leaching from animal farms. Such scattered pollution sources from operations as well as from fertilized fields will lead to more severe and complex antibiotic contamination in these rural areas compared to point-source-affected urban rivers; therefore, the antibiotic contamination in these regions requires additional attention and study.

The particular concern over VAs exposure is the increased presence of antibiotic resistant bacteria in the environment. Antibiotic-resistant bacteria might develop resistance within animal bodies from exposure to administered VAs [22], which can be excreted in feces and be subsequently released into the environment along with antibiotics [23,24]. Furthermore, long-term exposure to sub-therapeutic levels of antimicrobial agents in aquatic environments imposes selection stress on environmental microorganisms [25]. Once antibiotic-resistant microorganisms appear in the environment, they enlarge the resistance community through the transfer of corresponding ARGs among microbial populations [3,26]. In addition, antibiotic-resistant bacteria are likely to transport vertically and horizontally through physical or biological media [27]. Since livestock breeding and related agricultural activity is a potential source of antibiotics and resistant bacteria in the environmental, there is an urgent need to elucidate how antibiotics, and particularly antibiotic-resistant fecal indicator bacteria (*E. coli*) contamination in rural aquatic environments, might be impacted by intensive concentrated animal feeding operations (CAFOs).

China is the largest animal feeding country in the world and is experiencing the expansion and intensification of animal feeding operations in many areas. Due to the absence until now of relevant regulation, the residues of a variety of VAs have been consistently reported at high levels in animal waste [7,13,15]. Nevertheless, limited information is available providing baseline data about surface water pollution with antibiotics and antibiotic-resistant bacteria in typical livestock production regions. Therefore, this study investigated the Jiyun River, Beijing, which flows through Pinggu County, which is one of the primary meat-providing counties in Beijing, to (1) determine the occurrence and spatial distribution of antibiotics in the water system, (2) examine the antibiotic susceptibility of *E. coli* in this basin and (3) explore the correlation between *E. coli* resistance and antibiotic concentrations. We seek to provide useful information of the influence that is exerted on the aquatic ecosystem by VAs administration in the animal industry.

Materials and Methods

Study area, sampling sites and sample collection

The study area of Pinggu County is located east of Beijing, China. This county has the highest animal feeding density among the suburbs of Beijing and provided 3.4×10^5 and 7.11×10^6 swine and poultry products, respectively, to capital market in 2011 [28]. The main stream of the Jiyun River is Ju River, with a stretch of 54.4 km, which flows across Pinggu County over an area of 1352 km^2 and eventually enters Bohai Bay.

The Jiyun River has two tributaries: Cuo River and Jinji River. The sampling sites were selected along the upper, middle and lower reaches of the main stream and two tributaries. Eight of the 9 sites were directly adjacent to animal farms. These animal farms are small-scale operations, and none of the 8 farms was equipped with a professional lagoon system or fecal treatment procedure. The effluents of these farms are directly discharged into adjacent water bodies. The fecal wastes were simply piled up with maize straw in the air, but usually without any bulking agents, and then applied to nearby farmlands during fertilizer season. Six fresh fecal samples were collected from six of the eight animal operations with the permission of farmers, but samples from the other two farms were not available. From each farm, multiple points of fresh feces were obtained from the pigpen and were mixed to obtain a representative sample. Two intersection sites (JC and JCJJ) in the main stream Ju River separately receiving flow from Cuo River and Jinji River, respectively, and a site from the Haizi reservoir were also selected. The GPS coordinates of each location in this study are listed in Table S1. The sampling locations and sampling activity in this study did not require specific permissions and the sampling did not involve endangered or protected species. The study area and sample sites distributions are presented in Figure 1. The sampling was conducted on May 16, 2013, and a total of 12 water samples and 6 fresh fecal samples were obtained.

The samples were separated into two parts: one part received sodium azide immediately to inhibit microbial activity and prevent antibiotic biodegradation for further antibiotic analysis [29,30], and the other part was free of sodium azide for microbiological property determination and pure *E. coli* clone isolation. The samples were preserved in cold boxes, transported to the laboratory within one day and maintained at 4°C until use. The microbial samples were incubated and the pH values were measured immediately upon return to the laboratory. The antibiotics were detected beginning on the second day.

Sample preparation and analysis

Twelve target antibiotics in three different groups (tetracyclines (TCs): tetraycline (TC), oxytetracycline (OTC), chlortetracycline (CTC) and doxycycline (DOC); quinolones (QLs): ciprofloxacin (CFC), enrofloxacin (EFC) and ofloxacin (OFC); and sulfonamides (SAs): sulfadiazine (SDZ), sulfamethoxazole (SMX), sulfamonomethoxine (SMM), sulfameter (SM), sulfachinoxaline (SCX)) were analyzed in this study.

The samples were pretreated according to the method of [12,15] with slight modification. Briefly, a 2-L water sample was filtered through 0.45-μm pore glass fiber membrane filter. The filtered water sample was acidified to a pH of 3.0 with 30% H_2SO_4 (V/V), followed by the addition of Na_2-EDTA (0.5 g L^{-1}) as a chelating agent. The acidified water samples were concentrated by solid-phase extraction (SPE) over an Oasis HLB cartridge (6cc, 500 mg, Waters, USA) at a flow rate of 5 ml min^{-1}. The antibiotics that were retained on the HLB cartridges were eluted with 10 ml of methanol (0.1% formic acid) after being dried with nitrogen gas at a flow rate of 5 ml min^{-1}. The eluent was concentrated to near dryness under a gentle nitrogen stream in a 37°C water bath. A mixture of internal standard containing tetracycline-D6, enrofloxacin-D5 and sulfamethazine-D4 was added to compensate for the variations of TCs, FQs and SAs, respectively, during detection. Methanol was supplemented to a final volume of 1 ml. The final solution was filtered through a PTFE filter (Millipore) for liquid chromatography - tandem mass spectrometry (LC-MS-MS) analysis. For the solid samples, 3 g of fresh fecal samples were extracted with a 20-ml mixture of methanol and EDTA-McIlvaine buffer (V:V = 1:1) with 30 minutes of rotation and ultrasonication, respectively, three times. The extracts were centrifuged and the supernatant was collected and diluted with deionized water to maintain a methanol proportion below 10%. The substantial purification and concentration followed the steps of the water samples.

The extracted samples were analyzed by LC-MS-MS. The separation of the target compounds was performed on a C18 column (250×4.6 mm, 5 mm; Akzo Nobel, Sweden) in a HPLC system (DIONEX UltiMate 3000). A binary elution gradient consisting of A (0.1% formic acid water solution) and B (methanol) was used in the following program: 0–2 min: 15–30% B, 2–5 min: 30–40% B, 5–15 min: 40–70% B, 15–17 min: 70–100% B, and 17–21 min 100% B. A triple quadrupole mass spectrometer (API 3200, AB-SCIEX, Framingham, MA) that was equipped with an electrospray ionization (ESI) source in positive ion mode was used. The multiple reaction monitoring (MRM) mode was used for the

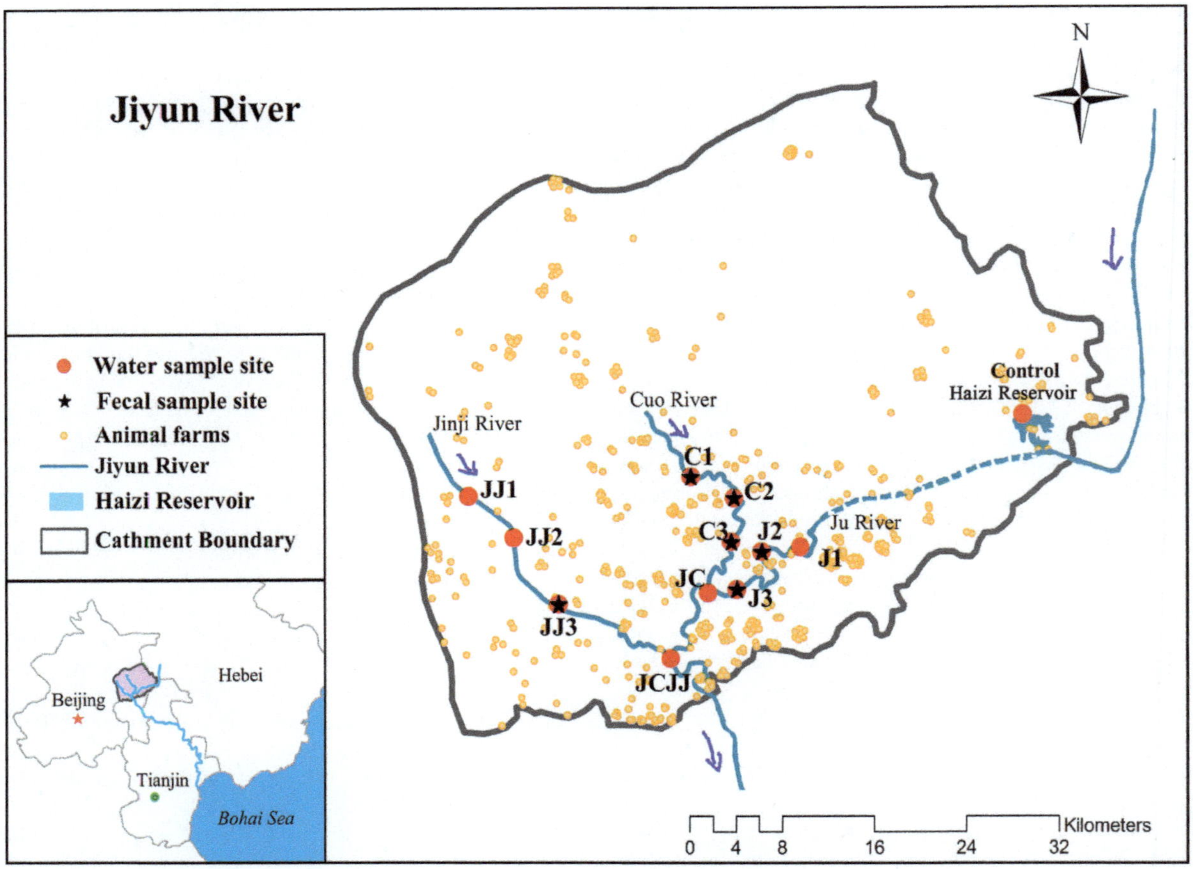

Figure 1. Map of sampling sites in the Jiyun River.

quantitation. The most intensive ion pairs together with retention time were used to identify the targeted antibiotics. Multipoint internal calibration curves were used to quantify the antibiotics. The recoveries of the 12 monitored compounds were in the range of 71 to 100%.

E. coli isolation and confirmation

E. coli in the water was detected using the enzyme substrate method [31]. Briefly, Colilert substrates (IDEXX Laboratories, Inc., Westbrook, ME) were added to 100 ml water samples and were mixed thoroughly until all of the substrate dissolved per the manufacturer's instructions. The dissolved water was sealed in an IDEXX Quanti-Tray/2000 (IDEXX Laboratories, Inc., Westbrook, ME) and incubated at 37°C for 24 h. The wells that changed to yellow and produced additional fluorescence were identified as harboring *E. coli.*

The *E. coli* isolates were obtained and confirmed according to previous descriptions [26,31] with some minor revision. Briefly, 10 µl of a 10^{-4}–10^{-5} dilution of liquid from 10–15 wells that were positive for *E. coli* growth on the plate for each site was streaked onto a Mueller-Hinton (MH) agar plate for incubation of 18 h at 37°C to obtain 10–50 clones on each plate. Approximately 4 isolates were randomly selected from well-separated colonies on each plate and purified on Mueller-Hinton (MH) agar plates using the crossed dilution method. To confirm these isolates, they were separately inoculated in *Escherichia coli* broth (EC broth) containing 4-methylumbelliferyl-, BD-glucuronide (EC-MUG) and tryptophan separately. The isolates that generated a blue fluorescence in EC-MUG and produced indole (identified by a rose-colored product after the addition of Kovacs agent) after 24 h of incubation at 37°C were confirmed as *E. coli*. In total, 218 isolates were collected from the study area. The confirmed isolates were re-streaked onto MH agar plates for subsequent susceptibility tests.

Antibiotics susceptibility test

In total, eight drugs were tested for *E. coli* susceptibility primarily based on their detection in this study and their clinical importance. CTC and SMM tests were performed via growth measurement in Mueller-Hinton Broth in the presence and absence of antibiotics. The other six antibiotics were analyzed via the disk diffusion method [32]. The CTC and SMM test concentration was set to 25 µg and 500 µg per ml MH broth, respectively, according to a combination of prescribed doses in CLSI guidelines and the levels that have been used in other studies [33–35]. The CTC stock was directly prepared in MH broth. The SMM was dissolved in MH broth with a few drops of NaOH solution. A pure *E. coli* culture was suspended in sterile sodium chloride solution to 0.5 McFarland turbidity level after the incubation for 18 hours at 37°C. Fifty µl of a 10^{-2} suspension was used to separately inoculate wells containing 50 µl of MH broth with or without target antibiotic. The preliminary test indicated that the amount of NaOH remaining in SMM wells had no effect on *E. coli* growth. Therefore, the first and second wells in each group of the 4 wells were always the negative control (MH broth) and the positive control (MH broth with *E. coli* inoculum), respectively. After incubation for 22 hours at 37°C, the 96-well plate was read spectrophotometrically at 600 nm. The isolates were recorded as resistant to a particular antibiotic if the growth, as measured through OD values, was inhibited by less than 15% compared to that of the positive control or were recorded as susceptible when the growth was reduced by at least half [27,33,36].

The six antibiotic disks (Oxoid, UK) included ampicillin (AMP, 10 μg), chloromycetin (C, 30 μg), levofloxacin (LEV, 5 μg), tetracycline (TE, 30 μg), gentamycin (CN, 10 μg) and sulfamethoxazole-trimethoprim (SXT, 25 μg/1.25 μg). A 0.5 McFarland turbidity level of the *E. coli* isolate suspension was obtained using sterile sodium chloride solution after the isolates were incubated on MH agar plates at 37°C. The suspension was evenly inoculated onto an MH agar plate using autoclaved gauze. Three different disks were placed onto each inoculated plate. The disks were far away from each other and from the plate rim to avoid overlapping inhibition zones. The inhibition zone diameter was precisely measured using a ruler after 16–18 h of incubation at 37°C. The diameter was compared to the diameter of the susceptible, moderate and resistant standards as listed in CLSI 2009. The antibiotic resistance frequency was calculated as the ratio of *E. coli* isolates that were resistant to antibiotics to the total number that were isolated from each site. The multi-antibiotic resistance (MAR) index was estimated by the equation $a/(b \times c)$, where a is the total antibiotic resistance score of all of the isolates from the sample, b is the number of tested drugs, and c is the number of isolates from the sample [26]. The significance of difference in average *E. coli* resistance frequencies and MAR among the three rivers was examined through a non-parametric k independent samples Kruskal-Wallis test using SPSS 17.0. The correlations between *E. coli* counts and resistance frequencies or MAR as well as among the percent resistances to the eight drugs in the seven sites were examined via a Spearman correlation analysis using SPSS 17.0. The two sites, including the control and C1, were removed during the correlation analysis because of low antibiotic levels and few *E. coli* isolates, respectively.

Results and Discussion

Occurrences and levels of antibiotics in the Jiyun River

The detection frequencies and concentration profiles of the 12 monitored antibiotics are summarized in Table 1. The selected VAs were extensively detected in the investigated water system at percentages ranging from 58% to 100%. The concentrations varied largely among these sites from less than 1 ng L^{-1} to 450 ng L^{-1}. Based on our determination, the SAs (SM, SMX and SDZ) and QLs (OFC, EFC and CFC) might be the predominant antibiotics in the water, as they were present in all of these investigated sites at an average concentration range of 3.79 to 92.97 ng L^{-1}. In comparison, the TCs were generally one or two orders of magnitude lower (mean 2.17–16.12 ng L^{-1}) than were the SAs (except for SCX) in concentration.

Figure 2 depicts the antibiotic distributions in the Jiyun River. The three rivers displayed varying antibiotic abundances with approximately 4- and 2-fold greater average antibiotic levels at the 3 sites in the Cuo River and Jinji River tributaries than that in the main stream Ju River. The total antibiotic contents at mid- and downstream (JJ2 and JJ3, and C2 and C3) in the two tributaries increased compared to the upstream site (JJ1 and C1). Our results contrast with those of other studies investigating antibiotic variation in point-source-affected water bodies, in which decreased antibiotic concentrations were detected in the water over distances from the site receiving a point-source effluent due to natural attenuation by adsorption, dilution, photolysis, hydrolysis and biodegradation [37,38].

The monitored sites, with the exception of Haizi reservoir, J1, JC and JCJJ, were immediately close to swine production farms, from which the sewage was directly discharged into receiving rivers, and the animal feces were used as a fertilizer in the nearby farmlands. Residual antibiotics in manure can concentrate, migrate in agricultural soil and finally end up in the aquatic environment through runoff when the feces are applied as fertilizer [20,21]. Therefore, the potential continuous input of antibiotics from the discharge of animal farms as well as from the runoff of fertilized farmland was inferred to account for the accumulated antibiotic contamination along the river. In addition, high amounts of antibiotics were also recorded at the two sites where the Ju River intersected with the two tributaries: JC (Ju River/Cuo River) and JCJJ (Ju River/Cuo River/Jinji River) (Figure 2). The antibiotic concentrations at the three sites J1, J2 and J3 were almost the lowest of the surveyed sites in the three rivers, except for JJ1. One possible reason is due to the dilution and rapid transfer of antibiotics in large water flow of the main stream. However, the specific reason requires further study. High concentrations of antibiotics at the two intersection sites likely indicate potential compound transfer with flow from the tributaries into the main stream.

As seen in Figure 2, most of the surveyed sites in the Jiyun River were generally determined to have higher levels of SAs (SM, SMM, SMX and SDZ) and QLs (OFC and CFC) than TCs. These findings agree with the results of many other studies [4,39,40] and may be chiefly explained by the discrepant partition characteristics among different antibiotic classes. The TCs have strong combination abilities with soil/sediment and their mineral or organic components via cation bridging and/or cation exchange [41,42], which causes their retention in the soil or dispersion to the sediment after being discharged into rivers. This result is supported by detection of TCs in the sediment of this river (Table S3). In contrast, most SA compounds predominantly exist in anionic species with negative charges at environmental pH values of >7 [43,44] and are consequently less frequently adsorbed onto solid-phase material. The pH values of these water samples were in the range of 7.29 to 9.24. This range facilitates the dissolution and migration of SAs. Thus, it is reasonable that the SAs were primarily observed in the river water.

A parallel survey of the antibiotic residues in six fecal samples of the animal operations that were directly adjacent to the water sample sites was conducted to explore the association of surface water antibiotic contamination with livestock production. The 12 target antibiotics were widely detected in these feces at levels of 1.03–56200 ng g^{-1}. The water and fecal samples from the study area were generally coincident in the antibiotic composition of the two classes of VAs: SAs and QLs (Figure 3). Both of the matrixes were dominated by SM and SMM, and SCX was seldom detected. In addition, these matrixes presented the same order of QLs: OFC>CFC>EFC. In contrast, although large amounts of TCs were detected in these animal feces, they were found at relatively low levels in the water. This finding may be explained by the strong adsorption of TCs in the environment as discussed above and is also supported by the detection of TCs in the river sediment (Table S3).

The detection concentrations of VAs in the current study are lower than or comparable to the levels in other surface waters that are proximate to animal operations in Jiangsu, China (560–2420 ng L^{-1}) [45] and in the US (1000–1500 ng L^{-1}) [46]. These reported results and the concentrations in this study were significantly higher than in rivers receiving effluent from WWTPs, such as the Victoria River, HongKong, China (<20 ng L^{-1}); the Elbe River, Germany (30–70 ng L^{-1}); and the Ebro River, Spain

Table 1. Frequencies and concentrations of the 12 target antibiotics in the Jiyun River (n = 12).

Class	Compound	Frequency(%) (%)	Range (Mean)(ng L^{-1}) (ng/L)	MDLs((ng L^{-1}))
Tetracyclines (TCs)	Tetraycline (TC)	83.33	n.d-11.00 (2.17)	2.14
	Oxytetracycline (OTC)	91.67	n.d-100.00 (16.12)	2.35
	Chlortetracycline (CTC)	83.33	n.d-40.60 (12.92)	2.87
	Doxycycline (DOC)	58.33	n.d-11.75 (2.84)	2.45
Quinolones (QLs)	Ciprofloxacin (CFC)	100.00	3.56–24.80 (11.61)	2.15
	Enrofloxacin (EFC)	100.00	0.55–13.41 (3.79)	0.25
	Ofloxacin (OFC)	100.00	1.34–102.00 (27.89)	1.10
Sulfonamides (SAs)	Sulfadiazine (SDZ)	100.00	0.03–385.70 (62.45)	0.01
	Sulfamethoxazole (SMX)	100.00	4.29–230.00 (54.65)	1.15
	Sulfamonomethoxin (SMM)	75.00	n.d-450.00 (147.64)	1.10
	Sulfameter (SM)	100.00	0.51–387.00 (92.97)	0.16
	Sulfachinoxalin (SCX)	91.67	n.d-13.95 (2.90)	0.57

n.d: non-detected.
MDLs: method detection limitations for the 12 compounds.

(0.2–35.6 ng L^{-1}) [47–49]. These results indicate that the influences of the rural aquatic ecosystem caused by livestock production are much greater than are those in urban areas caused by human populations.

Antibiotic-resistance for *E. coli* clones that were isolated from the Jiyun River

The detected *E. coli* counts in the Jiyun River were shown in Table S2. A total of 218 *E. coli* clones were isolated from the Jiyun River that not only exhibited a high percentage of resistance to the drugs that are frequently detected in these rivers but also to those that were not monitored but are of clinical concern. In total, 88% of the 218 isolates exhibited resistance to one or more of the tested drugs. The most frequent resistance appeared for CTC (61.01%), followed by AMP and TE (approximately 50%), CN (45.91%), SXT and SMM (39.09% and 36.24%), LEV (30.45%) and C (16.82%) (Figure 4). High levels of *E. coli* resistance to CTC, TE and AMP have been found in other aquatic environments [27,33] and in WWTP [50]. The great resistance frequency against AMP in the environment was possibly because of its comparatively older utilization history (over years) [51]. This resistance also reflects the common use of AMP in agricultural activities [52]. The high resistance rate to CTC maybe is due to the potential long-term exposure of *E. coli* to CTC in these aquatic environments. The frequent TE resistance despite its low concentration in water column may be caused by the possibility that the majority of TE-resistant *E. coli* have developed resistance against TE before they enter rivers because high levels of TCs were identified in the surveyed animal feces. In addition, the tet genes (e.g., tet(A), tet(C) and tet(G)) readily spreading among gram-negative bacteria such as *E. coli* through transposons and smaller plasmids might also facilitate the abundance of TE resistant *E. coli* [39]. The relatively lower resistance frequency against C agrees with the observations of an earlier previous study [53].

Approximately 72% of these bacteria exhibited multiple antibiotic resistances, and, in total, 48 resistance profiles were determined (Figure 5). The five primary profiles were AMP, CN, CTC, TE/CTC and TE/SMM/CTC. The total rate of resistance to the five profiles for *E. coli* that were isolated from the Jiyun River was approximately 24%. It is worth noting that 3.67% of the isolates from the 9 sites exhibited eight-drug resistance. Ongoing multiple antibiotic exposure must have occurred in the study area as it has been repeatedly demonstrated that bacterial antibiotic resistance diversity is closely related to antibiotic contamination frequency [50].

Table 2 lists the relationships of the percent resistances among the eight drugs. Interestingly, significant correlations were obtained among the three drugs SXT, C, and AMP ($p<0.01$) and between TE and CTC ($p<0.01$). Positive but statistically non-significant relationships were also identified for CN and LEV with SXT, LEV and C, LEV and AMP, CN and AMP, and SMM and SXT ($p>0.05$, $r>0.5$). Strongly positive correlations were also reported in other drugs, such as nalidixic acid MICs with cefotaxime and ciprofloxacin MICs [27] and vancomycin with daptomycin resistance in *Staphylococcus aureus* (*S. aureus*) [51]. These correlations indicate that the *E. coli* clones that were tested in this study probably developed co-selection by containing one drug resistance gene on the plasmids that were resistant to other antimicrobials [54]. These correlations also indicate possible similar resistance mechanisms for the different drugs, as reported for *S. aureus,* showing strongly positive correlation between vancomycin and daptomycin susceptibility likely due to the common physical barrier of a thickened cell [51].

Table 3 presents the antibiotic resistance rate and profiles of *E. coli* from the sites of the Jiyun River. Although comparatively less *E. coli* were detected in the two sites of Jinji River (Table S2), the isolated cultures exhibited high resistance rates and MAR (Table 3). Spearman correlation analysis revealed no statistical correlations between *E. coli* MPN and resistance rate ($r=0.07$, $p=0.85$) and between *E. coli* MPN and MAR ($r=-0.42$, $p=0.27$). It suggests the prevalence of antibiotic resistance for *E. coli* in the Jiyun River regardless of the *E. coli* contamination extent. The *E. coli* isolates from each of the three rivers displayed generally consistent increases in both the resistance rates and MAR indexes from up- to midstream. Statistically no-significant difference was found among the three rivers in resistance rates and resistance diversity (MAR indexes) based on a non-parametric k independent sample Kruskal-Wallis test ($p=0.50$ for resistance rate and $p=0.10$ for MAR). The *E. coli* isolates from different water systems showed varying drug numbers of resistance. The Ju

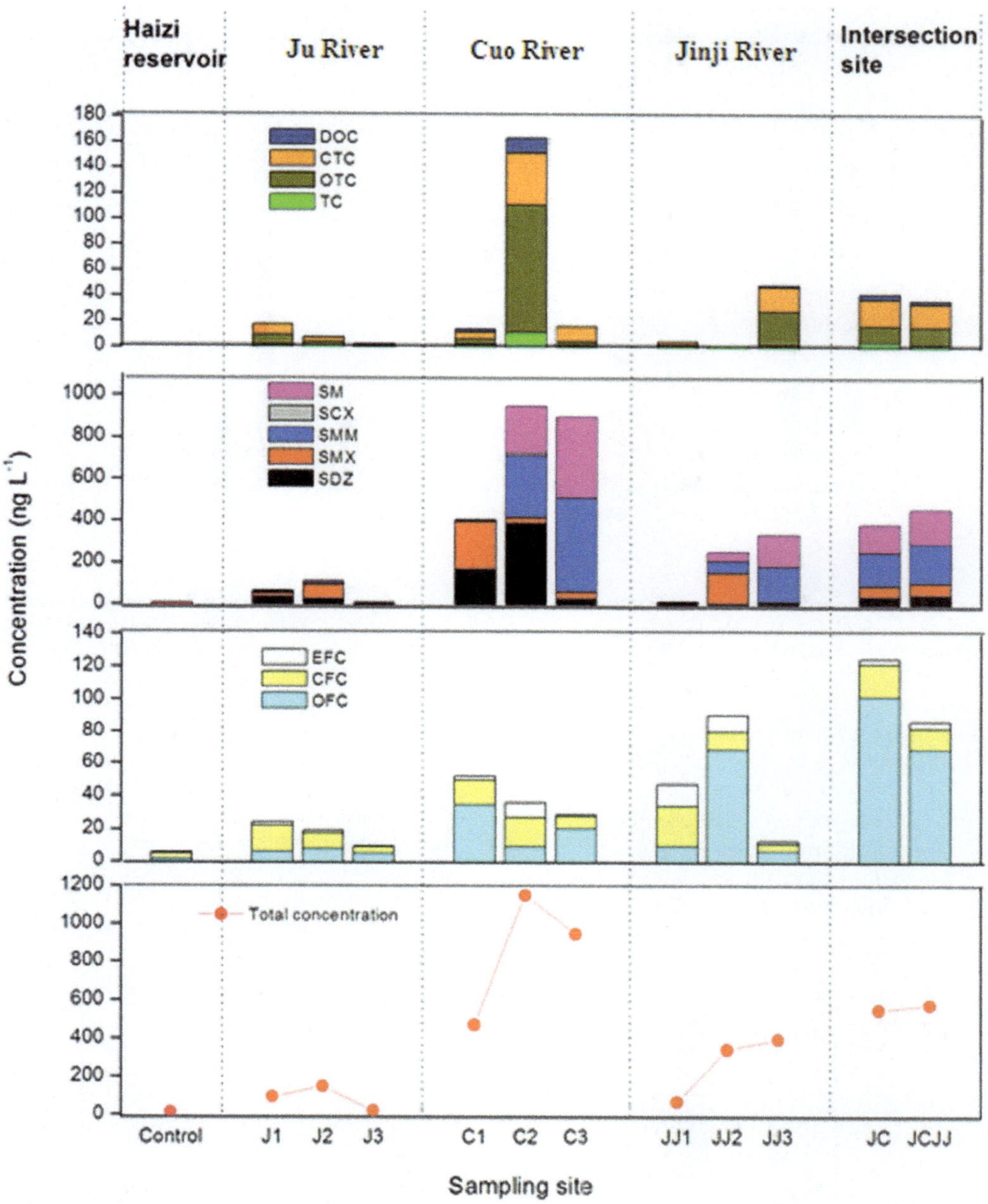

Figure 2. Distributions of the three classes of antibiotics in the 12 sampling sites. Control, Haizi reservoir; J1, J2 and J3, up-, mid- and downstream of Ju River; C1 C2, and C3, up-, mid- and downstream of Cuo River; JJ1, JJ2, and JJ3, up-, mid- and downstream of Jinji River; JC and JCJJ, intersection sites of Ju River with Cuo River and Jinji River, respectively. TC, tetracycline; OTC, oxytetracycline; CTC, chlortetracycline; DOC, doxycycline; CFC, ciprofloxacin; EFC, enrofloxacin; OFC, ofloxacin; SDZ, sulfadiazine; SMX, sulfamethoxazole; SMM, sulfamonomethoxine; SM, sulfameter; SCX, sulfachinoxaline.

River presented all of the resistance profiles from single- to eight-drug, dominated by one- to five-drug resistance. The Cuo River primarily displayed one-, two- and three-drug resistance. In contrast, the *E. coli* from Jinji River was more chiefly resistant to more than five drugs. In particular, 15% and 22% of *E. coli* isolates in JJ1 and JJ2, respectively, exhibited eight-drug resistance. Similar to the sites prior to intersection, the resistance profiles of *E. coli* at the two intersection sites also covered nearly all of the patterns from one- to seven-drug resistance.

The extent of the antibiotic-resistance of *E. coli* in the Jiyun River maybe is the highest currently reported and was significantly higher than the reports for other agricultural watersheds of British Columbia (Resistance rate: 20–50%) [27], the Beijing Wenyu River (Average rate: 48%, MAR: 0.11–0.14) [26], and the Dongjiang in Guangzhou (MAR: 0.17–0.50) [55]. Furthermore, the extent of resistance in all of these agriculture areas or both agriculture and human-affected areas was greater than the urban surface water in Japan (resistance rate: 37%) and the US (resistance rate: 5–38%) [31,56]. The generally greater average resistance rate and MAR in surface water of the study agricultural region relative to reported results is probably related to generally higher levels of antibiotic contamination in such areas. In addition to the cause of frequent and long-term exposure to various different levels of antibiotics, special water conditions, such as low

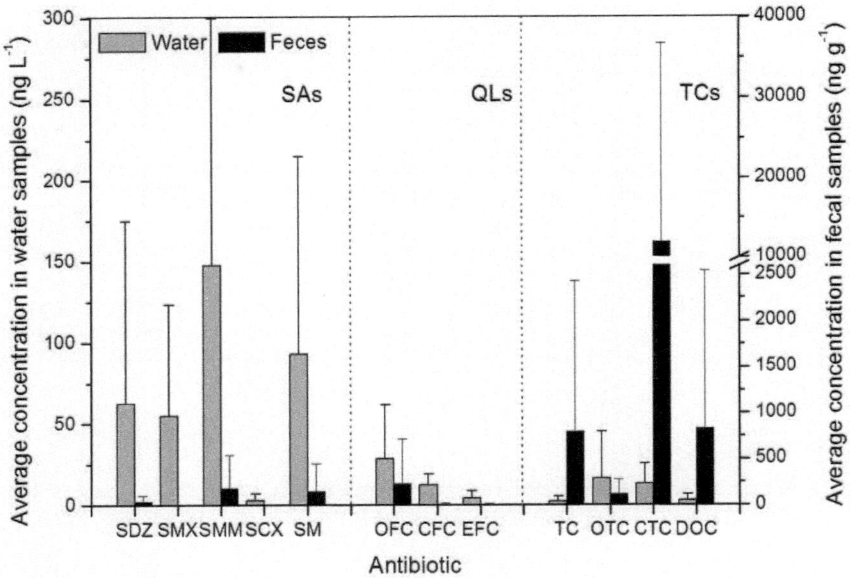

Figure 3. The means and standard deviations (SD) for antibiotics in water samples (n = 12) and in animal fecal samples (n = 6). TCs, tetracyclines; QLs, quinolones; SAs, sulfonamides; TC, tetracycline; OTC, oxytetracycline; CTC, chlortetracycline; DOC, doxycycline; CFC, ciprofloxacin; EFC, enrofloxacin; OFC, ofloxacin; SDZ, sulfadiazine; SMX, sulfamethoxazole; SMM, sulfamonomethoxine; SM, sulfameter; SCX, sulfachinoxaline.

flow and high nutrient concentrations, might also contribute to the high antibiotic resistance in the Jiyun Rivers [39,57,58].

Relationships between the resistance frequencies of *E. coli* and the corresponding antibiotics

The correlations between the resistance rate and the detection concentration for the five drugs SMM, CTC, TE, LEV and SXT were separately examined. In general, no clear relationship was observed for any of these drugs (Figure S1). The relationship between the extent of antibiotic resistance or ARG abundance with the antibiotic contents has also been examined in other studies, and contradictory observations have been reported. The abundances of resistance genes are positively related to the antibiotic concentrations in the Haihe and Huangpu Rivers in China [2,39]. However, no clear correlation was noted between

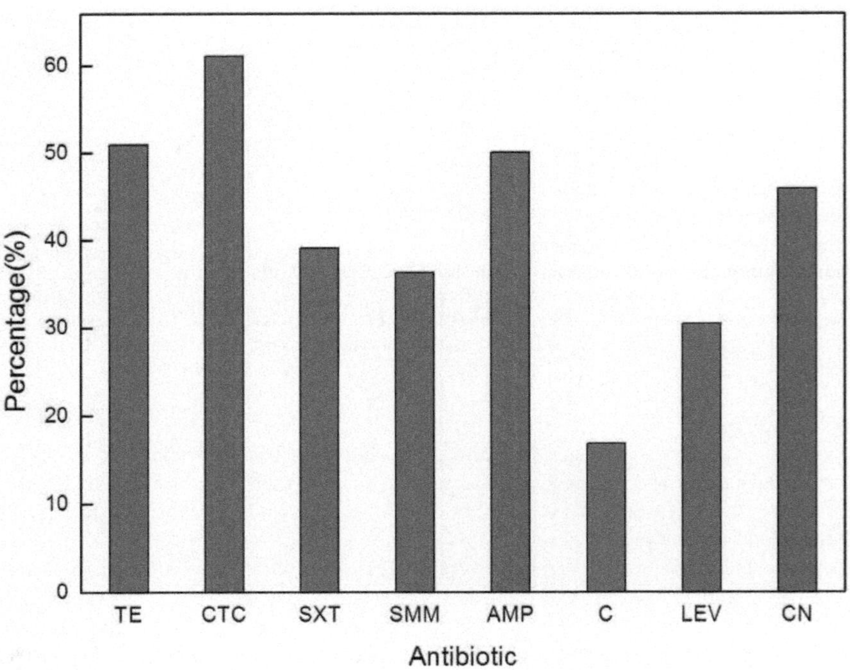

Figure 4. Resistance percentage to the eight tested drugs for all the *E. coli* isolates that were isolated from the Jiyun River (n = 218). TE, tetracycline; CTC, chlortetracycline; SXT, sulfamethoxazole-trimethoprim; SMM, sulfamonomethoxine; AMP, ampicillin; C, chloromycetin; LEV, levofloxacin; CN, gentamycin.

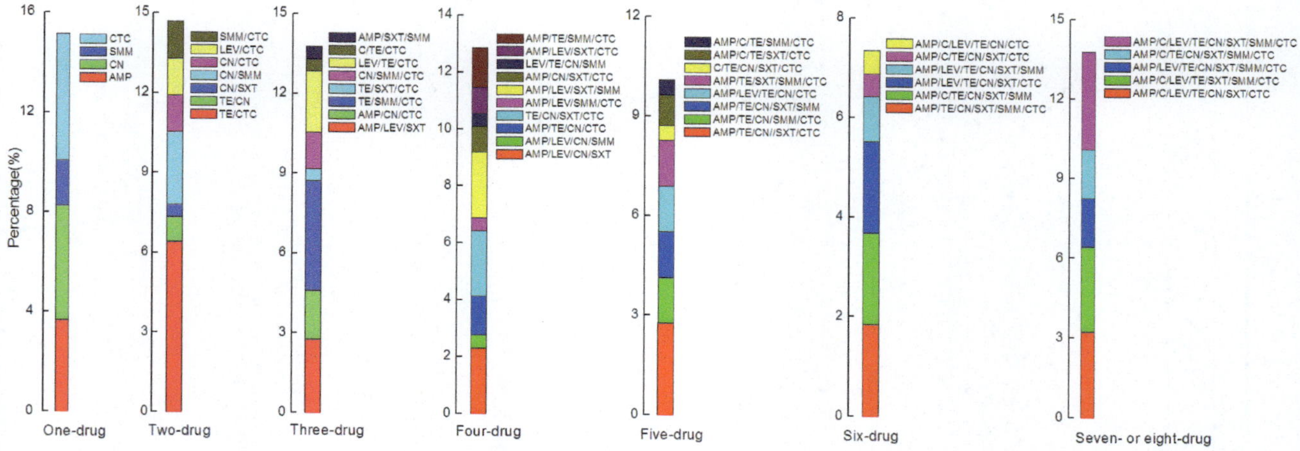

Figure 5. Percentage of *E. coli* number in each antibiotic-resistance pattern accounting for the total *E. coli* isolates from the Jiyun River. TE, tetracycline; CTC, chlortetracycline; SXT, sulfamethoxazole-trimethoprim; SMM, sulfamonomethoxine; AMP, ampicillin; C, chloromycetin; LEV, levofloxacin; CN, gentamycin.

fluoroquinolones and the corresponding resistant *E. coli* in the rivers of Osaka, Japan [3], in agreement with the findings in the present study. The lack of correlation between the antibiotic concentrations and the extent of resistance may be caused by several confounding effects. Antibiotic-resistant bacteria are linked not only to the exposure to corresponding antibiotics but also to other factors, such as the co-selection of toxic heavy metals [59] and the cross-resistance among different types of antibiotics [54], as well as certain water quality variables, such as nutrients, temperature, DO and salinity [27]. In addition to the direct formation of resistance for *E. coli* when exposed to the selective pressure of antibiotics and other factors, *E. coli* can also indirectly obtain resistance through the horizontal ARG transfer among microbial populations [3,26]. Antibiotics and resistant *E. coli* have different fates and persistence in the environment. For example, antibiotic resistance will remain for at least one month after the disappearance of antibiotics [60]. Considering the complexity of the antibiotic-resistance-formation mechanisms and the influences of various water quality parameters in aquatic environments, more in-depth and systematic studies should be conducted in the future. Regardless, it is clear that a large amount of VAs have been abused in animal production farms and reside in the surrounding aquatic environment, which is likely a primary driver for the emergence of resistant strains in the environment.

The observations in the present study are a cause for concern in terms of the following aspects. First, the Jiyun River is upstream of the Haihe River, which is the largest water system in north China and the main river flowing into Bohai Bay. Thus, high levels of antibiotics and resistant *E. coli* in the Jiyun River serve as potential contamination sources of the Bohai Bay. Second, in most instances, the surface water in rural areas is directly used for agricultural irrigation, which will likely result in the uptake and translocation of antibiotics by plants. After being irrigated with carbamazepine-containing wastewater for 60 and 110 days, carbamazepine was detected in soybean roots and beans [61]. Finally, given that more than 90% of the antibiotic-resistant *E. coli* in aquatic environments is conveyed by corresponding ARGs [26], this resistant *E. coli* might be the potential reservoir of various ARGs, which is particularly worth noting.

Antibiotics and antibiotic-resistant *E. coli* in the potential drinking water source

Only 12.15 ng L^{-1} antibiotics were detected at the site of the Haizi reservoir, which is the potential drinking water source of the Beijing population. However, 38 *E. coli* isolates were also retrieved from this site that displayed multiple antibiotic resistances (Table 3). The presence of *E. coli* and resistance in the well-protected Haizi reservoir may be caused by input from the upper stretch in Hebei province. Wildlife is also a suspected source because wildlife has been reported as the primary *E. coli* contributor in headwaters and can acquire antibiotic resistance through crossover with livestock animals [62–65]. In addition, stationary water flow in such reservoirs is also favorable for resistant bacterial growth and ARG transfer [57,58]. However, the specific reason is still unclear and requires further examination. The occurrence of antibiotic-resistant microorganisms and ARGs in drinking water sources or headwaters has been reported in other regions, such as the Huangpu River, China [66] and the Elk Creek Watershed, British Columbia [27]. The risk from *E. coli* could be completely eliminated if used for human consumption because the water is required to be treated by the Beijing Water Group Company to reach drinking standards. These results underscore the urgent need to explore the source and risks of resistant bacteria and ARG contamination in drinking water sources.

Conclusions

This study demonstrated the extensive use and partial abuse of VAs in livestock farming in the study area, which probably is responsible for the prevalence of VAs, particularly hydrophilic VA compounds, in the surrounding surface water. A total of 88% of *E. coli* isolates from these rivers were resistant to one or more antimicrobial agents. Significant correlations were found among the resistance rate of SXT, C and AMP as well as between TE and CTC, suggesting a possible cross-selection for resistance among these drugs. An increasing tendency was observed for total antibiotic contents along the two tributaries, which was most likely attributed to the potential continuous input from animal operations discharge and field runoff. Also, the antibiotics resistance frequency for *E. coli* from midstream is greater than those from upstream for the three rivers. *E. coli* isolates from

Table 2. Spearman correlation matrix of the percent resistances to the eight drugs in seven of the nine sampling sites.

	AMP	C	LEV	TE	CN	SXT	SMM	CTC
AMP	1.000	–	–	–	–	–		
C	**0.898****	1.000	–	–	–	–		
LEV	0.503	0.524	1.000	–	–	–		
TE	0.048	0.286	0.286	1.000	–	–		
CN	0.587	0.452	−0.048	0.333	1.000	–		
SXT	**0.934****	**0.905****	0.548	0.214	0.595	1.000		
SMM	0.311	0.405	0.000	0.071	0.262	0.524	1.000	
CTC	0.036	0.286	0.143	**0.952****	0.357	0.119	−0.071	1.000

The sites of Haizi reservoir and Cuo River upstream have been removed based on the low antibiotic levels and few *E. coli* isolates, respectively. AMP = ampicillin, C = chloromycetin, LEV = levofloxacin, TE = tetracycline, CN = gentamycin, SXT = sulfamethoxazole-trimethoprim, SMM = sulfamonomethoxine, CTC = chlortetracycline). ** indicate the significance level of $p<0.01$.

Table 3. Overall antibiotic-resistance frequency (%), MAR index and one-, two-, three-, four-, five-, six-drug, seven-drug and eight-drug resistance frequency (%) for *E. coli* isolated from each of the nine sampling sites.

Parameter	Control (n = 38)	J1 (n = 25)	J2 (n = 40)	C1 (n = 2)	C2 (n = 24)	JJ1 (n = 26)	JJ2 (n = 9)	JC (n = 29)	JCJJ (n = 25)
Frequency (%)	82	88	98	50	88	77	100	93	84
MAR index	0.47	0.37	0.40	0.13	0.31	0.46	0.79	0.45	0.29
One-drug (%)	3	23	26	0	10	20	0	7	43
Two-drug (%)	3	27	15	100	29	15	0	22	14
Three-drug (%)	23	5	13	0	52	0	0	19	5
Four-drug (%)	29	18	15	0	5	0	11	11	19
Five-drug (%)	10	9	18	0	0	5	22	19	10
Six-drug (%)	6	5	5	0	0	30	11	15	0
Seven-drug (%)	23	9	8	0	0	15	33	7	10
Eight-drug (%)	3	5	0	0	5	15	22	0	0

Control, Haizi reservoir; J1 and J2, up- and midstream of Ju River; C1 and C2, up- and midstream of Cuo River; JJ1 and JJ2, up- and midstream of Jinji River; JC and JCJJ, intersection sites of Ju River with Cuo River and Jinji River, respectively.

different water systems showed varying drug numbers of resistance. No obvious correlation was found between the antibiotic resistant rate of *E. coli* and the corresponding antibiotic concentrations, indicating that the resistance formation process must be affected by aquatic factors besides antibiotics. These results provide baseline data on the antibiotics and antibiotic-resistant bacteria contamination that are associated with widespread livestock production. For improved contamination control and environmental protection in rural areas, there is an urgent need to develop management protocols in the animal industry. In addition, the occurrence of resistance genes and potential threats to agricultural ecosystem safety should also be noted, as water is the main interface of antibiotics, resistant bacteria and ARG transfer in the ecosystem.

Supporting Information

Figure S1 Relationships between CTC resistance rate versus CTC concentration (a), TE resistance rate versus TC concentration (b), SMM resistance rate versus SMM concentration (c), SXT resistance rate versus SMX concentration (d) and LEV resistance rate versus OFC concentration (e) in seven of the nine sampling sites. The sites of Haizi reservoir and Cuo River upstream have been removed based on the low antibiotic levels and few *E. coli* isolates, respectively. TE, tetracycline; CTC, chlortetracycline; SXT, sulfamethoxazole-trimethoprim; SMM, sulfamonomethoxine; LEV, levofloxacin; TC, tetracycline; CTC, chlortetracycline; OFC, ofloxacin; SMX, sulfamethoxazole; SMM, sulfamonomethoxine.

Table S1 GPS coordinates (deg./min./sec.) of the 12 water sample sites in study region (Control, Haizi reservoir; J1, J2 and J3, up-, mid- and downstream of Ju River; C1 C2, and C3, up-, mid- and downstream of Cuo River; JJ1, JJ2, and JJ3, up-, mid- and downstream of Jinji River; JC and JCJJ, intersection sites of Ju River with Cuo River and Jinji River, respectively).

Table S2 Most possible number (MPN) of the *E. coli* in the Jiyun River (MPN/100 ml) (Control, Haizi reservoir; J1, and J2, up- and midstream of Ju River; C1 and C2, up- and midstream of Cuo River; JJ1 and JJ2, up- and midstream of Jinji River; JC and JCJJ, intersection sites of Ju River with Cuo River and Jinji River, respectively.).

Table S3 Detection frequencies, ranges and means of the 12 target antibiotics in sediment of the Jiyun River.

Author Contributions

Conceived and designed the experiments: XZ. Analyzed the data: XZ. Wrote the paper: XZ. Collected and analyzed samples: XZ BL JW MG LW. Directed the experiment design, implementation, and manuscript writing: YL CF.

References

1. Kolpin DW, Furlong ET, Meyer MT, Thurman EM, Zaugg SD, et al. (2002) Pharmaceuticals, hormones, and other organic wastewater contaminants in US streams, 1999–2000: A national reconnaissance. Environ Sci Technol 36: 1202–1211.
2. Luo Y, Mao DQ, Rysz M, Zhou QX, Zhang HJ, et al. (2010) Trends in antibiotic resistance genes occurrence in the Haihe River, China. Environ Sci Technol 44: 7220–7225.
3. Adachi F, Yamamoto A, Takakura KI, Kawahara R. (2013) Occurrence of fluoroquinolones and fluoroquinolone-resistance genes in the aquatic environment. Sci Total Environ 444: 508–514.
4. Yan CX, Yang Y, Zhou JL, Liu M, Nie MH, et al. (2013) Antibiotics in the surface water of the Yangtze Estuary: Occurrence, distribution and risk assessment. Environ Pollut 175: 22–29.
5. Levy SB. (1998) The challenge of antibiotic resistance. Sci. Am 278: 46–53.
6. Sarmah AK, Meyer MT, Boxall ABA. (2006) A global perspective on the use, sales, exposure pathways, ocurrence, fate and effects of veterinary antibiotics (VAs) in the environment. Chemosphere 65: 725–759.
7. Li YX, Zhang XL, Li W, Lu XF, Liu B, et al. (2013) The residues and environmental risks of multiple veterinary antibiotics in animal faeces. Environ Monit Assess 185: 2211–2220.
8. Cogliani C, Goossens H, Greko C. (2011) Restricting antimicrobial use in food animals: lessons from Europe. Microbe 6: 274.
9. DANMAP. Danmap 2011-use of antimicrobial agents and occurrence of antimicrobial resistance in bacteria from food animals, food and humans in Denmark. Available: http://danmap.org/Downloads/~/media/Projekt%20sites/Danmap/DANMAP%20reports/Danmap _2011.ashx/. Accessed 2013 July 26.
10. Hong PY, Al-Jassim N, Ansari MI, Mackie RI. (2013) Environmental and public health implications of water reuse: Antibiotics, antibiotic resistant bacteria, and antibiotic resistance genes. Antibiotics 2: 367–399.
11. Kumar KC, Gupta S, Chander Y, Singh AK. (2005) Antibiotic use in agriculture and its impact on the terrestrial environment. Adv Agron 87: 1–54.
12. Gutiérrez IR, Watanabe N, Harter T. (2010) Effect of sulfonamide antibiotics on microbial diversity and activity in a Californian Mollic Haploxeralf. J Soil Sediment 10: 537–544.
13. Martínez-Carballo E, González-Barreiro C, Scharf S, Gans O. (2007) Environmental monitoring study of selected veterinary antibiotics in animal manure and soils in Austria. Environ Pollut 148: 570–579.
14. Zhao L, Dong YH, Wang H. (2010) Residues of veterinary antibiotics in manure from feedlot livestocks in eight provinces of China. Sci Total Environ 408: 1069–1075.
15. Motoyama M, Nakagawa S, Tanoue R, Sato Y, Nomiyama K, et al. (2011) Residues of pharmaceutical products in recycled organic manure produced from sewage sludge and solid waste from livestock and relationship to their fermentation level. Chemosphere 84: 432–438.
16. Pan X, Qiang ZM, Ben WW, Chen MX. (2011) Residual veterinary antibiotics in swine manure from concentrated animal feeding operations in Shandong Province, China. Chemosphere 84: 695–700.
17. Dolliver H, Gupta S, Noll S. (2008) Antibiotic degradation during manure composting. J Environ Qual 37: 1245–1253.
18. Bao Y, Zhou Q, Guan L, Wang Y. (2009) Depletion of chlortetracycline during composting of aged and spiked manures. Waste Manage 29: 1416–1423.
19. Ali M, Wang JJ, DeLaune RD, Seo DC, Dodla SK, et al. (2013) Effect of redox potential and pH status on degradation and adsorption behavior of tylosin in dairy lagoon sediment suspension. Chemosphere 91: 1583–1589.
20. Kay P, Blackwell PA, Boxall A. (2005) Transport of veterinary antibiotics in overland flow following the application of slurry to arable land. Chemosphere 59: 951–959.
21. Sun P, Barmaz D, Cabrera ML, Pavlostathis SG, Huang CH. (2013) Detection and quantification of ionophore antibiotics in runoff, soil and poultry litter. J Chromatogr A 1312: 10–17.
22. Pezzotti G, Serafin A, Luzzi I, Mioni R, Milan M, et al. (2003) Occurrence and resistance to antibiotics of *Campylobacter jejuni and Campylobacter coli* in animals and meat in northeastern Italy. Int J Food Microbiol 82: 281–287.
23. Heuer H, Schmitt H, Smalla K. (2011) Antibiotic resistance gene spread due to manure application on agricultural fields. Curr opin microbiol 14: 236–243.
24. Marti R, Scott A, Tien YC, Murray R, Sabourin L, et al. (2013) Impact of manure fertilization on the abundance of antibiotic-resistant bacteria and frequency of detection of antibiotic resistance genes in soil and on vegetables at Harvest. Appl Environ Microbiol 79: 5701–5709.
25. Gullberg E, Cao S, Berg OG, Ilbäck C, Sandegren L, et al. (2011) Selection of resistant bacteria at very low antibiotic concentrations. PLOS pathogens 7: e1002158.
26. Hu JY, Shi JC, Chang H, Li D, Yang M, et al. (2008) Phenotyping and genotyping of antibiotic-resistant *Escherichia coli* isolated from a natural river basin. Environ Sci Technol 42: 3415–3420.
27. Maal-Eared R, Bartlett KH, Bowie WR, Hall ER. (2013) Phenotypic antibiotic resistance of Escherichia coli and *E. coli* O157 isolated from water, sediment and biofilms in an agricultural watershed in British Columbia. Sci Total Environ 443: 315–323.
28. China Agricultural Statistic Year Book. 2012.
29. Batt AL, Kim S, Aga DS. (2007) Comparison of the occurrence of antibiotics in four full-scale wastewater treatment plants with varying designs and operations. Chemosphere 68: 428–435.
30. Zhou LJ, Ying GG, Zhao JL, Yang JF, Wang L et al. (2011) Trends in the occurrence of human and veterinary antibiotics in the sediments of the Yellow River, Hai River and Liao River in northern China. Environ Pollut 159: 1877–1885.

31. Akiyama T, Savin MC. (2010) Populations of antibiotic-resistant coliform bacteria change rapidly in a wastewater effluent dominated stream. Sci Total Environ 408: 6192–6201.
32. CLSI. (2009) Performance Standards for Antimicrobial Susceptibility Testing; Sixteenth Informational Supplement, M100-S16. Clinical and Laboratory Standards Institute (National Committee for Clinical Laboratory Standards), Wayne, PA.
33. Parveen S, Murphree RL, Edmiston L, Kaspar CW, Portier KM. (1997) Association of multiple-antibiotic-resistance profiles with point and nonpoint sources of *Escherichia coli* in Apalachicola Bay. Appl. Environ. Microbiol 7: 2607–2612.
34. Whitlock JE, Jonesb DT, Harwooda VJ. (2002) Identification of the sources of fecal coliforms in an urban watershed using antibiotic resistance analysis. Water Res 36: 4273–4282.
35. CLSI. (2012) Performance Standards for Antimicrobial Susceptibility Testing; Sixteenth Informational Supplement, M100- S16. Clinical and Laboratory Standards Institute (National Committee for Clinical Laboratory Standards), Wayne, PA.
36. Walczak JJ, Xu SP. (2011) Manure as a Source of Antibiotic-Resistant *Escherichia coli* and *Enterococci*: a Case Study of a Wisconsin, USA Family Dairy Farm. Water Air Soil Pollut 219: 579–589.
37. Tamtam F, Mercier F, Le Bot B, Eurin J, Tuc Dinh Q, et al. (2008) Occurrence and fate of antibiotics in the Seine River in various hydrological conditions. Sci Total Environ 393: 84–95.
38. Luo Y, Xu L, Rysz M, Wang YQ, Zhang H, et al. (2011) Occurrence and Transport of Tetracycline, Sulfonamide, Quinolone, and Macrolide Antibiotics in the Haihe River Basin, China. Environ Sci Technol 45: 1827–1833.
39. Jiang L, Hu XL, Yin DQ, Zhang HC, Yu ZY. (2011) Occurrence, distribution and seasonal variation of antibiotics in the Huangpu River, Shanghai, China. Chemosphere 82: 822–828.
40. Zou SC, Xu WH, Zhang RJ, Tang JH, Chen YJ, et al. (2011) Occurrence and distribution of antibiotics in coastal water of the Bohai Bay, China: Impacts of river discharge and aquaculture activities. Environ Pollut 159: 2913–2920.
41. Pils JRV, Laird DA. (2007) Sorption of tetracycline and chlortetracycline on K- and Ca-saturated soil clays, humic substances, and clay-humic complexes. Environ Sci Technol 41: 1928–1933.
42. Xu XR, Li XY. (2010) Sorption and desorption of antibiotic tetracycline on marine sediments. Chemosphere 78: 430–436.
43. Gao J, Pedersen JA. (2005) Adsorption of sulfonamide antimicrobial agents to clay minerals. Environ Sci Technol 39: 9509–9516.
44. Kahle M, Stamm C. (2007) Time and pH-dependent sorption of the veterinary antimicrobial sulfathiazole to clay minerals and ferrihydrite. Chemosphere 68: 1224–1231.
45. Wei RC, Ge F, Huang SY, Chen M, Wang R. (2011) Occurrence of veterinary antibiotics in animal wastewater and surface water around farms in Jiangsu Province, China. Chemosphere 82: 1408–1414.
46. Campagnolo ER, Johnson KR, Karpati A, Rubin CS, Kolpin DW et al. (2002) Antimicrobial residues in animal waste and water resources proximal to large-scale swine and poultry feeding operations. Sci Total Environ 299: 89–95.
47. Wiegel S, Aulinger A, Brockmeyer R, Harms H, Löffler J, et al. (2004) Pharmaceuticals in the river Elbe and its tributaries. Chemosphere 57: 107–126.
48. Xu WH, Zhang G, Zou SC, Li XD, Liu YC. (2007) Determination of selected antibiotics in the Victoria Harbour and the Pearl River, South China using high-performance liquid chromatography-electrospray ionization tandem mass spectrometry. Environ Pollut 145: 672–679.
49. García-Galán MJ, Díaz-Cruz MS, Barceló D. (2011) Occurrence of sulfonamide residues along the Ebro river basin: removal in wastewater treatment plants and environmental impact assessment. Environ Int 37: 462–473.
50. Reinthaler FF, Posch J, Feierl G, Wüst G, Haas D, et al. (2003) Antibiotic resistance of *E. coli* in sewage and sludge. Water Res 37: 1685–1690.
51. Cui LZ, Tominaga EJ, Neoh HM, Hiramatsu KC. (2006) Correlation between reduced daptomycin susceptibility and vancomycin resistance in vancomycin-intermediate *Staphylococcus aureus*. Antimicrob Agents Ch 50: 1079–1082.
52. China MOA (Ministry of Agriculture of the People Republic of China) (2010) Chinese Veterinary Pharmacopoeia, the 4th edition. Beijing, China: Agriculture Press.
53. Rooklidge SJ. (2004) Environmental antimicrobial contamination from terracumulation and diffuse pollution pathways. Sci Total Environ 325: 1–13.
54. Bean DC, Livermore DM, Hall MC. (2009) Plasmids imparting sulfonamide resistance in *Escherichia coli*: implications for persistence. Antimicrob Agents Ch 53: 1088–1093.
55. Su HC, Ying GG, Tao R, Zhang RQ, Zhao JL, et al. (2012) Class 1 and 2 integrons, sul resistance genes and antibiotic resistance in *Escherichia coli* isolated from Dongjiang River, South China. Environ Pollut 169: 42–49.
56. Ham YS, Kobori H, Kang JH, Matsuzaki T, Iino M, et al. (2012) Distribution of antibiotic resistance in urban watershed in Japan. Environ Pollut 162: 98–103.
57. Schlüter A, Szczepanowski R, Pühler A, Top EM. (2007) Genomics of IncP-1 antibiotic resistance plasmids isolated from wastewater treatment plants provides evidence for a widely accessible drug resistance gene pool. FEMS Microbiol Rev 31: 449–477.
58. Rahube TO, Yost CK. (2010) Antibiotic resistance plasmids in wastewater treatment plants and their possible dissemination into the environment. Afr J Biotechnol 9: 9183–9190.
59. Ji XL, Shen QH, Liu F, Ma J, Xu G, et al. (2012) Antibiotic resistance gene abundances associated with antibiotics and heavy metals in animal manures and agricultural soils adjacent to feedlots in Shanghai; China. J of Hazard Mater 235–236: 178–185.
60. Rysz M, Alvarez PJJ. (2004) Amplification and attenuation of tetracycline resistance in soil bacteria: aquifer column experiments. Water Res 38: 3705–3712.
61. Wu CX, Spongberg AL, Witter JD, Fang M, Czajkowski KP. (2010) Uptake of pharmaceutical and personal care products by soybean plants from soils applied with biosolids and irrigated with contaminated water. Environ Sci Technol 44: 6157–6161.
62. Hagedorn C, Robinson SL, Filtz JR, Grubbs SM, Angier TA, et al. (1999) Determining sources of fecal pollution in a rural Virginia watershed with antibiotic resistance patterns in fecal streptococci. Appl Environ Microbiol 65: 5522–5531.
63. Sayah RS, Kaneene JB, Johnson Y, Miller RA. (2005) Patterns of antimicrobial resistance observed in Escherichia coli isolates obtained from domestic and wild-animal fecal samples, human septage, and surface water. Appl Environ Microbiol 71: 1394–1404.
64. Meays CL, Broersma K, Nordin R, Mazumder A, Samadpour M. (2006) Spatial and annual variability in concentrations and sources of Escherichia coli in multiple watersheds. Environ Sci Technol 40: 5289–5296.
65. Blanco G, Lemus JA, Grande J. (2009) Microbial pollution in wildlife: linking agricultural manuring and bacterial antibiotic resistance in red-billed choughs. Environ Res 109: 405–412.
66. Jiang L, Hu XL, Xu T, Zhang H, Sheng D, et al. (2013) Prevalence of antibiotic resistance genes and their relationship with antibiotics in the Huangpu River and the drinking water sources, Shanghai, China. Sci. Total. Environ. 458: 267–272.

Bacteriology and Changes in Antibiotic Susceptibility in Adults with Community-Acquired Perforated Appendicitis

Hong Gil Jeon[1], Hyeong Uk Ju[1], Gyu Yeol Kim[2], Joseph Jeong[3], Min-Ho Kim[4], Jae-Bum Jun[1]*

1 Department of Internal Medicine, Ulsan University Hospital, University of Ulsan College of Medicine, Ulsan, Republic of Korea, **2** Department of Surgery, Ulsan University Hospital, University of Ulsan College of Medicine, Ulsan, Republic of Korea, **3** Department of Laboratory Medicine, Ulsan University Hospital, University of Ulsan College of Medicine, Ulsan, Republic of Korea, **4** Biomedical Research Center, Ulsan University Hospital, University of Ulsan College of Medicine, Ulsan, Republic of Korea

Abstract

This study evaluated bacterial etiology and antibiotic susceptibility in patients diagnosed with community-acquired perforated appendicitis over a 12-year-period. We retrospectively reviewed records of adult patients diagnosed with perforated appendicitis at an 800-bed teaching hospital between January 2000 and December 2011. In total, 415 culture-positive perforated appendicitis cases were analyzed. *Escherichia coli* was the most common pathogen (277/415, 66.7%), followed by *Streptococcus* species (61/415, 14.7%). The susceptibility of *E. coli* to ampicillin, piperacillin/tazobactam, ceftriaxone, cefepime, amikacin, gentamicin, and imipenem was 35.1%, 97.1%, 97.0%, 98.2%, 98.9%, 81.8%, and 100%, respectively. The overall susceptibility of *E. coli* to quinolones (ciprofloxacin or levofloxacin) was 78.7%. During the study period, univariate logistic regression analysis showed a significant decrease in *E. coli* susceptibility to quinolones (OR = 0.91, 95% CI 0.84–0.99, $P = 0.040$). We therefore do not recommend quinolones as empirical therapy for community-acquired perforated appendicitis.

Editor: D William Cameron, University of Ottawa, Canada

Funding: The authors have no support or funding to report.

Competing Interests: The authors have declared that no competing interests exist.

* Email: uvgotletter@hanmail.net

Introduction

Acute appendicitis is one of the most common abdominal surgical emergencies; it is also typically a community-acquired infection. Despite the generally favorable outcome, complicated appendicitis, such as perforated appendicitis, is associated with increased morbidity compared with simple acute appendicitis [1,2]. Because *Escherichia coli* and *Bacteroides fragilis* are most commonly associated with appendicitis, antibiotic therapies are generally selected to target these bacteria [3,4].

For adult patients with community-acquired complicated intra-abdominal infections of mild-to moderate severity, the use of ticarcillin-clavulanate, cefoxitin, ertapenem, moxifloxacin, or tigecycline as single-agent therapy or combinations of metronidazole with cefazolin, cefuroxime, ceftriaxone, cefotaxime, levofloxacin, or ciprofloxacin are recommended by Infectious Diseases Society of America (IDSA) guidelines [5]. However, with increased *E. coli* resistance to quinolones, investigation of local microbiologic findings had been proposed when selecting empirical therapies [5].

Because previous literature has reported a proportionally greater ratio of extended-spectrum β-lactamase (ESBL) and quinolone-resistant *E. coli* among bacteria responsible for community-acquired abdominal infections in Asia compared to other regions, careful selection of empirical antibiotics is particularly important in Asia [6–8]. We therefore conducted a study of the local microbiological profile and changes in antibiotic resistance in community-acquired perforated appendicitis over the past 12 years. These results may help us to inform selection of empirical antibiotic treatments for community-acquired complicated appendicitis.

Materials and Methods

Patient selection and data collection

We retrospectively reviewed the records of adult patients (age ≥ 18 years) who were diagnosed to have perforated appendicitis at Ulsan University Hospital, an 800-bed teaching hospital, between January 2000 to December 2011. Hospital charts and follow-up records were reviewed.

Definitions

Perforated appendicitis was defined as either gross or microscopic evidence of appendiceal perforation. The appendix was not considered to be ruptured by the mere presence of suppurative peritoneal fluid or gangrenous appendicitis without microscopic evidence of perforation.

Community-acquired appendicitis was defined as appendicitis that occurred within 48 hours of hospital admission. Patients were excluded from the study if they had at least 1 of the following health care risk factors: 1) presence of an invasive device at time of admission, 2) history of MRSA infection or colonization, 3) history of surgery, hospitalization, dialysis, or residence in a long-term care facility in the 12 months preceding the culture date [5].

Systemic inflammatory response syndrome (SIRS), sepsis, severe sepsis, and septic shock were defined as described elsewhere [9].

Specimen culture, species identification, and susceptibility testing

Specimens were obtained by swabbing the suppurative peritoneal fluid or periappendiceal abscess. In some cases, specimens were obtained by swabbing the lumen of appendix or by retrieving the suppurative peritoneal fluid via syringe aspiration. The swab specimens were transported to the laboratory in a transport medium (Amies transport medium without Charcoal; Asan Pharmaceuticals Co., Ltd., Hwasung, Korea). The specimens were either dispatched to the microbiology laboratory directly or stored in the operating room until the next day if collected after the working hours. The specimens were inoculated on blood agar, chocolate agar, and MacConkey agar plates. Samples were not inoculated into anaerobic culture. An automated VITEK 2 system (bioMerieux, Inc. Durham, NC, USA) was used to identify pathogens and perform ESBL susceptibility testing. The Vitek 2 ESBL test has 6 wells containing cefepime at 1 μg/mL, cefotaxime at 0.5 μg/mL, and ceftazidime at 0.5 μg/mL alone and in combination with clavulanic acid (10 μg/mL, 4 μg/mL, and 4 μg/mL, respectively); growth rate in each well is quantitatively assessed with an optical scanner. The proportional growth reduction (over 50%) in wells containing cephalosporin plus clavulanic acid compared with those containing cephalosporin alone was considered evidence of ESBL production. Susceptibility testing results were interpreted according to the National Committee for Clinical Laboratory Standards (CLSI) guidelines published in 2009 [10]. However, cephalosporin susceptibility results of ESBL-positive strains were interpreted on the basis of the strains' respective minimal inhibitory concentration (MIC) breakpoints.

Ethics statement

This retrospective study was approved by the Institutional Review Board (IRB) committee of Ulsan University Hospital. Written consent given by the patients was waived by the approving IRB.

Statistical analysis

Statistical analyses were performed by using IBM SPSS Statistics for Windows, version 21 (IBM Corp., Armonk, NY, USA). The Chi-squared test was used to compare frequencies. A univariate logistic regression model was used to calculate odds ratios (OR), 95% confidence intervals, and p-values. The significance level was set at 0.05.

Results

Study population and clinical characteristics

Of 3,379 patients, 567 (16.7%) who received appendectomies during the study period were diagnosed with perforated appendicitis. Of these, we discarded 4 cases of health care-associated infection, 6 without confirmatory cultures, and 142 culture-negative cases; in total, we analyzed 415 culture-positive perforated appendicitis cases. The average length of hospitalization was 9.1±5.1 days. Patient ages ranged between 20 years and 94 years (mean 48.6±17.0 years), with 51.1% (212/415) men (Table 1). A majority of patients (404, 97.3%) underwent open appendectomy via a McBurney incision; laparotomy with a low midline incision was performed in 9 patients (2.2%) and laparoscopic appendectomy was performed in 2 patients (0.5%). The most common underlying disease was hypertension, reported in 56 patients (13.5%). Severe sepsis or septic shock was observed in 70 patients (16.8%), while 1 patient (0.2%) died of sepsis after mechanical ileus. Post-operative complications included wound infection in 18 patients (4.3%), abdominal abscesses or peritonitis in 7 patients (1.6%), and mechanical ileus in 6 patients (1.4%). A combination therapy comprising cephalosporin and metronidazole was the most frequent empirical antibiotic treatment.

Microbiological features

The most commonly isolated bacteria was *E. coli* (277 isolates, 66.7%), followed by *Streptococcus* spp. (61, 14.7%), *Enterococcus* spp. (32, 7.7%), *Klebsiella* spp. (25, 6.0%), and *Pseudomonas aeruginosa* (24, 5.8%) (Table 2). More than 2 organisms were isolated in 75 cases (18.0%).

Antibiotic susceptibilities of isolated organisms

Data on antibiotic susceptibilities of isolated organisms showed that *E. coli* had 78.7% susceptibility to quinolones (ciprofloxacin or levofloxacin). Susceptibilities to ampicillin, aztreonam, ampicillin/sulbactam, amoxicillin/clavulanic acid, piperacillin/tazobactam, cefazolin, cefoxitin, ceftriaxone, cefepime, trimethoprim/sulfamethoxazole, amikacin, gentamicin, tobramycin, and imipenem were 35.1%, 95.2%, 41.4%, 83.5%, 97.1%, 89.8%, 97.7%, 97.0%, 98.2%, 65.6%, 98.9%, 81.8%, 83.4%, and 100%, respectively (Table 3). ESBL-producing strains accounted for 3.9% of *E. coli* species. *Streptococcus* species showed 68.9% susceptibility to penicillin, and 100% susceptibility to ceftriaxone. *Enterococcus* species were 71.9% susceptible to penicillin. The susceptibilities of *P. aeruginosa* to piperacillin/tazobactam, cefepime, quinolones, amikacin, and imipenem were 95.2%, 100%, 87.5%, 100%, and 95.8%, respectively.

Comparisons of bacterial species and *E. coli* isolate antibiotic susceptibilities by clinical severity

We compared the bacterial species and antibiotic susceptibilities of *E. coli* isolates according to the clinically indicated severity (Table 4). The cases were redistributed into two major groups: "sepsis" and "severe sepsis." Infected patients without SIRS and the patients with sepsis were grouped together in the "sepsis" group, whereas the patients with severe sepsis and septic shock were grouped together in the "severe sepsis" group. A total of 345 patients (83.1%) were included in the sepsis group and 70 (16.9%) were included in the severe sepsis group. *E. coli* isolates were found more frequently in the severe sepsis group (74.3%) than in the sepsis group (65.2%), but the difference was not statistically significant. The isolation rates of the other species were also not significantly different between groups. There were no statistically significant differences in *E. coli* susceptibility to all antibiotics between groups.

Changes in *E. coli* antimicrobial susceptibility according to the year

Yearly changes in *E. coli* antimicrobial susceptibility during the study period were examined (Fig. 1). Univariate logistic regression analysis showed that *E. coli* susceptibility to quinolones significantly decreased, with annual susceptibility rates of 89.4%, 83.3%, 89.2%, 84.2%, 66.6%, 74.0%, 82.6%, 69.2%, 80.0%, 61.9%, 65.0%, and 85.0%, during the period of 2000 to 2011 (OR = 0.91, 95% CI 0.84–0.99, $P = 0.040$). In particular, *E. coli* susceptibility to cefoxitin ($P = 0.052$) and ceftriaxone ($P = 0.054$) decreased during the study period, but the change was not statistically

Table 1. Baseline characteristics of patients with perforated appendicitis.

Characteristics	Number (%)
Number of culture positive patients	415
Hospital day (Mean ± SD)	9.1±5.1
Age (Mean ± SD)	48.6±17.0
Sex	
male	212 (51.1)
female	203 (48.9)
Operation method	
laparoscopic	2 (0.5)
McBurney	404 (97.3)
laparotomy	9 (2.2)
Underlying disease	
hypertension	56 (13.5)
diabetes mellitus	26 (6.3)
hepatitis B virus	16 (3.9)
solid cancer	12 (2.9)
Initial manifestation	
infection without SIRS	73 (17.6)
sepsis	272 (65.7)
severe sepsis	67 (16.1)
septic shock	3 (0.7)
In-hospital mortality	
alive	414 (99.8)
death	1 (0.2)
Infectious complication	
wound infection	18 (4.3)
intra-abdominal abscess or peritonitis	7 (1.6)
mechanical ileus	6 (1.4)
Antibiotics	
1st (or 2nd) generation cephalosporin + metronidazole	215 (51.8)
3rd generation cephalosporin + metronidazole	193 (46.5)
ciprofloxacin + metronidazole	4 (1.0)
piperacillin/tazobactam	3 (0.7)

SIRS = systemic inflammatory response syndrome.

significant. Nor were any statistically significant changes observed in *E. coli* susceptibility to other antibiotics such as ampicillin ($P = 0.235$), aztreonam ($P = 0.168$), piperacillin/tazobactam ($P = 0.645$), cefazolin ($P = 0.126$), cefepime ($P = 0.393$), trimethoprim/sulfamethoxazole ($P = 0.732$), amikacin ($P = 0.835$), gentamicin ($P = 0.389$), and tobramycin ($P = 0.645$).

Discussion

This study evaluated microbiological profiles and antibiotic susceptibilities of pathogens isolated from cases of perforated appendicitis. The flora detected in complicated intra-abdominal infection differs between community-acquired and nosocomial infections. We considered appendicitis suitable for studying community-acquired bacterial infections since this illness is largely community-acquired. In fact, only 4 patients discarded from analysis owing to health care-associated infections. *P. aeruginosa* isolates in this study showed overall high levels of antibiotic susceptibility with no multidrug-resistant strains, supporting the idea that appendicitis is more commonly a community-acquired rather than nosocomial infection [11].

E. coli was the most common pathogen identified in this study (66.7% of all isolates), similar to findings in previous appendicitis literature [3,12]. Similarly, *Streptococcus* and *Enterococcus* species were the most frequently isolated gram-positive organisms [12,13]. The ratio of ESBL-producing *E. coli* was 3.9%, within previously reported ranges of 3.5–15.4% [8,14]. The isolation rate of *E. coli* was greater in the severe sepsis group, although this difference was not statistically significant. Some studies have reported that *P. aeruginosa* is a commonly isolated strain in appendicitis, with an isolation rate of 19–32%; however, this was not the case in the current study [12,15].

Although *E. coli* showed a high susceptibility rate of 97% to second- and third-generation cephalosporins that are most

Table 2. Distribution of bacterial species.

	Species	Number (%)[f]
Gram negative organism	*Escherichia coli*	277 (66.7)
	Klebsiella species[a]	25 (6.0)
	Pseudomonas aeruginosa	24 (5.8)
	Other gram negative organism[b]	45 (10.8)
Gram positive organism	*Streptococcus* species[c]	61 (14.7)
	Enterococcus species[d]	32 (7.7)
	Staphylococcus aureus	6 (1.4)
	Other gram positive organism[e]	23 (5.5)

[a]Includes: *K. pneumoniae, K. oxytoca.*
[b]Includes: *Achromobacter xylosoxidans, Acinetobacter lwoffii, Aeromonas hydrophila, Comamonas testosteroni, Hafnia alvei, Proteus mirabilis, Raoultella planticola, Serratia* species, *Enterobacter cloacae.*
[c]Includes: *S. alactolyticus, S. anginosus, S. cristatus, S. constellatus, S. gordonii, S. intermedius, S. mitis, S. salivarius, S. sanguinis*, Viridans Streptococci.
[d]Includes: *E. avium, E. faecalis, E. faecium, E. gallinarum, E. hirae, E. raffinosus.*
[e]Includes: *Gemella morbillorum, Lactococcus garvieae, Leuconostoc mesenteroides, Pediococcus pentosaceus.*
[f]Polymicrobial infection: 75 cases (18.0%).

Table 3. Antibiotic susceptibilities of isolated organisms that caused perforated appendicitis.

	E. coli (total)	*E. coli* (non-ESBL)	*E. coli* (ESBL)	*Streptococcus* species	*Enterococcus* species	*P. aeruginosa*
Antibiotic	(n = 277)	(n = 266)	(n = 11)	(n = 61)	(n = 32)	(n = 24)
Penicillin				42/61 (68.9)	23/32 (71.9)	
Ampicillin	97/276 (35.1)	97/265 (36.6)	0/11 (0)		24/27 (88.8)	
Aztreonam	220/231 (95.2)	217/220 (98.6)	3/11 (27.2)			
Ampicillin/sulbactam	84/203 (41.4)	84/200 (42.0)	0/3 (0)			
Amoxicillin/clavulanic acid	61/73 (83.5)	55/65 (84.6)	6/8 (75.0)			
Piperacillin/tazobactam	240/247 (97.1)	229/236 (97.0)	11/11 (100)			20/21 (95.2)
Cefazolin	248/276 (89.8)	248/265 (93.5)	0/11 (0)			
Cefoxitin	264/270 (97.7)	254/259 (98.0)	10/11 (90.9)			
Ceftriaxone	267/275 (97.0)	262/264 (99.2)	5/11 (45.4)	39/39 (100)		2/24 (8.3)
Cefepime	227/231 (98.2)	220/220 (100)	7/11 (63.6)			22/22 (100)
Quinolone	218/277 (78.7)	215/266 (80.8)	3/11 (27.2)		25/28 (89.2)	21/24 (87.5)
Trimethoprim/sulfamethoxazole	181/276 (65.6)	177/265 (66.7)	4/11 (36.3)	34/47 (72.3)	11/17 (64.7)	1/24 (4.1)
Amikacin	274/277 (98.9)	264/266 (99.2)	10/11 (90.9)			24/24 (100)
Gentamicin	226/276 (81.8)	221/265 (83.4)	5/11 (45.4)			24/24 (100)
Tobramycin	231/277 (83.4)	227/266 (85.3)	4/11 (36.3)			24/24 (100)
Vancomycin				60/61 (98.3)	30/32 (93.7)	
Imipenem	276/276 (100)	265/265 (100)	11/11 (100)	11/11 (100)	23/26 (88.5)	23/24 (95.8)

Table 4. Comparisons of bacterial species and antibiotic susceptibilities of *E. coli* between the sepsis group and the severe sepsis group.

	Sepsis[a] (%)	Severe sepsis[b] (%)	*P*-value
Species			
E. coli	225/345 (65.2)	52/70 (74.3)	0.142
P. aeruginosa	20/345 (5.8)	4/70 (5.7)	0.978
Streptococcus species	52/345 (15.1)	9/70 (12.9)	0.633
Enterococcus species	25/345 (7.3)	7/70 (10.0)	0.431
Antibiotics susceptibilities of *E. coli*			
Piperacillin/tazobactam	197/202 (97.5)	43/45 (95.6)	0.472
Cefoxitin	214/220 (97.3)	50/50 (100)	0.238
Ceftriaxone	215/223 (96.4)	52/52 (100)	0.166
Cefepime	184/187 (98.4)	43/44 (97.7)	0.760
Ciprofloxacin or levofloxacin	176/225 (78.2)	42/52 (80.8)	0.686

[a]Infection without SIRS (systemic inflammatory response syndrome) & sepsis.
[b]Severe sepsis & septic shock.

commonly used for empirical antibiotic treatment, the susceptibility decreased during the study period, albeit without statistical significance ($P = 0.052$ and $P = 0.054$, respectively). The susceptibility to quinolones was 78.7%, with a statistically significant ($P = 0.040$) decrease during the study period. Previous studies by Bochicchio et al (2006) and Rob et al (2013) reported that the susceptibility rate of *E. coli*, isolated from appendicitis samples, to quinolones was 71.4–85.6% [6,8]. The *E. coli* susceptibility to quinolones and cephalosporins reported by Rob et al (2013) was lower than that reported by Bochicchio et al (2006). This may be attributable to *E. coli*'s increased resistance to the antibiotics or the lowered MIC breakpoint for cephalosporins set by the CLSI guidelines. For most antibiotics, *E. coli* susceptibility rates observed in this study were similar to those reported by Bochicchio et al (2006), with the susceptibility rate to quinolones being slightly lower. Both previous studies found high susceptibilities to carbapenem, amikacin, and piperacillin/tazobactam; in this study, ESBL-producing organisms were particularly sensitive to piperacillin/tazobactam (12/12, 100%). The susceptibility of *Streptococcus* species to penicillin was 68.9%, and all strains were susceptible to ceftriaxone. *P. aeruginosa* isolated in this study was highly susceptible to amikacin, cefepime, piperacillin/tazobactam, and carbapenem, but was slightly less susceptible to quinolones (87.5%).

All patients undergoing operation for appendicitis should receive antimicrobial therapy [16]. Appropriate antimicrobial therapy includes agents effective against facultative and aerobic gram-negative organisms and anaerobic organisms. There are

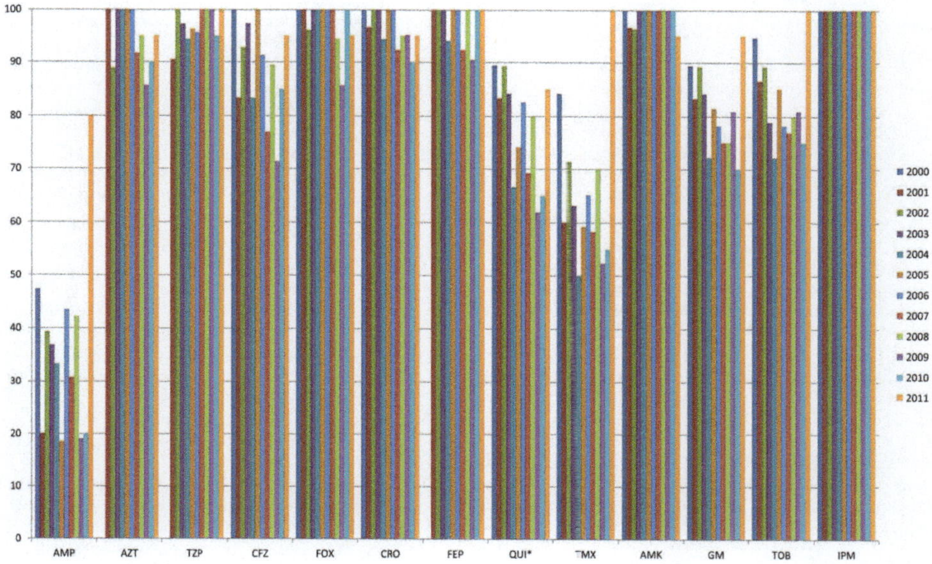

Figure 1. Change of antimicrobial susceptibility among *E. coli* during the 12-year-period. AMP, ampicillin; AZT, aztreonam; TZP, piperacillin/tazobactam; CFZ, cefazolin; FOX, cefoxitin; CRO, ceftriaxone; FEP, cefepime; QUI, quinolone; TMX, trimethoprim/sulfamethoxazole; AMK, amikacin; GM, gentamicin; TOB, tobramycin; IPM, imipenem. * During the study period, there was a significant decrease in antimicrobial susceptibility on univariate logistic regression analysis ($P = 0.040$).

data that inadequate empiric antibiotic therapy results in increased morbidity or treatment failure in complicated appendicitis [17,18]. If resistance to a given antibiotic is present in 10%–20% or more of isolates of a common intra-abdominal pathogen in the community, use of that agent should be avoided [5]. A report in Taiwan proposed that a quinolone be used to treat community-acquired complicated intra-abdominal infections, as *E. coli* was found to be 82–85% susceptible to ciprofloxacin and levofloxacin [19]. In this study, however, the resistance rate of *E. coli* to quinolones is >20%; therefore, its use as an empirical antibiotic is not advisable in Korea. Second- and third-generation cephalosporins appeared to be an appropriate treatment for this application according to our results. Although third-generation cephalosporins might be a better treatment choice because that *Streptococcus* species showed 100% susceptibility to ceftriaxone, further studies are needed to thoroughly trace variations in susceptibility, given that the decrease in *E. coli* susceptibility, observed during the study period, was not statistically significant. Piperacillin/tazobactam and carbapenem might be considered to treat *P. aeruginosa* or ESBL-producing organisms in patients with signs of severe sepsis such as organ dysfunction. However, these species were not frequently isolated in all patient groups including the severe sepsis group of the current study, and spectrum of these antibiotics may be too broad. *E. coli* also showed high susceptibility to amikacin, but concerns remain regarding use of aminoglycoside antibiotics owing to their nephrotoxicity and ototoxicity. Considering the high resistance of *E. coli* to ampicillin and ampicillin/sulbactam–and the questionable significance of enterococci as pathogens in complicated intra-abdominal infections–these antibiotics are not recommended for treating perforated appendicitis.

On the basis of evidence that culture testing of intraoperative specimens does not affect the prognosis of patients with perforated appendicitis, many institutions may not perform routine culture testing [20,21]. However, considering the current reality of increasing antibiotic resistance, routine culture testing might be useful to identify changes in susceptibility and to select appropriate antibiotics [5]. Anaerobic cultures are not necessary for patients with community-acquired intra-abdominal infection if empiric antimicrobial therapy active against common anaerobic pathogens is provide [5]. Although anaerobic bacteria culturing was not performed in this study, previous reports on anaerobic culture showed that *Bacteroides fragilis*, along with *E. coli*, was the most commonly isolated pathogen in appendicitis [14,20]. In past studies of appendicitis that conducted anaerobic susceptibility testing, *B. fragilis* was found to be more than 95% susceptible to metronidazole [4,14,22]. Anaerobic bacteria culturing could be considered for future studies if an increase in anaerobic bacterial resistance to metronidazole is observed.

The retrospective nature of the present study might have resulted in intrinsic bias and the data may not represent the entire population because data was collected from a single institution. However, considering that the quinolone resistance rate we observed was similar to that reported in previous studies conducted in Korea [23,24]–which involved community-acquired *E. coli* bacteremia originated from various infections including intra-abdominal infection–we speculated that quinolone resistance rate among *E. coli* causing intra-abdominal infection in Korea should be similar to the one determined in this study.

In conclusion, *E. coli* was the most commonly identified pathogen in patients with perforated appendicitis. The quinolone resistance rate was >20% in *E. coli* isolated from community-acquired perforated appendicitis. The isolates were decreasingly susceptible to quinolones during the study period. We advise against the use of quinolones as a first line antibiotic therapy in community-acquired perforated appendicitis in Korea.

Author Contributions

Conceived and designed the experiments: J-BJ. Performed the experiments: HGJ HUJ. Analyzed the data: GYK JJ M-HK. Contributed to the writing of the manuscript: J-BJ.

References

1. Lau WY, Wong SH (1981) Randomized, prospective trial of topical hydrogen peroxide in appendectomy wound infection. High risk factors. Am J Surg 142: 393–397.
2. Schmit PJ, Hiyama DT, Swisher SG, Bennion RS, Thompson JE Jr (1994) Analysis of risk factors of postappendectomy intra-abdominal abscess. J Am Coll Surg 179: 721–726.
3. Bennion RS, Baron EJ, Thompson JE Jr, Downes J, Summanen P, et al. (1990) The bacteriology of gangrenous and perforated appendicitis–revisited. Ann Surg 211: 165–171.
4. Lau WY, Teoh-Chan CH, Fan ST, Yam WC, Lau KF, et al. (1984) The bacteriology and septic complication of patients with appendicitis. Ann Surg 200: 576–581.
5. Solomkin JS, Mazuski JE, Bradley JS, Rodvold KA, Goldstein EJ, et al. (2010) Diagnosis and management of complicated intra-abdominal infection in adults and children: guidelines by the Surgical Infection Society and the Infectious Diseases Society of America. Clin Infect Dis 50: 133–164.
6. Bochicchio GV, Baquero F, Hsueh PR, Paterson DL, Rossi F, et al. (2006) In vitro susceptibilities of Escherichia coli isolated from patients with intra-abdominal infections worldwide in 2002–2004: results from SMART (Study for Monitoring Antimicrobial Resistance Trends). Surg Infect (Larchmt) 7: 537–545.
7. Paterson DL, Rossi F, Baquero F, Hsueh PR, Woods GL, et al. (2005) In vitro susceptibilities of aerobic and facultative Gram-negative bacilli isolated from patients with intra-abdominal infections worldwide: the 2003 Study for Monitoring Antimicrobial Resistance Trends (SMART). J Antimicrob Chemother 55: 965–973.
8. Lob SH, Badal RE, Bouchillon SK, Hawser SP, Hackel MA, et al. (2013) Epidemiology and susceptibility of Gram-negative appendicitis pathogens: SMART 2008–2010. Surg Infect (Larchmt) 14: 203–208.
9. Bone RC, Sibbald WJ, Sprung CL (1992) The ACCP-SCCM consensus conference on sepsis and organ failure. Chest 101: 1481–1483.
10. Clinical and Laboratory Standards Institute (2009) Performance standards for antimicrobial susceptibility testing. Nineteenth informational supplement. Document M100–S19. Wayne, PA: CLSI.
11. Montravers P, Lepape A, Dubreuil L, Gauzit R, Pean Y, et al. (2009) Clinical and microbiological profiles of community-acquired and nosocomial intra-abdominal infections: results of the French prospective, observational EBIIA study. J Antimicrob Chemother 63: 785–794.
12. Guillet-Caruba C, Cheikhelard A, Guillet M, Bille E, Descamps P, et al. (2011) Bacteriologic epidemiology and empirical treatment of pediatric complicated appendicitis. Diagn Microbiol Infect Dis 69: 376–381.
13. Chen CY, Chen YC, Pu HN, Tsai CH, Chen WT, et al. (2012) Bacteriology of acute appendicitis and its implication for the use of prophylactic antibiotics. Surg Infect (Larchmt) 13: 383–390.
14. Chan KW, Lee KH, Mou JW, Cheung ST, Sihoe JD, et al. (2010) Evidence-based adjustment of antibiotic in pediatric complicated appendicitis in the era of antibiotic resistance. Pediatr Surg Int 26: 157–160.
15. Fallon SC, Hassan SF, Larimer EL, Rodriguez JR, Brandt ML, et al. (2013) Modification of an evidence-based protocol for advanced appendicitis in children. J Surg Res 185: 273–277.
16. Andersen BR, Kallehave FL, Andersen HK (2001) Antibiotics versus placebo for prevention of postoperative infection after appendectomy. Cochrane Database Syst Rev: CD001439.
17. Yellin AE, Heseltine PN, Berne TV, Appleman MD, Gill MA, et al. (1985) The role of Pseudomonas species in patients treated with ampicillin and Sulbactam for gangrenous and perforated appendicitis. Surg Gynecol Obstet 161: 303–307.
18. Berne TV, Yellin AW, Appleman MD, Heseltine PN (1982) Antibiotic management of surgically treated gangrenous or perforated appendicitis. Comparison of gentamicin and clindamycin versus cefamandole versus cefoperazone. Am J Surg 144: 8–13.
19. Lau YJ, Chen YH, Huang CT, Lee WS, Liu CY, et al. (2012) Role of moxifloxacin for the treatment of community-acquired [corrected] complicated intra-abdominal infections in Taiwan. J Microbiol Immunol Infect 45: 1–6.

20. Kokoska ER, Silen ML, Tracy TF Jr, Dillon PA, Kennedy DJ, et al. (1999) The impact of intraoperative culture on treatment and outcome in children with perforated appendicitis. J Pediatr Surg 34: 749–753.
21. Foo FJ, Beckingham IJ, Ahmed I (2008) Intra-operative culture swabs in acute appendicitis: a waste of resources. Surgeon 6: 278–281.
22. Wojcik-Stojek B, Bulanda M, Martirosian G, Heczko P, Meisel-Mikolajczyk F (2000) In vitro antibiotic susceptibility of Bacteroides fragilis strains isolated from excised appendix of patients with phlegmonous or gangrenous appendicitis. Acta Microbiol Pol 49: 171–175.
23. Lee S, Han SW, Kim KW, Song do Y, Kwon KT (2014) Third-generation cephalosporin resistance of community-onset Escherichia coli and Klebsiella pneumoniae bacteremia in a secondary hospital. Korean J Intern Med 29: 49–56.
24. Park S, Park J, Lee S (2005) Analysis on the etiology and prognostic factors of community-acquired bacteremia in a community-based tertiary hospital. Infect Chemother 37: 255–264.

21

Seasonal Variability in Airborne Biotic Contaminants in Swine Confinement Buildings

Priyanka Kumari, Hong L. Choi*

Department of Agricultural Biotechnology, Research Institute for Agriculture and Life Science, Seoul National University, Gwanak, Seoul, Republic of Korea

Abstract

Little is known about the seasonal dynamics of biotic contaminants in swine confinement buildings (SCBs). The biotic contaminants of seven SCBs were monitored during one visit in the winter and one during the summer. Paired-end Illumina sequencing of the 16S rRNA gene, V3 region, was used to examine seasonal shifts in bacterial community composition and diversity. The abundances of 16S rRNA genes and six tetracycline resistance genes (*tet*B, *tet*H, *tet*Z, *tet*O, *tet*Q, and *tet*W) were also quantified using real-time PCR. Bacterial abundances, community composition and diversity all showed strong seasonal patterns defined by winter peaks in abundance and diversity. Microclimatic variables of SCBs, particularly air speed, PM2.5 and total suspended particles (TSP) were found significantly correlated to abundances, community composition, and diversity of bacterial bioaerosols. Seasonal fluctuations were also observed for four tetracycline resistance genes, *tet*H, *tet*O, *tet*Q, and *tet*W. The frequency of occurrences of these resistance genes were significantly higher in samples collected during winter and was also significantly correlated with air speed, PM2.5 and TSP. Overall, our results indicate that biotic contaminants in SCBs exhibit seasonal trends, and these could be associated with the microclimatic variables of SCBs. The correlations established in the current study could be helpful in establishing better management strategies to minimize the potential health impacts on both livestock and humans working in this environment.

Editor: Jack Anthony Gilbert, Argonne National Laboratory, United States of America

Funding: This work was supported by Korea Institute of Planning and Evaluation for Technology in Food, Agriculture, Forestry, and Fisheries (IPET) from the Ministry of Agriculture, Food, and Rural Affairs (MAFRA) through project no. 312036-03-2-HD030, and in part by Korea Institute of Energy Technology Evaluation and Planning (KETEP) from the Ministry of Trade, Industry and Energy (MOTIE) through project no. 2012-3020090040. The funders had no role in study design, data collection and analysis, decision to publish, or preparation of the manuscript.

Competing Interests: The authors have declared that no competing interests exist.

* Email: ulsoo8@snu.ac.kr

Introduction

The intensification of pig farming in confined buildings with high animal densities can lead to poor indoor air quality. Microbial decomposition of proteinaceous waste products in feces and urine results in elevated concentrations of volatile organic compounds, ammonia (NH_3), and sulfides [1], whereas feed materials, skin debris, bedding material, and dried manure generate airborne particulates that carry adsorbed microorganisms and endotoxins [2]. Poor indoor air quality in swine confinement buildings (SCBs) affects both animal and human health. For example, airborne particulates can deposit in nasal channels and the respiratory tract and cause damage to lung tissues [3]. Furthermore, some airborne bacteria and gases such as NH_3, hydrogen sulphide (H_2S) (from the manure), and carbon di-oxide (CO_2) (pig activity) can cause or trigger chronic respiratory tract inflammation in workers and pigs [4–11].

Microclimatic variables can influence the formation of aerosols containing microorganisms. There are great variations in the outside temperature in South Korea (-7 to 1°C in winter and from 22°C to 30°C in summer) [12], so maintaining an optimal indoor temperature in SCBs can be challenging. Typically in the winter, all of the openings are closed, and the ventilation rate has to be minimal to reduce the heat loss. This low ventilation rate could induce an increased concentration of airborne contaminants. In contrast, the ventilation is maximal during the summer, thus diminishing the indoor temperature and contributing to driving the indoor air outside the SCBs.

Bacteria constitute a huge proportion of organisms within bioaerosols in SCBs, with a mean concentration of 10^5 cfu m^{-3} [13,14]. While much has been done to monitor the indoor airborne biotic contaminants in SCBs [14–16], relatively little is known about the seasonal dynamics of airborne biotic contaminants and their interaction with microclimate parameters in SCBs. Nehme *et al.* [14,15] examined the seasonal variability of airborne bacterial and archaeal communities in SCBs and found that, although the microbial abundances were significantly higher during the winter, the biodiversity was similar in each SCB during both the winter and summer seasons. However, these studies used low-resolution molecular fingerprinting tools, which lacked the coverage and depth of high-throughput sequencing methods. In a recent study, Hong et al. [16] used 454-pyrosequencing to analyze the airborne biotic contaminants in pig and poultry confinement buildings sampled from different climate conditions and found that the different livestock as well as production phase were associated with distinct airborne bacterial communities; however, they did not evaluate the effect of microclimate variables on bacterial bioaerosol communities.

The usage of antibiotics in swine farms has promoted the development and abundance of antibiotic resistance in microbes [17,18], which can become aerosolized within the SCBs. Antibiotic resistant genes (ARGs) can be transferred to pathogens through transformation or phage-mediated transduction, and/or by conjugation, posing a serious threat to public health. Horizontal gene transfer plays important roles in the evolution and transmission of ARGs between bacterial species and includes the movement of ARGs from fecal bacteria to environmental bacteria, as well as the reverse; that is, emergence of novel mechanisms of acquired resistance in pathogens, ARGs that originally were present in harmless bacteria [19]. Tetracycline was chosen for this study because it is the most widely used broad spectrum antibiotic in livestock production worldwide, and is particularly prevalent in pig production [20]. The mechanism by which bacteria resist tetracycline antibiotics is heavily biased by ecological niche [21], and compared to the existing literature on tetracycline resistance genes (Tc^R) in soils and in water, relatively little is known regarding Tc^R genes in aerosols of SCBs [16,22]. Three Tc^R genes (*tet*B, *tet*H, *tet*Z) encoding efflux proteins (EFP), and three others (*tet*O, *tet*Q, *tet*W) encoding ribosomal protection proteins (RPP) were selected for this study because these genes have been detected in aerosol of SCBs [16] and because these Tc^R genes encode two main mechanisms of bacterial resistance to tetracycline, which have been found associated with bacteria of public health interest [23–25].

Therefore, the aim of this study was to answer the following questions using the Illumina Hiseq sequencing of the V3 region of the 16S rRNA gene and qPCR of both 16S rRNA and Tc^R genes: (1) How does the bacterial bioaerosol community composition and diversity vary in SCBs during the winter and summer seasons? (2) Does the abundance of 16S rRNA and Tc^R genes vary in SCBs during the winter and summer seasons? (3) What are the major microclimate variables affecting the abundance, community composition, and diversity of airborne biotic contaminants in SCBs?

Materials and Methods

Characteristics of animal confinement buildings

The study was conducted in the winter (January) and summer (June) of 2013 on seven commercial pig farms located in six South Korean provinces. All the commercial pig farms sampled in this study are privately owned. Permission to access privately owned farms was given by the farm owners and for future permissions we can contact the owners. All samples were collected in the growing/finishing houses of SCBs. The average outdoor temperature across all of South Korea ranges from −7 to 1°C in winter and 22°C to 30°C in summer. Temperature differences among the six provinces were less than 2°C [12]. Ventilation in SCBs was mechanical by exhaust fans on walls. The number of animals kept in each sampling room ranged from 140 to 480, and the stocking density varied from 0.88 to 1.41 m^2/head. The age of the pigs varied from 67–150 days in each sampling room at the time of sampling. The manure removal system was deep-pit with slats, and the cleaning cycle varied from 4–6 months.

Microclimate variables

The microclimate variables were measured from three points in the aisle outside the pens at every 8 h interval till 24 h (Fig. S1), and all the microclimate variables were reported as averages corresponding to each sampling period. Air temperature and relative humidity were measured with a hygrothermograph (SK-110TRH, SATO, Tokyo, Japan). Air speed was measured with an anemometer (model 6112, KANOMAX, Osaka, Japan). Particulate matter concentrations ($\mu g\ m^{-3}$) were measured using an aerosol mass monitor (GT-331, SIBATA, Soca-city, Japan). The mass concentrations of PM10 (PM average aerodynamic diameter ≤10 μm), PM2.5 (PM mean aerodynamic diameter ≤2.5 μm), PM1 (PM mean aerodynamic diameter ≤1 μm), and total suspended particles (TSP) were obtained simultaneously, at a flow rate of 2.83 l min^{-1}. The concentrations of NH_3, H_2S and CO_2 were measured by gas detector tubes (Gastec Co., Ltd., Kanagawa, Japan).

Sample collection and DNA extraction

Aerosol samples were collected from the middle point in the aisle outside the pens at a height of 1.4 m above the floor (Fig. S1). Air samples were captured on sterile 0.22-μm cellulose nitrate filters (Fisher Scientific, Pittsburgh, PA) via vacuum filtration with a flow rate of 4 l min^{-1} for 24 h. The cellulose nitrate filters were kept at 4°C until processing in the laboratory. Bacterial DNA was extracted directly from the filters using the PowerSoil DNA isolation kit (MoBio Laboratories, Carlsbad, CA). Individual filters were aseptically cut into small pieces, loaded into the bead tube of the DNA extraction kit, and heated to 65°C for 10 min followed by 2 min of vortexing. The remaining steps of the DNA extraction were performed according to the manufacturer's instructions. The purified DNA was resuspended in 50 μl of solution S6 (MoBio Laboratories) and stored at −20°C until PCR amplification. Blank filters were analyzed alongside sample filters to test for contamination, and following DNA extraction and amplification, blank filters were consistently found to be free of microbial contaminants.

Illumina sequencing and data processing

The extracted DNA was amplified using primers 338F (5′-XXXXXXXX-GTACTCCTACGGGAGGCAGCAG-3′) and 533R (5′-TTACCGCGGCTGCTGGCAC-3′) targeting the V3 region of bacterial 16S rRNA ('X' denotes 8-mer barcode sequence) [26]. Paired-end sequencing was performed at Beijing Genome Institute (BGI) (Hongkong, China) using 2×150 bp Hiseq2500 (Illumina) according to the manufacturer's instructions. Library preparation, sequencing, and initial quality filtering were performed as described previously [27]. The sequence data obtained by Illumina Hiseq2500 sequencing were processed using mothur [28]. Paired-end sequences were assembled, trimmed, and filtered in mothur. Next, the sequences were aligned against a SILVA alignment (http://www.arb-silva.de/). Putative chimeric sequences were detected and removed via the Chimera Uchime algorithm contained within mothur [29]. Sequences were denoised using the *'pre.cluster'* command in mothur, which applies a pseudo-single linkage algorithm with the goal of removing sequences that are likely due to sequencing errors [30]. All sequences were classified using the EzTaxon-e database (http://eztaxon-e.ezbiocloud.net/) [31], using the classify command in mothur at 80% Naïve Bayesian bootstrap cutoff with 1000 iterations. Sequence data were deposited in SRA at NCBI with the accession number of SRP039383.

Quantification of 16S rRNA and Tc^R genes

A real-time polymerase chain reaction was used to quantify bacterial 16S rRNA and six Tc^R genes (RPP class: *tet*O, *tet*Q and *tet*W; EFP class: *tet*B, *tet*H and *tet*Z, refer Levy et al. [32] for the details on nomenclature) using the SYBR Green approach with the primers described in Table S1. The copy numbers of 16S rRNA and Tc^R genes in bioaerosol samples were measured against the standard curves of plasmids containing copies of the respective genes, using a 10-fold serial dilution. All reactions were conducted in triplicate with the 20 µl qPCR mixtures containing 10 µl of 2× SYBR Green PCR Master Mix (Applied BioSystems, Foster City, CA, USA), 1.0 µl each of the 10 µM forward and reverse primers, and 7.0 µl of sterile, DNA-free water. Standard and bioaerosol (ca. 1.0 ng) DNA samples were added at 1.0 µl per reaction. The reaction was carried out on an ABI Prism 7300 sequence detector (Applied Biosystems, Foster City, CA, USA) with an initial step of 95°C for 10 min, followed by 40 cycles of denaturation (95°C for 10 s), and primer annealing and extension (60°C for 1 min). Dissociation curve analysis was performed to ensure the specificity of PCR, which included an increment of temperature from 60°C to 95°C at an interval of 0.5°C for 5 s. Gene copy numbers were determined using a regression equation for each assay and relating the cycle threshold (CT) value to the known numbers of copies in the standards. The correlation coefficients of standard curves ranged from 0.957 to 0.996.

Statistical processing and analysis of results

Rarefaction curves and diversity indices were generated using mothur, with the bacterial phylotype (OTU) defined here at 97% threshold of 16S rRNA gene sequence similarity. Phylogenetic diversity (PD) was calculated as Faith's PD [33], the total phylogenetic branch length separating OTUs in each rarefied sample. To allow for robust comparisons among samples containing different numbers of sequences, phylotype richness and phylogenetic diversity were calculated based on samples rarefied to contain 15,909 sequences. To test for differences in relation to season on OTU richness, PD, 16S rRNA and Tc^R genes abundances, we used a t-test for normal data and the Wilcoxon rank-sum test for non-normal data. We used the same procedure to test whether the relative abundance of the most abundant phyla differed across seasons.

To avoid including collinear variables in further analyses, we used a Spearman's correlation matrix and highly correlated (Spearman's $r \geq 0.8$) microclimatic variables were removed from further analysis. Non-metric multidimensional scaling (NMDS) was generated based on pairwise Bray-Curtis dissimilarities between samples using the vegan R package [34]. The analysis of similarity (ANOSIM) function in the vegan R package with 999 permutations was used to test for differences in bacterial communities among the winter and summer season. The vectors of microclimate variables were fitted onto ordination space (Bray–Curtis NMDS) to detect possible associations between patterns of community structure and microclimate variables using the 'envfit' function of the vegan R package, and statistical significance was evaluated among 999 random permutations. Analyses for Venn diagram generation were performed using the mothur, and the Venn diagram was plotted using *R* package VennDiagram [35]. Differentially abundant bacterial genera between the winter and summer seasons were identified using a parametric approach (Metastats) [36]. All statistical analysis, graphs, and ordinations were produced using R version 3.0.2 [37].

Results

The means of microclimate variables in the SCBs during the winter and summer seasons are presented in Table 1. All of the microclimate variables in the SCBs differed significantly between the winter and summer season samples ($P<0.05$; Table 1), except total suspended particles, NH_3 and H_2S ($P>0.05$, Table 1). Spearman's correlation matrix showed highest correlation between PM10 and TSP ($r = 0.93$; Table 2). Temperature and airspeed were the next most correlated variables ($r = 0.9$; Table 2)

Table 1. Seasonal means (±SD) of microclimate variables in swine confinement buildings.

Variable	Temperature, °C	Relative humidity, %	Air speed, m/s	PM10, µg m^{-3}	PM2.5, µg m^{-3}	PM1, µg m^{-3}	TSP, µg m^{-3}	NH_3, mg/L	H_2S, mg/L	CO_2, mg/L
Season										
Winter	20.8±5.5	68.5±18.6	0.02±0.02	470.8±468.8	48.4±27	25.7±20.1	1252.2±1320.5	24.4±22.7	0.36±0.34	2833.1±1433.6
Summer	31.5±4.2	86.3±13.7	0.18±0.06	77.5±19.5	17.1±8.2	7.6±4.2	130.8±34.2	14.9±12.1	0.41±0.43	1242.6±423.7
P-value[a]	<0.001	0.003	<0.001	0.04	0.03	0.03	0.07	0.21	0.82	0.03

[a]For each variable, *P*-value was used to determine the significance of means among the two seasons.

Table 2. Spearman rank correlations between measured microclimatic variables in SCBs.

Microclimate variables	Temperature	Relative humidity	Air speed	PM10	PM2.5	PM1	TSP	NH_3	H_2S	CO_2
Temperature	1									
Relative humidity	0.26	1								
Air speed	0.9***	0.3	1							
PM10	−0.52	−0.46	−0.39	1						
PM2.5	−0.69**	−0.35	−0.7**	0.72**	1					
PM1	−0.63*	−0.17	−0.64*	0.34	0.81***	1				
TSP	−0.39	−0.48	−0.23	0.93***	0.53	0.18	1			
NH_3	−0.06	0.02	−0.28	−0.13	0.12	−0.02	−0.15	1		
H_2S	0.21	−0.11	−0.06	−0.11	0.04	−0.11	−0.15	0.68**	1	
CO_2	−0.26	−0.02	−0.46	0.19	0.48	0.3	0.14	0.82***	0.44	1

*$P<0.05$;
**$P<0.01$;
***$P<0.001$.

followed by NH_3 and CO_2 (r = 0.82; Table 2) and PM2.5 and PM1 (r = 0.81; Table 2). Therefore, we removed PM10, temperature, PM1 and NH_3, and used the remaining six microclimate variables, (i.e. airspeed, relative humidity, PM2.5, TSP, H_2S and CO_2) for further analyses.

From the 14 samples, we obtained 13,597 OTUs at 97% similarity from 497,607 good-quality sequences. The average number of OTUs per sample was 2442±910 (standard deviation [SD]), ranging from 1287 to 4045 OTUs. To compare diversity levels and community profiles between samples controlling for differences in sequencing depth, samples were compared at the same sequencing depth (15,909 randomly selected sequences per sample). At this depth of coverage, bacterial richness ($P<0.01$; Fig. 1A) and phylogenetic diversity levels ($P = 0.01$; Fig. 1B) were significantly higher in winter in comparison to summer. Spearman's correlation coefficients showed a significant negative correlation between air speed and both OTU richness and PD of bacterial bioaerosol communities (Table 3). Whereas, PM2.5 and TSP were positively correlated to OTU richness and PD (Table 3).

The most abundant bacterial phyla were *Firmicutes*, representing 50.9% of all sequences, followed by *Bacteroidetes* (21.3%),

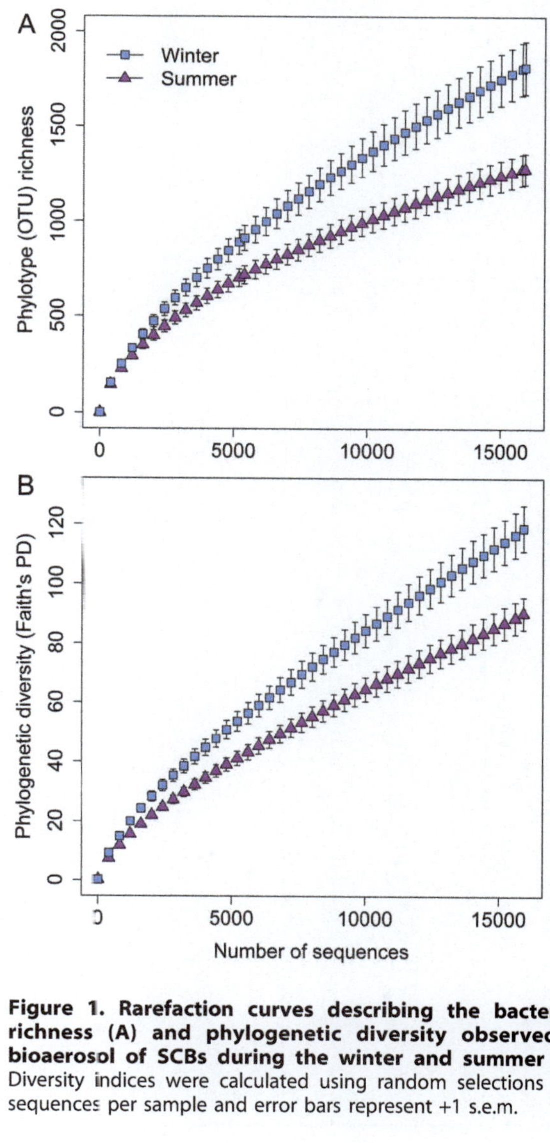

Figure 1. Rarefaction curves describing the bacterial OTU richness (A) and phylogenetic diversity observed in the bioaerosol of SCBs during the winter and summer seasons. Diversity indices were calculated using random selections of 15,909 sequences per sample and error bars represent +1 s.e.m.

Table 3. Spearman rank correlations between microclimatic variables, diversity and abundances of airborne biotic contaminants in SCBs.

Microclimate variables	OTU richness	Phylogenetic diversity	16S rRNA copy number	*tetB*	*tetH*	*tetZ*	*tetO*	*tetQ*	*tetW*
Relative humidity	−0.56*	−0.52	−0.4	−0.24	−0.26	−0.24	−0.47	−0.3	−0.39
Air speed	−0.7**	−0.71**	−0.78***	−0.14	−0.69**	−0.52	−0.61*	−0.68**	−0.73**
PM2.5	0.57*	0.57*	0.78***	0.24	0.62*	0.66*	0.58*	0.79***	0.79***
TSP	0.08	0.06	0.55*	0.45	0.56*	0.57*	0.64*	0.58*	0.6*
H_2S	0.35	0.31	0.09	−0.11	−0.25	0.1	−0.19	−0.03	−0.07
CO_2	0.42	0.4	0.56*	−0.24	0.32	0.63*	0.41	0.6*	0.49

*$P<0.05$;
**$P<0.01$;
***$P<0.001$.

Proteobacteria (18.5%), and, to a lesser degree, *Actinobacteria* (3.9%), *Tenericutes* (0.6%), and *Spirochaetes* (0.4%); 0.8% of all sequences were unclassified. We found significant differences in relative abundance across seasons for *Proteobacteria* ($P = 0.04$) (Fig. 2) and *Actinobacteria* ($P = 0.03$) (Fig. 2). The composition of the airborne bacterial communities was significantly influenced by seasons (ANOSIM statistic $R = 0.96$, $P<0.01$; Fig. 3). The samples collected during the winter season harbored bacterial communities distinct from those found in samples collected during summer. An environmental fitting analysis, using microclimatic variables, showed that airspeed ($r^2 = 0.70$, $P = 0.002$), PM2.5 ($r^2 = 0.39$, $P = 0.04$), TSP ($r^2 = 0.39$, $P = 0.04$) and CO_2 ($r^2 = 0.43$, $P = 0.04$) were significantly associated with bacterial community composition.

Venn diagrams (Fig. S2) illustrate that OTU overlap between seasons as well as show unique OTUs. Overall, 20% of OTUs were shared between the winter and summer season. These 2,705 shared OTUs represented the majority of sequences (457,100 sequences or 91% of the total data set) (Fig. S2). Given the distinct clustering pattern in bacterial community composition in SCBs by season, we investigated in more detail what bacterial genera determine more strongly the distinct community composition in each season. Some bacterial genera were found to be dominant in both winter and summer seasons. *Lactobacillus* (18.3% and 16.3% on average) and *Prevotella* (19.6% and 6.2%) were the most dominant genera in both seasons. However, there were some differences in abundant genera between the two seasons (Table 4). The bacterial bioaerosol of SCB in the winter season was dominated by a single genus – *Prevotella* – at nearly 19.6% (Metastats $P = 0.01$). *Faecalibacterium* (0.8%), *Blutia* (0.7%), and *Catenibacterium* (0.7%) were also more abundant in the winter than in the summer (all $P \leq 0.01$). Genera that were more abundant in the summer included *Sphingomonas* (3.7%), *Capnocytophaga* (3%), *Haemophilus* (2.8%), and *Streptococcus* (2.6%) (all $P \leq 0.01$).

The abundances of 16S rRNA genes in the bioaerosols of SCBs were significantly higher in the winter (mean $= 1.4 \times 10^8$ bacteria m^{-3}, $n = 7$, $P<0.01$, Fig. 4) than in the summer (mean $= 1.2 \times 10^7$ bacteria m^{-3}, $P<0.01$, $n = 7$, Fig. 4). Six classes of Tc^R genes (*tet*B, *tet*H, *tet*Z, *tet*O, *tet*Q, and *tet*W) were further quantified using qPCR. All six classes of Tc^R genes were detected in high abundance in both winter and summer bioaerosol samples (Fig. 4). Tc^R genes encoding RPP (*tet*O, *tet*Q and *tet*W) were present in significantly higher copy numbers than Tc^R encoding EFP (*tet*B, *tet*H and *tet*Z; *t*-test, *P*-value = 0.04). Seasonal trends were also detected in four Tc^R genes, which include *tet*H, *tet*O, *tet*Q, and *tet*W, and their abundances peaked in bioaerosol samples collected during the winter (Fig. 4). Among the six microclimate variables, air speed was found negatively correlated ($P<0.05$) with the abundances of 16S rRNA, *tet*H, *tet*O, *tet*Q, and *tet*W genes (Table 3), however, PM2.5 and TSP were found positively correlated ($P<0.05$) with these genes (Table 3).

Discussion

In this study, Illumina sequencing was used to provide a comprehensive insight into the bacterial bioaerosol community composition and diversity in SCBs. The use of high-throughput molecular sequencing methods revealed indoor microbial biodiversity that was previously difficult or impossible to observe [38].

Our results indicate that bacterial bioaerosol diversity was significantly higher in winter and this result is in contrast with the findings of Nehme et al. [14], who examined the influence of seasonal variation on bacterial biodiversity in SCBs and found that

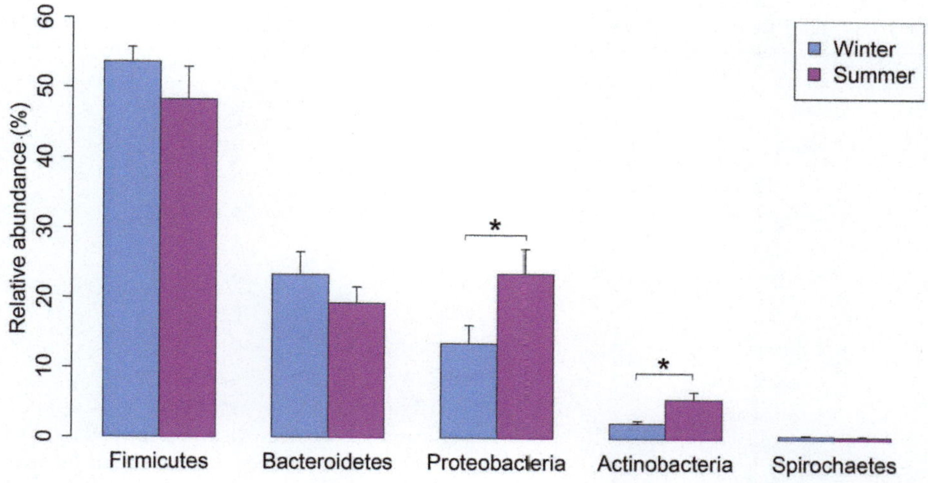

Figure 2. Relative abundance (means ± SE) of the most abundant bacterial phyla detected in SCBs bioaerosol during the winter and summer seasons. * indicates significantly different at >0.05.

biodiversity was unchanged between different seasons of the year. One of the possible explanations for this discrepancy in results could be that Nehme et al. [14] analyzed very limited number of sequences for estimating the bacterial bioaerosol diversity. Furthermore, Nehme et al. [14] used denaturing gradient gel electrophoresis and 16S-cloning-and-sequencing approach to characterize the bacterial bioaerosol community, which lack resolution and throughput, respectively, compared to next-generation sequencing based methods [39–41]. We found that both air speed and particulate matter strongly influenced the bacterial bioaerosol diversity. Airspeed has been shown to impact the diversity of indoor bacterial communities [42], and these results suggest that the higher diversity levels in the winter season are likely to be a function of ventilation rate and particulate matter, as during summer under high ventilation rates, more airborne particulate matter carrying bacteria would be transferred out of the SCBs.

Seasonality also influenced bacterial bioaerosol community composition and most of the observed variation was explained by the microclimate variables. In several previous studies, it has been reported that microclimate variables are the most important factor which affects the indoor bacterial bioaerosol community composition and diversity [42,43]. This relationship could be due to a direct link between the growth and survival of certain taxa and microclimate conditions in SCBs, or an increase in the dispersal of microbes from animals or animal feces under these conditions. Consistent with previous studies [14,16,44], our bioaerosol samples of SCBs were dominated by the phyla *Firmicutes*, *Bacteroidetes*, *Proteobacteria*, and *Actinobacteria*. The genera *Lactobacillus* and *Prevotella*, which accounted for the predominant bacterial genera in our samples, are commonly associated

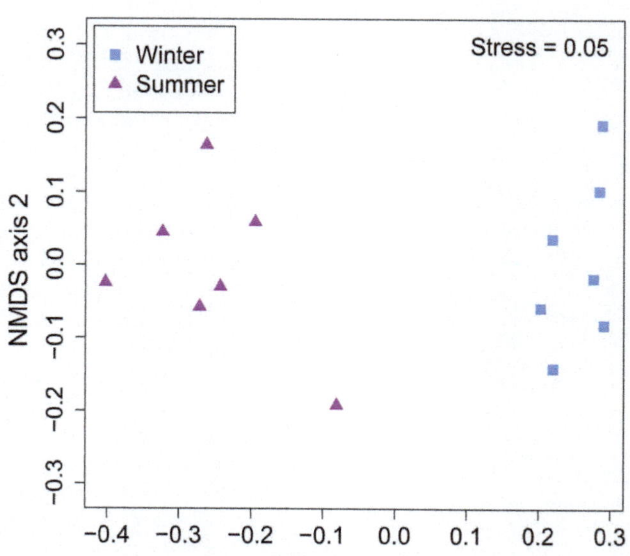

Figure 3. NMDS of Bray-Curtis pairwise dissimilarity of bacterial bioaerosol community in SCBs during the winter and summer seasons.

Table 4. Differentially abundant bacterial genera in swine confinement buildings sampled during winter and summer seasons.

Phylum	Genus	Winter (%)	Summer (%)	p-value
Bacteroidetes	*Capnocytophaga*	0.00	**3.06**	<0.01
Bacteroidetes	*Chitinophaga*	0.01	**0.68**	<0.01
Bacteroidetes	*Dyadobacter*	0.00	**0.37**	<0.01
Bacteroidetes	*Prevotella*	**19.57**	6.24	0.01
Firmicutes	*Blautia*	**0.75**	0.19	0.01
Firmicutes	*Bulleidia*	**0.56**	0.33	<0.01
Firmicutes	*Butyricicoccus*	**0.49**	0.19	0.01
Firmicutes	*Catenibacterium*	**0.74**	0.28	<0.01
Firmicutes	*Faecalibacterium*	**0.87**	0.20	<0.01
Firmicutes	*Oscillibacter*	**0.38**	0.12	0.01
Firmicutes	*Ruminococcus*	**0.33**	0.11	<0.01
Firmicutes	*Streptococcus*	0.37	**2.64**	0.01
Firmicutes	*Subdoligranulum*	**0.69**	0.28	0.01
Proteobacteria	*Escherichia*	0.02	**1.26**	<0.01
Proteobacteria	*Haemophilus*	0.00	**2.82**	<0.01
Proteobacteria	*Neisseria*	0.00	**0.32**	0.01
Proteobacteria	*Sphingomonas*	0.17	**3.70**	0.01

Differences in relative abundance of bacterial genera between seasons are represented with Metastats p-values. Significantly different ($P \leq 0.01$) and relatively abundant genera (>0.3%) were displayed.

with the fecal microbiota; this result further validates the claims that swine feces and soiled bedding material are the primary source of the bacteria in the bioaerosols of SCBs [14,16,45]. Although many of the genera described were detected in both seasons, many others were only detected in one of the seasons, thus justifying the view that the bacterial bioaerosol communities in both the winter and summer season samples are different.

The high abundance of 16S rRNA and Tc^R genes detected in the bioaerosol samples suggests an alternative airborne transmission route, which can lead to their persistence and dispersal into the nearby environment. The level of contamination detected in this study was far higher than the proposed limit of bacterial contamination associated with human respiratory symptoms [8]. Similar to several previous studies, our results indicate that swine workers are exposed to a higher level of airborne bacteria than the occupational recommendations [13,14,44]. The prevalence of Tc^R genes encode RPP in the present study is not surprising, since these genes were found to be predominant in the gastrointestinal tracts of pigs and steers [46], and their elevated possibilities of transfer from one bacteria to another because of their close relationship with mobile genetic elements such as plasmids, conjugative transposons, integrons, and consequently their wide host range [23,24]. Among all Tc^R genes, *tet*Q had the highest abundance in bioaersol samples (the average abundance was $8.89 \times 10^5 \pm 1.45 \times 10^6$ copies m^{-3}) followed by *tet*Z, *tet*O, *tet*W and *tet*H, with *tet*B having the least abundance. The relatively high level of *tet*Q is not surprising because it is seen equally in both Gram-positive and Gram-negative bacterial genera [23], and most of which have been shown to dominate the bioaerosols of SCBs such as *Clostridium*, *Lactobacillus*, *Staphylococcus*, *Streptococcus*, *Prevotella* etc. [14,16,44]. The lack of *tet*B and *tet*H is also not surprising because they are found only associated with Gram-negative bacteria [23], which are less common in bioaerosols of SCBs. However, higher average abundance of *tetZ* compared to *tet*O and *tet*W was not expected because it has relatively narrow host range (Only detected in *Lactobacillus* and *Corynebacterium*; [23]). These results are consistent with the recent findings that both ecology and bacterial phylogeny are the primary determinant of ARG content in environment [21,47]. qPCR is more indicative of the potential for aerosol-mediated transfer of antibiotic resistance between environments than culture-based methods, and results can be more easily compared among studies. Nonetheless, the method is limited by it its ability to detect only a fragment of genes targeted. Truncated sequences and non-expressed sequences cannot be resolved from expressed gene sequences using qPCR,

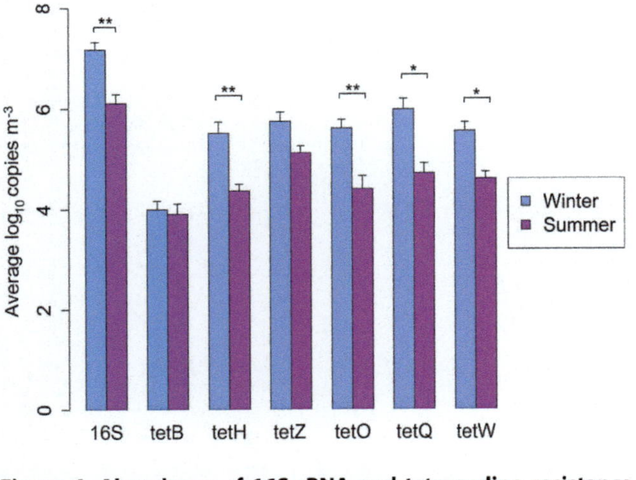

Figure 4. Abundance of 16S rRNA and tetracycline resistance genes in the bioaerosols of SCBs. Asterisks above solid lines indicate significant differences between the winter and summer seasons samples of SCBs. * indicates significantly different at >0.05, ** indicates at >0.01.

so the levels reported could overestimate the number of functional, full length genes.

Our findings also revealed that 16S rRNA and most of the Tc^R genes showed higher abundance in winter bioaerosol samples. Seasonal trends in microbial 16S rRNA gene concentration has already been reported by Nehme et al. [15,29], who showed that total microbial 16S rRNA genes concentrations in bioaersol of SCBs were significantly higher in winter. Seasonal fluctuation in Tc^R genes abundances have been reported several times in wastewater treatment plants and livestock lagoons [48,49]; however, this is the first time a seasonal trend has been reported in bioaerosol samples of SCBs. The abundance of these genes were negatively correlated to airspeed and positively correlated to PM2.5 and TSP. The reduced ventilation in SCBs to avoid heat loss in winter could be responsible for the increased concentration of TSP and PM2.5 in SCBs, which in turn could increase the abundance of 16S rRNA and Tc^R genes. In this survey only a limited number of ARGs were investigated in bioaersols, however, a variety of ARGs encoding different antibiotic resistance could be present in bioaerosols of SCBs, where different classes of antibiotics (e.g., macrolides, lincosamides) are frequently used in addition to tetracycline. So there is a need for further study to explore more diverse ARGs in bioaersols of SCBs.

The detection of 16S rRNA and Tc^R genes in high abundance is of particular concern, because Tc^R airborne pathogenic bacteria present in SCBs have been shown to colonize the nasal flora of pig farmers [50], and this could pose potential occupational health problem. Indoor air ventilated from the SCBs to the external environment can cause detrimental effects to the ambient air quality. For instance, multiple drug resistances bacteria were recovered in an air plume up to 150 m downwind from a SCB at higher percentages than upwind [51]. Furthermore, presence of some airborne pathogens have been detected over long distances from their emission site which were found capable to infect healthy animals intramuscularly or intratracheally [52]. Most of the previous studies consistently indicated an association between environmental exposure to SCBs and respiratory symptoms indicative of asthma of their neighbors [53–55]. Surprisingly, Smit et al. [56] found an inverse association between indicators of air pollution from livestock farms and respiratory morbidity among neighboring resident. However, before drawing firm conclusions from this study, these results should be confirmed with more objective disease information.

Though, this study had fewer samples than previous studies, our results however indicate that seasons have an influence on the biotic contaminants abundance, community composition and diversity, in indoor air of SCBs. Seasonality was significantly associated with microclimate variables, indicating that indoor environmental conditions play an important role in structuring airborne biotic contaminants in SCBs. Based on the results of this study, better management practices and regulations can be designed to minimize the potential health impact on both the farm workers and the public residing in close proximity to these buildings.

Supporting Information

Figure S1 Indoor plan view of sampling points (circles) in swine confinement buildings. The microclimate variables were measured from 3 points (P1, P2 and P3) and aerosol samples were collected from the middle point (P2) in the aisle outside the pens.

Figure S2 Venn diagrams showing the overlap of OTUs (at 97% similarity) between winter and summer seasons. All samples in each season were pooled and then the percentage of shared and season-specific OTUs was calculated.

Table S1 Q-PCR primers used to quantify the abundance of 16S rRNA genes and tetracycline resistance genes.

Acknowledgments

Priyanka Kumari was supported by SNU Global and BK21 plus scholarships. We are grateful to Professor Naomichi Yamamoto for his insightful comments on the manuscript and to two anonymous reviewers for helpful comments.

Author Contributions

Conceived and designed the experiments: HLC PK. Performed the experiments: PK. Analyzed the data: PK. Contributed reagents/materials/analysis tools: HLC PK. Wrote the paper: HLC PK.

References

1. O'Neill D, Phillips V (1992) A review of the control of odour nuisance from livestock buildings: Part 3, properties of the odorous substances which have been identified in livestock wastes or in the air around them. J Agr Eng Res 53: 23–50.
2. Cambra-López M, Aarnink AJ, Zhao Y, Calvet S, Torres AG (2010) Airborne particulate matter from livestock production systems: A review of an air pollution problem. Environ Pollut 158: 1–17.
3. Carpenter G (1986) Dust in livestock buildings—review of some aspects. J Agr Eng Res 33: 227–241.
4. Donham K, Haglind P, Peterson Y, Rylander R, Belin L (1989) Environmental and health studies of farm workers in Swedish swine confinement buildings. Br J Ind Med 46: 31–37.
5. Heederik D, Brouwer R, Biersteker K, Boleij JS (1991) Relationship of airborne endotoxin and bacteria levels in pig farms with the lung function and respiratory symptoms of farmers. Int Arch Occup Environ Health 62: 595–601.
6. Choudat D, Goehen M, Korobaeff M, Boulet A, Dewitte JD, et al. (1994) Respiratory symptoms and bronchial reactivity among pig and dairy farmers. Scand J Work Environ Health 20: 48–54.
7. Mackiewicz B (1998) Study on exposure of pig farm workers to bioaerosols, immunologic reactivity and health effects. Ann Agr Environ Med 5: 169–176.
8. Cormier Y, Tremblay G, Meriaux A, Brochu G, Lavoie J (1990) Airborne microbial contents in two types of swine confinement buildings in Quebec. Am Ind Hyg Assoc J 51: 304–309.
9. Israel-Assayag E, Cormier Y (2002) Adaptation to organic dust exposure: a potential role of l-selectin shedding? Eur Respir J 19: 833–837.
10. Dosman J, Lawson J, Kirychuk S, Cormier Y, Biem J, et al. (2004) Occupational asthma in newly employed workers in intensive swine confinement facilities. Eur Respir J 24: 698–702.
11. Charavaryamath C, Singh B (2006) Pulmonary effects of exposure to pig barn air. J Occup Med Toxicol 1:10.
12. Ministry of Environment Republic of Korea (2008) Environmental Statistics Yearbook, No. 12 (in Korean).
13. Chang C, Chung H, Huang C-F, Su H-JJ (2001) Exposure of workers to airborne microorganisms in open-air swine houses. Appl Environ Microbiol 67: 155–161.
14. Nehme B, Létourneau V, Forster RJ, Veillette M, Duchaine C (2008) Culture-independent approach of the bacterial bioaerosol diversity in the standard swine confinement buildings, and assessment of the seasonal effect. Environ Microbiol 10: 665–675.
15. Nehmé B, Gilbert Y, Létourneau V, Forster RJ, Veillette M, et al. (2009) Culture-independent characterization of archaeal biodiversity in swine confinement building bioaerosols. Appl Environ Microbiol 75: 5445–5450.
16. Hong PY, Li X, Yang X, Shinkai T, Zhang Y, et al. (2012) Monitoring airborne biotic contaminants in the indoor environment of pig and poultry confinement buildings. Environ Microbiol 14: 1420–1431.

17. Blake D, Hillman K, Fenlon D, Low J (2003) Transfer of antibiotic resistance between commensal and pathogenic members of the Enterobacteriaceae under ileal conditions. J Appl Microbiol 95: 428–436.
18. Zhu Y-G, Johnson TA, Su J-Q, Qiao M, Guo G-X, et al. (2013) Diverse and abundant antibiotic resistance genes in Chinese swine farms. Proc Natl Acad Sci USA 110: 3435–3440.
19. Baquero F, Martínez J-L, Cantón R (2008) Antibiotics and antibiotic resistance in water environments. Curr Opin Biotechnol 19: 260–265.
20. Delsol A, Anjum M, Woodward M, Sunderland J, Roe J (2003) The effect of chlortetracycline treatment and its subsequent withdrawal on multi-resistant *Salmonella enterica* serovar Typhimurium DT104 and commensal *Escherichia coli* in the pig. J Appl Microbiol 95: 1226–1234.
21. Gibson MK, Forsberg KJ, Dantas G (2014) Improved annotation of antibiotic resistance determinants reveals microbial resistomes cluster by ecology. ISME J doi:10.1038/ismej.2014.106.
22. Ling AL, Pace NR, Hernandez MT, LaPara TM (2013) Tetracycline resistance and class 1 integron genes associated with indoor and outdoor aerosols. Environ Sci Technol 47: 4046–4052.
23. Roberts MC, Schwarz S, Aarts HJ (2012) Erratum: Acquired antibiotic resistance genes: an overview. Front Microbiol 3: Article 384.
24. Chopra I, Roberts M (2001) Tetracycline antibiotics: mode of action, applications, molecular biology, and epidemiology of bacterial resistance. Microbiol Mol Biol Rev 65: 232–260.
25. Santamaría J, López L, Soto CY (2011) Detection and diversity evaluation of tetracycline resistance genes in grassland-based production systems in Colombia, South America. Front Microbiol 2: Article 252.
26. Huse SM, Dethlefsen L, Huber JA, Welch DM, Relman DA, et al. (2008) Exploring microbial diversity and taxonomy using SSU rRNA hypervariable tag sequencing. PLoS Genet 4: e1000255.
27. Zhou H-W, Li D-F, Tam NF-Y, Jiang X-T, Zhang H, et al. (2011) BIPES, a cost-effective high-throughput method for assessing microbial diversity. ISME J 5: 741–749.
28. Schloss PD, Westcott SL, Ryabin T, Hall JR, Hartmann M, et al. (2009) Introducing mothur: open-source, platform-independent, community-supported software for describing and comparing microbial communities. Appl Environ Microbiol 75: 7537–7541.
29. Edgar RC, Haas BJ, Clemente JC, Quince C, Knight R (2011) UCHIME improves sensitivity and speed of chimera detection. Bioinformatics 27: 2194–2200.
30. Huse SM, Welch DM, Morrison HG, Sogin ML (2010) Ironing out the wrinkles in the rare biosphere through improved OTU clustering. Environ Microbiol 12: 1889–1898.
31. Kim O-S, Cho Y-J, Lee K, Yoon S-H, Kim M, et al. (2012) Introducing EzTaxon-e: a prokaryotic 16S rRNA gene sequence database with phylotypes that represent uncultured species. Int J Syst Evol Microbiol 62: 716–721.
32. Levy SB, McMurry LM, Barbosa TM, Burdett V, Courvalin P, et al. (1999) Nomenclature for new tetracycline resistance determinants. Antimicrob Agents Chemother 43: 1523–1524.
33. Faith DP (1992) Conservation evaluation and phylogenetic diversity. Biol Conserv 61: 1–10.
34. Oksanen J, Kindt R, Legendre P, O'Hara B, Stevens MHH, et al. (2007) The vegan package. Community ecology package.
35. Chen H, Boutros PC (2011) VennDiagram: a package for the generation of highly-customizable Venn and Euler diagrams in R. BMC Bioinformatics 12: 35.
36. White JR, Nagarajan N, Pop M (2009) Statistical methods for detecting differentially abundant features in clinical metagenomic samples. PLoS Comput Biol 5: e1000352.
37. R Development Core Team (2008) R: A language and environment for statistical computing. Vienna, Austria: R Foundation for Statistical Computing.
38. Sogin ML, Morrison HG, Huber JA, Welch DM, Huse SM, et al. (2006) Microbial diversity in the deep sea and the underexplored "rare biosphere". Proc Natl Acad Sci USA 103: 12115–12120.
39. Bent SJ, Pierson JD, Forney LJ (2007) Measuring species richness based on microbial community fingerprints: the emperor has no clothes. Appl Environ Microbiol 73: 2399–2401.
40. MacLean D, Jones JD, Studholme DJ (2009) Application of'next-generation'sequencing technologies to microbial genetics. Nat Rev Microbiol 7: 287–296.
41. Bartram AK, Lynch MD, Stearns JC, Moreno-Hagelsieb G, Neufeld JD (2011) Generation of multimillion-sequence 16S rRNA gene libraries from complex microbial communities by assembling paired-end Illumina reads. Appl Environ Microbiol 77: 3846–3852.
42. Kembel SW, Meadow JF, O'Connor TK, Mhuireach G, Northcutt D, et al. (2014) Architectural design drives the biogeography of indoor bacterial communities. PLoS One 9: e87093.
43. Kembel SW, Jones E, Kline J, Northcutt D, Stenson J, et al. (2012) Architectural design influences the diversity and structure of the built environment microbiome. ISME J 6: 1469–1479.
44. Kristiansen A, Saunders AM, Hansen AA, Nielsen PH, Nielsen JL (2012) Community structure of bacteria and fungi in aerosols of a pig confinement building. FEMS Microbiol Ecol 80: 390–401.
45. Nonnenmann M, Bextine B, Dowd S, Gilmore K, Levin J (2010) Culture-independent characterization of bacteria and fungi in a poultry bioaerosol using pyrosequencing: a new approach. J Occup Environ Hyg 7: 693–699.
46. Aminov R, Garrigues-Jeanjean N, Mackie R (2001) Molecular ecology of tetracycline resistance: development and validation of primers for detection of tetracycline resistance genes encoding ribosomal protection proteins. Appl Environ Microbiol 67: 22–32.
47. Forsberg KJ, Patel S, Gibson MK, Lauber CL, Knight R, et al. (2014) Bacterial phylogeny structures soil resistomes across habitats. Nature 509: 612–616.
48. McKinney CW, Loftin KA, Meyer MT, Davis JG, Pruden A (2010) *tet* and *sul* antibiotic resistance genes in livestock lagoons of various operation type, configuration, and antibiotic occurrence. Environ Sci Technol 44: 6102–6109.
49. Yang Y, Li B, Ju F, Zhang T (2013) Exploring variation of antibiotic resistance genes in activated sludge over a four-year period through a metagenomic approach. Environ Sci Technol 47: 10197–10205.
50. Létourneau V, Nehmé B, Mériaux A, Massé D, Cormier Y, et al. (2010) Human pathogens and tetracycline-resistant bacteria in bioaerosols of swine confinement buildings and in nasal flora of hog producers. Int J Hyg Environ Health 213: 444–449.
51. Gibbs SG, Green CF, Tarwater PM, Mota LC, Mena KD, et al. (2006) Isolation of antibiotic-resistant bacteria from the air plume downwind of a swine confined or concentrated animal feeding operation. Environ Health Perspect 114: 1032.
52. Otake S, Dee S, Corzo C, Oliveira S, Deen J (2010) Long-distance airborne transport of infectious PRRSV and *Mycoplasma hyopneumoniae* from a swine population infected with multiple viral variants. Vet Microbiol 145: 198–208.
53. Kilburn KH (2012) Human impairment from living near confined animal (hog) feeding operations. J Environ Public Health 2012: 565690.
54. Schinasi L, Horton RA, Guidry VT, Wing S, Marshall SW, et al. (2011) Air pollution, lung function, and physical symptoms in communities near concentrated swine feeding operations. Epidemiology 22: 208–215.
55. Pavilonis BT, Sanderson WT, Merchant JA (2013) Relative exposure to swine animal feeding operations and childhood asthma prevalence in an agricultural cohort. Environ Res 122: 74–80.
56. Smit LA, Hooiveld M, van der Sman-de Beer F, Opstal-van Winden AW, Beekhuizen J, et al. (2013) Air pollution from livestock farms, and asthma, allergic rhinitis and COPD among neighbouring residents. Occup Environ Med doi:10.1136/oemed-2013-101485.

22

Systematic Review of the Literature and Evidence-Based Recommendations for Antibiotic Prophylaxis in Trauma: Results from an Italian Consensus of Experts

Daniele Poole[1]*, Arturo Chieregato[2], Martin Langer[3], Bruno Viaggi[2], Emiliano Cingolani[4], Paolo Malacarne[5], Francesca Mengoli[6], Giuseppe Nardi[4], Ennio Nascimben[7], Luigi Riccioni[4], Ilaria Turriziani[6], Annalisa Volpi[8], Carlo Coniglio[6], Giovanni Gordini[6], on behalf of the Trauma Update Working Group

1 U.O. Anestesia e Rianimazione, Ospedale S. Martino, Belluno, Italy, **2** SOD Anestesia e Area Intensiva CTO, AOU Careggi, Firenze, Italy, **3** Dipartimento di Anestesia e Rianimazione, Fondazione IRCCS Istituto Nazionale dei Tumori e Università degli Studi di Milano, Milano, Italy, **4** UOC Shock e Trauma, AO San Camillo-Forlanini, Roma, Italy, **5** U.O. Anestesia e Rianimazione - P.S., Azienda Ospedaliera Universitaria Pisana, Pisa, Italy, **6** U.O. Rianimazione 118, Ospedale Maggiore, Bologna, Italy, **7** Neurorianimazione Ospedale S. Maria di Ca' Foncello, Treviso, Italy, **8** Anestesia e Rianimazione 1, Azienda Ospedaliero-Universitaria di Parma, Parma, Italy

Abstract

Background: Antibiotic prophylaxis is frequently administered in severe trauma. However, the risk of selecting resistant bacteria, a major issue especially in critical care environments, has not been sufficiently investigated. The aim of the present study was to provide guidelines for antibiotic prophylaxis for four different trauma-related clinical conditions, taking into account the risks of antibiotic-resistant bacteria selection, thus innovating previous guidelines in the field.

Methods: The MEDLINE database was searched for studies comparing antibiotic prophylaxis to controls (placebo or no antibiotic administration) in four clinical traumatic conditions that were selected on the basis of the traumatic event frequency and/or infection severity. The selected studies focused on the prevention of early ventilator associated pneumonia (VAP) in comatose patients with traumatic brain injury, of meningitis in severe basilar skull fractures, of wound infections in long-bone open fractures. Since no placebo-controlled study was available for deep surgical site-infections prevention in abdominal trauma with enteric contamination, we compared 24-hour and 5-day antibiotic prophylaxis policies. A separate specific research focused on the question of antibiotic-resistant bacteria selection caused by antibiotic prophylaxis, an issue not adequately investigated by the selected studies. Randomised trials, reviews, meta-analyses, observational studies were included. Data extraction was carried out by one author according to a predefined protocol, using an electronic form. The strength of evidence was stratified and recommendations were given according to the Grading of Recommendations Assessment, Development and Evaluation (GRADE) criteria.

Results: Uncertain evidence deserving further studies was found for two-dose antibiotic prophylaxis for early VAP prevention in comatose patients. In the other cases the risk of resistant-bacteria selection caused by antibiotic administration for 48 hours or more, outweighed potential benefits.

Conclusions: When accounting for antibiotic-resistant bacteria selection we found no evidence in favour of antibiotic prophylaxis lasting two or more days in the studied clinical conditions.

Editor: Zhi Zhou, Purdue University, United States of America

Funding: The authors have no support or funding to report.

Competing Interests: The authors have declared that no competing interests exist.

* Email: daniele.poole@alice.it

Introduction

In the attempt to investigate antibiotic prophylaxis in several trauma-related clinical conditions, taking into account the potential ecological risk, innovating previous approaches, in February 2013 a meeting of experts was held in Bologna in the context of the 8th edition of the *Trauma Update and Organization* conference.

Infection is the consequence of contamination due to the contact between normally sterile tissues and microorganisms from the external environment or from internal sites. The inoculation occurs when the trauma breaks natural barriers against infections.

All attempts to reduce the bacterial charge in the tissues should be made. Cleaning, irrigation, disinfection of wounds, followed by the appropriate early surgical treatment when indicated, are crucial for infection prevention. Antimicrobial prophylaxis is only one element integrating these interventions and should be limited to the period immediately after the trauma.

One major drawback of exposure to antibiotics is the emergence of drug-resistant bacteria strains, as the result of selection pressure on resistant colonizing strains or on bacteria with newly acquired mutations. Unfortunately, most studies dealing with antibiotic prophylaxis have not adequately investigated this issue, which is particularly relevant in critical care settings where bacterial resistance to antibiotics has become a major problem. Here, we focused on studies comparing antibiotic prophylaxis and placebo in reducing infections in several trauma-related conditions taking into account the risk of antibiotic-resistant bacteria selection to produce a fair risk/benefit ratio assessment, in contrast with other guidelines addressing similar issues [1,2].

In the present paper, we report the contents of the review, the analysis and discussion of the literature, and the conclusions of the conference.

Methods

The works was organized in three sequential steps:

Step 1: The Coordinating Committee formed by five intensivists selected an initial list of topics. Head, thorax, abdomen, and limbs trauma were considered. Because of time constraints (there were only two days to discuss the issues during the convention) only four clinical conditions were selected based on traumatic event frequency and/or infection severity. The scenarios included antibiotic prophylaxis for 1) early VAP in comatose patients, 2) meningitis in severe skull fractures, 3) wound infection in long-bone open fractures (we initially focused on osteomyelitis but found no study specifically investigating this issue), 4) deep surgical-site infections in abdominal trauma with hollow viscus perforation. We also considered but discarded pneumonia in thoracic contusion, empyema after chest tube insertion, surgical-site infection for long-bone closed fractures, infection after soft tissues open injuries, abdominal trauma without bowel perforation, and inhalation pneumonia in pre-hospital intubation.

The Coordinating Committee then generated the queries for the literature search and selected the articles fulfilling the criteria established for each subject. The fundamental criterion, shared by all the investigated issues, was the comparison between adult patients subjected to antibiotic prophylaxis and controls. When such studies were not available comparison between groups receiving different pharmacological protocols were considered.

The MEDLINE database was searched using the free PubMed provider. Randomised trials, reviews, meta-analyses, or observational studies were selected. Pediatric and animal studies, and articles not written in English, were excluded. The MEDLINE search covered a period between January 1970 and January 2014. This search was integrated with the consultation of reference lists of retrieved articles.

The Coordinating Committee members unanimously selected the articles after reviewing the abstracts, with the help of one of the authors with specific expertise in statistics and research methodology. Full-texts were then retrieved and a further selection was carried out.

The statistical and methodological expert also performed data extraction according to a predefined plan (the protocol was not pre-registered), using a dedicated electronic form that automatically performed all planed computations and plot generation. The electronic form was developed *ad hoc* for this revision, and tested and refined using the studies on early VAP prevention with antibiotic prophylaxis.

Data collected included study inclusion criteria, antibiotic class and dose, duration of prophylaxis, antibiotic-resistant bacteria selection when investigated, infection rates, and mortality rates when available, along with information needed for the GRADE evidence quality assessment.

Data were presented as event rates in treatment arms and controls, absolute risks, absolute risk reductions, and relative risks. Adjusted odds ratios were reported when dealing with multivariate analysis results in observational studies. Numbers needed to treat were also calculated. Confidence intervals were computed for all the above measures. Relative risks were graphically represented in Forest plots.

Design heterogeneity among selected studies advised against their combination in meta-analyses. For the same reason statistical approaches to investigate presence of asymmetry and potential publication bias were not carried out [3].

The queries and the results of the summarised results were submitted to the External Scientific Advisory Panel, formed by four intensivists with expertise in the field of infections and trauma in critical care settings, who discussed them together with Coordinating Committee members.

Evidence provided by randomised controlled trials (RCTs) and observational studies was ranked on the basis of the Grading of Recommendations Assessment, Development and Evaluation (GRADE) criteria. [4–6] The GRADE provides a framework to guide reviewers for the assessment of the strength of evidence, taking into account that different reviewers or reviewing groups can evaluate the same evidence differently. [7] Moreover, it eliminates automatic links between the strength of the recommendation and the strength of the evidence, leaving to the reviewers the possibility of grading recommendations based on other variables relevant to clinical decisions, besides evidence.

The GRADE stratifies quality of evidence on a four-level scale, ranging from high to very low, [7], in favour or against intervention. In the middle of this scale an absence of evidence level was added. This event occurs frequently in studies with small sample size, where the alternative hypothesis and statistical power have not been adequately taken into account. In our opinion, negative results from such studies should thus not be considered indicative of no effect, but rather of an unresolved issue [8].

The reporting of RCTs was evaluated on the basis of currently shared standards. [9] We considered quasi-randomised studies, in which treatment allocation is based on non-random methods such as alternation, leading to severe bias in regards to sequence generation and allocation concealment.

Our evaluation of meta-analyses was restrictive, according to the principle that heterogeneity among studies in terms of inclusion criteria, study design, study treatment, type of outcome, which are not accounted for by statistical methods focused on heterogeneity of outcome rates, [10] should contraindicate the merging of their results. [11] When dealing with observational studies we discarded al investigations that reported only crude-data comparisons without performing adjustment for potential confounders.

Step 2: The results were presented to the physicians invited to the symposium - intensivists, infectious disease specialists, surgeons, emergency physicians, radiologists, and physiatrists – and the issues were collectively discussed. The discussion was audio-recorded and at a later stage revised by members of the Coordinating Committee to identify and highlight the main opinions that emerged from the discussion.

Step 3: the scientific advisors and the Coordinating Committee members formed the jury that provided recommendations that were presented as preliminary results on the final day of the conference, but discussed and refined in the following months by e-mail. A final meeting was held one year after the conference, to discuss diverging opinions until unanimity was reached. To develop the recommendations the panel followed the GRADE framework. This is based on four points: quality of evidence, balance between desirable and undesirable effects, values and preferences, uncertainty about whether the intervention represents a wise use of resources [12].

The GRADE initially attributes a high default level of evidence to RCTs and a low level to observational studies. [7] RCTs can then be up- or downgraded according to the following criteria: study limitations due to insufficient internal validity, indirectness of evidence, imprecision of results, inconsistency between studies, and selective publication of studies. [13–17] While the first three are related to single studies, the last two points take into account the relation with other studies in terms of inconsistent results and publication bias, respectively. We added an *other* criterion which included other weak features of the study, such as the unbalance in important variables between study arms in RCTs that may occur when the sample size is small.

For RCTs the first criterion, i.e. bias, takes into account allocation concealment and blinding, completeness of follow-up and adherence to the intention-to-treat principle, premature study interruptions for benefit, and selective reporting of outcomes. [17] The last criterion takes into account the existence and availability of a study protocol or on the clear omission to report important outcomes. [17] We were not able to comply with this criterion because, for most RCTs, protocols were not available. For each of these items, concerns can be serious or very serious, down-scoring the evidence by one and two points respectively. For example, a RCT with an initial high-evidence grading can be downgraded to moderate or low, or even very low, if there are other weak points.

Limits in observational studies can include lack of controls, as in case series, or, when controls are present, lack of adjustment for important covariates, which may be unevenly distributed between study groups [17].

Initially, evidence is rated only one point up when the relative risk is between 0.5 and 0.2 or between 2 and 5, and two points up when lower than 0.2 or higher than 5, obviously when dealing with statistically significant results. However, even in the presence of large effects, when imprecision is strong (i.e. wide 95% confidence intervals) and negligible effects cannot be ruled out, upgrading is not performed. The rating can also be increased when a dose-response is evident or plausible potential confounders would only increase the confidence in the results [7,18].

The strength of recommendations can be strong or weak, in favour or against interventions. [19] Only when an intervention with a strong likelihood of effectiveness had not been sufficiently investigated (i.e. absence of evidence as defined above) we considered the possibility of not giving any recommendation, although we usually gave a weak negative recommendation in most cases of absence of evidence, in the light of the limited resources available.

The conference dealt with four different issues of broad clinical interest but focusing especially on relevance for critically ill patients and from the point of view of intensive care unit (ICU) and emergency department physicians.

To investigate the main undesirable effects of prophylaxis, we preliminarily performed a systematic research on the effect of short antibiotic administration on bacterial antibiotic-resistance selection, to provide a fair assessment of the balance between benefits and risks. The results of this investigation integrated, when appropriate, the evidence concerning antibiotic prophylaxis in the clinical conditions we explored.

A transparent report of the reviewing process and of evidence grading is provided for each query in the supporting information.

The review was conducted complying with the Preferred Reporting Items for Systematic Reviews and Meta-Analyses (PRISMA) statement recommendations (Table S6 in File S6) [20].

Results

We tried to find an answer to the following question: *Does short antibiotic prophylaxis cause resistant bacteria selection?* To do this we explored the influence of short antibiotic course on resistant bacteria selection at two levels: at the patient's level, investigated before and after exposure, and at the centre level, comparing restrictive and liberal antibiotic prescription strategies between centres or within a single centre before and after the policy change.

We also assumed that there is strong evidence that prolonged antibiotic exposure (say, 7–10 days) increases the risk of colonization with resistant bacterial strains, [21] and without formally investigating this issue with the GRADE approach, we attributed to it a default high level of evidence.

Selection of antibiotic-resistant bacteria in patients exposed to prophylaxis

The contemporaneous combination of the mutant bacteria and antibiotic exposure determines the spread of antibiotic-resistant bacteria. [22] Although patients admitted to the hospital after trauma are usually young and healthy, given the presence of resistant bacteria also within the community, the possibility that they may be already colonized by such strains cannot be ruled out. [23,24] Moreover, colonization in non-colonized patients can occur because of newly acquired bacterial mutations, selection of bacterial resistant strains among heterogeneous populations, [25] and horizontal transfer, [26] the latter being particularly relevant in ICU environments. [27] We made an extensive literature revision focusing on antibiotic-resistant colonization in patients undergoing short antibiotic treatment, but RCTs in this field were unavailable. Thus, we had to rely on the few observational studies found (Table S5 in File S5).

In a study carried out on a cohort of 2,641 patients receiving antibiotic prophylaxis for cardiac surgery, of 1,094 undergoing cultures 426 were shown to harbour bacteria resistant to the antibiotic they were exposed to. Logistic regression, adjusting for many possible confounders, indicated an increased risk of resistance selection in case of antibiotic prophylaxis lasting more than 48 hours. [28] Overall the description of the statistical procedures was scanty, hindering quality assessment. One major limitation was that only part of the patients were submitted to cultures and that physicians in the centre were "encouraged" to keep antibiotic prophylaxis duration within 48 hours, potentially introducing selection biases that logistic regression may have not accounted for. Moreover, colonization status was not assessed before prophylaxis was administered, so it is not known whether the patients had previous colonization or acquired the resistant pattern during therapy or during the postoperative stay in a high-risk environment as the ICU.

In a before/after study involving 5 wards, the rate of acquisition within 30 days from initial antibiotic exposure, was 3% for MRSA (22/724), 2% for vancomycin-resistant enterococci (VRE) (10/430), and 1% (11/840) for ciprofloxacin-resistant *Pseudomonas Aeruginosa* (CRPA). [29] Forty per cent of newly acquired colonization with resistant bacteria was detected 48 hours from the first antibiotic dose. The initial colonization rate was 16% of 864 patients for MRSA, 50% for VRE, and 3% for ciprofloxacin-resistant *Pseudomonas Aeruginosa*. Cephalosporins, quinolones, and macrolides were the antibiotics more frequently administered (31, 29, 19% of cases, respectively). Absence of control (i.e. patient not exposed to antibiotics) although not irrelevant was not a major issue, in our opinion, because we do not expect factors other than antibiotic pressure to affect the colonization rate within a 48-hour period. A low colonization rate, as in this study, is still clinically relevant if we account for the duration and possibility of horizontal spread of resistant bacterial colonization, especially in environments such as the ICU [27,30].

Another study investigated 29 patients undergoing cardiac surgery and 10 coronary angioplasty, receiving prophylaxis for three days with a first or a second-generation cephalosporin. [31] Cultures were collected at time 0 and on the third day. The investigators detected colonization by methicillin-resistant coagulase-negative staphylococci in 36 of 43 cultured sites (84%) previously not colonized and a quantitative colonization increase in 17 of 28 sites (61%) harbouring low-level colonization before antibiotic administration. The main limitation of the study was that multiple samples were collected from each patient that may have borne the same resistant bacteria, inflating the final count of true events. A reliable quantitative estimation of resistant bacteria selection was hence impossible. This study was also limited by its small sample size.

Finally, in a study investigating the effect of a five-day course with clarithromycin, 5 patients colonised by susceptible enterococcus strains few days after the end of treatment developed resistant types, which in 3 cases persisted for one year or more. [32] Although, almost a case report, this study is indicative of how antibiotic pressure can select resistant strains and how long the effects can last over time.

Effect of liberal vs. aggressive antibiotic prescriptive strategies

The limitation of antibiotics use is widely recognized as an effective strategy for the reduction in bacterial resistance. [33,34] However, the role of antibiotic prophylaxis policies, corresponding to strong limitations in antibiotic use, have never been investigated in relation to antibiotic-resistance acquisition. Indeed a recent review concerning the influence of antibiotic exposure on MRSA diffusion, the shortest period of antibiotic administration considered was 7 days. [21] In a before/after study dealing with late-generation cephalosporin administration policy, surgical prophylaxis was among the exceptions to restriction, indicating that the authors did not feel it was significant for the purposes of their policy [35].

Conclusions: there is low evidence indicating that antibiotic prophylaxis lasting 48 hours or more is capable of determining the colonization with resistant bacterial strains, that is the *undesirable effect* reported in the text and tables for each of the interventions under scrutiny.

Question 1: is antibiotic prophylaxis for the prevention of ventilator-associated lower respiratory tract infections effective in intubated patients with traumatic head injury (TBI) and coma, i.e. Glasgow Coma Scale (GCS) ≤8?

Results: the MEDLINE search provided only three studies, two RCTs and one observational study, [36–38] all complying with our inclusion criteria. Although, the studies did not deal exclusively with TBI patients, the results can be reasonably extended to this category of patients. The two RCTs are described in Table 1, Table S1 in File S1, and Figure S1 and tested respectively a 3-day and a single-day antibiotic prophylaxis. Both RCTs reported a statistically significant reduction of early-VAP, but no reduction of late-onset pneumonia and ICU mortality. Besides other flaws, major limitations of both RCTs were the absence of blinding, the small sample size and the lack of power to detect increases in antibiotic-resistant infections in the treatment group (outcome investigated only in the first RCT). [36] Moreover, the second RCT was not specifically focused on comatose patients. [37] The premature interruption of the first RCT after the first interim analysis when only 38 patients were randomised, on the basis of the reduction of early-onset pneumonia incidence in the treatment arm, while no difference in mortality was detected, is a questionable choice. Finally, the beneficial effect of the postponement of pneumonia in a phase of minor vulnerability of the brain in severe brain injury is only theoretical and was not demonstrated with a robust outcome, such as the long-term mortality/severe disability rate. The main undesirable effect, i.e. antibiotic-resistant bacteria selection, according to the evidence we collected, was likely to occur after three days of antibiotic administration, while probably irrelevant when only two doses are given.

The results of a recent observational study, including 129 patients suggested that a single dose of ceftriaxone was an independent predictor of reduced incidence of early-onset pneumonia. [38] The regression analysis was performed including a propensity score and the variable "antibiotic prophylaxis" in a regression model, but the number of outcomes (overall 15 cases of early-onset VAP) did not justify the inclusion of the two above variables according established rules. [39,40] Moreover, the sample size and the number of events were small, and the model was underpowered to include potential important covariates capable of influencing the significance of ceftriaxone administration in the predictive model. The observational study appeared to be, hence, unreliable and was not taken into account in our final evaluation.

Conclusions: The two RCTs had substantial different interventions and outcomes, thus we provided recommendations for two different queries (Table 2).

Intervention 1. Three-day course with a beta-lactamase-protected penicillin for the prevention of ventilator-associated early-VAP in intubated patients with TBI and coma:

Level of evidence in favour: Very low evidence.

Level of evidence against: Low evidence.

Recommendation: Weak against intervention.

Intervention 2. 2-dose wide spectrum antibiotic administration for the prevention of early-VAP in intubated patients with TBI and coma.

Level of evidence in favour: Very low evidence.

Level of evidence against: No evidence.

Recommendation: Weak against intervention.

Table 1. Antibiotic prophylaxis for early-VAP prevention in comatose patients: patients, interventions, and outcomes.

Sample	Treatment	Control	Desirable effect		Undesirable effect		Desirable effect			
			Outcome	Outcome quality	Outcome	Outcome quality	Treatment arm % (95%-CI)	Controls % (95%-CI)	Difference % (95%-CI)	RR (95%-CI)
Ventilated patients, cerebral haemorrhage, GCS ≤8	Ampicillin-sulbactam 3-day course - 19 patients	No placebo -19 patients	Early VAP	Weak, surrogate of death/severe disability	Antibiotic- resistant bacteria selection not Adequately investigated	Robust, Clinically relevant for the patient/crucial for the health-care organization	21.1 (8.5 to 43.3)	57.9 (36.3 to 76.9)	-36.8 (−59.6 to −5.8)	0.36 (0.14 to 0.94)
Ventilated patients, cerebral haemorrhage, GCS ≤12	2-dose cefuroxime - 50patients	No placebo - 50 patients	Early VAP	Weak, surrogate of death/severe disability	Antibiotic- resistant bacteria selection not investigated	Robust, Clinically relevant for the patient/crucial for the health-care organization	16 (8.3 to 28.5)	36 (24.1 to 49.9)	-20 (−35.8 to −2.8)	0.44 (0.21 to 0.93)

Question 2: is antibiotic prophylaxis effective in patients with basilar skull fractures from non-penetrating head trauma to decrease the occurrence of meningitis?

Results: of 146 studies found only four were RCTs. One was discarded because only 10 patients were randomised while 129 were the object of a retrospective evaluation and therefore included among observational studies. [41] The remaining were single-centre trials, including 109, 37, and 52 patients respectively, all providing negative results, but the three studies were underpowered to demonstrate clinically relevant differences (Table 3, Table S2 in File S2, Figure S2). [42–44] The 37 patients enrolled in the second study with basilar fractures were a subset of 196 with either basilar or skull open fractures. Thus the findings of interest for our review were the result of a subgroup analysis, and as such biased. [45] The other two trials also suffered from methodological bias. Most important long-lasting antibiotic prophylaxis was administered (5-day ceftriaxone, 3-day ceftriaxone or ampicillin-sulbactam, and 8-day average penicillin course, respectively) and no monitoring of antibiotic-resistant bacteria selection was investigated. This was a major undesirable effect of such prolonged antibiotic prophylaxis, as supported by the literature revision we carried out.

The negative evidence regarding this issue are summarised in a recent Cochrane review and meta-analysis [46] that also includes the study of the randomised ten patients mentioned above and a RCT published as correspondence. [47] Although the studies were not heterogeneous in terms of meningitis odds ratios, they were heterogeneous in terms of meningitis absolute rates, inclusion criteria, duration of antibiotic treatment, ceftriaxone dose in the two studies that used this antibiotic, [43,44] and number of patients lost to follow-up, advising against their inclusion in a meta-analysis. Two other non-recent meta-analyses were biased by the merging of RCTs and observational studies [48,49].

We found four observational studies matching the inclusion criteria, none of which performed adjusted comparisons, thus not providing any acceptable evidence in favour of or against antibiotic prophylaxis [41,50–52].

Conclusions: The three studies under scrutiny all investigated prolonged antibiotic administration, that was thus the object of our query (Table 4).

Intervention. Three or more days of antibiotic prophylaxis for the prevention of meningitis in patients with basal-skull fractures:

Level of evidence in favour: No evidence.

Level of evidence against: Low evidence.

Recommendation: Strong against intervention.

Question 3: is antibiotic prophylaxis beneficial for the reduction of wound infection rate in patients with long-bone open fractures?

Results: of 174 studies fulfilling the search criteria, only three trials, [53–55] one observational study, [56] and one meta-analysis [57] matched the inclusion criteria. The studies were carried out in between the 1970s and the 1980s. In the first RCT, patients with open-fractures were only a subset of the randomised sample, [53] the findings thus being the result of a subgroup analysis and as such inadequate to provide conclusive answers. [45] The second study [54] was of higher level, but still the reporting was insufficient to allow a thorough assessment of its quality according to current standards. [9] In these two studies, the diagnosis required positive cultures. The third study was very low quality,

Table 2. Antibiotic prophylaxis for early-VAP prevention in comatose patients - Level of evidence and recommendations.

Patients and intervention	Desirable effect	Undesirable effect*	Benefit/risk profile	Values and preferences	Resource use	Recommendation	Rationale
Ventilated patients, cerebral haemorrhage, GCS≤8 – Ampicillin-sulbactam 3-day course	Very low evidence in favour of intervention +□□□	Low evidence against intervention ++□□	Unfavourable	Not available	Unwise	Weak against intervention	The benefit/risk ratio was unfavourable. Resistant bacteria selection was considered a likely event for a three-day antibiotic course, a more robust and thus prevalent outcome compared to early-VAP reduction, a surrogate of middle term death and severe disability.
Ventilated patients, cerebral haemorrhage, GCS≤8-2-dose cefuroxime	Very low evidence in favour of intervention +□□□	No evidence against intervention □□□□	Uncertain	Not available	Not assessable	Weak against intervention	The intervention requires further investigation focused on hard outcomes such as death and severe disability. The selection of resistant bacteria appears to be unlikely with two doses of antibiotic although it cannot be ruled out. Notwithstanding the large effect size (RR 0.44), given the high degree of imprecision a limited reduction of early VAP (lower 95%-CI = 0.93) is possible, corresponding to a 2.8 absolute risk reduction and a NNTB of 36, a small effect at the price of an excessively high number of patients to expose to antibiotics, especially in the context of the ICU.

*Not investigated in the studies, evidence could be found from external sources. NNTB = number needed to treat for benefit.

Table 3. Antibiotic prophylaxis for the prevention of meningitis in basilar skull fractures: patients, interventions, and outcomes. NC = not computable.

			Desirable effect		Undesirable effect		Desirable effect			
Sample	**Treatment**	**Control**	**Outcome**	**Outcome quality**	**Outcome**	**Outcome quality**	**Treatment arm % (95%-CI)**	**controls % (95%-CI)**	**Difference % (95%-CI)**	**RR (95%-CI)**
Open skull fractures or basilar skull fractures	3-day course ceftriaxone or ampicillin-sulphadiazine - 25 patients	No placebo - 12 patients	Meningitis reduction	Robust, Prognostically relevant outcome	Antibiotic- resistant bacteria selection not investigated	Robust, Clinically relevant for the patient/crucial for the health-care organization	0 (0 to 13.3)	8.3 (1.5 to 35.4)	-8.3 (-35.4 to 6.6)	NC
Acute traumatic pneumo-cephalus verified by CT scan	5-day course ceftriaxone - 53 patients	No placebo - 56 patients	Meningitis reduction	Robust, Prognostically relevant outcome	Antibiotic- resistant bacteria selection not investigated	Robust, Clinically relevant for the patient/crucial for the health-care organization	18.9 (10.6 to 31.4)	21.4 (12.7 to 33.8)	-2.6 (-17.5 to 12.7)	0.88 (0.42 to 1.86)
Traumatic rhinorrhoea or otorrhoea	Average 7.7 days course penicillin - 26 patients	Placebo - 26 patients	Meningitis reduction	Robust, Prognostically relevant outcome	Antibiotic-resistant bacteria selection not investigated	Robust, Clinically relevant for the patient/crucial for the health-care organization	0 (0 to 12.9)	3.8 (0.7 to 18.9)	-3.8 (-18.9 to 9.4)	NC

with no blinding and no placebo, and a partial report of the results. It was carried out on open fractures due to gunshot, with infections diagnosis not based on cultures. [55] This study was negative, since 2 cases out of 32 were reported in the single-day cefazolin group and 2 out of 35 in the group not receiving antibiotic. We considered this study methodologically too weak to be considered a reliable source of evidence.

The antibiotic tested in the first two RCTs (Table 5, Table S3 in File S3, Figure S3) were instead dicloxacillin or benzyl penicillin, [53] and cloxacillin, [54] respectively. In the first study treatment was continued for 48 hours and, in the second, for ten days. The influence of prophylaxis on bacterial resistance to antibiotics was not investigated in any of these studies. We assumed that a 10-day antibiotic administration would certainly greatly increase the risk of antibiotic-resistant bacteria selection. We attributed, instead, a lower level of evidence to a 48-hour treatment (Table 6).

A 13.3 and 22.6% wound-infection rate decrease was found in the antibiotic-prophylaxis arms in the two studies, corresponding respectively to an almost statistically significant result and a fully statistically significant result. Interestingly, in the second study, infections occurring within six weeks were considered early infections, including infections that, reasonably, should have not been influenced by early antibiotic prophylaxis. In the first study, instead, the duration of the "window" period was not specified.

The observational study that matched the inclusion criteria suggested a protective effect of antibiotic prophylaxis, but since it was based only on the crude infection rate, it was not included in the analysis [56].

The meta-analysis included, besides the three RCT mentioned above, two other studies that we excluded because one was written in German [58] and the other including hand, foot, and fingers fracture, pediatric patients and gunshot injuries, without performing any subgroup analysis on long-bone fractures. [59] The meta-analysis concluded that antibiotic prophylaxis is effective in reducing wound infections. However, study designs, inclusion criteria, treatment protocols, and quality of the trials, were so heterogeneous to question the reliability of their merging in a single sample. We hence considered this study a potential source of misleading information.

We did not find any study comparing no antibiotic prophylaxis with 24-hours antibiotic prophylaxis, an option recommended by other guidelines presumably on the basis of indirect evidence [1,2,60,61].

Conclusions: Two different antibiotic strategies were adopted in the two studies that were included in the final evaluation, leading to two answers to two different queries (Table 6).

Intervention 1. 2-day antibiotic administration for the prophylaxis of wound infections (not specifically osteomyelitis).

Level of evidence in favour: No evidence.

Level of evidence against: Low evidence.

Recommendation: Weak against intervention.

Intervention 2. Prolonged (ten-day) antibiotic prophylaxis for the prevention of wound infections (not specifically osteomyelitis).

Level of evidence in favour: Low evidence.

Level of evidence against: High evidence.

Recommendation: Strong against intervention.

Table 4. Antibiotic prophylaxis for the prevention of meningitis in basilar skull fractures - Level of evidence and recommendations.

Patients and intervention	Desirable effect	Undesirable effect*	Benefit/risk profile	Values and preferences	Resource use	Recommendation	Rationale
Open skull fractures or basilar skull fractures - 3-day course ceftriaxone or ampicillin/sulphadiazine	No evidence in favour of intervention □□□□	Low evidence against intervention ++□□	Unfavourable	Not available	Unwise	Strong against intervention	
Acute traumatic pneumocephalus verified by CT scan - 5-day course ceftriaxone	No evidence in favour of intervention □□□□	Low evidence against intervention ++□□	Unfavourable	Not available	Unwise	Strong against intervention	Results of the RCTs were negative. The studies were seriously biased The risk of resistant bacteria selection is a likely and relevant undesirable effect when prolonged antibiotic prophylaxis is administered.
Traumatic rhinorrhoea or otorrhoea - Average 7.7 days course penicillin	No evidence in favour of intervention □□□□	Low evidence against intervention ++□□	Unfavourable	Not available	Unwise	Strong against intervention	

*Not investigated in the studies, evidence could be found from external sources.

Question 4: is antibiotic prophylaxis indicated to reduce the risk of deep surgical site infections in patients with abdominal trauma and enteric abdominal contamination submitted to emergent surgery?

Results: out of 504 articles resulting from the literature search, none compared treatment and no-treatment arms. Our results were consistent with those of two recent Cochrane meta-analyses that did not find any RCTs investigating the effectiveness of antibiotic prophylaxis compared to placebo in reducing infections following penetrating abdominal trauma [62,63].

Since no direct evidence exists with regard to the effectiveness of antibiotic prophylaxis in abdominal trauma compared to placebo, under the assumption that antibiotic prophylaxis is effective in preventing surgical site infections, [64] we oriented the query towards the duration of antibiotic prophylaxis as it was studied by three RCTs focused on penetrating abdominal trauma and deep surgical infections occurrence, [65–67] and excluded a fourth one because it did not report the rate of deep surgical site infections. [68] The main limitation of the studies was that only a minority of the patients had associated hollow viscus injury. Moreover, the infection rate for those who bore intestinal perforations was not reported, making it impossible to analyse subgroup performance. Thus, it was only possible to study the number of events in relation to the overall sample, including those without perforation. Another weakness was that no evidence concerning blunt trauma, a common event in abdominal trauma, was available.

The first two studies dealt almost exclusively with deep surgical site infections, i.e. intra-abdominal abscess, peritonitis, and necrotizing fasciitis, while the third one reported disaggregated data concerning deep and superficial surgical site infections. Thus, we were able to analyse data concerning deep surgical infections. The studies recruited 317, 515, and 300 patients respectively, and infections rates were 10 vs. 8% in the 5-day and 24-hour treatment arms in the first two studies, and 6% in both arms in the third one (Table 7). Although the studies were underpowered to detect clinically meaningful differences in the outcome rate (the power to detect a 5% difference ranged between 40 and 60% in the three RCTs), looking at the range of the absolute-difference 95%-confidence intervals, there is little chance that a clinically relevant superiority of 5-day prophylaxis is present (Table 7, Table S4 in File S4, Figure S4). Moreover, on the basis of our specific evidence assessment, five days of antibiotic administration have a high probability of determining resistant-bacterial selection unlike a 24-hour therapy.

In the studies, surgical management was carried out within few hours from the injury, and early surgery appears to be a reasonable condition for short (i.e. ≤24 hours) antibiotic prophylaxis. We adopted the Infective Disease Society of America (IDSA) definition of early surgery (i.e. carried out within 12 hours from trauma). [69] There is currently no clear proof in favour of this cut-off but it, nevertheless, seems reasonable.

Conclusions: The review was focused on different antibiotic strategies under the assumption that antibiotic prophylaxis is effective in reducing infections compared to placebo (Table 8).

Intervention. Administration of 24-hour is equivalent to 5-day antibiotic prophylaxis for surgical abdominal trauma with hollow viscus injury and surgical repair within 12 hours of trauma:

Level of evidence in favour: Very low evidence.

Level of evidence against: No evidence.

Recommendation: Weak in favour intervention.

Table 5. Antibiotic prophylaxis for the prevention of wound infections in long-bone open fractures: patients, interventions, and outcomes.

			Desirable effect		Undesirable effect		Desirable effect			
Sample	Treatment	Control	Outcome	Outcome quality	Outcome	Outcome quality	Treatment arm % (95%-CI)	Controls % (95%-CI)	Difference %(95%-CI)	RR (95%-CI)
Long-bone open fractures	48-hour course, dicoxacillin or penicillin - 60 patients	Placebo - 30 patients	Wound infection (not specifically osteomyelitis)	Weak, Minor prognostic relevance	Antibiotic- resistant bacteria selection, not investigated by the study	Robust, Clinically relevant for the patient/crucial for the health-care organization	6.7 (2.6 to 15.9)	20 (9.5 to 37.3)	−13.3 (−31.1 to 0.7)	0.33 (0.1 to 1.09)
Long-bone open fractures	cloxacillin 10-day course - 43 patients	Placebo - 44 patients	Wound infection (not specifically osteomyelitis)	Weak, Minor prognostic relevance	Antibiotic- resistant bacteria selection, not investigated by the study	Robust, Clinically relevant for the patient/crucial for the health-care organization	4.7 (1.3 to 15.5)	27.3 (16.3 to 41.8)	−22.6 (−37.6 to −7.3)	0.17 (0.04 to 0.72)

Table 6. Antibiotic prophylaxis for the prevention of wound infections in long-bone open fractures - Level of evidence and recommendations.

Patients and intervention	Desirable effect	Undesirable effect*	Benefit/risk profile	Values and preferences	Resource use	Recommendation	Rationale
Long-bone open fractures - 48-hour course dicoxacillin or penicillin	No evidence in favour of intervention □□□□	Low evidence against intervention ++□□	Unfavourable	Not available	Unwise	Weak against intervention	The study was slightly negative but the power was insufficient to detect clinically significant differences. Being the result of a subgroup analysis the effectiveness of the intervention is only hypothetical. There is evidence of risk for resistant bacteria selection even with a two-day antibiotic course. Wound infections (excluding osteomyelitis) are not as relevant an outcome as resistant bacteria selection, given the potentially high number needed to treat (Table S3 in File S3).
Long-bone open fractures - 10-day course cloxacillin	Low evidence in favour of intervention ++□□	High evidence against intervention ++++	Unfavourable	Not available	Unwise	Strong against intervention	It is common knowledge that 10-day antibiotic prophylaxis has a very high chance of determining resistant bacteria selection, which outweighs wound. infections prevention, generating an unfavourable benefit/risk profile.

*Not investigated in the studies, evidence could be found from external sources.

Table 7. Antibiotic prophylaxis for the prevention of deep surgical-site infections in abdominal trauma with enteric contamination: patients, interventions, and outcomes.

			Desirable effect		Undesirable effect		Desirable effect			
Sample	**Treatment**	**Control**	**Outcome**	**Outcome quality**	**Outcome**	**Outcome quality**	**Treatment arm % (95%-CI)**	**Controls % (95%-CI)**	**Difference % (95%-CI)**	**RR (95%-CI)**
Penetrating abdominal trauma	24-hour cefoxitin or cefotetan - 265 patients	5-day cefoxitin or cefotetan - 250 patients	Deep surgical site infections	Robust, Prognostically relevant outcome	Antibiotic-resistant bacteria selection, not investigated in the study	Robust, Clinically relevant for the patient/ crucial for the health-care organization	7.9 (5.2 to 11.8)	10 (6.9 to 14.3)	−2.1 (−7.2 to 2.9)	0.79 (0.46 to 1.38)
Penetrating abdominal trauma	24-hour ampicillin-sulbactam - 158 patients	5-day ampicillin-sulbactam - 159 patients	Deep surgical site infections	Robust, Prognostically relevant outcome	Antibiotic-resistant bacteria selection, not investigated in the study	Robust, Clinically relevant for the patient/crucial for the health-care organization	8.2 (4.9 to 13.6)	10.1 (6.3 to 15.7)	−1.8 (−8.4 to 4.7)	0.82 (0.41 to 1.64)
Penetrating abdominal trauma	24-hour cefoxitin - 148 patients	5-day cefoxitin - 152 patients	Deep surgical site infections	Robust, Prognostically relevant outcome	Antibiotic-resistant bacteria selection, not investigated in the study	Robust, Clinically relevant for the patient/crucial for the health-care organization	6.1 (3.2 to 11.2)	5.9 (3.1 to 10.9)	0.2 (−5.5 to 5.9)	1.03 (0.42 to 2.52)

Discussion

The main limitation of our study is that the literature search was limited to the MEDLINE database and only to publications in English, because of financial constraints. However, the issues we have dealt with have been largely investigated in previous reviews, and the guidelines included in the literature we have scrutinized wee used as an alternative source of references. We are therefore confident that we did not miss important publications in this field.

It is our firm belief that our recommendations, and guidelines in general, should not be clinical rules that impose specific clinical behaviours on doctors, but only indications applicable to an average patient, and should never replace the complex and individualised decisional process that physicians follow for individual patients [70].

In studying the role of antibiotic prophylaxis in trauma we found that in many cases the studies did not provide sufficient information for GRADE categories to be fully applied. Specifically, the assessment of the overall benefit/risk profile requires the comparison between desirable and undesirable effects. In the case of prolonged antibiotic prophylaxis (in several scrutinized studies antibiotic were administered for three days or more), the undesirable effect is the selection of antibiotic-resistant bacteria, an outcome not investigated in most of the studies. [71] Thus we had to integrate studies investigating exclusively infection prevention with evidence from other sources dealing with selection of resistant bacteria, a novel approach compared to other experiences that have proposed a different interpretation of evidence in this field [1,2].

The risk of resistant mutant selection due to short antibiotic pressure has not been investigated with RCTs. We had therefore to rely on before-after studies investigating either the colonization of single patients after treatment or the effect of antibiotic policies changes in single centres.

We were able to find observational studies suggesting that a 2/3-day antibiotic course could already result in the selection of antibiotic-resistant bacteria. We graded this evidence as low or very low, due to several limitations of the studies. However, in the face of similar strength of evidence in favour and against the intervention, as in the case of the 3-day antibiotic prophylaxis for wound infections in open-fractures or early-VAP prevention in comatose patients, we considered the antibiotic-resistant resistant selection the main concern in terms of clinical impact and social relevance of the outcome., Antibiotic-resistant bacteria spread constitutes an emergency for the global health-care system, given its high burden of mortality and resource consume. [72–74] Since resistant mutations are widespread in hospitals and extending within the community, the solution to this problem appears to be the restoration of susceptible bacteria by strongly limiting antibiotic use. [22] Our concern is that the common use of antibiotics in prophylaxis, even for short periods, might significantly worsen this problem. Our recommendations go in this direction.

In conclusion, we would like to mention a striking report from a neurosurgical centre that in the late 1960s had to face an outbreak of resistant *Klebsiella Aerogenes*, which at its peak caused urinary infections in 1 of 4 patients, respiratory infections in 1 of 8, and 9 cases of meningitis 8 of which were fatal. [75–77] Though antibiotic treatment was targeted on sensitivity tests and despite the major increase in antibiotic consumption in the ward between 1966 ad 1969 the outbreak was out of control. As the authors state, "in desperation", antibiotic use whether for treatment or prophylaxis was abandoned. This resulted in an immediate reduction of infections and the elimination of the Klebsiella from the ward without any serious adverse event being reported as the consequence of this strategy. This report although deviating from the paradigm of

Table 8. Antibiotic prophylaxis for the prevention of deep surgical-site infections in abdominal trauma with enteric contamination - Level of evidence and recommendations.

Query	Desirable effect	Undesirable effect*	Benefit/risk profile	Values and preferences	Resource use	Recommendation	Rationale
Penetrating abdominal trauma - 24-hour cefoxitin or cefotetan vs. 5-day cefoxitin or cefotetan	Very low evidence in favour of intervention ✚□□□	No evidence against intervention □□□□	Favourable	Not available	Wise	Weak in favour of intervention	Only part of the patients had intestinal perforation, but no specific data is available for this subset; this hampers conclusive answers regarding patients with perforation, the focus of the review. However, the risk of resistant bacteria selection is itself a sufficient
Penetrating abdominal trauma - 24-hour ampicillin-sulbactam vs. 5-day ampicillin-sulbactam	Very low evidence in favour of intervention ✚□□□	No evidence against intervention □□□□	Favourable	Not available	Wise	Weak in favour of intervention	condition to contraindicate the 5-day antibiotic course. We assume that single-day prophylaxis is effective on the basis of indirect evidence from elective abdominal-surgery antibiotic prophylaxis studies and less dangerous in terms of antibiotic-resistant bacteria selection than a 5-day treatment. This issue however deserves further investigations.
Penetrating abdominal trauma - 24-hour cefoxitin vs. 5-day cefoxitin	Very low evidence in favour of intervention ✚□□□	No evidence against intervention □□□□	Favourable	Not available	Wise	Weak in favour of intervention	

*Not investigated in the studies, no evidence could be found from external sources.

evidence base medicine (we are dealing with a single-centre epidemiological descriptive report with no statistical analysis performed and no precise design) nevertheless provides a demonstration of the effectiveness of a drastic antibiotic containment strategy, a potential choice when "the going gets tough".

Supporting Information

Figure S1 Absolute proportions differences and relative risks for the studies concerning the first query.

Figure S2 Absolute proportions differences and relative risks for the studies concerning the second query.

Figure S3 Absolute proportions differences and relative risks for the studies concerning the third query.

Figure S4 Absolute proportions differences and relative risks for the studies concerning the fourth query.

File S1 MEDLINE database search, flow diagram illustrating the literature selection process, and Table S1 illustrating evidence assessment for the first query.

File S2 MEDLINE database search, flow diagram illustrating the literature selection process, and Table S2 illustrating evidence assessment for the second query.

File S3 MEDLINE database search, flow diagram illustrating the literature selection process, and Table S3 illustrating evidence assessment for the third query.

File S4 MEDLINE database search, flow diagram illustrating the literature selection process, and Table S4 illustrating evidence assessment for the fourth query.

File S5 MEDLINE database search, flow diagram illustrating the literature selection process, and Table S5 illustrating evidence assessment for antibiotic-resistant bacteria (undesirable effect).

File S6 Preferred Reporting Items for Systematic Reviews and Meta-Analyses (PRISMA) statement recommendations checklist (Table S6).

Acknowledgments

The authors would like to thank Fulvia Zerri for her help in the organization of the meetings and for her extraordinary support to the works of the panel.

Author Contributions

Conceived and designed the experiments: DP AC ML BV. Analyzed the data: DP. Contributed to the writing of the manuscript: DP AC ML BV. Interpretation of data: DP AC ML BV EC PM FM GN EN LR IT AV CC GG. Drafting the work and revising it critically for important intellectual content: DP AC ML BV EC PM FM GN EN LR IT AV CC GG.

References

1. Hauser CJ, Adams CA Jr, Eachempati SR, Council of the Surgical Infection S (2006) Surgical Infection Society guideline: prophylactic antibiotic use in open fractures: an evidence-based guideline. Surg Infect (Larchmt) 7: 379–405.
2. Luchette FA, Bone LB, Born CT, DeLong WG, Hoff WS, et al. (2000) EAST Practice Management Guidelines Workgroup: Practice management guidelines for prophylactic antibiotic use in open fractures. Eastern Association for the Surgery of Trauma.
3. Ioannidis JP, Trikalinos TA (2007) The appropriateness of asymmetry tests for publication bias in meta-analyses: a large survey. CMAJ 176: 1091–1096.
4. Atkins D, Eccles M, Flottorp S, Guyatt GH, Henry D, et al. (2004) Systems for grading the quality of evidence and the strength of recommendations I: critical appraisal of existing approaches The GRADE Working Group. BMC Health Serv Res 4: 38.
5. Guyatt G, Oxman AD, Akl EA, Kunz R, Vist G, et al. (2011) GRADE guidelines: 1. Introduction-GRADE evidence profiles and summary of findings tables. J Clin Epidemiol 64: 383–394.
6. Guyatt GH, Oxman AD, Vist GE, Kunz R, Falck-Ytter Y, et al. (2008) GRADE: an emerging consensus on rating quality of evidence and strength of recommendations. BMJ 336: 924–926.
7. Balshem H, Helfand M, Schunemann HJ, Oxman AD, Kunz R, et al. (2011) GRADE guidelines: 3. Rating the quality of evidence. J Clin Epidemiol 64: 401–406.
8. Altman DG, Bland JM (1995) Absence of evidence is not evidence of absence. BMJ 311: 485.
9. Moher D, Schulz KF, Altman DG, Consort G (2001) The CONSORT statement: revised recommendations for improving the quality of reports of parallel-group randomized trials. Ann Intern Med 134: 657–662.
10. Higgins JP, Thompson SG, Deeks JJ, Altman DG (2003) Measuring inconsistency in meta-analyses. BMJ 327: 557–560.
11. Feinstein AR (1995) Meta-analysis: statistical alchemy for the 21st century. J Clin Epidemiol 48: 71–79.
12. Andrews JC, Schunemann HJ, Oxman AD, Pottie K, Meerpohl JJ, et al. (2013) GRADE guidelines: 15. Going from evidence to recommendation-determinants of a recommendation's direction and strength. J Clin Epidemiol 66: 726–735.
13. Guyatt GH, Oxman AD, Kunz R, Brozek J, Alonso-Coello P, et al. (2011) GRADE guidelines 6. Rating the quality of evidence–imprecision. J Clin Epidemiol 64: 1283–1293.
14. Guyatt GH, Oxman AD, Kunz R, Woodcock J, Brozek J, et al. (2011) GRADE guidelines: 8. Rating the quality of evidence–indirectness. J Clin Epidemiol 64: 1303–1310.
15. Guyatt GH, Oxman AD, Kunz R, Woodcock J, Brozek J, et al. (2011) GRADE guidelines: 7. Rating the quality of evidence–inconsistency. J Clin Epidemiol 64: 1294–1302.
16. Guyatt GH, Oxman AD, Montori V, Vist G, Kunz R, et al. (2011) GRADE guidelines: 5. Rating the quality of evidence–publication bias. J Clin Epidemiol 64: 1277–1282.
17. Guyatt GH, Oxman AD, Vist G, Kunz R, Brozek J, et al. (2011) GRADE guidelines: 4. Rating the quality of evidence–study limitations (risk of bias). J Clin Epidemiol 64: 407–415.
18. Guyatt GH, Oxman AD, Sultan S, Glasziou P, Akl EA, et al. (2011) GRADE guidelines: 9. Rating up the quality of evidence. J Clin Epidemiol 64: 1311–1316.
19. Andrews J, Guyatt G, Oxman AD, Alderson P, Dahm P, et al. (2013) GRADE guidelines: 14. Going from evidence to recommendations: the significance and presentation of recommendations. J Clin Epidemiol 66: 719–725.
20. Liberati A, Altman DG, Tetzlaff J, Mulrow C, Gotzsche PC, et al. (2009) The PRISMA statement for reporting systematic reviews and meta-analyses of studies that evaluate health care interventions: explanation and elaboration. PLoS Med 6: e1000100.
21. Tacconelli E, De Angelis G, Cataldo MA, Pozzi E, Cauda R (2008) Does antibiotic exposure increase the risk of methicillin-resistant Staphylococcus aureus (MRSA) isolation? A systematic review and meta-analysis. J Antimicrob Chemother 61: 26–38.
22. Levy SB (2002) The 2000 Garrod lecture. Factors impacting on the problem of antibiotic resistance. J Antimicrob Chemother 49: 25–30.
23. Naimi TS, LeDell KH, Como-Sabetti K, Borchardt SM, Boxrud DJ, et al. (2003) Comparison of community- and health care-associated methicillin-resistant Staphylococcus aureus infection. JAMA 290: 2976–2984.
24. Valverde A, Grill F, Coque TM, Pintado V, Baquero F, et al. (2008) High rate of intestinal colonization with extended-spectrum-beta-lactamase-producing organisms in household contacts of infected community patients. J Clin Microbiol 46: 2796–2799.
25. Schentag JJ, Hyatt JM, Carr JR, Paladino JA, Birmingham MC, et al. (1998) Genesis of methicillin-resistant Staphylococcus aureus (MRSA), how treatment of MRSA infections has selected for vancomycin-resistant Enterococcus faecium,

and the importance of antibiotic management and infection control. Clin Infect Dis 26: 1204–1214.
26. Lipsitch M, Samore MH (2002) Antimicrobial use and antimicrobial resistance: a population perspective. Emerg Infect Dis 8: 347–354.
27. Bonten MJ, Slaughter S, Ambergen AW, Hayden MK, van Voorhis J, et al. (1998) The role of "colonization pressure" in the spread of vancomycin-resistant enterococci: an important infection control variable. Arch Intern Med 158: 1127–1132.
28. Harbarth S, Samore MH, Lichtenberg D, Carmeli Y (2000) Prolonged antibiotic prophylaxis after cardiovascular surgery and its effect on surgical site infections and antimicrobial resistance. Circulation 101: 2916–2921.
29. Tacconelli E, De Angelis G, Cataldo MA, Mantengoli E, Spanu T, et al. (2009) Antibiotic usage and risk of colonization and infection with antibiotic-resistant bacteria: a hospital population-based study. Antimicrob Agents Chemother 53: 4264–4269.
30. Filius PM, Gyssens IC, Kershof IM, Roovers PJ, Ott A, et al. (2005) Colonization and resistance dynamics of gram-negative bacteria in patients during and after hospitalization. Antimicrob Agents Chemother 49: 2879–2886.
31. Kernodle DS, Barg NL, Kaiser AB (1988) Low-level colonization of hospitalized patients with methicillin-resistant coagulase-negative staphylococci and emergence of the organisms during surgical antimicrobial prophylaxis. Antimicrob Agents Chemother 32: 202–208.
32. Sjolund M, Wreiber K, Andersson DI, Blaser MJ, Engstrand L (2003) Long-term persistence of resistant Enterococcus species after antibiotics to eradicate Helicobacter pylori. Ann Intern Med 139: 483–487.
33. Gould IM (1999) A review of the role of antibiotic policies in the control of antibiotic resistance. J Antimicrob Chemother 43: 459–465.
34. Tacconelli E (2009) Antimicrobial use: risk driver of multidrug resistant microorganisms in healthcare settings. Curr Opin Infect Dis 22: 352–358.
35. Rahal JJ, Urban C, Horn D, Freeman K, Segal-Maurer S, et al. (1998) Class restriction of cephalosporin use to control total cephalosporin resistance in nosocomial Klebsiella. JAMA 280: 1233–1237.
36. Acquarolo A, Urli T, Perone G, Giannotti C, Candiani A, et al. (2005) Antibiotic prophylaxis of early onset pneumonia in critically ill comatose patients. A randomized study. Intensive Care Med 31: 510–516.
37. Sirvent JM, Torres A, El-Ebiary M, Castro P, de Batlle J, et al. (1997) Protective effect of intravenously administered cefuroxime against nosocomial pneumonia in patients with structural coma. Am J Respir Crit Care Med 155: 1729–1734.
38. Valles J, Peredo R, Burgueno MJ, Rodrigues de Freitas AP, Millan S, et al. (2013) Efficacy of single-dose antibiotic against early-onset pneumonia in comatose patients who are ventilated. Chest 143: 1219–1225.
39. Babyak MA (2004) What you see may not be what you get: a brief, nontechnical introduction to overfitting in regression-type models. Psychosom Med 66: 411–421.
40. Peduzzi P, Concato J, Kemper E, Holford TR, Feinstein AR (1996) A simulation study of the number of events per variable in logistic regression analysis. J Clin Epidemiol 49: 1373–1379.
41. Ignelzi RJ, VanderArk GD (1975) Analysis of the treatment of basilar skull fractures with and without antibiotics. J Neurosurg 43: 721–726.
42. Klastersky J, Sadeghi M, Brihaye J (1976) Antimicrobial prophylaxis in patients with rhinorrhea or otorrhea: a double-blind study. Surg Neurol 6: 111–114.
43. Demetriades D, Charalambides D, Lakhoo M, Pantanowitz D (1992) Role of prophylactic antibiotics in open and basilar fractures of the skull: a randomized study. Injury 23: 377–380.
44. Eftekhar B, Ghodsi M, Nejat F, Ketabchi E, Esmaeeli B (2004) Prophylactic administration of ceftriaxone for the prevention of meningitis after traumatic pneumocephalus: results of a clinical trial. J Neurosurg 101: 757–761.
45. Assmann SF, Pocock SJ, Enos LE, Kasten LE (2000) Subgroup analysis and other (mis)uses of baseline data in clinical trials. Lancet 355: 1064–1069.
46. Ratilal BO, Costa J, Sampaio C, Pappamikail L (2011) Antibiotic prophylaxis for preventing meningitis in patients with basilar skull fractures. Cochrane Database Syst Rev: CD004884.
47. Hoff JT, Brewin A, U HS (1976) Letter: Antibiotics for basilar skull fracture. J Neurosurg 44: 649.
48. Brodie HA (1997) Prophylactic antibiotics for posttraumatic cerebrospinal fluid fistulae. A meta-analysis. Arch Otolaryngol Head Neck Surg 123: 749–752.
49. Villalobos T, Arango C, Kubilis P, Rathore M (1998) Antibiotic prophylaxis after basilar skull fractures: a meta-analysis. Clin Infect Dis 27: 364–369.
50. Dagi TF, Meyer FB, Poletti CA (1983) The incidence and prevention of meningitis after basilar skull fracture. Am J Emerg Med 1: 295–298.
51. Zrebeet HA, Huang PS (1986) Prophylactic antibiotics in the treatment of fractures at the base of the skull. Del Med J 58: 741–748.
52. Frazee RC, Mucha P Jr, Farnell MB, Ebersold MJ (1988) Meningitis after basilar skull fracture. Does antibiotic prophylaxis help? Postgrad Med 83: 267–268, 273–264.
53. Bergman BR (1982) Antibiotic prophylaxis in open and closed fractures: a controlled clinical trial. Acta Orthop Scand 53: 57–62.
54. Braun R, Enzler MA, Rittmann WW (1987) A double-blind clinical trial of prophylactic cloxacillin in open fractures. J Orthop Trauma 1: 12–17.
55. Dickey RL, Barnes BC, Kearns RJ, Tullos HS (1989) Efficacy of antibiotics in low-velocity gunshot fractures. J Orthop Trauma 3: 6–10.
56. Patzakis MJ, Wilkins J, Moore TM (1983) Use of antibiotics in open tibial fractures. Clin Orthop Relat Res: 31–35.
57. Gosselin RA, Roberts I, Gillespie WJ (2004) Antibiotics for preventing infection in open limb fractures. Cochrane Database Syst Rev: CD003764.
58. Rojczyk M (1983) Treatment results in open fractures, aspects of antibiotic therapy. Hefte Unfallheilkd 162: 33–38.
59. Patzakis MJ, Harvey JP Jr, Ivler D (1974) The role of antibiotics in the management of open fractures. J Bone Joint Surg Am 56: 532–541.
60. Hoff WS, Bonadies JA, Cachecho R, Dorlac WC (2011) East Practice Management Guidelines Work Group: update to practice management guidelines for prophylactic antibiotic use in open fractures. J Trauma 70: 751–754.
61. Hospenthal DR, Murray CK, Andersen RC, Bell RB, Calhoun JH, et al. (2011) Guidelines for the prevention of infections associated with combat-related injuries: 2011 update: endorsed by the Infectious Diseases Society of America and the Surgical Infection Society. J Trauma 71: S210–234.
62. Brand M, Goosen J, Grieve A (2009) Prophylactic antibiotics for penetrating abdominal trauma. Cochrane Database Syst Rev: CD007370.
63. Brand M, Grieve A (2013) Prophylactic antibiotics for penetrating abdominal trauma. Cochrane Database Syst Rev 11: CD007370.
64. Bratzler DW, Houck PM, Surgical Infection Prevention Guidelines Writers W, American Academy of Orthopaedic S, American Association of Critical Care N, et al. (2004) Antimicrobial prophylaxis for surgery: an advisory statement from the National Surgical Infection Prevention Project. Clin Infect Dis 38: 1706–1715.
65. Fabian TC, Croce MA, Payne LW, Minard G, Pritchard FE, et al. (1992) Duration of antibiotic therapy for penetrating abdominal trauma: a prospective trial. Surgery 112: 788–794; discussion 794–785.
66. Kirton OC, O'Neill PA, Kestner M, Tortella BJ (2000) Perioperative antibiotic use in high-risk penetrating hollow viscus injury: a prospective randomized, double-blind, placebo-control trial of 24 hours versus 5 days. J Trauma 49: 822–832.
67. Bozorgzadeh A, Pizzi WF, Barie PS, Khaneja SC, LaMaute HR, et al. (1999) The duration of antibiotic administration in penetrating abdominal trauma. Am J Surg 177: 125–131.
68. Cornwell EE III, Dougherty WR, Berne TV, Velmahos G, Murray JA, et al. (1999) Duration of antibiotic prophylaxis in high-risk patients with penetrating abdominal trauma: a prospective randomized trial. J Gastrointest Surg 3: 648–653.
69. Solomkin JS, Mazuski JE, Bradley JS, Rodvold KA, Goldstein EJ, et al. (2010) Diagnosis and management of complicated intra-abdominal infection in adults and children: guidelines by the Surgical Infection Society and the Infectious Diseases Society of America. Clin Infect Dis 50: 133–164.
70. Hurwitz B (1999) Legal and political considerations of clinical practice guidelines. BMJ 318: 661–664.
71. Guyatt GH, Oxman AD, Kunz R, Atkins D, Brozek J, et al. (2011) GRADE guidelines: 2. Framing the question and deciding on important outcomes. J Clin Epidemiol 64: 395–400.
72. de Kraker ME, Davey PG, Grundmann H, group Bs (2011) Mortality and hospital stay associated with resistant Staphylococcus aureus and Escherichia coli bacteremia: estimating the burden of antibiotic resistance in Europe. PLoS Med 8: e1001104.
73. de Kraker ME, Wolkewitz M, Davey PG, Koller W, Berger J, et al. (2011) Burden of antimicrobial resistance in European hospitals: excess mortality and length of hospital stay associated with bloodstream infections due to Escherichia coli resistant to third-generation cephalosporins. J Antimicrob Chemother 66: 398–407.
74. de Kraker ME, Wolkewitz M, Davey PG, Koller W, Berger J, et al. (2011) Clinical impact of antimicrobial resistance in European hospitals: excess mortality and length of hospital stay related to methicillin-resistant Staphylococcus aureus bloodstream infections. Antimicrob Agents Chemother 55: 1598–1605.
75. Price DJ, Sleigh JD (1970) Control of infection due to Klebsiella aerogenes in a neurosurgical unit by withdrawal of all antibiotics. Lancet 2: 1213–1215.
76. Price D, Sleigh JD (1970) Infection due to Klebsiella aerogenes in a neurosurgical unit. J Clin Pathol 23: 378.
77. Price DJ, Sleigh JD (1972) Klebsiella meningitis–report of nine cases. J Neurol Neurosurg Psychiatry 35: 903–908.

iBrick: A New Standard for Iterative Assembly of Biological Parts with Homing Endonucleases

Jia-Kun Liu[1], Wei-Hua Chen[1], Shuang-Xi Ren[1], Guo-Ping Zhao[1,2,3,4]*, Jin Wang[1]*

1 CAS Key Laboratory of Synthetic Biology, Institute of Plant Physiology and Ecology, Shanghai Institutes for Biological Sciences, Chinese Academy of Sciences, Shanghai, China, **2** State Key Lab of Genetic Engineering & Center for Synthetic Biology, Department of Microbiology and Microbial Engineering, School of Life Sciences, Fudan University, Shanghai, China, **3** Shanghai-MOST Key Laboratory of Disease and Health Genomics, Chinese National Human Genome Center at Shanghai, Shanghai, China, **4** Department of Microbiology and Li Ka Shing Institute of Health Sciences, The Chinese University of Hong Kong, Prince of Wales Hospital, Shatin, New Territories, Hong Kong SAR, China

Abstract

The BioBricks standard has made the construction of DNA modules easier, quicker and cheaper. So far, over 100 BioBricks assembly schemes have been developed and many of them, including the original standard of BBF RFC 10, are now widely used. However, because the restriction endonucleases employed by these standards usually recognize short DNA sequences that are widely spread among natural DNA sequences, and these recognition sites must be removed before the parts construction, there is much inconvenience in dealing with large-size DNA parts (*e.g.*, more than couple kilobases in length) with the present standards. Here, we introduce a new standard, namely iBrick, which uses two homing endonucleases of I-SceI and PI-PspI. Because both enzymes recognize long DNA sequences (>18 bps), their sites are extremely rare in natural DNA sources, thus providing additional convenience, especially in handling large pieces of DNA fragments. Using the iBrick standard, the carotenoid biosynthetic cluster (>4 kb) was successfully assembled and the actinorhodin biosynthetic cluster (>20 kb) was easily cloned and heterologously expressed. In addition, a corresponding nomenclature system has been established for the iBrick standard.

Editor: Mark Isalan, Imperial College London, United Kingdom

Funding: This work was supported by the National Basic Research Program of China [2012CB721102], Natural Science Foundation for the Youth [31300095] and National Natural Science Foundation of China [31121001]. The funders had no role in study design, data collection and analysis, decision to publish, or preparation of the manuscript.

Competing Interests: A patent of "A method for construction and assembly of DNA in vitro" (application no. in State Intellectual Property Office of the P.R.C: 01410235720.6) has been filed by CAS Key Laboratory of Synthetic Biology with Jin Wang, Guo-Ping Zhao, Jia-Kun Liu and Wei-Hua Chen as inventors.

* Email: cmg.wangjin@gmail.com (JW); gpzhao@sibs.ac.cn (GPZ)

Introduction

Synthetic biology is a newly emerging discipline that learns from the engineering principles and uses complex combinations of DNA components to construct biological systems with new properties [1,2]. The development of efficient methods for *de novo* synthesis of biological parts from a pool of oligos has greatly promoted the progress of synthetic biology. Nowadays, more and more companies are providing commercial services for *de novo* DNA synthesis, and most of them are using the polymerase chain assembly (PCA) [3–7] method to synthesize the designed DNA pieces up to several kilobases. Once small pieces are obtained, multiple approaches, *e.g.* BioBricks standard [8–10], Gibson assembly [11–13] and Master Ligation [14], are now available for assembling them into genes, pathways, modules or even a whole genome.

Among these standards (or techniques), BioBricks standard, which was originally introduced by Knight *et al.* (http://hdl.handle.net/1721.1/21168), is a milestone in the field of synthetic biology. A BioBricks part is a piece of DNA sequence flanked by XbaI and SpeI restriction sites on their 5′ and 3′ ends, respectively. Because XbaI and SpeI digested DNAs have compatible cohesive ends, they can be ligated head-to-tail to generate a hybrid DNA piece with a scar, and the resulted DNA piece is flanked by XbaI and SpeI again, which can be taken as a new BioBricks part for further iterative assembly. Techniques used in BioBricks assembly are merely DNA restriction/ligation and are easy to learn and to be widely adopted. On the other hand, once a BioBricks part is created and characterized, it can be reused in a variety of projects with different purposes. Built on the basis of BioBricks standard, the International Genetically Engineered Machine Competition (iGEM) has not only inspired young scientists in the field of synthetic biology but also promoted the sharing of biological resources.

However, the original BioBricks assembly scheme (BBF RFC 10, http://hdl.handle.net/1721.1/45138) produces an internal scar of 8 base pairs (bps) and has an in-frame stop codon. Therefore, it is not suitable for creation of fusion proteins, and certainly limits its landscape of potential applications. In fact, a series of improvements have been made to overcome the shortcomings of the original BioBricks standard thereafter. For example, BBF RFC 25 (http://hdl.handle.net/1721.1/45140) adds an NgoMIV site and an AgeI site in the prefix and suffix of RFC 10, respectively. As a consequence, BBF RFC 25 produces a scar of ACCGCC, encoding Thr-Gly, and supports the construction of plasmids for expression of in-frame fusion proteins. In addition, BglBricks [9], which is assigned BBF RFC 21 and uses BamHI and BglII instead of XbaI and SpeI, has addressed several key problems associated with the original BioBricks standard and is therefore a standard of great potential.

With the progress of synthetic biology, the demand of synthesis of long DNA sequences increases. Although the cost of DNA synthesis has significantly dropped in the past few years, *de novo* synthesis of large pieces of DNA sequences is still too expensive for most labs to afford, and it is also time-consuming. On the other hand, there exist numerous natural genetic materials, *e.g.* antibiotic biosynthetic gene clusters, many of which can actually be directly used for analysis without codon optimization or other manipulations. However, as these large pieces usually contain many recognition sites of type IIP and IIS restriction enzymes (REs) which excludes the direct usage of the present type II RE-based BioBricks standards [15–17]. Although internal restriction sites can be eliminated by synonymous base mutagenesis, it is practically unfeasible with regard to the cost of time and money. Moreover, base substitution may bring uncertainties in the phenotypes of the DNA sequences probably due to the influences of the synonymous single nucleotide polymorphisms (SNPs) [18–20]. Although other methods, such as PSA [21], Gibson Assembly and Master Ligation can be adopted, they do not allow convenient reuse of the constructed materials.

Here, we introduce a new standard designated iBrick for iterative assembly of biological parts, which employs two homing endonucleases (HEs) of I-SceI [22–24] and PI-PspI [25], which produce compatible non-palindromic cohesive ends of "TTAT". Parts in iBrick standard are prefixed with I-SceI and suffixed with PI-PspI (**Fig. 1**). As the cohesive ends produced *via* I-SceI and PI-PspI digestion are non-palindromic, the parts can be assembled in a defined order in iBrick standard (**Fig. 2**), similar to those using the BioBricks standards. In addition, because HEs used in iBrick recognize DNA sequences longer than 18 bps, which are much longer than that of the conventional type II REs and are thus extremely rare in natural DNA sequences, there is usually no need for modification of the DNA sequences regardless of their length. With the use of iBrick standard, we showcase here successful assembly of the carotenoid biosynthetic cluster (*crtEBI*) and convenient manipulation of the actinorhodin biosynthetic cluster (*act*) from *Streptomyces coelicolor* for heterologous expression in a thermophilic *Streptomyces* strain.

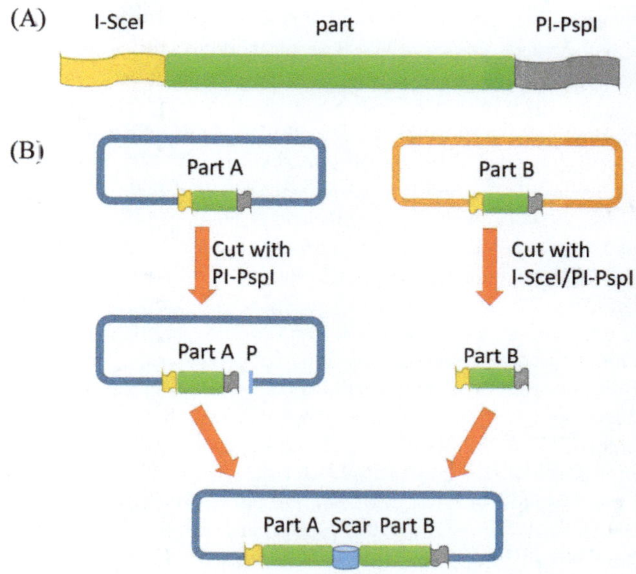

Figure 2. Schematic diagram of iBrick forward assembly. (A) Construction of iBrick parts, which should be prefixed and suffixed in iBrick standard. (B) Forward assembly procedure. Part B is firstly released with I-SceI and PI-PspI digestion, and is then inserted into the PI-PspI digested vector, obtaining an assembled part AB. Alternatively, the vector can also be cut with I-SceI followed by insertion of part B to obtain part BA, which is designated the reverse assembly.

Materials and Methods

Bacterial strains, media, reagents and primers

E. coli strain DH5α was used for cloning of plasmids with pUC replication origin (pUCori); strain EPI300 (Epicentre) was used for cloning oriV-containing plasmids; and strain ET12567 was used for conjugation with *Streptomyces sp.* 4F [26]. *E. coli* strains were grown in Luria-Bertani medium [27], while *Streptomyces sp.* 4F was grown on mannitol soya flour [28] plates for spore preparation and on R2YE plates [29] for actinorhodin production.

Enzymes used in this study (unless specified) were purchased from NEB. GeneRuler 1-kb DNA ladder was purchased from Thermo Fermentas and λ-HindIII ladder was from TAKARA. High fidelity DNA polymerase KOD Plus Neo was purchased from TOYOBO. The lycopene standard was purchased from Sigma-Aldrich.

Primers used in this study were listed in **Table S1**.

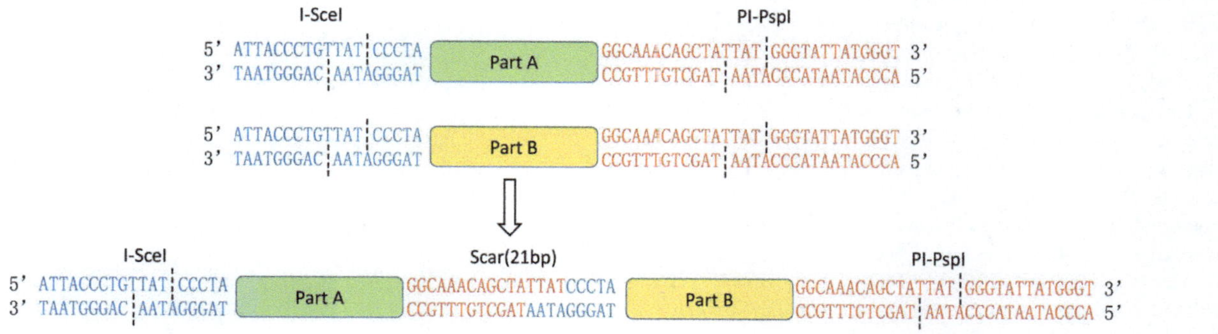

Figure 1. DNA sequences of the prefix, suffix and scar in iBrick assembly. Parts A and B are prefixed with I-SceI (shown in blue) and suffixed with PI-PspI (shown in red). After cleavage, compatible cohesive end ("TTAT") can be ligated together to produce a 21-bp scar between parts. When translated in frame, this 21-bp DNA sequence encodes 7 amino acids.

Protocol for construction of the iBrick plasmids

To *de novo* construct the base plasmids for iBrick assembly, beta-lactamase encoding gene [21], T4-g32 terminator, pUCori together with the *crtE* gene were used to construct the base plasmid pIB1A1_0-C000001. Specifically, *bla* was PCR amplified and the forward primer used contained the T4-g32 terminator sequence. The amplicon was then digested with NotI and phosphorylated with T4 polynucleotide kinase (PNK). Similary, pUCori was PCR amplified from pUC18 plasmid [30] and then the PCR product was cut with NotI at one end before it was ligated to *bla* to construct plasmid pJK5. Notably, several rare cutting enzymes that recognize 8-bp sites, *i.e.* PacI, NotI, SwaI and AscI, were used to link the replication origin, ampicillin resistance cassette and the terminator sequences in pJK5 (**Fig. S1**). Then, the *crtE* gene of the lycopene biosynthetic cluster was amplified from BioBrick BBa_K274200, flanked with SwaI/I-SceI sites at the 5′ end and PI-PspI/AscI sites at the 3′ end, and was then digested with SwaI and AscI and inserted into the same sites in pJK5 to obtain pIB1A1_0-C000001, in which two HE sites of I-SceI and PI-PspI were successfully introduced. Therefore, vector pIB1A1_0 prepared from I-SceI and PI-PspI double digestion of pIB1A1_0-C000001 was used as the base vector for creating small parts for iBrick assembly.

In addition to the ampicillin resistance cassette, other antibiotic resistance genes, including kanamycin, chloramphenicol and hygromycin were also amplified with the same flanking restriction sites of PacI and NotI, and were further used to substitute the *bla* gene in pIB1A1_0-C000001 to obtain base plasmids of pIB1K1_0-C000001, pIB1C1_0-C000001 and pIB1H1_0-C000001, respectively (**Fig. S4**).

The replication origin of pCC2FOS (Epicentre) containing both the F factor-based partitioning system and single-copy origin of ori2 and the inducible oriV [31] was PCR amplified and digested with NotI/SwaI before cloning into pIB1A1_0-C000001, obtaining pIB2A1_0-C000001. Similarly, the *bla* gene was replaced by *kan*, *apr* and *hyg* to obtain base plasmids of pIB2K1_0-C000001, pIB2Am1_0-C000001and pIB2H1_0-C000001, respectively (**Fig. S4**). Vectors prepared from these base plasmids were used as base vectors for cloning of large iBrick parts.

Procedure of assembly and verification of the lycopene biosynthetic cluster (*crt*)

In addition to *crtE*, other lycopene biosynthesis genes, *i.e. crtB* and *crtI*, and the arabinose inducible *pBAD* promoter were PCR amplified and digested with I-SceI and PI-PspI before being individually ligated to pIB1A1_0, obtaining iBrick plasmids harboring parts of *crtB*, *crtI* and *pBAD* promoter. These plasmids were named as pIB1A1_0-C000002, pIB1A1_0-C000003 and pIB1A1_0-P000001, respectively. Parts were verified through DNA sequencing with primer iPrimer1 and primer iPrimer2.

To assemble the *crt* cluster, part IBP_C000003 (*crtI*) was firstly released from pIB1A1_0-C000003 with I-SceI and PI-PspI digestion and was then cloned into the PI-PspI site of pIB1A1_0-C000002 to obtain pIB1A1_0-X000001, where the head of *crtI* was ligated to the tail of *crtB*. Similarly, *crtE* was cut from pIB1A1_0-C000001 with I-SceI and PI-PspI and cloned into the PI-PspI site of pIB1A1_0-P000001 to obtain pIB1A1_0-X000002, which contains *crtE* and *pBAD*. Finally, *crtBI* genes were cut from pIB1A1_0-X000001 and inserted to the PI-PspI site of pIB1A1_0-X000002, obtaining pIB1A1_0-X000003, which contains the *pBAD*-driven *crt* expression cluster. To avoid self-ligation, the PI-PspI-digested vector should be dephoshorylated, *e.g.* by Antarctic Phosphatase (NEB) in this study.

To test the carotenoid expression, pIB1A1_0-X000003 was transformed into *E. coli* strain DH5α and the obtained strain was cultured in LB medium supplemented with arabinose at a final concentration of 10%, employing DH5α transformed with plasmid pJK5 (**Fig. S1**) as a negative control. After shaking at 37°C for 12 h, cells were harvested and lysed with equal volumes of acetone. Chromatography was performed with Agilent LC 1200/Accurate Mass 6520A QTOF (Agilent Technologies) with a well-plate auto-sampler and the analytical column was Agilent Zorbax XDB-C18 4.6*50 mm, 1.8 μm. The system was operated in isocratic mode with a flow of 0.3 ml/min of 42% ACN mixed with 42% MeOH and 16% dichloromethane. As lycopene is a little, highly symmetric and non-polar molecule, APCI (Atmospheric-pressure chemical ionization) was applied to ionize lycopene.

Protocol for manipulation and heterologous expression of the actinorhodin biosynthetic cluster (*act*)

Base plasmid pIB2K1_0-C000001 was PCR amplified with primers of iBrick-actF and iBrick-actR to obtain PCR products with homologous sequences of the left and right arms of *act* cluster at both ends. Cosmid N07_85 containing the whole *act* cluster was transformed into *E. coli* strain BW25113 (pIJ790) [32], and the resulted strain was then used to prepare electro-competent cells for transformation of the above PCR products to perform gap repair assay [33]. Transformants were selected on LB plate containing kanamycin, and the obtained candidate plasmids were then verified by both their sizes and their restriction patterns, with correct clones named as pIB2K1_0-K000001. The integration cassette containing the phiC31 integrase and the oriT was firstly PCR amplified from pSET152 [34] and cloned into the I-SceI and PI-PspI sites in pIB1A1_0-C000001, obtaining pIB1A1_0-X000004. Then, the integration cassette was released by I-SceI and PI-PspI digestion and introduced into the same sites of pIB2K1_0-C000001 to obtain pIB2K1_0-X000004. To conjugate a plasmid into *Streptomyces*, *E. coli* strain ET12567 (pUZ8002) [32], which carries the kanamycin and chloramphenicol resistances, is often used. And the kanamycin resistance in pIB2K1_0-X000004 was further replaced with apramycin resistance gene (*apr*) and the obtained plasmid was called pIB2Am1_0-X000004. Finally, the *act* cluster was released from pIB2K1_0-K000001 by I-SceI and PI-PspI digestion and inserted into the PI-PspI site of pIB2Am1_0-X000004 to get pIB2Am1_0-X000005, an iBrick plasmid containing the *act* expression cluster and an integration cassette. Plasmid pIB2Am1_0-X000005 was then conjugated into *Streptomyces* 4F strain for heterologous expression of actinorhodin following the procedures previously described [14].

Results

Assembly of *crt* cluster with iBrick parts

To construct the biosynthetic cluster of carotenoid with the iBrick standard, three genes of *crtEBI* were individually PCR amplified from BioBrick BBa_K274200 and then cloned into an iBrick base vector pIB1A1_0 to obtain pIB1A1_0-C000001 - pIB1A1_0-C000003, respectively. Each part contains a separate ribosome binding site (RBS) and is prefixed with I-SceI and suffixed with PI-PspI. Similarly, the arabinose inducible promoter *pBAD* was prepared as an iBrick part of IBP_P000001. Then, these four parts were assembled together to produce pIB1A1_0-X000003, in which, the carotenoid biosynthetic cluster (*crt*) is expressed under the control of *pBAD* (**Fig. 3A**). Totally, two rounds of hierarchical forward assembly, including three manipulation steps were performed with efficiency no less than 75% in

Table 1. Ligation efficiency of iBrick assembly.

Fragments	Obtained plasmids (kb)	Obtained clones*	Ratio (positive/tested)
Lycopene			
Vector+*pBAD-crtE*	pIB1A1_0-X000001 (4)	2264	75% (9/12)
Vector+*crtBI*	pIB1A1_0-X000002 (4.2)	2294	75% (9/12)
Vector+*pBAD-crtEBI*	pIB1A1_0-X000003 (6.5)	1984	75% (186/248)
Actinorhodin			
Vector+*act*	pIB1A1_0-X000005 (31)	592	25% (3/12)

* Clones on plates were evenly divided into 4 even sections with only one counted and the total number was then roughly calculated as four times of the number.

each step (**Table 1**), which should be sufficient enough for most assembly tasks.

The assembled *crt* cluster was then tested in *E. coli* strain DH5α to examine its biological function, employing an isogenic strain but harboring an intermediate plasmid (pJK5, without the *crt* cluster, **Fig. S1**) as a negative control. After arabinose induction, clones with the whole cluster turned red while those without *crt* stayed white (data not shown), and the product of *crt* cluster was further confirmed to be lycopene with LC-MS analysis (**Fig. 3**).

Cloning and heterologous expression of *act* cluster using iBrick standard

Vector pIB2K1_0, which contains a FOSori and an inducible high copy number origin oriV, is used as the base vector to clone large iBrick parts, *e.g.* the *act* cluster in this study (>20 kb). Gap repair technology [33] was employed to directly clone the *act* cluster into vector pIB2K1_0 to obtain pIB2K1_0-K000001. The phiC31 integrase, *attP* site and the oriT element, which are necessary for conjugation and integration in *Streptomyces*, were together prepared as an iBrick part (IBP_X000004). Part IBP-K000001 containing the *act* cluster was then ligated to the tail of part IBP_X000004, obtaining an iBrick plasmid (pIB2Am1_0-X000005, **Fig. 4A–4C**) for heterologous expression of actinorhodin in a fast growing thermophilic actinomycete, *Streptomyces* 4F [26]. After incubation for three days on R2YE plate, 4F harboring the *act* cluster successfully expressed the blue pigmented actinorhodin, demonstrating the robustness and convenience of iBrick standard in manipulation of large DNA pieces (**Fig. 4D**). Notably, although the ligation efficiency (25%) for construction of pIB2Am1_0-X000005 is remarkably lower than those for small pieces, it should still be sufficient enough for most projects (**Table 1**).

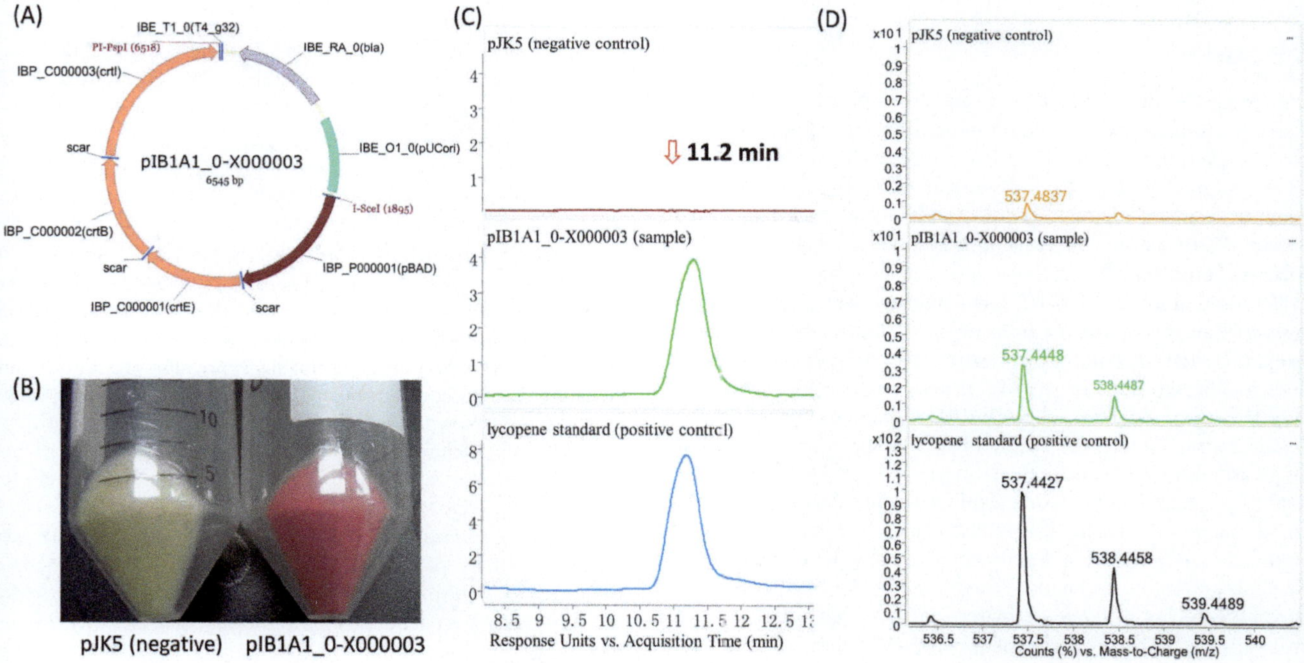

Figure 3. Assembly and expression of the *crt* cluster in *E. coli*. (**A**) Constructed expression plasmid pIB1A1_0-X000003 for *crt* cluster expression. Plasmid map was drawn with Vector NTI Software (Life Technologies). (**B**) Production of carotenoids in *E. coli*. Strain harboring pJK5 was used as a negative control. (C&D) Detection of the product of pIB1A1_0-X000003 with Agilent LC 1200/Accurate Mass 6520A QTOF (Agilent Technologies). (**C**) When analyzed with LC, product of pIB1A1_0-X000003 had the same retention time of 11.2 min as the lycopene standard. No obvious peaks were observed at this retention time in the negative control. (**D**) Mass analysis proved the fermentation product had the same molecular weight as the standard. Tiny peaks were observed in the negative control, which might be caused by contamination. In both C and D, the scales of the diagrams were adjusted to be the same in the tested sample (pIB1A1_0-X000003) and the negative control.

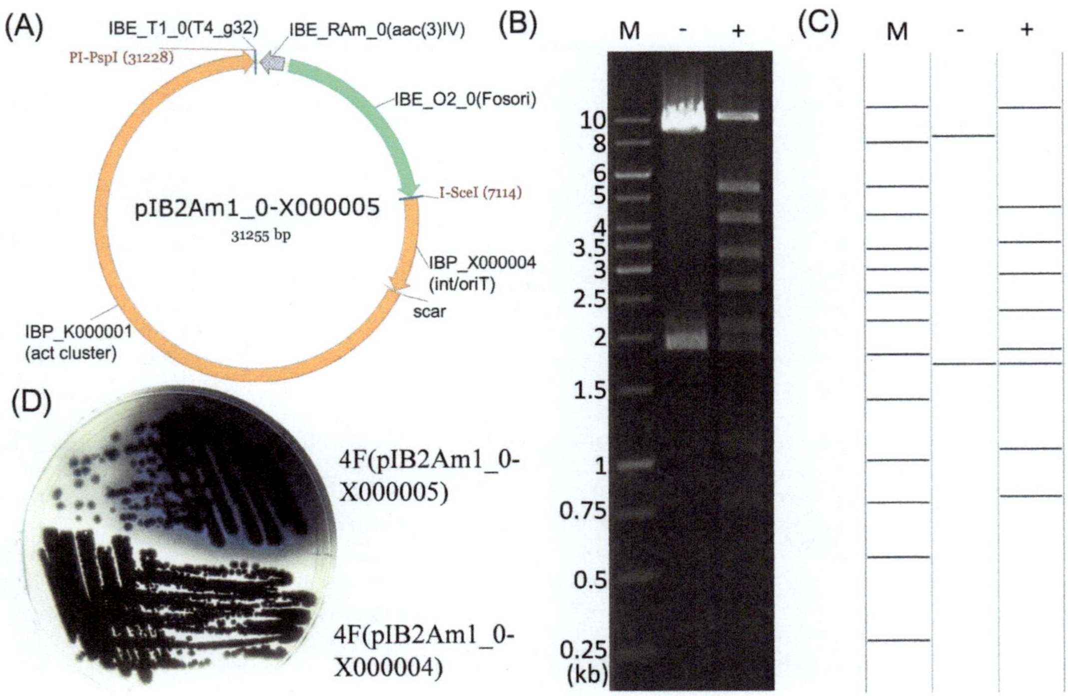

Figure 4. Heterologous expression of the *act* cluster in *Streptomyces* 4F. (A) Constructed expression plasmid pIB2Am1_0-X000005. (B&C) Verification of pIB2Am1_0-X000005 with both the experimental digestion (B) and the electronic restriction (C) with XcmI. M, 1 kb DNA ladder with sizes labeled; -, pIB2Am1_0-X000004; +, pIB2Am1_0-X000005. For "−", theoretical sizes are 8.32 and 1.88 kb; for "+", sizes are 9.93, 5.28, 4.18, 3.38, 2.67, 2.08, 1.88, 1.08 and 0.78 kb. Same patterns were obtained for the experimental group and the electronic analysis group. (D) Heterologous expression of *act* cluster in *Streptomyces* 4F, employing pIB2Am1_0-X000004 as a negative control. Strains were cultured on R2YE plate at 30°C for 2 days.

Discussion

The idea of BioBricks standard was developed to introduce the engineering principles, including standardization, decoupling and abstraction into the field of synthetic biology [10]. Since the birth of the original BioBricks standard (BBF RFC 10), a lot of efforts have been made to improve the standard and enrich the parts library. Participants include scientists around the world, *e.g.* iGEM students, graduate students, postdocs and principal investigators. Till today, at least 103 BBF RFC schemes have been publicly available on the website of BBF and over 20,000 BioBricks parts have been constructed, many of which have been deposited in the iGEM Registry and are widely reused. More importantly, the iGEM Registry not only collects the physical parts from worldwide iGEMers but also provides an open platform for judging the popularity and robustness of each part, which is quite instructive for new users (http://parts.igem.org/Main_Page). With the development of BioBricks, many detailed rules have been gradually set up for the whole system, including the nomenclature, symbols and so on, and a lot of them have been tested in practical applications by different research groups. Therefore, iBrick standard accepts a majority of the present rules and will only make necessary supplements and improvements.

Description of iBrick standard

Unlike type IIP REs that are commonly used in BioBricks standards, I-SceI and PI-PspI are HEs that produce nonpalindromic ends and therefore two enzymes instead of four are sufficient for orderly assembly of iBrick parts. In addition, as both I-SceI and PI-PspI recognize long DNA sequences (>18 bp), there are rare such sites in natural DNA sequences and therefore basically there is no need for modification of internal DNA sequences (*e.g.* removal of internal restriction sites) during the parts assembly processes.

Among hundreds of thousands of parts to be cloned, components other than protein-coding genes are limited for module or vector constructions, *e.g.* the replication origins, the antibiotic resistance cassettes for selection and the transcriptional terminators, and therefore these components are prepared separately and are prefixed and suffixed with typical type IIP REs in iBrick standard. Rare-cutting REs that recognize 8-bp DNA sequences, including PacI, NotI, SwaI and AscI are used to link these components, which may greatly facilitate the exchange of these components during vector preparation. Specifically, prefixes/suffixes chosen are SwaI/NotI for the replication origins, NotI/PacI for the antibiotic resistance cassettes and AscI/PacI for the transcriptional terminators (**Figs. S1 and S2**). Other DNA sequences, including CDS, promoters, clusters, or pathways are defined as iBrick parts and should be prefixed with I-SceI and suffixed with PI-PspI. In addition, the above mentioned components, including replication origins, antibiotic resistance cassettes and transcriptional terminators can also be prepared as iBrick parts with I-SceI as the prefix and PI-PspI as the postfix as well. However, these components should only be used in construction of vectors with special purposes, *e.g.* vectors containing two or more different replication origins for different hosts or with two or more different antibiotic resistance cassettes for different selection purposes, and in such circumstances, the second component can be added as an iBrick part following the iBrick standard.

The replication origin of the vector must have its capacity in corresponding to the size of the parts being cloned. In this study, the pUCori and a fosmid-derived origin (FOSori) [31] are used in corresponding to the cloning of small *versus* large parts, respectively.

The iBrick standard much resembles other BioBrick standards with respect to the usage of classic restriction and ligation techniques. Definition of iBrick constituents follows that of BioBrick, including parts, vectors and plasmids (**Table 2**). An iBrick part can be prepared through either direct PCR amplification or other techniques to introduce the prefix and suffix at both ends of the part, which is then digested with I-SceI and PI-PspI before being ligated to an iBrick vector with compatible cohesive ends. An iBrick plasmid thus prepared is used to reserve the iBrick part, which should normally be verified either by DNA sequencing or functional analyses prior to be assigned a serial number and stored in the registry. To carry out an assembly, an iBrick part can be firstly released from an iBrick plasmid *via* I-SceI and PI-PspI digestion and then ligated to an iBrick vector, to produce a new iBrick plasmid (**Fig. 1**). The iBrick vector is usually made from an iBrick plasmid with either I-SceI digestion or PI-PspI digestion, depending on the direction of the assembly procedure, *i.e.* PI-PspI digestion allows a forward assembly while I-SceI digestion should be used in the reverse assembly. As only one enzyme cutting is used to prepare the vector, dephosphorylation by alkaline phosphatase is necessary to avoid self-ligation of the vector.

Take the forward assembly for example (**Fig. 2**), the 3′ terminus of part A is ligated to the 5′ terminus of part B to obtain part AB and a 21-bp scar of "5′-GGCAAACAGCTATTATCCCTA-3′" is produced in between. The newly obtained part AB is flanked with I-SceI at the 5′ end and PI-PspI at the 3′ end and therefore partAB can be taken as a new iBrick part for subsequent assembly, allowing iterative assembly of multiple parts. In most cases, the scar is merely a piece of short DNA sequence that links iBrick parts A and B. However, when iBrick standard is used for construction of plasmids to express fusion proteins, the translation stop codon in part A should be omitted and the scar should be translated in frame, which encodes a linker peptide of Gly-Lys-Gln-Leu-Leu-Ser-Leu. Although the scar contains no rare codons when expressed in *E. coli*, the protein linker contains three closely spaced leucines that may cause problems with respect to protein solubility, probably due to the large hydrophobic nature of leucine side chains. Considering the fact that HEs always tolerate a small number of base substitutions within their target DNA sequences without a loss in their enzyme efficiencies [35,36], one possible solution would be to slightly alter their targets, which would still be recognized and digested by I-SceI and PI-PspI but generate a new scar that encodes a more optimal protein linker for the solubility of the fusion protein.

Nomenclatures of iBrick standard

To avoid confusion, a nomenclature system is established for iBrick. The iBrick system can be divided into parts (including parts, devices and even systems), vectors and plasmids. For the parts, a specific name is given to each of them as "IBP_$N", where "IB" stands for iBrick, "P" stands for Parts, an alphabetic letter, represented by "$" indicates the categories of the parts (**Table 2**) and "N" is a unique identification given with a six-figure Arabic number.

Besides "parts", other components that compose a vector are called "elements", including replication origins, antibiotic resistance cassettes, transcriptional terminators, *etc*. An *element* is different from a *part* in their prefixes and suffixes and they are distinguished in their nomenclatures, *i.e.* "IBE_$N_n", where the letter "E" stands for Elements; "$" stands for an alphabetic letter that indicates the specific type of the elements; "N" is a unique

Table 2. Letter allocations for parts and elements in iBrick standard.

Abbreviations	Categories
iBrick Parts (Prefixed with I-SceI and suffixed with PI-PSPI)	
Abbreviations	**Categories**
IBP_C	CDS
IBP_K	Cluster/Pathway
IBP_P	Promoter
IBP_S	RBS Sequences
IBP_X	Combined parts
IBP_U	Unclassified
IBP_I	BioBrick iGEM parts
IBP_B	BglBrick parts
IBP_O	Replication Origin, built in iBrick standard
IBP_R	Antibiotic Resistance, built in iBrick standard
IBP_T	Transcriptional Terminator, built in iBrick standard
iBrick Elements (Prefixed/suffixed with type IIP enzymes)	
Abbreviations	**Categories**
IBE_O	Replication Origin (SwaI/NotI)
IBE_R	Antibiotic Resistance (NotI/PacI)
IBE_T	Transcriptional Terminator (PacI/AscI)

identification given with an Arabic number indicating the subcategory of a given type; and "n" indicates the version of the element (**Tables 2&3**). For example, an inducible FOSori is designated as IBE_O2_0 for an iBrick Element of replication origin (O) –fosmid ori (replication origin category 2) of version 0; and the original kanamycin resistance cassette is given the nomenclature of IBE_RK_0 for an iBrick Element of antibiotic resistance cassette – (kanamycin resistance gene, K) of version 0. Normally, elements are stored as iBrick base-plasmids and can be simply prepared through restriction of the plasmids with type IIP REs followed by gel purification. Based on the nomenclature of elements, vectors are designated as pIB(O)(R)(T)_n, similar to that of BioBrick [8], where "O" indicates the replication origin; "R" indicates the antibiotic resistance (**Table 4**); "T" indicates the type of terminator chosen; and "n" indicates the version of the vector.

A plasmid constructed by a vector and a part is given a conjoint name combining the vector and the part, *e.g.* part IBP_C000001 ligated to vector pIB1K1_0 results in a plasmid pIB1K1_0-C000001 (**Table 5**). When multiple parts are assembled together, a new serial number must be given to indicate the combination of these parts. Besides, the same parts with altered arrangements must be assigned with a different identification number.

Possible applications of iBrick in future

The DNA sequences recognized by homing nucleases used in iBrick are extremely rare in natural DNA sequences, bringing much convenience in preparation of iBrick parts, especially for those large DNA pieces. In this study, the entire *act* biosynthetic cluster was directly taken from a cosmid with gap repair and stored as an iBrick part. Using iBrick assembly, this large part can be easily operated, *e.g.* ligated with an integrating cassette for heterologous expression. Because many secondary metabolites biosynthetic clusters are mined and produced in either its original host or a phylogenetically close host, there is usually no need to perform codon optimization. For example, polyketide biosynthetic clusters from actinomycetes are preferably characterized and optimized for industrial production in an actinomycete host rather than in an *E. coli* strain. With the availability of a huge number of bacterial genomic data (including both finished and unfinished), the number of predicted clusters for secondary metabolites grows rapidly. However, a lot of these clusters are cryptic and promoter engineering might be necessary for further characterization. Simply, an inducible strong promoter can be added in front of the whole cluster to drive its expression, instead of substitution of its original promoter. Therefore, the whole cluster can actually be cloned as a large iBrick part and then ligated with another part of an inducible promoter, using iBrick assembly. Moreover, with the development of one-step cloning techniques of large DNA pieces, *e.g.* TAR [37] and RecE-mediated LLHR [38], large iBrick parts can also be cloned directly from genomic DNAs through combination of these efficient cloning methods. Besides, for cloning of those extremely large DNA pieces (>100 kb), which might be too large to be directly cloned, iBrick standard might be effective in combination with other assembly techniques, *e.g.* the Site-Specific Recombination based Tandem Assembly method (SSRTA) [39]. SSRTA uses the phiBT1 integrase mediated efficient recombination between paired mutated *attB* and *attP* sites [40]. Therefore, a large part can firstly be divided into several subparts, which can be directly cloned with one-step strategies; then, adaptors containing paired *attB* and *attP* sites can be ligated to the subparts that are released from I-SceI and PI-PspI digestion and SSRTA can subsequently be employed to accomplish the assembly of the large part.

Considering the large size of some parts, a combined origin with ori2 and oriV [31] is used in iBrick standard. When the plasmid is produced in a strain containing the *araC–*P_{BAD} driven *trfA* gene, the copy number can be induced from single copy to 10–200 copies per cell, simply through addition of arabinose [31]. Because many parts may express toxic products that are harmful to the hosts, control of both the copy number of the plasmid and the promoter activities are extremely useful for cloning and preservation of the toxic parts without killing the hosts. When the plasmids need to be prepared in a large population, inducers can be added when the cell density reaches high and cells can be harvested for plasmids preparation in a short time. Besides, the induction of the copy number of plasmids containing oriV can also be achieved through co-transformation of pJK46 (**Fig. S3**), which expresses TrfA in any *E. coli* strains.

Developments of iBrick in future

Due to the long recognition sequences, the 21-bp scar produced in iBrick is much longer than routine BioBricks standards (*e.g.* 6 bps in BglBricks [9]) and iBrick is therefore more suitable in assembling large parts instead of small parts (*e.g.* transcriptional terminators and RBSs). During the process of construction and storage of iBrick plasmids that contain scars, no recombination between these scars has been observed in *E. coli* hosts. Besides, the constructed plasmid containing scars seems quite stable in Streptomyces 4F. Taken together, although the *in vivo* recombination between scars cannot be totally expelled, the efficiency should be extremely low. Besides, the

Table 3. iBrick elements used in this study.

Systematic name	Common name	Version	Type	Source
IBE_O1_0	pUCori	0	replication origin	pUC18 [30]
IBE_O2_0	inducible FOSori	0	replication origin	pCC2FOS (Epicentre)
IBE_RA_0	*bla*	0	ampicillin resistance	pUC18
IBE_RK_0	*kan*	0	kanamycin resistance	pET28a (Novagen)
IBE_RC_0	*cat*	0	chloramphenicol resistance	pCC2FOS
IBE_RAm_0	*aac3(IV)*	0	apramycin resistance	pSET152 [34]
IBE_RH_0	*hyg*	0	hygromycin resistance	pML814 [44]
IBE_T1_0	T4_g32	0	transcriptional terminator	T4 phage [45]

Table 4. Letter abbreviations for antibiotic resistance markers in iBrick standard*.

Code	Antibiotics
A	ampicillin
Am	apramycin
C	chloramphenicol
E	erythromycin
G	gentamycin
H	hygromycin
K	kanamycin
N	neomycin
Na	nalidixic acid
R	rifampicin
S	spectinomycin
St	streptomycin
T	tetracycline
Tm	trimethoprim
Ts	thiostrepton
Z	zeocin

* Most of the letter abbreviations are the same as those described by Shetty *et al.* [8].

recombination efficiency in other hosts, including *Saccharomyces* and cell lines, is still unclear at the moment and needs be investigated in future studies.

A non-cleaving mutant of I-SceI was ever successfully linked to FokI, a type IIS RE, generating a new type of IIS REs of CdnDs [41]. Interestingly, CdnDs recognize long DNA sequences just as I-SceI but cut DNA outside of their recognition sites, thus promising in application of scar-less assembly of large DNA pieces. To develop a scar-less assembly scheme for iBrick standard, PI-PspI needs to be engineered in a similar way to acquire a hybrid type IIS RE that recognizes PI-PspI recognition sequences but cuts DNA outside.

Because different enzymes are used in iBrick standard and BioBrick RFC 10 (the most popular BioBricks standard), the two standards are not compatible with each other at the moment. Considering a large number of parts have been collected by BioBrick Registry using the RFC 10 standard, this incompatibility needs to be solved by persistent developments of iBrick standard in future.

The optimal reaction temperature for PI-PspI is 65°C and cannot be inactivated by heating. Therefore, to make a double digestion, I-SceI is firstly added and the reaction is performed at 37°C in PI-PspI buffer, then PI-PspI is added directly and the digestion is transferred to 65°C. Besides, PI-PspI seems extremely stable, and the procedure combining gel electrophoresis and column purification is preferable for removal of PI-PspI than other treatments, *e.g.* SDS, proteinase K or trypsin (data not shown). Therefore, efforts are now making to engineer PI-PspI to lower its optimal reaction temperature and make it easier to be inactivated, *e.g.* by heat inactivation. Moreover, to better support iBrick standard, establishment of a bioinformatics

Table 5. iBrick parts used in this study.

Systematic name	Common name	Vector	Type	Source
IBP_C000001	*crtE*	pIB1A1_0	CDS	BioBrick BBa_K274200
IBP_C000002	*crtB*	pIB1A1_0	CDS	BioBrick BBa_K274200
IBP_C000003	*crtI*	pIB1A1_0	CDS	BioBrick BBa_K274200
IBP_K000001	*act* cluster	pIB2K1_0	Cluster	fosmid N07_85 (this lab)
IBP_P000001	*pBAD*	pIB1A1_0	Promoter	pKD46 [43]
IBP_X000001	*crtBI*	pIB1A1_0	Combined	IBP_C000002, IBP_C000003
IBP_X000002	*pBAD-crtE*	pIB1A1_0	Combined	IBP_C000001, IBP_P000001
IBP_X000003	*pBAD-crtEBI*	pIB1A1_0	Combined	IBP_P000001, IBP_C000001-3
IBP_X000004	phiC31 int, *oriT*, *attP*	pIB1A1_0/pIB2K1_0/pIB2Am1_0	Combined	pSET152 [34]
IBP_X000005	phiC31 int, *oriT*, *attP*, *act* cluster	pIB2Am1_0	Combined	IBP_X000004, IBP_K000001

platform as well as development of useful tools are now under way.

Supporting Information

Figure S1 Plasmid map of pJK5. The construction procedures can be found in *Materials and Methods*. Plasmid pJK5 was used as a negative control in analysis of carotenoid production.

Figure S2 Plasmid map of pIB1A1_0-C000001. Plasmid pIB1A1_0-C000001 is used as a base plasmid for construction of iBrick parts. Base vector pIB1A1_0 can be prepared *via* I-SceI and PI-pspI double digestion of the base plasmid, followed by gel electrophoresis and gel purification.

Figure S3 Plasmid map of pJK46. The *trfA* gene was firstly amplified from pRK415 [42] with primers of trfA-F2 and trfA-R4 and the amplicon was then inserted into the HincII site of pUC18. After being verified by DNA sequencing, *trfA* gene was then released from pUC18 through EcoRI and SmaI digestion, and introduced to the same sites of pKD46 [43] to obtain pJK46.

Figure S4 Frequently used base vectors. The construction processes can be found in *Materials and Methods*. Base vectors are used for preparation of iBrick parts and subsequent assembly.

Table S1 Primers used in this study.

Acknowledgments

We thank Chao LEI and Yi-Ning LIU (IPPE, SIBS) for their help in detection of the lycopene production.

Author Contributions

Conceived and designed the experiments: JW GPZ WHC SXR. Performed the experiments: JKL. Analyzed the data: JW GPZ JKL WHC. Contributed reagents/materials/analysis tools: JW GPZ JKL WHC SXR. Contributed to the writing of the manuscript: JW GPZ.

References

1. Lu TK, Khalil AS, Collins JJ (2009) Next-generation synthetic gene networks. Nature Biotechnology 27: 1139–1150.
2. Endy D (2005) Foundations for engineering biology. Nature 438: 449–453.
3. Marchand JA, Peccoud J (2012) Building block synthesis using the polymerase chain assembly method. Methods Mol Biol 852: 3–10.
4. TerMaat JR, Pienaar E, Whitney SE, Mamedov TG, Subramanian A (2009) Gene synthesis by integrated polymerase chain assembly and PCR amplification using a high-speed thermocycler. J Microbiol Methods 79: 295–300.
5. Mamedov TG, Padhye NV, Viljoen H, Subramanian A (2007) Rational *de novo* gene synthesis by rapid polymerase chain assembly (PCA) and expression of endothelial protein-C and thrombin receptor genes. J Biotechnol 131: 379–387.
6. Booth PM, Buchman GW, Rashtchian A (1994) Assembly and cloning of coding sequences for neurotrophic factors directly from genomic DNA using polymerase chain reaction and uracil DNA glycosylase. Gene 146: 303–308.
7. Stemmer WP, Crameri A, Ha KD, Brennan TM, Heyneker HL (1995) Single-step assembly of a gene and entire plasmid from large numbers of oligodeoxyribonucleotides. Gene 164: 49–53.
8. Shetty RP, Endy D, Knight TF Jr. (2008) Engineering BioBrick vectors from BioBrick parts. J Biol Eng 2: 5.
9. Lee TS, Krupa RA, Zhang F, Hajimorad M, Holtz WJ, et al. (2011) BglBrick vectors and datasheets: A synthetic biology platform for gene expression. J Biol Eng 5: 12.
10. Rokke G, Korvald E, Pahr J, Oyas O, Lale R (2014) BioBrick assembly standards and techniques and associated software tools. Methods Mol Biol 1116: 1–24.
11. Gibson DG, Young L, Chuang RY, Venter JC, Hutchison CA, 3rd, et al. (2009) Enzymatic assembly of DNA molecules up to several hundred kilobases. Nat Methods 6: 343–345.
12. Gibson DG, Smith HO, Hutchison CA 3rd, Venter JC, Merryman C (2010) Chemical synthesis of the mouse mitochondrial genome. Nat Methods 7: 901–903.
13. Gibson DG, Glass JI, Lartigue C, Noskov VN, Chuang RY, et al. (2010) Creation of a bacterial cell controlled by a chemically synthesized genome. Science 329: 52–56.
14. Chen WH, Qin ZJ, Wang J, Zhao GP (2013) The MASTER (methylation-assisted tailorable ends rational) ligation method for seamless DNA assembly. Nucleic Acids Res 41: e93.
15. Engler C, Kandzia R, Marillonnet S (2008) A one pot, one step, precision cloning method with high throughput capability. PLoS One 3: e3647.
16. Engler C, Marillonnet S (2011) Generation of families of construct variants using golden gate shuffling. Methods Mol Biol 729: 167–181.
17. Engler C, Gruetzner R, Kandzia R, Marillonnet S (2009) Golden gate shuffling: a one-pot DNA shuffling method based on type IIs restriction enzymes. PLoS One 4: e5553.
18. Hunt R, Sauna ZE, Ambudkar SV, Gottesman MM, Kimchi-Sarfaty C (2009) Silent (synonymous) SNPs: should we care about them? Methods Mol Biol 578: 23–39.
19. Komar AA (2007) Silent SNPs: impact on gene function and phenotype. Pharmacogenomics 8: 1075–1080.
20. Komar AA (2007) Genetics. SNPs, silent but not invisible. Science 315: 466–467.
21. Blake WJ, Chapman BA, Zindal A, Lee ME, Lippow SM, et al. (2010) Pairwise selection assembly for sequence-independent construction of long-length DNA. Nucleic Acids Res 38: 2594–2602.
22. Kuijpers NG, Chroumpi S, Vos T, Solis-Escalante D, Bosman L, et al. (2013) One-step assembly and targeted integration of multigene constructs assisted by the I-SceI meganuclease in *Saccharomyces cerevisiae*. FEMS Yeast Res 13: 769–781.
23. Viret JF (1993) Meganuclease I-SceI as a tool for the easy subcloning of large DNA fragments devoid of selection marker. Biotechniques 14: 325–326.
24. Plessis A, Perrin A, Haber JE, Dujon B (1992) Site-specific recombination determined by I-SceI, a mitochondrial group I intron-encoded endonuclease expressed in the yeast nucleus. Genetics 130: 451–460.
25. Xu MQ, Southworth MW, Mersha FB, Hornstra LJ, Perler FB (1993) *In vitro* protein splicing of purified precursor and the identification of a branched intermediate. Cell 75: 1371–1377.
26. Chen W, Qin Z (2011) Development of a gene cloning system in a fast-growing and moderately thermophilic *Streptomyces* species and heterologous expression of *Streptomyces* antibiotic biosynthetic gene clusters. BMC Microbiol 11: 243.
27. Naatsaari L, Mistlberger B, Ruth C, Hajek T, Hartner FS, et al. (2012) Deletion of the Pichia pastoris KU70 homologue facilitates platform strain generation for gene expression and synthetic biology. PLoS One 7: e39720.
28. Jamsai D, Orford M, Nefedov M, Fucharoen S, Williamson R, et al. (2003) Targeted modification of a human beta-globin locus BAC clone using GET Recombination and an I-Scei counterselection cassette. Genomics 82: 68–77.
29. Kieser, Buttner MJ, Chater KF, Hopwood DA (2000) Practical *Streptomyces* Genetics. Norwich Research Park, Norwich NR4 7UH, England: The John Innes Foundation Press.
30. Yanisch-Perron C, Vieira J, Messing J (1985) Improved M13 phage cloning vectors and host strains: nucleotide sequences of the M13mp18 and pUC19 vectors. Gene 33: 103–119.
31. Wild J, Hradecna Z, Szybalski W (2002) Conditionally amplifiable BACs: switching from single-copy to high-copy vectors and genomic clones. Genome Res 12: 1434–1444.
32. Gust B, Chandra G, Jakimowicz D, Yuqing T, Bruton CJ, et al. (2004) Lambda red-mediated genetic manipulation of antibiotic-producing *Streptomyces*. Adv Appl Microbiol 54: 107–128.
33. Marsischky G, LaBaer J (2004) Many paths to many clones: a comparative look at high-throughput cloning methods. Genome Res 14: 2020–2028.
34. Bierman M, Logan R, O'Brien K, Seno ET, Rao RN, et al. (1992) Plasmid cloning vectors for the conjugal transfer of DNA from *Escherichia coli* to *Streptomyces* spp. Gene 116: 43–49.
35. Colleaux L, D'Auriol L, Galibert F, Dujon B (1988) Recognition and cleavage site of the intron-encoded omega transposase. Proc Natl Acad Sci U S A 85: 6022–6026.
36. Monteilhet C, Perrin A, Thierry A, Colleaux L, Dujon B (1990) Purification and characterization of the in vitro activity of I-Sce I, a novel and highly specific endonuclease encoded by a group I intron. Nucleic Acids Res 18: 1407–1413.
37. Kouprina N, Larionov V (2006) TAR cloning: insights into gene function, long-range haplotypes and genome structure and evolution. Nat Rev Genet 7: 805–812.
38. Fu J, Bian X, Hu S, Wang H, Huang F, et al. (2012) Full-length RecE enhances linear-linear homologous recombination and facilitates direct cloning for bioprospecting. Nat Biotechnol 30: 440–446.
39. Zhang L, Zhao G, Ding X (2011) Tandem assembly of the epothilone biosynthetic gene cluster by *in vitro* site-specific recombination. Sci Rep 1: 141.

40. Zhang L, Wang L, Wang J, Ou X, Zhao G, et al. (2010) DNA cleavage is independent of synapsis during *Streptomyces* phage phiBT1 integrase-mediated site-specific recombination. J Mol Cell Biol 2: 264–275.
41. Lippow SM, Aha PM, Parker MH, Blake WJ, Baynes BM, et al. (2009) Creation of a type IIS restriction endonuclease with a long recognition sequence. Nucleic Acids Res 37: 3061–3073.
42. Keen NT, Tamaki S, Kobayashi D, Trollinger D (1988) Improved broad-host-range plasmids for DNA cloning in gram-negative bacteria. Gene 70: 191–197.
43. Datsenko KA, Wanner BL (2000) One-step inactivation of chromosomal genes in *Escherichia coli* K-12 using PCR products. Proc Natl Acad Sci U S A 97: 6640–6645.
44. Amon J, Brau T, Grimrath A, Hanssler E, Hasselt K, et al. (2008) Nitrogen control in *Mycobacterium smegmatis*: nitrogen-dependent expression of ammonium transport and assimilation proteins depends on the OmpR-type regulator GlnR. J Bacteriol 190: 7108–7116.
45. Miller ES, Kutter E, Mosig G, Arisaka F, Kunisawa T, et al. (2003) Bacteriophage T4 genome. Microbiol Mol Biol Rev 67: 86–156.

Permissions

All chapters in this book were first published in PLOS ONE, by The Public Library of Science; hereby published with permission under the Creative Commons Attribution License or equivalent. Every chapter published in this book has been scrutinized by our experts. Their significance has been extensively debated. The topics covered herein carry significant findings which will fuel the growth of the discipline. They may even be implemented as practical applications or may be referred to as a beginning point for another development.

The contributors of this book come from diverse backgrounds, making this book a truly international effort. This book will bring forth new frontiers with its revolutionizing research information and detailed analysis of the nascent developments around the world.

We would like to thank all the contributing authors for lending their expertise to make the book truly unique. They have played a crucial role in the development of this book. Without their invaluable contributions this book wouldn't have been possible. They have made vital efforts to compile up to date information on the varied aspects of this subject to make this book a valuable addition to the collection of many professionals and students.

This book was conceptualized with the vision of imparting up-to-date information and advanced data in this field. To ensure the same, a matchless editorial board was set up. Every individual on the board went through rigorous rounds of assessment to prove their worth. After which they invested a large part of their time researching and compiling the most relevant data for our readers.

The editorial board has been involved in producing this book since its inception. They have spent rigorous hours researching and exploring the diverse topics which have resulted in the successful publishing of this book. They have passed on their knowledge of decades through this book. To expedite this challenging task, the publisher supported the team at every step. A small team of assistant editors was also appointed to further simplify the editing procedure and attain best results for the readers.

Apart from the editorial board, the designing team has also invested a significant amount of their time in understanding the subject and creating the most relevant covers. They scrutinized every image to scout for the most suitable representation of the subject and create an appropriate cover for the book.

The publishing team has been an ardent support to the editorial, designing and production team. Their endless efforts to recruit the best for this project, has resulted in the accomplishment of this book. They are a veteran in the field of academics and their pool of knowledge is as vast as their experience in printing. Their expertise and guidance has proved useful at every step. Their uncompromising quality standards have made this book an exceptional effort. Their encouragement from time to time has been an inspiration for everyone.

The publisher and the editorial board hope that this book will prove to be a valuable piece of knowledge for researchers, students, practitioners and scholars across the globe.

List of Contributors

Dongdong Zhang, Haiyang Jiang, Kai Wen, Jianzhong Shen and Xingyuan Cao
College of Veterinary Medicine, China Agricultural University, Beijing, China

Xiaoqi Tao
College of Veterinary Medicine, China Agricultural University, Beijing, China
College of Food Science, Southwest University, Chongqing, China

Lijun Gu
Research Center for Asian Infectious Diseases, the Institute of Medical Science, the University of Tokyo, Tokyo, Japan
Japan-China Joint Laboratory of Molecular Immunology and Molecular Microbiology, Institute of Microbiology, Chinese Academy of Sciences, Beijing, P. R. China

Ai Kawana-Tachikawa
Division of Infectious Diseases, Advanced Clinical Research Center, the Institute of Medical Science, the University of Tokyo, Tokyo, Japan

Teiichiro Shiino and Wataru Sugiura
AIDS Research Center, National Institute of Infectious Diseases, Tokyo, Japan

Hitomi Nakamura
Department of Infectious Disease Control, International Research Center for Infectious Diseases, the Institute of Medical Science, the University of Tokyo, Tokyo, Japan
Division of Infectious Diseases and Applied Immunology, Research Hospital, The Institute of Medical Science, The University of Tokyo, Tokyo, Japan

Michiko Koga and Tadashi Kikuchi
Division of Infectious Diseases, Advanced Clinical Research Center, the Institute of Medical Science, the University of Tokyo, Tokyo, Japan
Division of Infectious Diseases and Applied Immunology, Research Hospital, The Institute of Medical Science, The University of Tokyo, Tokyo, Japan

Eisuke Adachi and Tomohiko Koibuchi
Division of Infectious Diseases and Applied Immunology, Research Hospital, The Institute of Medical Science, The University of Tokyo, Tokyo, Japan

Takaomi Ishida
Research Center for Asian Infectious Diseases, the Institute of Medical Science, the University of Tokyo, Tokyo, Japan
Japan-China Joint Laboratory of Molecular Immunology and Molecular Microbiology, Institute of Microbiology, Chinese Academy of Sciences, Beijing, P. R. China

George F. Gao
CAS Key Laboratory of Pathogenic Microbiology and Immunology, Institute of Microbiology, Chinese Academy of Sciences, Beijing, P. R. China

Masaki Matsushita
Biotech Research and Development, Wakunaga Pharmaceutical Corporation, Hiroshima, Japan

Aikichi Iwamoto
Research Center for Asian Infectious Diseases, the Institute of Medical Science, the University of Tokyo, Tokyo, Japan
Division of Infectious Diseases, Advanced Clinical Research Center, the Institute of Medical Science, the University of Tokyo, Tokyo, Japan
Department of Infectious Disease Control, International Research Center for Infectious Diseases, the Institute of Medical Science, the University of Tokyo, Tokyo, Japan
Division of Infectious Diseases and Applied Immunology, Research Hospital, The Institute of Medical Science, The University of Tokyo, Tokyo, Japan

Noriaki Hosoya
Division of Infectious Diseases, Advanced Clinical Research Center, the Institute of Medical Science, the University of Tokyo, Tokyo, Japan
Department of Infectious Disease Control, International Research Center for Infectious Diseases, the Institute of Medical Science, the University of Tokyo, Tokyo, Japan

Qi Wang, Cui Hua Liu, Fei Liu and Bing Xi Li
CAS Key Laboratory of Pathogenic Microbiology and Immunology, Institute of Microbiology, Chinese Academy of Sciences, Beijing, China

Susanna K. P. Lau and Patrick C. Y. Woo
State Key Laboratory of Emerging Infectious Diseases, Department of Microbiology, The University of Hong Kong, Hong Kong Special Administrative Region, China

Yanlin Zhao
National Center for Tuberculosis Control and Prevention, Chinese Center for Disease Control and Prevention, Beijing, China

Hong Min Li and Yong Liang Hu
Institute for Tuberculosis Research, the 309 Hospital, Beijing, China

Ke Zhao, Wenzhen Kang, Yuan Li, Qing Liu, Wei Jiang, Yan Zhuang, Zisheng Guo1, Zhuoran Yu, Xinhong Li, Chunfu Wang, Na Yao and Yongtao Sun
Department of Infectious Diseases, Tangdu Hospital, Fourth Military Medical University, Xi'an, Shaanxi Province, China

Qingquan Liu
Department of Endocrinology, Tangdu Hospital, Fourth Military Medical University, Xi'an, Shaanxi Province, China

Stella M. Bernardo and Samuel A. Lee
Section of Infectious Diseases, New Mexico Veterans Healthcare System, Albuquerque, NM, United States of America
Division of Infectious Diseases, University of New Mexico Health Science Center, Albuquerque, NM, United States of America

Christopher P. Allen, Anna Waller, Susan M. Young, Tudor Oprea and Larry A. Sklar,
Center for Molecular Discovery, University of New Mexico, Albuquerque, NM, United States of America

Eric T. Sumrall, Elizabeth B. Gallo and Iruka N. Okeke
Department of Biology, Haverford College, Haverford, Pennsylvania, United States of America

Aaron Oladipo Aboderin
Department of Medical Microbiology and Parasitology, Obafemi Awolowo University, Ile-Ife, Osun State, Nigeria

Adebayo Lamikanra
Department of Pharmaceutics, Obafemi Awolowo University, Ile-Ife, Osun State, Nigeria

Min Zhang, Mingquan Shang, Weiwei Yang, Junli Chen and Hong Shang
Key Laboratory of AIDS Immunology of National Health and Family Planning Commission, Department of Laboratory Medicine, The First Affiliated Hospital, China Medical University, Shenyang, China
Collaborative Innovation Center for Diagnosis and Treatment of Infectious Diseases, Hangzhou, China

Zhe Wang
Disease Prevention and Control Center of Henan Province, Zhengzhou, China

Aram Heo and Woojun Park
Laboratory of Molecular Environmental Microbiology, Department of Environmental Science and Ecological Engineering, Korea University, Seoul, Republic of Korea

Hyun-Jin Jang and Jung-Suk Sung
Department of Life Science, Dongguk University, Seoul, Republic of Korea

Petros Isaakidis, Mrinalini Das, Lorraine Rebello and Christopher Peskett
Médecins Sans Frontières, Mumbai, India

Ajay M V Kumar
International Union Against Tuberculosis and Lung Disease (The Union), South-East Asia Regional Office, New Delhi, India

Minni Khetarpal
City TB Office, Bawala Wadi, Chinckpokli, Mumbai, India

Arun Bamne
Brihanmumbai Municipal Corporation, Mumbai, India

Balkrishna Adsul
Mumbai District AIDS Control Society (MDACS), Mumbai, India

Mamta Manglani
Pediatric Centre of Excellence for HIV Care, L.T.M. Medical College & General Hospital, Sion, Mumbai, India

Kuldeep Singh Sachdeva
Central TB Division, Directorate General of Health Services, Ministry of Health and Family Welfare, New Delhi, India

Malik Parmar
World Health Organisation - Country Office for India, New Delhi, India

Avinash Kanchar and B.B. Rewari
Department of AIDS Control, National AIDS Control Organisation, New Delhi, India

Alaka Deshpande
MGM Institute of Health Sciences, Navi Mumbai, India

Camilla Rodrigues and Anjali Shetty
P.D. Hinduja National Hospital and Medical Research Centre, Mumbai, India

Peter Saranchuk
Southern Africa Medical Unit (SAMU), Médecins Sans Frontières, Cape Town, South Africa

Chia-Ter Chao
Division of Nephrology, Department of Internal Medicine, National Taiwan University Hospital Jin-Shan Branch, New Taipei City, Taiwan
Graduate Institute of Toxicology, National Taiwan University, Taipei, Taiwan

Szu-Ying Lee
Division of Nephrology, Department of Internal Medicine, National Taiwan University Hospital Yun-Lin Branch, Yun-Lin County, Taiwan

Wei-Shun Yang
Division of Nephrology, Department of Internal Medicine, National Taiwan University Hospital Hsin-Chu Branch, Hsin-Chu County, Taiwan

Huei-Wen Chen, Kuan-Yu Hung and Jenq-Wen Huang
Division of Nephrology, Department of Internal Medicine, National Taiwan University Hospital, Taipei, Taiwan

Cheng-Chung Fang
Department of Emergency Medicine, National Taiwan University Hospital, Taipei, Taiwan

Chung- Jen Yen
Department of Geriatrics and Gerontology, National Taiwan University Hospital, Taipei, Taiwan

Chih-Kang Chiang
Department of Integrative diagnostics and Therapeutics, National Taiwan University Hospital, Taipei, Taiwan

Petra Videnska, Md. Masudur Rahman, Marcela Faldynova, Vladimir Babak, Marta Elsheimer Matulova, Darina Cejkova and Ivan Rychlik
Veterinary Research Institute, Brno, Czech Republic

Estella Prukner-Radovcic and Ivan Krizek
Faculty of Veterinary Medicine, University of Zagreb, Zagreb, Croatia

Sonja Smole-Mozina and Jasna Kovac
Biotechnical Faculty, University of Ljubljana, Ljubljana, Slovenia

Ama Szmolka and Bela Nagy
Institute for Veterinary Medical Research, Hungarian Academy of Sciences, Budapest, Hungary

Karel Sedlar
Department of Biomedical Engineering, Brno University of Technology, Brno, Czech Republic

Cheryl C. Y. Loh, Bruce Russell and Kitti W. K. Chan
Department of Microbiology, Yong Loo Lin School of Medicine, National University of Singapore, Singapore, Singapore

Rossarin Suwanarusk and Laurent Rénia
Singapore Immunology Network, Agency for Science Technology and Research, Biopolis, Singapore, Singapore

Yan Quan Lee and Kevin S. W. Tan
Department of Microbiology, Yong Loo Lin School of Medicine, National University of Singapore, Singapore, Singapore
NUS Graduate School for Integrative Sciences and Engineering, National University of Singapore Singapore, Singapore

Kit-Ying Choy and Larry M. C. Chow
Department of Applied Biology and Chemical Technology, The State Key Laboratory of Chirosciences, The Hong Kong Polytechnic University, Hung Hom, Kowloon, Hong Kong SAR, PR China

Martin J. Lear
Department of Chemistry, Graduate School of Science, Tohoku University, Aza Aramaki, Aoba-ku, Sendai, Japan

François H. Nosten
Centre for Tropical Medicine, Nuffield Department of Medicine, University of Oxford, Oxford, United Kingdom
Shoklo Malaria Research Unit, Mahidol-Oxford Tropical Medicine Research Unit, Faculty of Tropical Medicine, Mahidol University, Mae Sot, Thailand
Centre for Tropical Medicine, University of Oxford, Churchill Hospital, Oxford, United Kingdom

Yingying Su, Fujie Zhang, Huixin Liu, Lin Zhu, Jing Wu and Ning Wang
National Center for AIDS/STD Control and Prevention, Chinese Center for Disease Control and Prevention, Beijing, China

M. Kumi Smith
Department of Epidemiology, University of North Carolina, Chapel Hill, North Carolina, United States of America

Melissa Ehman
Tuberculosis Control Branch, Division of Communicable Disease Control, Center for Infectious Diseases, California Department of Public Health, Richmond, California, United States of America
Institute of Global Health, Global Health Sciences, University of California San Francisco, San Francisco, California, United States of America

Jennifer Flood and Pennan M. Barry
Tuberculosis Control Branch, Division of Communicable Disease Control, Center for Infectious Diseases, California Department of Public Health, Richmond, California, United States of America

Srey Viseth Horm, Sareth Rith, Sirenda Vong, Arnaud Tarantola, Sowath Ly, Paul F. Horwood and Philippe Buchy
Institut Pasteur du Cambodge, Réseau International des Instituts Pasteur, Phnom Penh, Cambodia

Sek Mardy
Institut Pasteur du Cambodge, Réseau International des Instituts Pasteur, Phnom Penh, Cambodia
World Health Organization, Phnom Penh, Cambodia

Sovann Ly, Seng Heng and Sok Touch
Communicable Disease Control Department, Ministry of Health, Phnom Penh, Cambodia

Paul Kitsutani,
Influenza Division, National Center for Immunization and Respiratory Disease, Center for Disease Control and Prevention, Atlanta, Georgia, United States of America

Vannra Ieng, Nima Asgari and Maria Concepcion Roces
World Health Organization, Phnom Penh, Cambodia

Borann Sar and Nora Chea
Centers for Disease Control and Prevention, Cambodia Office, Phnom Penh, Cambodia

Buth Sokhal and Sam An Ung
National Institute of Public Health, Phnom Penh, Cambodia

Ian Barr, Anne Kelso, Aeron Hurt and Naomi Komadina
WHO Collaborating Centre for Reference and Research on Influenza, Melbourne, Australia

Ans Timmermans, Chanthap Lon and David Saunders
Armed Forces Research Institute of Medical Sciences, Bangkok, Thailand

Zhiping Wang
Environmental Biotechnology Laboratory, The University of Hong Kong, Hong Kong
School of Environmental Science and Engineering, Shanghai Jiao Tong University, Shanghai, China

Feng Guo and Tong Zhang
Environmental Biotechnology Laboratory, The University of Hong Kong, Hong Kong

Lili Liu
Environmental Biotechnology Laboratory, The University of Hong Kong, Hong Kong
State Environmental Protection Key Laboratory of Environmental Risk Assessment and Control on Chemical Process, East China University of Science and Technology, Shanghai, China

Hye Won Jeong and Jung Yeon Heo
Department of Internal Medicine, Chungbuk National University College of Medicine, Heungduk-gu, Cheongju, Republic of Korea

Jung Soo Park
Department of Emergency Medicine, Chungbuk National University College of Medicine, Heungduk-gu, Cheongju, Republic of Korea

Woo Joo Kim
Division of Infections Disease, Department of Internal Medicine, Korea University College of Medicine, Guro-dong, Guro-gu, Seoul, Republic of Korea
Transgovernmental Enterprise for Pandemic Influenza in Korea, Seoul, Republic of Korea

Yini Dang, Xiaoying Zhou and Guoxin Zhang
Department of Gastroenterology, The First Affiliated Hospital of Nanjing Medical University, Nanjing, China, and First Clinical Medical College of Nanjing Medical University, Nanjing, China

Jan D. Reinhardt
Institute for Disaster Management and Reconstruction, Sichuan University, Chengdu, China, and Hong Kong Polytechnical University, Hung Hom, Hong Kong, China
Department of Health Sciences, University of Lucerne, Lucerne, Switzerland Swiss Paraplegic Research, Nottwil, Switzerland

Xuelian Zhang, Yanxia Li, Bei Liu, Jing Wang, Chenghong Feng, Min Gao and Lina Wang
State Key Laboratory of Water Environment Simulation, School of Environment, Beijing Normal University, Beijing, China

Hong Gil Jeon, Hyeong Uk Ju and Jae-Bum Jun
Department of Internal Medicine, Ulsan University Hospital, University of Ulsan College of Medicine, Ulsan, Republic of Korea

Gyu Yeol Kim
Department of Surgery, Ulsan University Hospital, University of Ulsan College of Medicine, Ulsan, Republic of Korea

Joseph Jeong
Department of Laboratory Medicine, Ulsan University Hospital, University of Ulsan College of Medicine, Ulsan, Republic of Korea

Min-Ho Kim
Biomedical Research Center, Ulsan University Hospital, University of Ulsan College of Medicine, Ulsan, Republic of Korea

Priyanka Kumari and Hong L. Choi
Department of Agricultural Biotechnology, Research Institute for Agriculture and Life Science, Seoul National University, Gwanak, Seoul, Republic of Korea

Daniele Poole
U.O. Anestesia e Rianimazione, Ospedale S. Martino, Belluno, Italy

Arturo Chieregato and Bruno Viaggi
SOD Anestesia e Area Intensiva CTO, AOU Careggi, Firenze, Italy

Martin Langer
Dipartimento di Anestesia e Rianimazione, Fondazione IRCCS Istituto Nazionale dei Tumori e Universitàdegli Studi di Milano, Milano, Italy

Emiliano Cingolani, Giuseppe Nardi and Luigi Riccioni
UOC Shock e Trauma, AO San Camillo-Forlanini, Roma, Italy

Paolo Malacarne
U.O. Anestesia e Rianimazione - P.S., Azienda Ospedaliera Universitaria Pisana, Pisa, Italy

Francesca Mengoli, Ilaria Turriziani, Carlo Coniglio and Giovanni Gordini
U.O. Rianimazione 118, Ospedale Maggiore, Bologna, Italy

Ennio Nascimben
Neurorianimazione Ospedale S. Maria di Ca' Foncello, Treviso, Italy

Annalisa Volpi
Anestesia e Rianimazione 1, Azienda Ospedaliero-Universitaria di Parma, Parma, Italy

Jia-Kun Liu, Wei-Hua Chen, Shuang-Xi Ren and Jin Wang
CAS Key Laboratory of Synthetic Biology, Institute of Plant Physiology and Ecology, Shanghai Institutes for Biological Sciences, Chinese Academy of Sciences, Shanghai, China

Guo-Ping Zhao
CAS Key Laboratory of Synthetic Biology, Institute of Plant Physiology and Ecology, Shanghai Institutes for Biological Sciences, Chinese Academy of Sciences, Shanghai, China
State Key Lab of Genetic Engineering & Center for Synthetic Biology, Department of Microbiology and Microbial Engineering, School of Life Sciences, Fudan University, Shanghai, China
Shanghai-MOST Key Laboratory of Disease and Health Genomics, Chinese National Human Genome Center at Shanghai, Shanghai, China
Department of Microbiology and Li Ka Shing Institute of Health Sciences, The Chinese University of Hong Kong, Prince of Wales Hospital, Shatin, New Territories, Hong Kong SAR, China

Index